Lecture Notes in Geoinformation and Cartography

For further volumes:
http://www.springer.com/series/7418

Stan Geertman · Fred Toppen
John Stillwell
Editors

Planning Support Systems for Sustainable Urban Development

Springer

Editors
Stan Geertman
Fred Toppen
Faculty of Geosciences
Utrecht University
Utrecht
The Netherlands

John Stillwell
School of Geography
University of Leeds
Leeds
UK

ISSN 1863-2246 ISSN 1863-2351 (electronic)
ISBN 978-3-642-37532-3 ISBN 978-3-642-37533-0 (eBook)
DOI 10.1007/978-3-642-37533-0
Springer Heidelberg New York Dordrecht London

Library of Congress Control Number: 2013941532

Printed on acid-free paper

Springer is part of Springer Science+Business Media (www.springer.com)

Preface

The international CUPUM (Computers in Urban Planning and Urban Management) conference has been one of the premier international conferences for the exchange of ideas and applications of computer technologies to address a range of social and environmental problems relating to urban areas. The first conference took place in 1989 in Hong Kong. Since then this bi-annual conference has been hosted in every continent of the world (with the sole exception of Africa) (see below). In 2013, delegates gathered for the 13th time in the city of Utrecht in the centre of The Netherlands.

The CUPUM Board has promoted the publication of a book with the collection of the scientific papers that were submitted to the conference. Those papers went through a competitive review process that resulted in the selection of what the reviewers deemed to be the best CUPUM papers. All these papers fit the main overarching theme of this specific CUPUM conference: *Planning Support Systems for Sustainable Urban Development*. Therein, we acknowledge that sustainable urban development is an issue that deserves wide and continuous attention. Moreover, it reflects the growing worldwide recognition of the importance of Planning Support Systems (PSS) with their background in GIScience as instruments for planning and policy support in general and for achieving sustainable urban development in particular.

Organizing the programme of an international conference and editing a volume of scientific papers requires time, effort, and support. First of all, as book editors, we would like to thank the authors for their high-quality contributions.

Second, we would like to thank the reviewers for their difficult task to assess and select the best papers for inclusion in this book. The double-blind review process was not an easy task, for sure, and it is always difficult when a number of potential authors are going to be disappointed. The group of reviewers who assisted us in this difficult selection process and provided the selected authors with valuable comments on their drafts were the members of the Board of Directors of CUPUM and the Advisors to this Board (see below). By fulfilling the review process double-blind and demanding at least three reviews per submission we believe that the review process has been conducted in a fair and equal way.

Third, we would like to thank our scientific sponsors (the Faculty of Geosciences and the Urban and Regional Research Centre URU of the Utrecht University) for their contribution in time and resources to this publication. In addition, we would like to thank Springer Publishers for their willingness to publish these contributions in their academic series *Springer Lecture Notes in Geoinformation and Cartography*. This is the first time that a selection of best papers from the CUPUM conference has been published by Springer. Last but definitely not least, we would like to thank Ingeborg Koppenol, our office manager, who checked and double checked all chapters and in doing so, ensured that a book would actually be forthcoming.

Past CUPUM Conferences

#	Years	Place	Country
I	1989	Hong Kong	HKG
II	1991	Oxford	GBR
III	1993	Atlanta	USA
IV	1995	Melbourne	AUS
V	1997	Mumbai	IND
VI	1999	Venice	ITA
VII	2001	Honolulu	USA
VIII	2003	Sendai	JPN
IX	2005	London	GBR
X	2007	Iguazu Falls	BRA
XI	2009	Hong Kong	CHN
XII	2011	Lake Louise (Calgary/Banff)	CAN
XIII	2013	Utrecht	NED

Board of Directors CUPUM

Name	Institute	Country	E-mail
Andrew Allan	University of South Australia	AUS	Andrew.Allan@unisa.edu.au
Michael Batty	University College London	GBR	m.batty@ucl.ac.uk
Joseph Ferreira	Massachusetts Institute of Technology	USA	jf@mit.edu
Stan Geertman	Utrecht University	NED	S.C.M.Geertman@uu.nl
John Douglas Hunt	University of Calgary	CAN	jdhunt@ucalgary.ca
Karl Kim	University of Hawaii	USA	karlk@hawaii.edu
Kazuaki Miyamoto	Tokyo City University	JPN	miyamoto@tcu.ac.jp
Paola Rizzi	Università degli Studi di Sassari	ITA	rizzi@uniss.it
Antonio N. Rodrigues da Silva	University of Sao Paulo	BRA	anelson@sc.usp.br
Anthony G. O. Yeh	University of Hong Kong	HKG	hdxugoy@hkucc.hku.hk

Advisors to the Board of Directors

Name	Institute	Country	E-mail
Ray Wyatt	University of Melbourne	AUS	raywyatt3@gmail.com
Michael Wegener	University of Dortmund	GER	mw@spiekermann-wegener.de
Dick Klosterman	What if? Incorporated, Hudson, Ohio	USA	dick.klosterman@gmail.com

Utrecht, The Netherlands
Leeds, UK, 2013

Stan Geertman
Fred Toppen
John Stillwell

Contents

Introduction to 'Planning Support Systems for Sustainable Urban Development'

Stan Geertman, John Stillwell and Fred Toppen

1 Introduction to Planning Support Systems

Planning Support Systems (PSS) are geo-information-technology-based instruments that are dedicated to supporting those involved in planning in the performance of their specific tasks (Batty 1995; Klosterman 1997). The term PSS appeared on the planning scene in the mid-1980s thanks to its progenitor, Britton Harris, although the concept of building instruments dedicated to the support of planning activities dates back much further. In a sense, it can be argued that PSS are related to Geographic Information Systems (GIS)—insofar as most PSS are very likely to be based on a GIS as the means of storing, managing and displaying spatial data—but while GIS are general-purpose tools that are applicable for many different spatial problems, each PSS is distinctive in its focus on supporting a specific planning task and may therefore contain a particular set of component parts. Although PSS are also related to so-called Spatial Decision Support Systems (SDSS), the two types of system differ in that PSS generally pay particular attention to long-range problems and strategic issues whereas SDSS have evolved from DSS used in business management and are generally designed to support shorter-term policymaking by isolated individuals or by business organizations (Clarke 1990). An alternative way of distinguishing SDSS and PSS is to recognise that SDSS aim to support operational decision making whereas the primary focus of PSS is on strategic planning activities. Given this function, a typical PSS will

S. Geertman (✉) · F. Toppen
Faculty of Geosciences, Utrecht University, Heidelberglaan 2 3508 Utrecht, The Netherlands
e-mail: s.c.m.geertman@uu.nl

F. Toppen
e-mail: f.j.toppen@uu.nl

J. Stillwell
School of Geography, University of Leeds, Leeds LS2 9JT, UK
e-mail: j.c.h.stillwell@leeds.ac.uk

S. Geertman et al. (eds.), *Planning Support Systems for Sustainable Urban Development*, Lecture Notes in Geoinformation and Cartography,
DOI: 10.1007/978-3-642-37533-0_1,

integrate planning-related theory, data, information, knowledge, methods and instruments within a single framework, with access provided through a shared graphical user interface (Geertman and Stillwell 2003, 2009).

Until relatively recently, attitudes toward PSS development tended to be rather negative. Klosterman (1998, p. 35) complained that *"instruments for planning support are no better developed now than they were ten years ago"*, and he was equally pessimistic about the adoption of new instruments and computer applications in planning practice in the near future. Harris (1999, p. 7) noticed that *"planners and designers have remained at best distrustful, or at worst downright antagonistic, toward computer-based models of support"*. With the turn of the century, however, the attitude toward instruments for planning support seems to be changed. At present, much more positive attention is being focused on planning support and its technological instruments than has been the case in the past several decades—a trend evident from the sheer volume of studies being undertaken, the dedicated conferences now taking place, and the diversity of papers and books that take PSS as their primary focus (e.g., Brail and Klosterman 2001; Geertman and Stillwell 2003, 2009; Brail 2008). Many leading academic authors consider PSS to be valuable support tools that will better enable planners to handle the complexity of planning processes, leading in turn to plans of higher quality and savings in terms of both time and resources (e.g., Stillwell et al. 1999).

2 PSS for Sustainable Urban Development

Since 2006, more than 50 % of the world population is living in urbanized areas (Castells 1996, 2001; Rollnick 2006), while around 1900 this was about 10 % (Priemus 2003). The increasing concentration of living and associated activities like working and recreation has consequences that vary in space and time. Besides the positive requirements such as the preservation of open spaces and the provision of preconditions for affordable public transport, there are also a range of negative implications that are outcomes of the urbanisation process. In rural areas, these problems can entail depopulation, fragmentation and difficulties in retaining economic vitality and liveability; in intermediate or transitional areas, the typical problems include an increase in traffic congestion, overconsumption of land, urban sprawl, *et cetera*; and for urban area, the problems may involve, *inter alia*, either over- and depopulation as well as socio-economic deprivation and environmental deterioration.

The notion of 'sustainability' can be envisaged as a present-day perspective to overcome these negative implications of continuing urbanization trends, and endeavour to undertake developments that will protect the environment, produce high quality spaces, generate economic prosperity and encourage social equity/ cohesion (e.g. Goedman and Zonneveld 2007). Remarkably, it is only quite recently that this notion has begun to be actually applied in the field of spatial planning of urban development, in particular through the use of ideas such as smart growth, network cities, ecological footprints, new urbanism, eco-cities, *et cetera* (e.g. Knaap 2002; Pivo 2002). Given its short history and, consequently, the rather basic state of knowledge about sustainability methods, it is a tremendous challenge to operationalize the notion of 'sustainable urban development' (Healey 2006, 2007; Tanguay et al. 2010; UN Habitat 2009).

Given the place, time and scale dependency of the implications indicated above, to attain a process of sustainable urban development requires informed policy measures that will vary in time and place and context too. This implies the need for appropriate knowledge production and consumption. Therein, the concept of 'governance' will play an ever increasing role in development planning. Herewith, it is acknowledged that besides government organisations, there are a host of associated institutions or stakeholders such as the private sector or the wider public, who will need to have a say in policy making and will try to influence the decisions that are to be taken. Different parties will always have different and often conflicting views. The level of sustainability will function as a yardstick for future urbanisation developments, although our present knowledge concerning the notion of 'sustainable urban development' and its implications for governance are still very limited.

It is certain that information and knowledge will play a prominent role in governance to promote sustainable urban development. The problems challenging sustainable urban development are both multidimensional and 'wicked'; there are very complex issues in which spatial, social, economic, and environmental dimensions interact and/or compete with each other. To be able to cope with this complexity, it is necessary to have available a range of support tools of which the present generation of PSS is a fine example. PSS are envisioned to support stakeholders to improve their knowledge base and to integrate and balance the various demands on space (Geertman and Stillwell 2003). At present, there is a lack of evidence that PSS are indispensable in contributing to the proper guidance of urban sustainable development processes although there are some promising signs and results from important case studies (Geertman and Stillwell 2009). Moreover, at the moment, there is a lack of sufficient insights into appropriate methodologies/approaches to achieve planning support for practitioners with the help of generated knowledge and PSS (Geertman 2006). It is an important goal of this book to show different ways in which PSS can support the process of sustainable urban development.

3 Structure of the Book

This book has been divided into five parts, the contents of which are summarised in the remainder of the chapter. There is always some danger in imposing structure on a wide range of contributions, but we have chosen a five part categorisation based on the purpose or use of each PSS. Part 1 is a collection of contributions that illustrate PSS that are used for spatial analysis and urban modelling. Part 2 contains a series of chapters about PSS that find their application specifically in environmental planning and modelling. In Part 3, traffic and network modelling is at the core of the contributions. Part 4 contains a range of PSS that are characterised by their distributed nature and which we identify as web-based support systems. Finally, Part 5 focuses on chapters which are dedicated to the function of planning and policy support. Therein, it is not so much the function of a specific system itself as much as the role that the kind of system (can) play in the support of planning and policy processes.

3.1 Part 1: Spatial Analysis and Urban Modelling

This second chapter "What-Ifs, If-Whats and Maybes: Sketch of Ubiquitous Collaborative Decision Support Technology", is an expanded and elaborated version of the keynote address, delivered by Harry Timmermans at the 13th CUPUM Conference in Utrecht, the Netherlands, July 2–5, 2013. In this chapter, written by *Soora Rasouli* and *Harry Timmermans,* the authors put forward the question whether dominant SDSS and PSS for sustainable urban development which are based on specific areal units are suitable for adequately predicting consumer response. Instead, they argue that the space–time behaviour of urban citizens should be represented by disaggregated models to enable greater sensitivity to behavioural heterogeneity and to higher levels of spatial and temporal resolution. This requires more detailed datasets but also more recognition of the need to address the issue of uncertainty in data and modelling systems in a more systematic and fundamental way.

In line with the previous chapter, *Toshihiro Osaragi* and *Teruo Nishimatsu* propose in "A Model of Land Use Change in City Areas Based on the Conversion of Unit Lots" a disaggregate land-use model, which is not based on the cell as a modelling unit but on the building lot. A modified version of the conventional Markov chain-style land-use model is used to create an approach based on the probability of conversion of unit lots. In order to do this, lot conversions were investigated to detect the variables that best describe the structure of the lot conversion. Based on the outcomes of this analysis, sub-models were created and validated that describe essential processes in land-use change like changes in land category, lot division, demolition of buildings and the reconstruction of buildings. These sub-models were then combined to create the final land-use model, after which its accuracy and temporal stability was validated. Moreover, the range of region sizes over which the model provided acceptable estimates was investigated. In general, the model calculations provided good results.

In their contribution "Simulating the Dynamics Between the Development of Creative Industries and Urban Spatial Structure: An Agent-Based Model", *Helin Liu* and *Elisabete Silva* present a simulation framework and an agent-based model for simulating the dynamics between the development of creative industries and urban spatial structure. In general, creative industries are envisaged as contributing to urban innovation and urban sustainability. However, in practice, the effectiveness of planning creative industries is unsatisfactory and frequently turns out to be economically and socially unsustainable, generating negative competition, requiring extensive government investment on regeneration, and resulting in vacancy rates or short-term usage and abandonment, once the government subsidy stops. The solution proposed in this contribution is to view the relationship among the agents involved as a dynamic complex process. In a spatially explicit model, simplified urban space is used to represent real urban land space. The agents involved include the creative firms, the creative workers and the local government. The resulting urban spatial structure is examined from two aspects: the spatial density distribution and the spatial clustering pattern of both the creative firms and creative workers. With the help of the framework, it is expected that more effective and sustainable urban development can be achieved. The framework is tested using the city of Nanjing, located in the Yangtze River Delta in China, as a case study. This work shows that the framework provides an interface allowing scenarios to be easily generated, although the model application also shows shortcomings such as the difficulty of fully incorporating agents' potentially irrational behaviour or subjective choices into the model.

In "LACONISS: A Planning Support System for Land Consolidation" by *Demetris Demetriou, John Stillwell* and *Linda See*, a prototype planning support system is presented called LACONISS, that integrates GIS, artificial intelligence techniques and multi-criteria decision methods to assist the planning process of land consolidation in a rural context. This process is supported in four distinctive modules: the measurement of land fragmentation; the design of alternative land distribution plans; the assessment of these alternative plans based on a set of criteria; and the production of a land partitioning plan, which is the final land reallocation plan. The spatial PSS has been implemented and applied in a case study area in Cyprus from which it shows that the system can contribute to both rural (and urban) sustainable development since it supports and automates the land reallocation process. This reallocation process can be considered as the principal component of any land consolidation project. One future challenge is how the PSS might be converted into a generic, commercialised or open source software system where the user could customise the application based on land consolidation legislation and practices of a particular country.

Kazuaki Miyamoto, Nao Sugiki, Noriko Otani and *Varameth Vichiensan* focus in "Qualitative and Quantitative Comparisons of Agent-Based and Cell-Based Synthesis Estimation Methods of Base-Year Data for Land-Use Micro Simulations" on the comparison of two approaches to estimate base-year micro-data for land-use microsimulations: cell-based population synthesis versus agent-based methods. The two methods are compared using both qualitative and quantitative methods and it is concluded that neither one is superior in each aspect. Cell-based methods are preferred when the microsimulation deals with data that are sufficiently simple, while agent-based methods are preferred when accurate and/or numerous micro-data attributes are demanded. In short, the findings suggest a way for selecting a better method based on the conditions of the microsimulation model and the purpose of its application.

In their contribution "Application of Land Use Model Combined with GIS and RS Technology in Supporting Urban Spatial Planning", *Rui Zhou, Hailong Su, Xinjun Wang, Yuanman Hu* and *Fenge Zhang* illustrate the application of the combined application of a land-use model, a GIS and a remote sensing instrument to simulate the future land use and landscape pattern change in Xinzhuang town in Jiangsu Province in China. The application shows that this combined land-use modelling approach can successfully simulate the spatial–temporal changes of future land-use under different scenarios. These outcomes can serve as an early warning system for understanding the future effects of land-use changes. Furthermore, the results can also be considered as a strategic guide for urban land-use planning in that the PSS can help local authorities to better understand the complex land-use system and undertake improved urban development and land-use management. Above all, it shows how a land-use change model can be a helpful and scientific tool for supporting urban land-use planning and policy making.

Scott Lieske, Roger Coupal, Jeffrey Hamerlinck, Donald McLeod and *Anna Scofield* present a geographical information science-based PSS in "Planning Support Systems for Fiscally Sustainable Planning" with which residential growth can be projected and the fiscal consequences of development can be assessed. To accomplish this, future urban growth is modelled, spatially and temporally disaggregated to indicate future residential growth at different planning horizons. Spatial indices are then calculated for different planning horizons and incorporated into an econometric model which calculates the effect of change in urban form on public service expenditures. Fiscal modelling and growth modelling are linked via land-use and density specifications. Applications of the system demonstrate that, with the help of the proposed PSS, savings on public service expenditures can be realised. As such, the results, both mapped and numeric, prove to be very useful for local government planning and decision making.

3.2 Part 2: Environmental Planning and Modelling

This part of the book begins with a chapter on "Generalisation of Planning Data as a Contribution to Strategic Environmental Assessments (SEA): The Example of a City-Wide Biotope-Type Assessment for Berlin", by *Michael Föerster, Antje Köeppen, Jöhanna Ferretti, Johann Koeppel* and *Birgit Kleinschmit* in which they focus on so-called GIS-based Strategic Environmental Assessments (SEA) that generally require the processing of a variety of environmental and planning data. More often than not, it is necessary to combine thematic datasets of different spatial resolutions and in order to overcome this difficulty, the authors present a method for the generalization of over-detailed planning data, in this case a city-wide biotope-type assessment of Berlin in Germany. The method includes the aggregation of an ecological assessment of a biotope-type map to a more generalized block-based inventory of open and green spaces in map form by areal weighting. The results show that the combination of the two datasets can be very sufficiently utilized to detect areas of high ecological significance in early stages of environmental planning. From this, the recommendation is to include the method as a part of a PSS used for SEA.

In "Using MapTable® to Learn About Sustainable Urban Development", *Peter Pelzer, Gustavo Arciniegas, Stan Geertman* and *Jaap de Kroes* introduce a map-based touch table, called MapTable, and explain its application as a PSS for sustainable urban development. More specifically, the chapter evaluates the extent to which this touch table can be of help to incorporate environmental values in the development of a sustainable neighbourhood in the region of Utrecht in the Netherlands. The conclusion is that the touch table facilitates learning processes by providing a platform for communication among stakeholders from different backgrounds. The application of the map-based touch table shows how it supports the integration of environmental values into spatial planning practice. Still several questions remain, of which one stands out and requires special attention in further research. This is the question of the role of the touch table in the facilitation of the communication among stakeholders from different backgrounds like urban designers and spatial planners.

Brian Deal, Varkki Pallathucheril and *Tom Heavisides* present their PSS called Land-use Evolution and impact Assessment Model (LEAM) in "Ecosystem Services, Green Infrastructure and the Role of Planning Support Systems". This is a dynamic-spatial modelling technology with which probable future development patterns can be simulated given varying planning policy or investment scenarios and the likely impacts on regional infrastructure and resources. In the chapter, the authors describe a pragmatic first step in an effort to link land-use modelling, green infrastructure resources and ecosystem valuation in their PSS in helping to sustain critical green infrastructure resources in the State of Illinois in the USA. The PSS loosely couples a dynamic-spatial model to a beneficiaries-based ecosystem valuation model and presents the resulting information in the Web-based GeoPortal. The GeoPortal facilitates exploring potential development scenarios within a river basin while tracking their implications to the regional green infrastructure system, including the related costs. The long-term vision for this work is to provide relevant information about ecosystem services and green infrastructure at varying scales to leverage land-use decision making at different political levels.

"Urban CO_2 Planning: A Decision Support System" by *Ivan Blecic, Arnaldo Cecchini, Matthias Falk, Serena Marras, David Pyles, Donatella Spano* and *Giuseppe Trunfio* presents a DSS with which carbon exchanges of alternative land-use scenarios can be estimated with the help of a modelling framework. It is envisioned that the complexities of urban systems and the interaction of carbon emissions with the environment and local meteorology make it a non-trivial task to support urban planning and management for a more sustainable metabolism of future cities with the help of a modelling framework. The proposed framework encompass three components: (1) a cellular automata (CA) model for the simulation of urban land-use dynamics; (2) a transportation model to capture the long-term average vehicle load on the road network; and (3) a soil–vegetation–atmosphere transfer model tightly coupled with a weather forecasting model for simulating interactions related to carbon fluxes between the city, environment and local weather. With the help of this modelling framework, the impacts of future planning alternatives on carbon emissions are evaluated for the City of Florence in Italy. As a result, the authors recognize the difficulty of calibration but also see the application of the framework as providing a promising perspective for developing decision support tools for sustainable urban planning.

In "A GIS-Based Performance Metrics for Designing a Low Energy Urban Agriculture System", *Steven Jige Quan, John David Minter* and *Perry Pei-Ju Yang* introduce a low energy urban agriculture system in Atlanta in the USA that was conceptualized by a performance-based urban design model, a model that suggests a metabolic process of energy, material and water flows in an urban setting. It explores how dimensions of representation, performance and scenario planning can be integrated into a PSS that connects urban form, ecological performance and future scenarios for design decision making. Based on the test case in Atlanta, the proposed PSS model demonstrates how a GIS platform is designed for accommodating various sources of information, including energy, material and water flows in the city. This is deemed essential information for a local food production system that aims to grow local food using food waste and thus reduce the food desert. In the PSS model, the authors establish a set of tools to manage the representational information, to use them for conducting performance energy flows and waste stream flows at four spatial scales, and to evaluate the overall performance of the system under different development scenarios. The findings are encouraging; they go beyond traditional solution-based urban design and provide a more comprehensive understanding of how an urban agriculture system functions and performs. The authors conclude that such results could not be achieved without a PSS framework.

Energy modelling in the built environment at different scales is the focus of "A Comprehensive Review of Existing Urban Energy Models in the Built Environment" by *Saleh Mohammadi, Bauke de Vries* and *Wim Schaefer.* Until recently, modelling approaches have focused on energy flow simulation in urban areas from generation to distribution and finally consumption. The authors provide an overview and an assessment of three main approaches to simulate energy flow in urban areas. The evaluation is done based on the energy flow direction and the application focus of these models in the urban energy system. Accordingly, three main distinct categories of models are identified, including: supply, demand and integrated models. Each of these categories uses different assumptions and simulation techniques and relies on different levels of data. As such, there is a wide range of urban energy models that vary greatly in terms of spatial and temporal scales, energy side focus, analysis variables and methodological approaches. Based on a uniform set of criteria, a critical review of these models is performed and the conclusion is that some key issues are missing in the existing urban energy models, like integrated multi-layer models, spatial implications of renewable energy technologies, and integrated simulation and optimization methods.

3.3 Part 3: Traffic and Network Modelling

The first chapter in this section of the book concerns a demonstration of "A Procedure Using GIS to Analyze the Access by Non-Motorized Transport to Transit Stations" by *Fernanda Monteiro* and *Vânia Campos*. The underlying idea is that for achieving sustainable mobility, it is important to encourage the use of non-motorized and public transport. The quality of urban space around public transport stations can be an important factor to encourage the access of non-motorized transportation, especially by foot or by bicycle. Thus, it is envisaged that the best way to increase the use of public transport, instead of automobile use, is to improve pedestrian facilities and cycling conditions to access the stations. The proposed procedure evaluates walk and bicycle paths near the stations based on indicators mainly related to the physical characteristics of these paths. To define these indicators, a bibliographical review and a survey research in subway stations in Rio de Janeiro in Brazil were developed. From the survey, the distances travelled by users to the stations were evaluated and the attractiveness factors of walk and bicycles trips around them were identified. Based on these indicators, a spatial analysis of paths around a transit station were performed with the help of GIS. The maps presented indicate the quality levels of the walking and bicycle paths and indicate where potential improvements can be made.

In "Locations with Frequent Pedestrian-Vehicle Collisions: Their Transportation and Neighborhood Environment Characteristics in Seattle and King County, Washington", *Junfeng Jiao, Anne Moudon* and *Yuan Li* describe an application for the identification of locations where vehicle–pedestrian collisions are frequent, and for the isolation of road and neighbourhood environment characteristics of locations that are associated with frequent collision occurrences. It is considered that successfully identifying these characteristics will help researchers to develop a better PSS to locate, predict or remove collision hot spots within a city and provide safer environments for the residents. The analyses demonstrate a consistent effect of the road and neighbourhood environment on the frequency of pedestrian collisions; the more intensely developed and travelled environments were also the locations where pedestrian collisions were more frequent. A direct association is noted too between proxy measures of pedestrian volumes, such as the density of residential units and jobs, and bus boarding and alighting counts with the number of collisions. This same direct association is noted with higher vehicular volumes. These proxy measures of pedestrian activity and exposure to vehicles confirm a lack of pedestrian safety in metropolitan areas. In fact, the relatively small number of locations with high and very high pedestrian-vehicle collisions should facilitate targeted, effective and inexpensive pedestrian safety improvement programmes.

The authors of "A Computer-Aided Approach for Planning Sustainable Trips to Large Trip Generators: The Case of Cycling Routes Serving University Campuses ", *Thais de Cássia Martinelli Guerreiro* and *Antônio Nélson Rodrigues da Silva* present an approach for the planning of sustainable trips to large trip generators, in their case cycling routes serving Sao Paulo's university campus. Using this approach, the design and evaluation of potential impacts of new cycling routes connecting a university campus with the city's street network can be performed. GIS tools are used to assess the potential impacts of proposed routes on bicycle, car and walking trips. The results show that the proposed methodology is adequate for exploratory studies and transferable to a PSS which could be designed for a variety of goals, such as the design of paths, management of bicycles flows, *et cetera*. Although this study focuses on the particular case of a university campus, it is concluded that the approach can be easily generalized and applied to other large trip generators.

In their contribution "Walk Route: A New Methodology to Find the Optimal Walking Route in the City of Atlanta", *Subhrajit Guhathakurta, Ge Zhang, Manoj Panguluru* and *Ramachandra Sivakumar* demonstrate a new methodology for finding optimal walking routes according to user-specified conditions. The pedestrian network planning methodology reflects the influence of environmental factors facilitating or impeding pedestrians' propensity to walk. Several characteristics of the built environment have been identified to influence walking behaviour. The Analytical Hierarchy Process (AHP) is used to weigh each attribute, which is then combined to calculate a walk score. Thereafter, the optimum route is determined by applying a routing algorithm based on the walkability cost of the network. In a case study of the city of Atlanta in the USA the application of this methodology is demonstrated. The methodology is designed to address multiple objectives: (1) it offers a means for pedestrians to identify routes that meet their requirements; (2) it provides a framework to measure the quality of the pedestrian infrastructure in the city; and (3) it serves as a decision support tool for urban planning to make neighbourhoods and places more walkable. Although this study focused on the pedestrian's experience, the methodology could also be modified and applied as a route optimization model for people with disabilities.

3.4 Part 4: Web-Based Support Systems

In the initial chapter of Part 4, "Access to UK Census Data for Spatial Analysis: Towards an Integrated Census Support Service", *John Stillwell, Justin Hayes, Rob Dymond-Green, James Reid, Oliver Duke-Williams, Adam Dennett* and *Jo Wathan* explain the development of a web-based 'one stop shop' where users can access a range of data from different sources, including those associated to the United Kingdom (UK) census of population. Censuses remain a crucially important source of demographic and socio-economic data both for academic research as well as planning and policy making for urban sustainability in the UK. Data from recent UK censuses have been used to produce a variety of products in digital form at different spatial scales, ranging from the counts of individuals or households with particular characteristics captured by the census questionnaire and often referred to as aggregate statistics, to samples of the population at individual or household level, known collectively as micro data. They are used to create geo-demographic profiles of local populations, to identify health, educational and occupational characteristics, to measure levels of deprivation and affluence, as well as to establish the characteristics of households and the housing which they inhabit. Access to these often large and complex data sets for social science research has been facilitated through the development of a set of services funded by the Economic and Social Research Council (ESRC) over the last decade. In the chapter, the authors describe the transition of the present online census support service to a more integrated service of Census Support within a new UK Data Service.

In "The Online What if? Planning Support System", *Christopher Pettit, Richard Klosterman, Marcos Nino-Ruiz, Ivo Widjaja, Patrizia Russo, Martin Tomko, Richard Sinnott* and *Robert Stimson* present an online version of a well-known PSS called Online What if (OWI). The chapter describes the development and applications of OWI that is being developed as a component of the Australian Urban Research Infrastructure Network (AURIN—www.aurin.org.au) project. AURIN has been established to provide advanced information infrastructure to support interdisciplinary research and promote sustainable urban development in Australia. OWI enables a range of end users to create and explore what if? land-use change scenarios for an area's future land use, population and employment patterns. The authors discuss OWI in the context of a demonstrator case study in Hervey Bay, Queensland, and introduces future applications of this public participation GIS (PPGIS) tool to support the sustainable planning of cities in Australia. Moreover, it illustrates the potential that advanced eResearch infrastructure can provide for helping planners, decision makers and private citizens to explore alternative sustainable futures for cities and regions around the world.

"A Web-Based Fuzzy CA Model for Urban Growth Simulation" by *Yan Liu* describes a methodology to implement a fuzzy constrained urban CA model within a web-based GIS environment. At present, CA models are increasingly adapted by geographers and planners to simulate the spatial and temporal process of urban growth. However, in practice, the operation of CA models and the calibration of the parameters mostly are just known to the modellers. This is largely due to the constraint that most CA models are developed on desktop computer programs. Consequently, there is little input from the user to test or visualize the actual operation or to evaluate the applicability of the model under different conditions. In contrast, in the web-based CA model presented in the chapter, users can visualize and test the operation of the model; they can modify or calibrate the model's parameters and evaluate its simulation accuracies; and can feed the model with 'what-if' conditions to generate alternative outcomes. The model is applied in a case study for the Gold Coast City in Southeast Queensland, Australia to simulate the spatial–temporal process of urban growth. Based on its application it is concluded that the web-based CA modelling platform provides a useful channel to foster public participation in urban planning and management. Moreover, it will contribute significantly to the development of CA modelling as well as to the understanding of urban growth dynamics.

In "Flexible Geospatial Platform for Distributed and Collaborative Urban Modelling", *Yi Zhu, Mi Diao, Joseph Ferreira, Weifeng Li* and *Shan Jiang* present an infrastructure for a flexible GeoPortal platform for supporting distributed urban modelling with spatially detailed land-use and transportation interactions. The modelling infrastructure supports the loosely-coupled large-scale land-use, transportation and environment modelling work with distributed research groups and public agencies. The infrastructure is based on the PostgreSQL database structure and two visualization tools which can be used to coordinate the results from a variety of analytical models, to provide web-based visualization of data and to save and share useful spatial patterns via WMS records or URL links. It provides collaborative tools to support parallel development and interconnection of complex model components in a manner that facilitates the visualization, discussion and sharing of intermediate results. Large-scale land-use, transportation and environment modelling contains many components which are themselves complex, in various stages of development and distributed among different research groups. Many of the datasets needed for model calibration are incomplete, more aggregated than desired or yet to be acquired. The GeoPortal platform is aimed at accelerating the realistic evolution of modelling data and cross-group learning as well as enhancing sense-making during early stages of model exploration and integration.

In the final contribution in this section, "The Participatory Cube: A Framework for Analysis of Online Participation Platforms", *Alenka Poplin, Gilberto Corso Pereira* and *Maria Célia Furtado Rocha* describe a framework for the analysis of online participation platforms. These platforms should enable interested citizens to contribute their knowledge and opinions and thus to create more sustainable living for everybody. The authors propose the so-called participatory cube as a framework for the analysis of the large variety of participatory applications available online. This framework consists of three axes which form a cube and which represent three dimensions identified as the most relevant for the analysis. These dimensions are: decision power; interactivity of communication; and access to space of communication. The framework enables the specific characteristics of online participation platforms to be analysed and assessed. This is applied to a range of case studies, originating from Germany and Brazil, and encompassing technologies that support civic engagement. From this comparison the authors conclude that additional research is needed to understand which approaches can enable satisfactory participation processes and how this success and satisfaction can be measured.

3.5 Part 5: Planning and Policy Support

The authors of "Application of Socio-Technical Research Methods in Understanding the Genesis and Potential Sustainability of Planning Support Systems", *Wayne Williamson* and *Bruno Parolin,* ask themselves the question: 'How does the application of socio-technical research methods enhance our understanding of the genesis and potential sustainability of PSS?' To answer this question they compare two methodologies. First, they have collected data from interviews and the application of the so-called Actor-Network Theory (ANT); second, they request the completion of an online questionnaire by planners and GIS staff working in local and state government and private practice. The questionnaire data were analysed using a structural equation modelling (SEM) technique under the Unified Theory of Acceptance and Use of Technology (UTAUT) framework. Results of applying ANT provide useful insights into the social and technical interactions that are required to build and implement a PSS. The UTAUT results stress that in order for Information and Communication Technology (ICT) applications to be widely accepted by planners, the organizations in which they work need to address performance expectancy and facilitating conditions as priorities. The insights gained from the ANT and UTAUT analyses appear to be complementary and suggest that the introduction of software is not a value neutral innovation that is just adopted by an organization, but rather, software can be viewed as a non-human actor that translates and is translated throughout the adoption process. Moreover, it shows there is a need for an on-going translation process which will determine the success or failure of the adoption process. In that, the process of introducing ICT tools into an organization is a process of network formation in which actors seek to persuade others to become enrolled and promote the acceptance of their view of the way the problem can best be solved.

Yanliu Lin and *Stan Geertman,* in "Governance Approaches in the Regeneration of Immigrant Communities: Potential Roles of Planning Support Systems", start by considering the political issue of massive migration from rural into urban areas in developing countries in particular. For the governments involved, it is almost impossible to provide or subsidize sufficient housing for these low-income migrant groups with the result that immigrant communities develop that are characterized by high concentrations of poverty and social and economic deprivation, and possibly also economic, physical, and social exclusion. Traditional planning approaches fail to sustainably regenerate these communities due to complex social, economic, spatial issues, and due to conflicts between key stakeholders. New governance approaches have been advocated in which the interplay between state, market and civil society results in several modes of governance in these contexts. The authors argue that each mode of governance has its positive and negative consequences for the regeneration of immigrant communities and that PSS can assist in dealing with the negative consequences. Therein, they conclude that three types of PSS—namely 'informing PSS', 'communicating PSS', and 'analysing PSS'—can play a distinctive role in dealing with the specific deficiencies of each mode of governance in the context of immigrant communities.

In "Supporting Planning Processes by the Use of Dynamic Visualisation", *Stefano Pensa, Elena Masala* and *Isabella Lami* set themselves the goal to build a method for using an intuitive visual language which will enable communication among decision and policy makers, as well as other kinds of public or private actors for allowing the planning process to be effective in including participation and collaboration as approaches to a sustainable urban development. In order to build common knowledge and facilitate reasoning among the different actors, the authors propose a method for sharing information through the use of dynamic maps. Therefore, they propose the Interactive Visualization Tool (InViTo), a visual method for managing spatial data in real-time, based on parametric three-dimensional modelling. InViTo is based on Grasshopper, a free plug-in of McNeel's 'Rhinoceros', a 3D modelling software package most often used in architecture design. InViTo is a tool conceived as a PSS for aiding the actors involved in sharing information and raising awareness of spatial issues. The system has been used in different case studies and has shown its effectiveness in creating awareness of spatial problems and enhancing discussions.

In the final chapter of the book, "Beauty and Brains: Integrating Easy Spatial Design and Advanced Urban Sustainability Models", the authors *Eduardo Dias, Marianne Linde, Azarakhsh Rafiee, Eric Koomen and Henk Scholten* describe the challenges of creating a PSS acceptable by designers and report the success of combining two existing planning support tools: 'Phoenix', and 'Urban Strategy'. The systems were developed in different contexts by and for different groups, but their combination unlocked the key to user acceptance and adoption. Their low learning curve attracts even technophobic stakeholders and makes use of recent advances in cloud computing for storage and processing to deliver immediate feedback. It allows anyone, but in particular designers, to sketch their set of solutions with natural movements and receive prompt feedback on key indicators of sustainability performance, enabling iterative improvements of designs. Based on past experiences and future case studies, the authors envisage a bright future for these integrated tools in supporting the professional tasks of urban designers, in short 'Geodesign'.

4 Conclusions

The suite of contributions that we have introduced and which follow in the remainder of the book represents a representative cross-section of PSS for sustainable urban development. This collection highlights the diversity and emphasises the dynamics of on-going PSS developments, in this case dedicated to the planning and policy making issues that surround sustainable urban development. It is worth reiterating that achieving sustainable urban development is a hugely difficult task to accomplish because of the multidimensional nature of the social, economic and environmental dimensions of sustainability and the way in which they interact and/or compete with each other across space and over time. Endeavouring to cope with this complexity requires support tools of which the present generation of PSS contains some excellent example.

We hope that the contents of the book chapters about PSS for Sustainable Urban Development assembled in this collection will encourage and enthuse you to recognise the current state of the art in PSS development and to appreciate the value of the new and innovative developments that are on-going. We also hope that those amongst you who are practitioners will consider ways in which PSS can be developed and nurtured for use in facilitating your own decision making and planning activities.

References

Batty M (1995) Planning support systems and the new logic of computation. Reg Dev Dialogue 16(1):1–17

Brail R (ed) (2008) Planning support systems for cities and regions. Lincoln Institute for Land Policy, Cambridge, Massachusetts

Brail R, Klosterman R (eds) (2001) Planning support systems, integrating geographic information systems, models and visualization tools. Esri Press, Redlands California

Castells M (1996) The rise of the network society. Blackwell, Oxford

Castells M (2001) The internet galaxy; reflections on the internet, business and society. Oxford University Press, Oxford

Clarke M (1990) Geographical information systems and model-based analysis: towards effective decision support systems. In: Scholten H, Stillwell J (eds) Geographical information systems for urban and regional planning. Kluwer, Dordrecht, pp 165–175

Geertman S (2006) Potentials for planning support: a planning conceptual approach. Environ Plan B: Plan Design 33(6):863–881

Geertman S, Stillwell J (eds) (2003) Planning support systems in practice, 1st edn. Springer, Berlin

Geertman S, Stillwell J (eds) (2009) Planning support systems: new methods and best practice. Springer publishers, New York

Goedman J, Zonneveld W (2007) In search for the sustainable urban region. International conference ENHR2007, Rotterdam

Harris B (1999) Computing in planning: professional and institutional requirements. Environ Plan B: Plan Design 26(3):321–331

Healey P (2006) Collaborative planning: shaping places in fragmented societies, 2nd edn. Palgrave Macmillan, Basingstoke

Healey P (2007) Urban complexity and spatial strategies: towards a relational planning for our times. Routledge, London

Klosterman RE (1997) Planning support systems: a new perspective on computer-aided planning. J Plan Educ Res 17(1):45–54

Klosterman RE (1998) Computer applications in planning. Environ Plan B: Plan Design (Anniversary Issue) 25:32–36

Knaap G (2002) Talking smart in the united states. Paper prepared for the international meeting on multiple intensive land use. Gouda

Pivo G (2002) The six C's of sustainable urbanisation (online), 2010. Available on the World Wide Web: http://www.cogeneration.net/sustainable_urbanisation.htm

Priemus H (2003) Business plan/research spearhead 'sustainable urban areas'. Delft University of Technology, Delft

Rollnick R (2006) The third World urban forum 'Spaceship Earth' spiralling irreversibly into the Urban Era. United Nations chronicle online edition

Stillwell J, Geertman S, Openshaw S (1999) Geographical information and planning. Springer, Berlin

Tanguay GA, Rajaonson J, Lefebvre J et al (2010) Measuring the sustainability of cities: an analysis of the use of local indicators. Ecol Ind 10(2):407–418

UN Habitat (2009) Planning sustainable cities: global report on human settlements 2009. Earthscan, London

Part I
Spatial Analysis and Urban Modelling

What-Ifs, If-Whats and Maybes: Sketch of Ubiquitous Collaborative Decision Support Technology

Soora Rasouli and Harry Timmermans

Abstract Since its introduction, geographic information science has witnessed a tremendous growth and can build on enormous achievements (e.g. Cheng et al. 2012). Current geographic information systems and the decision support systems and models that have been accompanying these systems have a strong 'geography' identity, typical of the era in which geographic information system were introduced. Systems are mostly based on spatial entities (mostly grids or polygons). To the extent that commercial and open source geographic information systems have been enriched with models, a similar strong geographic flavor can be discerned. Most models of spatial choice behavior are related to the aggregate spatial interaction models, models of land use change are often based on cellular automata. The question then becomes whether dominant spatial decision support systems, fundamentally based on aggregate spatial interaction, cellular automata and similar models, are suitable for adequately predicting consumer response. We content that in light of the increasing complexity of the decision making process and increasing personalization of decisions and lifestyles, these systems and their underlying models have increasingly become inadequate and obsolete. The field should shift to the development of more integral microscopic models of choice behavior, allowing more integral policy performance assessments. Moreover, mobile computing should allow and stimulate the development of real-time information and decision support systems that support the management of urban functions and include persuasive computing. Uncertainty analysis should play an integral role in these developments.

This chapter is an expanded and elaborated version of the keynote address, delivered by Timmermans at the 13th CUPUM Conference in Utrecht, July 3, 2013.

S. Rasouli · H. Timmermans (✉)
Urban Planning Group, Eindhoven University of Technology,
5600 MB, Eindhoven, The Netherlands
e-mail: h.j.p.timmermans@tue.nl

S. Geertman et al. (eds.), *Planning Support Systems for Sustainable Urban Development*, Lecture Notes in Geoinformation and Cartography,
DOI: 10.1007/978-3-642-37533-0_2, © Springer-Verlag Berlin Heidelberg 2013

1 Introduction: Shifting Contexts

Since its introduction, geographic information science has witnessed a tremendous growth and can build on enormous achievements (e.g. Cheng et al. 2012). Scholars from a variety of disciplines and backgrounds have become involved in this area of research and development, and made significant contributions to the field. The number and more importantly the quality of international journals in geographic information science have experienced wealthy growth and steady improvement. International conferences, including CUPUM, have continued their success in attracting leading scholars in the field, presenting their latest research findings and reflecting on existing areas of research to formulate evolving research agendas. The application of geographic information systems and associated decision and planning support systems has shown profound evidence of international dissemination and valorization (e.g. Geertman and Stillwell 2009).

This is not to say that all tools, models and integrated decision support systems have found ample application (Vonk et al. 2005, 2007a, b; Te Brömmelstroet and Schrijnen 2010). It is probably fair to argue that in many parts of the world geographical information systems with their mapping opportunities have gradually replaced conventional maps and databases. Information science has entered many private and public firms and organizations. It has affected daily work in a positive way, increasing productivity, although issues of synchronization, the existence of different data formats and software platforms remain (e.g. De Paoli and Miscione 2011). The application success of decision support systems shows considerably more variability. In some fields of application, such as transportation planning, the use of forecasting and impact assessment models is relatively common. Other fields, such as retail planning, have shown significant fluctuation (e.g. Huang and Levison 2011; Shen et al. 2011; Rasouli and Timmermans 2013), while the use of decision support systems in other domains is still virtually lacking (e.g. Johnson and Sieber 2011). Regional and international differences prevail.

In general, the acceptance of new technology depends on perceived needs, awareness of the technology and the extent to which the technology successfully addresses the needs (e.g. Bailey et al. 2011; Slotterback 2011; Rae and Wong 2012). In that context, it is relevant to memorize that geographic information systems and the decision support systems and models that have been accompanying these systems have a strong 'geography' identity, typical of the era in which geographic information system were introduced. Systems are mostly based on spatial entities (mostly grids or polygons). To the extent that commercial and open source geographic information systems have been enriched with models, a similar strong geographic flavor can be discerned (e.g. Voudouris 2011). Most models of spatial choice behavior are related to the aggregate spatial interaction models (e.g. Silveira and Dentinho 2010), models of land use change are often based on cellular automata. In particular, the number of applications of cellular automata model is staggering (e.g. Furtado et al. 2012; García et al. 2011, 2012; Long et al. 2012; Mahiny and Clarke 2012; Norte Pinto and Pais Antunes 2010; Plata-Rocha et al.

2011; Tang 2011; Thapa and Murayama 2011; van Vliet et al. 2012; White et al. 2012; Liu 2012).

This dominance of aggregate models is understandable considering the constant flirting of geography and regional science with aggregate models, originally developed in physics in combination with the geography background of early developers, and technical considerations such as the fact that the spatial representation of the data in geographic information systems is congruent with the principles underlying these aggregate models. The dominant inclusion of aggregate models in geographical information systems is not only understandable from the evolution of the key underlying disciplines, it is also understandable from a computer science perspective in that limited memory and computing power necessitated constraints on data and model complexity in the early years. Conveniently, the latter was amplified by the methodological principle of parsimony as the hallmark of scientific research and model development (e.g. Klosterman 2012).

The functionality of geographic information systems and associated models served the primary needs of urban planning (e.g. Vonk et al. 2007a, b). In many countries, the process of urban planning is orchestrated by public planning authorities, which on the one hand stimulate particular developments and on the other hand constrain others by formulating norms and guidelines. In addition, detailed land use planning shapes and signals legal authority for citizens and firms. In particular at higher levels of planning hierarchies, an academic approach to plan development and assessment has implied a relatively structured planning process involving the formulation of plan scenarios, the prediction of their likely effects in terms of a set of performance indicators (what-if questions) and an evaluation of plan scenarios and alternatives (e.g. Ligmann-Zielinska and Jankowski 2010; cf. Grêt-Regamey and Crespo 2011). Exogenous trends and scenarios are typically described for official statistical areas (neighborhoods, municipalities, provinces), explaining the use of the areal classification for data representation and model operation. This focus has been further legitimized by the fact that several specific planning domains have been strongly influenced by fundamentally spatial concepts. An example is the functional hierarchy of shopping centers which stimulates a neighborhood-based provision of stores serving local communities for their daily shopping needs, complemented by larger regional shopping centers and the city center itself serving a larger sphere of influence. Another example is the neighborhood concept, which assumes social coherence, and a sense of place, branded in local plans.

Spatial decision support systems provide the functionalities for evidence-based planning and design, with space typically represented at some level of aggregation (grid-polygon). The system produces the values of alternative plans on a set of performance indicators deemed relevant to judge the quality of the plans in meeting a set of corresponding plan objectives. The effects of plan interventions in many cases depend on how individuals and households are reacting to such interventions and adapt their current behavior. For example, the feasibility of new shopping malls and their external effects on the turnover of existing shopping centers depends on how households reorganize their shopping habits and substitute currently used centers for the new malls. Similarly, the success of the concept of high density, mixed land use neighborhoods in curbing mobility rates, depends on how much people's activity-travel decisions are influenced by such spatial characteristics.

2 From Aggregate to Microscopic Models

The question then becomes whether dominant spatial decision support systems, fundamentally based on aggregate spatial interaction, cellular automata and similar models, are suitable for adequately predicting consumer response. We content that in light of the increasing complexity of the decision making process and increasing personalization of decisions and lifestyles, these systems and their underlying models have increasingly become inadequate and obsolete. As argued by Chapin (1968) already decades ago, most land uses and transportation systems do no exist for their own sake, but because they are means for conducting mandatory and discretionary activities. People have certain needs and desires and in order to achieve these, they need to become involved in a set of activities. Because the facilities (or land use) enabling them to become involved in these activities are spatially distributed as a result of a (planned) sorting process, people need to travel. Activities constitute the link between the city and the people, while travel is the mechanism that makes the city working as an integral system. Chapin thus argued that the study of activity patterns, recurring in space and time, might ultimately yield better theoretical explanations and improved predictive models for urban planners and designers. Modelling consumer reactions to changing exogenous circumstances and plan interventions in terms of changes in their daily activity travel patterns thus seems paramount to fulfil the articulated need of policy assessment in some policy domains. Such microscopic models should logically outperform conventional aggregate models in that they capture particular behavioural mechanisms, behavioural heterogeneity and complexity that the spatial interaction models fundamentally ignore. The extent to which the microscopic models outperform aggregate models depends on the relative prevalence of such systematic errors and the extent to which these errors are counterbalanced elsewhere in the modelling process.

For a long time, such microscopic models were deemed impossible. However, increasing computing power, and developments in data fusion and modelling frameworks have allowed the first large-scale microscopic activity-based models of travel demand (e.g. Arentze and Timmermans 2000; Bhat and Singh 2000; Bowman and Ben-Akiva 2001; Miller and Roorda 2003; Raney et al. 2003), which simulate at high spatial and temporal resolutions which activities are conducted, where, when and for how long, with whom, and the transport mode involved.

3 The Shift to Integral Performance Assessments

The leading theme of this conference is sustainability. Our contention is that a shift from aggregate spatial interaction models to microscopic spatial choice and decision-making models would enhance our ability to better support urban decision making processes. Energy efficiency, social cohesion, cultural and economic prosperity, health and safe built environments characterize sustainable urban development. All these topics are strongly intertwined with how individuals and households organize their daily lives, reflected in their activity-travel patterns in different urban settings. Assuming that the achievement of these goals depends, at least partially, on a good understanding of urban activity-travel patterns, the development of valid and reliable activity-based models seems paramount to better support urban planning processes. It does not only have the advantage that actual decision makers as opposed to spatial units, which do not make any decisions, are modelled, the attention to space–time behaviour at a high level of spatial and temporal resolution also implies that in addition to the usual economic and social performance indicators, environmental impacts can be simulated.

Traffic contributes significantly to urban emissions and thus the impact of daily travel on emissions can be predicted. In this context, it should also be emphasized that current legislation articulates the importance of high spatial resolution at the level of a building plan. This level of resolution is also required if we move from emissions to air quality and exposure. The current models express exposure based on zonal population statistics. However, as a direct result of their daily activity-travel behaviour, the population moves across the city during the day. Therefore, exposure assessments, based on zonal population statistics are fundamentally flawed. Other examples of improved sensitivity of microscopic models to policy indicators where aggregate models fail include the amount of social exclusion, quality of life, and time pressure.

4 Ubiquitous Pervasive Information

The relationship between the urban systems and activity-travel patterns is of central importance in this context. As argued, several urban planning and design concepts saluted to this intricate connection. A better understanding of activity-travel patterns should ultimately lead to improved evidence-based planning, but due to dynamics in both the urban system and people's preferences, and the omnipresence of imperfect information, by definition urban systems will always be out of equilibrium. Wardop's user equilibrium assumes that network equilibrium occurs when no traveller can unilaterally change his route choice to reduce travel times. It may be a theoretically appealing concept, but has no behavioural foundation in the sense that the implicit notion of complete and full information does not have any credibility.

Information and communication technology can play a critical role in reducing user disutility of out-of-equilibrium artificial urban systems. First, by providing the right information at the right time, the uncertainty and possibly ineffective behavioural response and decisions that stem from incomplete and imperfect information may be reduced. In some cases, for example whether there is a direct bus connection between A and B, static information may suffice. In other cases, whether it is better right now to choose the car or the bus between A and B, real time traffic information will be more effective. Second, in addition to user benefits, public and private firms and organisations may also profit. For example, park guidance systems will allow park managers to maximize the use of their parking facilities. Users also benefit in this case assuming that indeed they prefer to find a vacant parking place as quickly as possible. Hence, the inherent disequilibrium may at least partially be addressed by providing the right information to the right people such as to optimize system level performance. Thirdly, whereas these examples involve neutral information about the (time-varying) state of the urban system, information and communication technology may also be used to try and persuade people to behaviour in a certain way. The underlying objective may be to improve personal service, but also to optimize service-level performance or some combination of these. Examples such as minimizing system-wide congestion, discouraging trucks traveling through environments with many children, and reducing emissions quickly come to mind.

This argument acknowledges the shift in policy from urban development to urban management and control. Such management and control stems from the notion that urban systems almost by definition are in a constant state of disequilibrium. Development strategies are ineffective in reducing the short-term negative effects of disequilibrium. The use of information and communication technology has become increasingly more prevalent in this context. Our contention is that ICT will be the new layer for the "smart" city of the future. Mobile technology, combined with intelligent systems, will create ubiquitous environments and make information omnipresent. Intelligence, accuracy, personalization, persuasion, real time are some of the buzzwords in this development (e.g. Lu and Liu 2012).

Mobile tools and mobile e-services will mark the next major developments in spatial decision support systems. The emergence of grid computing and service-oriented architectures, computing is becoming increasingly less confined to traditional computing platforms. Grid computing promises the accessibility of vast computing and data resources across geographically dispersed areas. Mobile wireless devices significantly enhanced this capability to deliver access to high-performance computing under demanding circumstances. Cloud computing makes storage of large amounts of data a lesser problem. E-services guarantee access to software. New dedicated languages and platforms will allow users to easily perform a set of related analyses, with dedicated support of how to conduct such analyses.

This shift to mobile platforms should not necessarily imply different content as evidenced by services and apps supplied by particular municipalities which have continued to provide access to city statistics, maps and current plans: the communication tool differs—the contents does not. However, Web2.0 technology does offer new opportunities. In particular, social media provide interesting new opportunities. Members of a community can exchange information, trace particular others, identify the most current location of particular friends, etc. These functionalities offer the co-production of maps, databases, exchange of real time information, a platform for organizing meetings and location-based services, etc. Applications rapidly emerge: updates of databases, underlying navigation systems, group-tracing systems, updates and experiences of actual, real time transport systems, community-based portals in the context of plan development, and many similar examples could be mentioned.

5 Implications for Research Agenda

To this point, we have contemplated that effective decision and planning support systems for sustainable urban development should address space–time behavior of its citizens as this constitutes the logical basis for developing an integrated model system that derives emission, exposure, noise, social exclusion and other performance indicators from the way people organize their daily lives in a particular urban setting as reflected in their daily activity-travel patterns. The main consequence for the development of geographic information systems is that areal units no longer make up the core of the system because individuals, households and activity locations should be represented at higher levels of spatial resolution. This need for higher resolution is reinforced by the requirements of especially environmental policies. In addition, with time as the linking pin, the need for a temporal organization of data emerges, creates a new challenge as most mainstream geographical information systems are not organized around temporal data.

Models themselves need major elaboration and new model types need to be developed. Whereas operational activity-based models of travel demand are not widely available, considerable progress is still required in short-term traffic forecasting, individual use of travel information, and particularly in modeling traveller response to persuasive information as a function of different underlying goals (personal preferences, system performance or some combination). Travellers will be aware of the fact that other travellers may also receive recommendations and hence have to make decisions, considering their beliefs of how other travellers will react. Such models of strategic decisions under conditions of uncertainty are still at the very early stages of development (Han and Timmermans 2006a, b). Assuming that such models will be developed with a sufficient degree of accuracy, then distributed recommendations have to be provided in case of a single control agent. What-if questions will be complemented by if-what questions! Such information provision systems are currently not available either, at least not any systems that go beyond simple practical solutions.

Models need data for estimation and validation. This new generation of models is no exception. In fact, data requirements are formidable, considering the real-time and high spatial and temporal resolution of the envisioned models. Traditionally, the data for activity-travel models stem from one-day travel surveys. It does not need many arguments to reason the pure inadequacy of such data for the next generation of models. We need "technology for technology". Fortunately, several recent technological developments may provide a solution. In recent years, transportation research has witnessed a tremendous growth of interest into the use of modern information and communication technology (ICT) for data collection. Modern mobile technology will become a superior alternative for these traditional surveys in collecting data about individual travel patterns. Mobiles devices range from laptop computers, Personal Digital Assistants (PDAs) and mobile phones to mobile game computers. In the near future, new technologies such as Augmented Reality may become widely available. Also changes in wireless networking will offer new opportunities: cellular telephony is moving from low-bandwidth to higher bandwidth Universal Mobile Telecommunications System (UMTS) to support more advanced services (e.g. graphics). Wireless networks provide a higher bandwidth but are primarily used for laptops and PDA's. Hotspots support access to the Internet. Bluetooth provides a low-bandwidth, short-range protocol for communication between devices (e.g. mobile phone and head set). An upcoming technology is Worldwide Interoperability for Microwave Access (WI-MAX) that provides a long range, fast connection and potentially competes with the UMTS standard. These developments lead to ubiquitous travel environments: information can be shared in a network environment from some geo-sensors providing users with information readily available anywhere, for any person and at any time. This information can be descriptive, but can also relate to recommendations. The same technology however can also be used to track individuals. When used in isolation, tracking technology such as GPS is insufficient as it only provides information about route choice behaviour and the current position of the person tracked. However, in principle such data can be enhanced with other spatial and non-spatial data. Moreover, GPS traces can be interpreted using particular algorithms to derive information about other facets of activity-travel patterns.

It should also be articulated that cars of the future are driving computers. Data on routes, emissions, energy consumptions, driving style, use of travel information and navigation systems, etc. can be directly extracted from the computer. Similarly, smart cards offer huge potential datasets on the use of public transportation systems. Devices recording energy consumption of home appliances should be very useful for modelling energy consumption. Admittedly, the different sources of data relate to different people. However, modern data fusion techniques can be used to create synthetic populations. These will not be perfect, but perfect for the task at hand, certainly offering much richer and larger data sets than the data that we currently are forced to use.

Indeed, these datasets will be huge, so huge and complex that it becomes difficult to process using current database management tools. A GPS device with a 1 s time interval will record 35 million data points over a 100 day time period. The Smart Card data in Seoul generates 10 million observations per day. The so-called "Big data" are difficult to handle with existing geographical information systems and associated relational and object-oriented databases, desktop statistical, estimation and mapping/visualization packages. Planning authorizes are currently exploring the opportunities of parallel and cloud computing using tens, hundreds, or even more servers. They did not touch these increased data set sizes. Hence, there seems a bright future for intelligent supervised machine learning and data mining algorithms.

The combination of these imperfect, imputed data and the anticipated use of microscopic simulation models with their inherent model uncertainty, bring about the need to systematically and more fundamentally address the issue of uncertainty in complex model systems (Rasouli and Timmermans 2012a, b). To differentiate policy impact from model uncertainty, a formal uncertainty analysis is needed. To support individuals in making decisions by suggesting certain actions, it is critical to analyze the impact of data uncertainty on the suggested actions, particularly in the parameter space where one advice shifts to another. Furthermore, the future itself is inherently uncertain. Uncertainty in scenarios should not be met with a fatalistic attitude. Rather, new concepts, tools and approaches should be developed to explicitly deal with the fundamentally probabilistic, conditional forecasts in policy assessments. Approaches should enhance our understanding of critical paths in the dynamics of spatial systems, possible bifurcations points and critical transition states? Rather than producing a single forecast for different scenarios, we should perform sensitivity analyses for critical parameter subspaces, use ensembles of models to derive probabilistic forecasts (e.g. Rasouli and Timmermans 2012c), identify trajectories that imply a resilient system, simulate internal system adjustment to external perturbations, etc. Such endeavours might help in arbitrating intelligently between different plan options.

To end this presentation, in the beginning we have argued for the need of increased disaggregation to make our models more sensitive to behavioural heterogeneity and the required sensitivity to higher spatial and temporal resolution. However, the results of the uncertainty analysis stipulated above with certainty indicate that the degree of model uncertainty in model forecasts and impact assessment will, albeit not linearly, increase with increased spatial and temporal resolution. Moreover, one would also expect input uncertainty to be higher at higher spatial and temporal resolution. In other words, the development of increasingly more disaggregated model of spatial choice behaviour should go hand in hand with a systematic comparison of model performance against simpler, more aggregate models. Keeping in mind limited resources and relevant margins in policy formulation and decisions, some optimum should be expected.

References

Arentze TA, Timmermans HJP (2000) Albatross. European Institute of Retailing and Services Studies, Eindhoven

Bailey K, Blandford B, Grossardt T, Ripy J (2011) Planning, technology, and legitimacy: structured public involvement in integrated transportation and land-use planning in the United States. Environ Plan B: Plan Design 38(3):447–467

Bhat C, Singh S (2000) A comprehensive daily activity-travel generation model system for workers. Transp Res A 34:1–22

Bowman J, Ben-Akiva ME (2001) Activity-based disaggregate travel demand system with activity schedules. Transp Res A 35:1–28

Chapin FS (1968) Activity systems and urban structure. J Am Inst Planners 34:11–18

Cheng T, Haworth J, Manley E (2012) Advances in geocomputation (1996–2011). Comput Environ Urban Syst 36:481–487

De Paoli S, Miscione G (2011) Relationality in geoIT software development: how data structures and organization perform together. Comput Environ Urban Syst 35:173–182

Furtado BA, Ettema D, Ruiz RM, Hurkens J, van Delden H (2012) A cellular automata intraurban model with prices and income-differentiated actors. Environ Plan B: Plan Design 39(5): 897–924

García AM, Santé I, Crecente R, Miranda D (2011) An analysis of the effect of the stochastic component of urban cellular automata models. Comput Environ Urban Syst 35:289–296

García AM, Santé I, Boullón M, Crecente R (2012) A comparative analysis of cellular automata models for simulation of small urban areas in Galicia, NW Spain. Comput Environ Urban Syst 36:291–301

Geertman SCM, Stillwell J (2009) Planning support systems: new methods and best practice (advances in spatial science). Springer Publishers, New York

Grêt-Regamey A, Crespo R (2011) Planning from a future vision: inverse modeling in spatial planning. Environ Plan B: Plan Design 38(6):979–994

Han Q, Timmermans HJP (2006a) Towards models of strategic spatial choice behaviour: theory and application issues. GeoJournal 67:195–206

Han Q, Timmermans HJP (2006b) Interactive learning in transportation networks under conditions of uncertainty, bounded rationality and strategic choice behaviour: a quantal response model. Transp Res Rec 1964:27–34

Huang A, Levinson D (2011) Why retailers cluster: an agent model of location choice on supply chains. Environ Plan B: Plan Design 38(1):82–94

Johnson PA, Sieber RE (2011) Negotiating constraints to the adoption of agent-based modeling in tourism planning. Environ Plan B: Plan Design 38(2):307–321

Klosterman RE (2012) Simple and complex models. Environ Plan B: Plan Design 39(1):1–6

Ligmann-Zielinska A, Jankowski P (2010) Exploring normative scenarios of land use development decisions with an agent-based simulation laboratory. Comput Environ Urban Syst 34:409–423

Liu Y (2012) Modelling sustainable urban growth in a rapidly urbanising region using a fuzzy-constrained cellular automata approach. Int J Geogr Inf Sci 26:151–167

Long Y, Shen Z, Mao Q (2012) Retrieving spatial policy parameters from an alternative plan using constrained cellular automata and regionalized sensitivity analysis. Environ Plan B: Plan Design 39(3):586–604

Lu Y, Liu Y (2012) Pervasive location acquisition technologies: opportunities and challenges for geospatial studies. Comput Environ Urban Syst 36:105–108

Mahiny AS, Clarke KC (2012) Guiding SLEUTH land-use/land-cover change modeling using multicriteria evaluation: towards dynamic sustainable land-use planning. Environ Plan B: Plan Design 39(5):925–944

Miller EJ, Roorda M (2003) Prototype model of household activity-travel scheduling. Transp Res Rec 1831:114–121

Norte Pinto N, Pais Antunes A (2010) A cellular automata model based on irregular cells: application to small urban areas. Environ Plan B: Plan Design 37(6):1095–1114

Plata-Rocha W, Gómez-Delgado M, Bosque-Sendra J (2011) Simulating urban growth scenarios using GIS and multicriteria analysis techniques: a case study of the Madrid region. Spain. Environ Plan B: Plan Design 38(6):1012–1031

Rae A, Wong C (2012) Monitoring spatial planning policies: towards an analytical, adaptive, and spatial approach to a 'wicked' problem. Environ Plan B: Plan Design 39(5):880–896

Raney B, Cetin A, Voellmy M, Vrtic M, Axhausen K, Nagel K (2003) An agent-based microsimulation model of Swiss travel: first results. Netw Spat Econ 3:23–41
Rasouli S, Timmermans HJP (2012a) Uncertainty in travel demand forecasting models. Transp Lett 4:55–73
Rasouli S, Timmermans HJP (2012b) Using ensembles of decision trees to predict transport mode choice decisions: effects on predictive success and uncertainty estimates. In: Proceedings 17th international conference Hong Kong Society of transportation, Hong Kong, China
Rasouli S, Timmermans HJP (2012c) Uncertainty, uncertainty, uncertainty: revisiting the study of dynamic complex spatial systems. Environ Plan A 44:1781–1784
Rasouli S, Timmermans HJP (2013) Assessment of model uncertainty in destination and travel forecasts of models of complex spatial shopping behaviour. J Retail Consum Serv 20:139–146
Shen Z, Yao XA, Kawakami M, Chen P (2011) Simulating spatial market share patterns for impacts analysis of large-scale shopping centers on downtown revitalization. Environ Plan B: Plan Design 38(1):142–162
Silveira P, Dentinho T (2010) Spatial interaction model of land use—an application to Corvo Island from the 16th, 19th and 20th centuries. Comput Environ Urban Syst 34:91–103
Slotterback CS (2011) Planners' perspectives on using technology in participatory processes. Environ Plan B: Plan Design 38(3):468–485
Tang J (2011) Modeling urban landscape dynamics using subpixel fractions and fuzzy cellular automata. Environ Plan B: Plan Design 38(5):903–920
Te Brömmelstroet M, Schrijnen PM (2010) From planning support systems to mediated planning support: a structured dialogue to overcome the implementation gap. Environ Plan B: Plan Design 37(1):3–20
Thapa RB, Murayama Y (2011) Urban growth modeling of Kathmandu metropolitan region, Nepal. Comput Environ Urban Syst 35:25–34
van Vliet J, Hurkens J, White R, van Delden H (2012) An activity-based cellular automaton model to simulate land-use dynamics. Environ Plan B: Plan Design 39(2):198–212
Vonk GA, Geertman SCM, Schot PP (2005) Bottlenecks blocking widespread usage of planning support systems. Environ Plan A 37(5):909–924
Vonk GA, Geertman SCM, Schot PP (2007a) New technologies stuck in old hierarchies. An analysis of diffusion of geo-information technologies in Dutch public organizations. Public Adm Rev 67(4):745–756
Vonk GA, Geertman SCM, Schot PP (2007b) A SWOT analysis of planning support systems. Environ Plan A 39(7):1699–1714
Voudouris V (2011) Towards a conceptual synthesis of dynamic and geospatial models: fusing the agent-based and object—field models. Environ Plan B: Plan Design 38(1):95–114
White R, Uljee I, Engelen G (2012) Integrated modelling of population, employment and land-use change with a multiple activity-based variable grid cellular automaton. Int J Geogr Inf Sci 26:1251–1280

A Model of Land Use Change in City Areas Based on the Conversion of Unit Lots

Toshihiro Osaragi and Teruo Nishimatsu

Abstract Many studies on land use change have been performed in order to provide fundamental information for future land use plans. In the estimation process of transition probabilities of land use, it should be noted that the spatial unit of land use conversions in city area is a lot. The present study proposes a land use model based on the transition probability of lots, which can be built in planning support systems for sustainable urban development. Lot conversions such as changes in land category, division of lots, the demolition/reconstruction of buildings, changes in building types are described in the models, and we investigate what variables are best to describe the structure of lot conversions. Using the model estimated from time-series GIS data, we discuss the effects of land use policy by simulating land use change in the future, and demonstrate that it is necessary to control the potential for development of lots by adopting adequate zoning regulations and widening roads, by taking into account effects on lot conversions.

1 Introduction

1.1 Land Use Model for Sustainable Urban Development

During the past decades the concept of sustainable development has established itself as a new requirement for urban and metropolitan level public action, which involves the design and practice of urban and land-use planning (Gauthier 2007).

T. Osaragi (✉) · T. Nishimatsu
Department of Mechanical and Environmental Informatics, Graduate School of Information Science and Engineering, Tokyo Institute of Technology, 2-12-1-W8-10 O-okayama, Meguro-ku, Tokyo 152-8552, Japan
e-mail: osaragi@mei.titech.ac.jp

T. Nishimatsu
e-mail: nishimatsu@os.mei.titech.ac.jp

S. Geertman et al. (eds.), *Planning Support Systems for Sustainable Urban Development*, Lecture Notes in Geoinformation and Cartography,
DOI: 10.1007/978-3-642-37533-0_3, © Springer-Verlag Berlin Heidelberg 2013

The concern over sustainable urban development will continue to grow, especially in the developing countries which are undergoing rapid urbanization. The main issue of sustainable urban development is to search for better urban forms that can help to sustain development, especially the minimization of unnecessary agricultural land loss (Li and Yeh 2000). In practice, land-use planning proved to be one of the most important areas in which conceptions of sustainable development are contested (Owens and Cowell 2002).

In the meantime, a sustainable urban development would require considerably more ambitious policies than today in order to limit energy consumption, reduce pollution and protect natural areas and arable land (Gauthier 2007). Namely, re-use of urban areas and more effective utilization of building sites is a possible strategy to this end, not only in developing countries but also in developed countries. Land use planning for a sustainable urban development must be oriented towards the conversion of agricultural land into urban land uses, but should not be based solely on such urbanization processes.

Changes that occur on urban land do not occur within the neat, uniform squares of a raster file; they occur on lots, whose shapes are often anything but neat and square. When an existing building on a lot is demolished, a new building is built for a new use, meaning that the land use has been changed (Fig. 1). This is because the ease of the permit process for demolishing buildings affects the direction and speed of the change in land use. Also, it is quite common for lots to be divided, with portions being sold off for conversion to residential use (Fig. 2). There is a need for a debate not only about change in land category (undeveloped land/building lot), but about division of lots, demolition/reconstruction of buildings, and change in building type. Thus, in contrast to land use changes in suburbs, where changes in land category are the main topic of concern, a model for land use in existing city areas that include lot division, building demolition/reconstruction, and building-type change is required, when we discuss sustainable urban development.

Fig. 1 Example of demolition and reconstruction of buildings observed in existing city area within 4 years

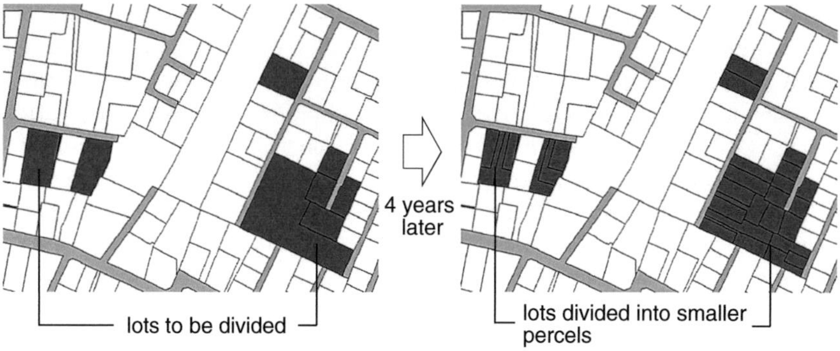

Fig. 2 Example of division of lots observed in existing city area

1.2 Behavioral Approach for Land Use Model

There are two approaches in the field of land use models: to consider the overall volume of land use in each region, or to directly model land use changes by individual unit lots. The former is called the aggregate model, and the latter is the disaggregate model. Nowadays, there is demand for development of a behavioral model that asks "How are land use decisions made?" behind the increasingly popular use of the disaggregate model in disparate fields of research.

Microsimulation models are recently broad trend, in which the dynamic behavior of individual agents is explicitly simulated over both time and space to generate aggregate system behavior. Agent-based models are ones in each individual actor within the system of interest are modeled as autonomous "agents", each one of which possesses identity, attributes and the capability to "behave", i.e. to make decisions and to act within the system (Miller et al. 2004).

Typical examples are models based on multi-agent systems (MAS) and cellular automata (CA), which are in vogue in many branches of interdisciplinary research fields, including urban planning and urban management, and built-in decision/planning support systems (McClean and Watson 1995; Li and Yeh 2000; Torrens 2002; Arentze and Timmermans 2003; Saarloos et al. 2005). Land use and land cover change (LUCC) models based on MAS/CA were offered as a means of complementing other techniques, after examining the strengths and weaknesses of myriad ordinal LUCC modeling approaches.

1.3 Limitation of Previous Approach

Kapelan et al. (2005) identified the principal strengths and weaknesses of planning support tools (PST) for sustainable urban development while considering the broader issue of how an integrated urban planning support system (PSS) can be developed. They identified a total of over 150 tools for sustainable urban development using 14 criteria [We consider here both PSS and decision support systems (DSS) interchangeable, since the difference between them lies mainly in the application of these systems.]. The issue of calibration and validation of mathematical models, especially impact assessment models in PSTs for sustainable urban development, is often neglected even though it is very important. A model, no matter how complex and detailed, is meaningless if its outputs cannot be trusted. Therefore, a PST should be fully calibrated using sufficient quality/quantity of observed data (Kapelan et al. 2005). Thus, the limitations of contemporary MAS/CA models include its poor statistical significance in parameter setting, statistical verification and calibration of models. Also, CA models include its poor ability to handle actual irregular shape of urban lots. Coupling MAS/CA with a statistical model, they can serve as an analytical engine to provide a flexible framework for the programming and running of dynamic spatial models.

PSS are generally regarded as systems in which technologies dedicated to the planning profession are brought together. Harris and Batty (1993) associated the concept of PSS with combining a range of computer-based methods and models into an integrated system that is used to support a particular planning function (Geertman and Stillwell 2004). Namely, various land use models for urban area are needed for integrated PPS for sustainable urban development.

Although this chapter addresses fundamental problems hidden in the conventional models proposed up to the present, the challenges shown in this chapter are common and partly overlap those of the MAS/CA models, which are often used in PPS. Namely, it has been widely recognized in geography that there exist common scale-related problems related to the verification and validation of MAS/CA models. Most MAS/CA models are, by nature, interdisciplinary. It is obvious, therefore, that we should draw on and combine knowledge from many disciplines in order to develop creative new tools for PSS for sustainable urban development.

1.4 Our Approach to Micro Land Use Model

In recent years, land use databases are being assembled at a rapid pace, as GIS technology is steadily being upgraded. It is now easy to obtain data from multiple time points. Given this background, a large body of basic research using Markov chain models exists in LUCC studies (Drewett 1969; Bourne 1971; Bell 1974; Bell and Hinojosa 1977; Robinson 1978; Jahan 1986; Ishizaka 1992; Muller and Middleton 1994; Theobald and Hobbs 1998; Qihao 2002). Markov models may be combined with CA for LUCC modeling, as evidenced by joint CA-Markov models (Li and Reynolds 1997; Balzter et al. 1998). Several methods were proposed to sharpen the Markov chain models for LUCC analysis.

However, there still remained several types of errors that resulted from the estimation method of land use transition probability (Osaragi and Aoki 2006). The transition probability matrix, which is at the heart of conventional Markov-chain land use models, makes its estimates on the basis of the overall sum of the areas of lots whose uses were changed between two time points, using raster database, for instance. This method is quite simple, but tends to give a misleading picture of the structure of actual land use conversions. There is a great difference between the matrices generated by estimates based on area and those generated by estimates based on unit lots. In addition, this method does not provide a dependable analysis of how likely it is for a change to occur on any individual lot. Nevertheless, we should take into consideration that the basic unit of land use change is not a cell but a building lot. Adopting the conventional method entails the risk of overlooking the true transition structure (Yoshikawa 1994).

This research uses a modified version of the conventional Markov chain-style land use model to create a model based on the probability of conversion of unit lots. In this approach, fixed asset tax records (non-spatial data) kept by local authorities were combined with spatial data such as lot location and shape in a single file to create a source of data about individual lots. Submodels were constructed to describe actions such as changes in lot division, building demolition, or building reconstruction. These were combined into the land use model proposed here. The accuracy of the model was then investigated. Future changes in land use were simulated with the model to predict the effects of factors such as zoning changes and street widening.

2 Approach to and Modeling of Land Use Conversions

In city areas, land use changes generally arise when older style buildings are demolished and replaced with contemporary buildings. This is quite different from land use changes in suburban areas, where a major concern is land development. Land use issues should also be considered from a disaggregate viewpoint—it is perfectly natural to examine these issues from the perspective of those people whose activities are tied to the land (such as homeowners and business owners), who make decisions about the exact uses of their own plots. In other words, analysis of the basic unit for changes in land use—the lot—should be considered from the viewpoint of behavioral science. This chapter expands the conventional concept of changes in land use in city areas to include the combination of changes both to land and buildings.

Figure 3 is a diagram of the model. The following should be noted: (1) The probability of division is estimated individually for each lot. The number of parcels resulting from each of the divided lots is then estimated. (2) If there is a building on a lot, it is predicted whether or not the building will be demolished. When the building is predicted not to be demolished, it is assumed that no change will occur in land use, thereby determining the use of the lot at the next time step. (3) If there is no pre-existing building on the lot, it is predicted whether or not the land category of the lot will be changed; for lots used for forest or agriculture, the prediction process is stopped and no prediction of any change is made. (4) If the

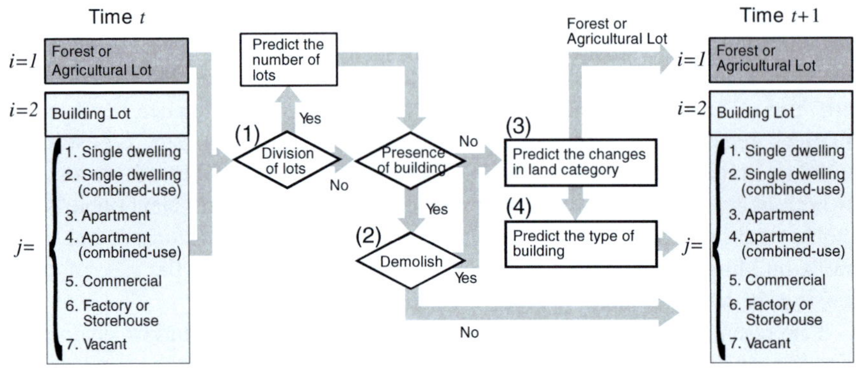

Fig. 3 Land use model based on unit lots

land category of the lot is for buildings, it is predicted that the building will be remodeled and upgraded. The mathematical form of this model is as follows:

$$x_{1*}(t+1) = Dt_{11}x_{1*}(t) + Dt_{21}x_{27}(t) + \sum_{l=1}^{6} [(1-r)D + R]\, t_{21}x_{2l}(t) \qquad (1)$$

$$\begin{aligned} x_{2j}(t+1) = {} & Dt_{12}k_{7j}x_{1*}(t) + Dt_{22}k_{7j}x_{27}(t) + rx_{2j}(t) \\ & + \sum_{l=1}^{6} [(1-r)D + R]\, t_{22}k_{lj}x_{2l}(t) \end{aligned} \qquad (2)$$

where $D = (1-d) + d\sum_n nd_n,\ R = rd\left(\sum_n nd_n - 1\right)$,

$x_{ij}(t)$ is the number of lots whose land category is i and land use (building type) is j at time t (see Fig. 3),

r is the remaining probability of building,

d is the probability of lot subdivision,

d_n is the probability of lot subdivision into n lots,

t_{i1i2} is the transition probability of land category from i_1 to i_2,

k_{j1j2} is the transition probability of building type from j_1 to j_2.

3 Estimating the Probability of Lot Conversion

The design of the land-use model takes into account that the collection of individual micro data (i.e. data which because of their micro location can be associated with individual buildings or small groups of buildings) is neither possible nor desirable. The land-use model therefore works with synthetic micro data which can be retrieved from generally accessible public data (Moeckel et al. 2002). This study uses fixed asset tax record data in Mitaka City, Tokyo from the period 1994–1998, in which around 60,000 individual lots are included. The probabilities of lot conversion estimated below vary greatly with conditions, including the attributes of lots, location and prevailing regulations on land use. Therefore, this study investigates what variables are best to include in models in order to describe the structure of lot conversions. Generally, the more explanatory variables, the more detailed look one can get at the structure of conversions; however, the model becomes unstable if classifications are too fine. A contingency table analysis model based on Akaike Information Criterion (AIC) was used to incorporate variables into the model in order of statistical significance (Akaike 1972, 1974). Since statistically significant variables mean nothing if they cannot be used in urban policy, variables were sought in a preliminary study (Table 1).

3.1 Probability of Lot Division

Sometimes portions of lots are sold off while houses are still standing on the unsold parts of the lots. However, lots with no house are more likely to be divided (Fig. 2).

Table 1 Variables describing the characteristics of lots

Attributes	
Area of lot	(1) 100 m^2, (2) 500 m^2, (3) 1,000 m^2, (4) 1,000 m^2 or more
Width of adjacent street	(1) 4 m, (2) 6 m, (3) 10 m, (4) 10 m or more
Current presence of building	(1) Yes (a building is standing), (2) No (no building)
Type of building on a lot	(1) Single dwelling, (2) Single dwelling (combined-use), (3) Apartment, (4) Apartment (combined-use), (5) Commercial, (6) Factory/storehouse
Age of building on a lot	(1) 10, (2) 20, (3) 30, (4) 40, (5) 50, (6) 60, (7) 70, (8) 80, (9) 81 years or more
History of lot division	(1) Yes (a lot was divided before), (2) No (never divided)
Combined-use here means that in addition to dwelling space, the building also contains commercial-use facilities with over 30 m^2 of floor space	
Regulations	
Land use zoning	(1) Low-rise residential, (2) Middle-to-high-rise residential, (3) Residential, (4) Quasi-residential, (5) Neighborhood commercial, (6) Commercial, (7) Quasi-industrial/industrial
Floor area zoning	(1) 50 %, (2) 80 %, (3) 100 %, (4) 150 %, (5) 200 %, (6) 300 %, (7) 500–600 %
Building area zoning	(1) 30 %, (2) 40 %, (3) 50 %, (4) 60 %, (5) 80 %
Location	
Districts	(1) Commercial, (2) Industrial, (3) Residential, (4) Residential with commercial, (5) Regular residential
Distance to the nearest bus stop	(1) 100 m, (2) 200 m, (3) 300 m, (4) 300 m or more
Distance to the nearest railway station	(1) 500 m, (2) 1,000 m, (3) 1,500 m, (4) 1,500 m or more

Current presence of a building was added to the list of explanatory variables for this reason. As the results of a contingency table analysis, four statistically significant variables (1:[Current presence of a building], 2:[Area of lot], 3:[Land category], 4:[Width of adjacent street]) were incorporated in the model. That is, the probability of lot division was estimated for each combination of these variables, i.e., 64 (=2 × 4 × 2 × 4) probabilities of lot division were estimated. [Current presence of a building] has the highest explanatory power; lot division was more probable when there was no building. The explanatory variable with the second highest influence was [Area of lot]. The larger the area, the higher the probability of division. [Distance to the nearest bus stop] was also employed to calculate the probability of lot division in the simulation, whose execution will be described below.

3.2 Probability of Building Demolition

In contrast to other probabilities of conversion, the probability of demolition of a building must be estimated while accounting for the fact that it actually existed until some point in time without being demolished (Osaragi 2004, 2005). This is employed as an explanatory variable on the basis of earlier investigations on the probability of demolition of buildings. As the results, three variables (1:[Type of building on a lot], 2:[Land use zoning], 3:[Age of building]) were incorporated in the model.

3.3 Probability of Change in Land Category

Conversion of residential land to forest or agricultural land is extremely rare, and its probability can be considered to be fixed without dependence on any other conditions. The variable of [History of division] has been added to the list of explanatory variables in order to account for the probability of the opposite process, division of a forest or agricultural lot and conversion for sale as residential plots. As the results, three statistically significant variables (1:[History of division], 2:[Area of lot], 3:[Width of adjacent street]) were selected for explanatory variables. [History of division] is the most powerful explanatory variable; the probability for a forest or agricultural lot to be converted to residential is extremely high. [Land use zoning] was also employed as an additional variable in the simulation in order to discuss the power of controllable variables in the change of land category.

3.4 Probability of Change of Building Type

It was studied what factors influence the change of certain classifications of buildings. 1:[Area of lot] and 2:[History of lot division] had the highest explanatory power in conversions from “Single dwelling” to other uses. However, in the simulation described below, the explanatory variable of zoning designation is used, which has about the same level of strength as land use regulations. 1:[History of lot division] and 2:[Area of lot] both have high explanatory values for apartment buildings. The greater the area of a lot, the higher the probability that it will be converted to use for an apartment building. The probability is also increased if the lot has ever been divided. Since there are low numbers (samples) of combined-use single dwellings (*combined-use* here means that in addition to dwelling space, the building also contains commercial-use facilities with over 30 m^2 of floor space), combined-use apartment houses, commercial establishments and factories, only the factor of 1:[History of lot division] is employed to explain conversions from these classifications. For conversions of vacant lot, three variables are adopted; i.e., 1:[History of lot division], 2:[Area of lot], 3:[Land use zoning], 4:[Width of adjacent street] after examining their consistency with the probabilities of other conversions. Locations adjoining wide streets and commercially zoned locations show high probabilities of conversion of vacant lot to commercial use.

4 Validation of Model

4.1 Accuracy of Estimates by Model

All of the lots were sorted into two groups at random, comprising 1/4 and 3/4 of the original set. The probability of conversion was calculated using the 3/4 portion, and Eqs. (1) and (2) were employed to test the probabilities for the lots in the remaining portion to test the accuracy of the estimates by the model. As described above, the prediction was carried out for each individual lot, but urban planning often calls for information relating to area, so the validation was carried out with respect to area. Figure 4 shows a partial comparison of estimated area and the actual area in 1998. The estimates were generally consistent with the actual areas.

4.2 Stability Across Time

While cellular modeling techniques offer greater flexibility for representing spatial and temporal dynamics, these dynamics are based on stationary transition probabilities (Parker et al. 2003). The models proposed thus far assume that the transition probability will remain stable in the future, except in a few instances where it has been tested (Bourne 1971; Bell 1974; Bell and Hinojosa 1977). Therefore, it is necessary to confirm the stability in the transition probability in terms of time.

The estimating parameters cannot be rigorously tested for their ability to predict the future, so they were set using data from 1994 to 1996, and then used in an

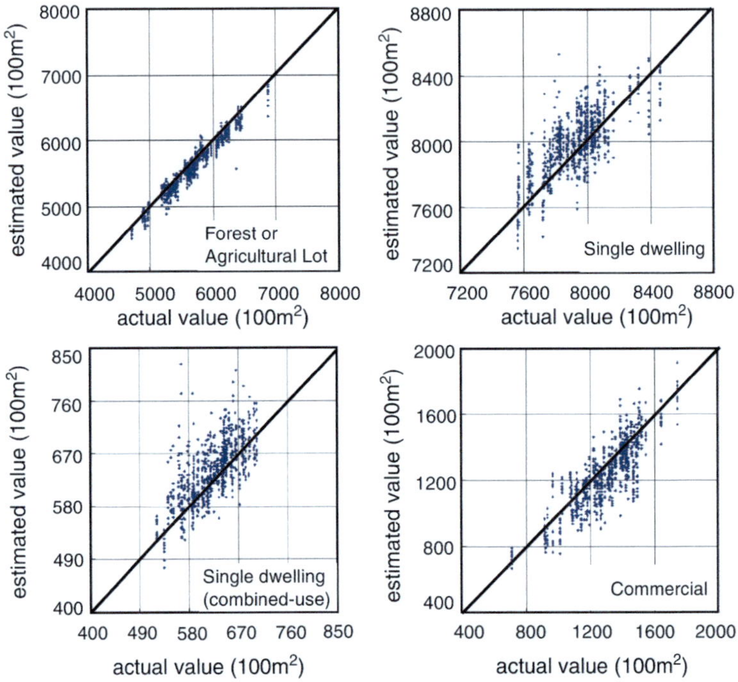

Fig. 4 Comparison of estimated area and the actual area of each lot

attempt to predict the changes in the 1996–1998 period. In addition to the magnitude of the errors, the method was also examined for its prediction of the direction of changes in land use. Figure 5 provides the fraction represented by the error in the estimates with respect to the actual values. The commercial use was slightly underestimated, but the model showed good overall consistency with the data.

It is generally assumed in Markov-chain models that the transition probability matrix is constant. This assumption is not very plausible for changes in land use, however, which are closely connected to middle/long term socioeconomic activities. The researcher or planner must revise both the database itself and the expected accuracy of the data there when preparing models that are expected to replicate actual events. For example, methods can be used to incorporate the varying probability of conversion over time.

4.3 Spatial Extent Predictable with Model

It was also investigated whether the same parameters can be applied in a comparatively small area. The parameters found using the data for the entire city of Mitaka were employed to predict changes in land use in 11 different *chō* (a subunit

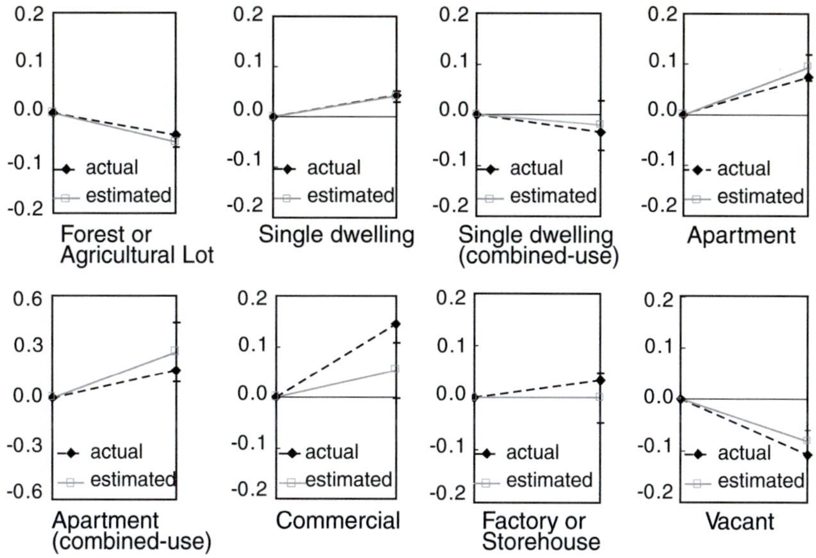

Fig. 5 Fraction represented by the error in the estimates

of a city approximately equal to a few tens of blocks). Some results were shown in Fig. 6. These predictions showed a generally good match with actual land use changes, but a greater extent of scatter was found in the estimates of combined-use apartment houses in Kitano-chō and factory/storehouses in Shinkawa-chō. When predictions are attempted over smaller spaces, these have an increased influence over the estimates for the larger region. Estimates at the aggregate area level are then prone to error.

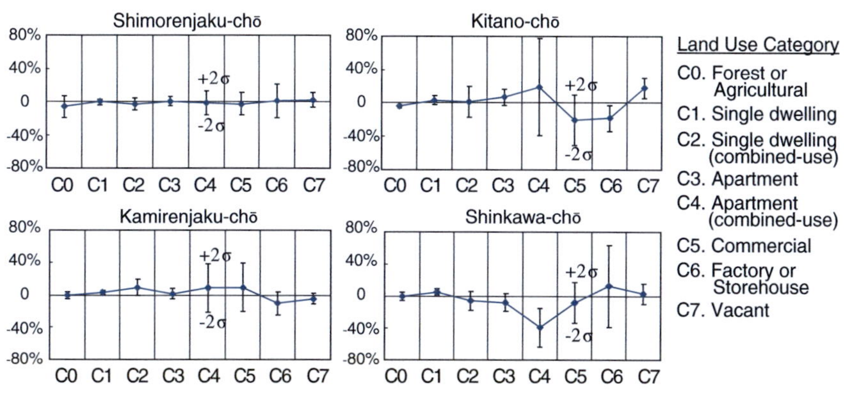

Fig. 6 Size of error according to the size of area for prediction

5 Simulations for Predicting Future Land Use

5.1 Analyses, from the Point of View of Future Land Use

The estimated parameters were used to predict the uses of lots in the entire area of Mitaka City over the next 20 years (Fig. 7). In terms of area, conversion to residential use is predicted to continue, with a reduction for forest- and agriculture-land and a large increase in single-dwelling and multiple-dwelling (apartment) residential land. Commercial land is expected to shrink slightly, along with land used for industrial purposes and material storage (Fig. 7, bottom right). However, the area zoned for combined commercial and multi-dwelling residential area was predicted to increase, indicating a high potential for commercial plots to be converted to such combined use. The mean lot area was also investigated (Fig. 7, upper right). The overall area for single-dwelling and multiple-dwelling residential land was predicted to increase, but the size of individual residential plots was predicted to decrease, like those of commercial plots. Lots for combined-use apartment buildings will become slightly larger and are less prone to be divided during reconstruction of buildings than ordinary apartment buildings. Lots for combined-use multiple-dwelling complexes have a high potential to be rebuilt with larger buildings at the time of change in land use category.

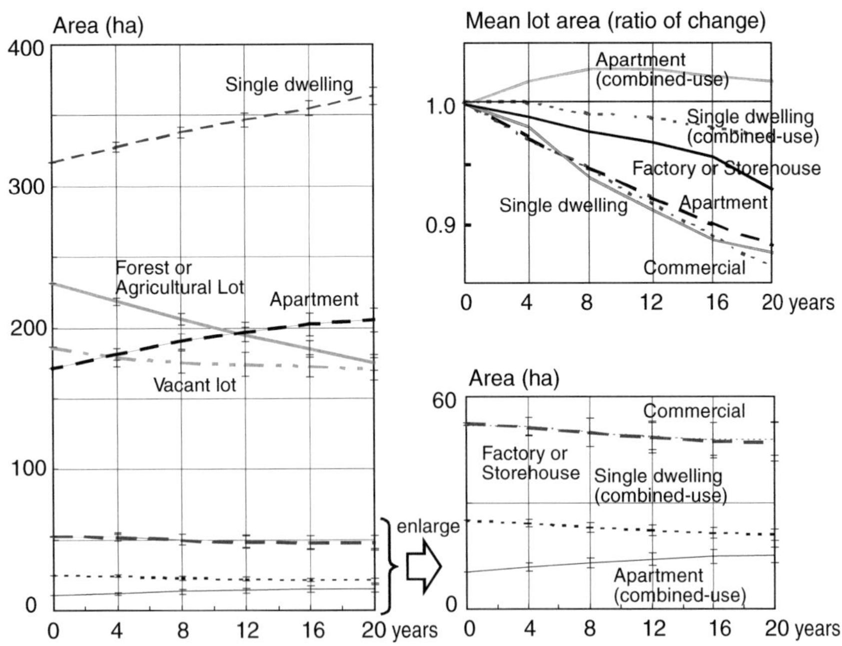

Fig. 7 Prediction of the uses of lots in the entire area over the next 20 years

5.2 Analysis of Influence of Land Use Plans

5.2.1 Direction and Speed of Conversions of Forest or Agricultural Land

Conversion of forest and agricultural land was investigated (Fig. 8). The area under these two uses was predicted to be reduced to about 75 % of its current total over the next 20 years. The after-conversion uses will be residential or vacant, and much of the vacant lots will be converted to residential land. The after-conversion uses were examined for variations with before-conversion uses (Fig. 8, right side). Conversions of agricultural and forest land in middle-to-high-rise residential districts and in residential districts were quite fast, consuming about half the available land in the 20-year period. These rates of conversion of the agricultural and forest lots in middle-to-high-rise residential districts and in residential districts, multiple-dwelling, commercial use or vacant status were three, five and two times as much as average rates, respectively. A clear influence was predicted for district land use designation on conversions.

5.2.2 Direction and Speed of Conversions of Vacant Lots

The fate of vacant lots was also predicted (Fig. 9). The area under this designation was predicted to fall by 35 % over 20 years. By conversion type, relatively few

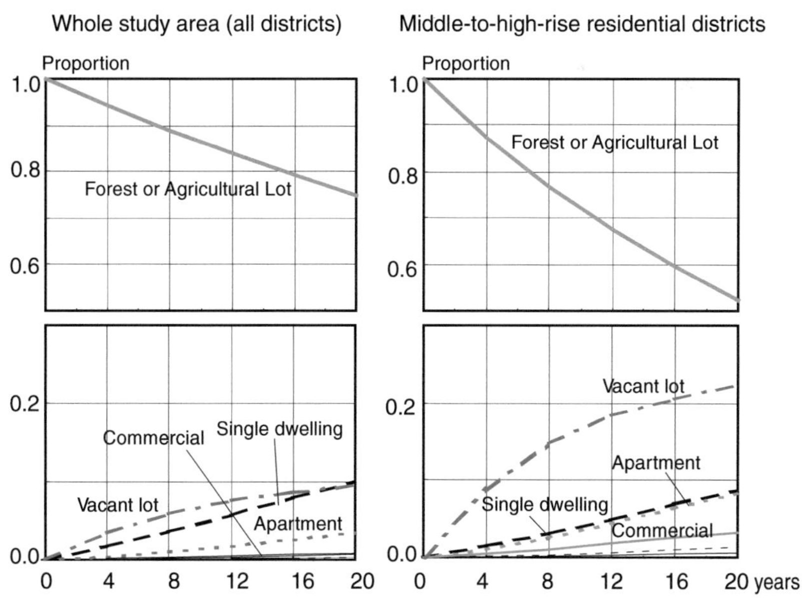

Fig. 8 Conversions of forest or agricultural land

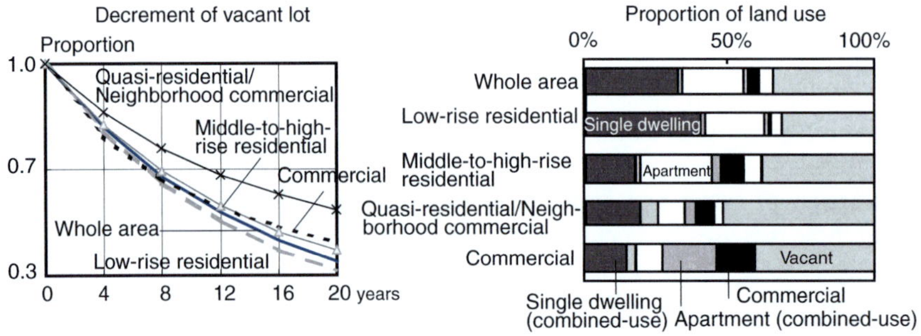

Fig. 9 Conversions of vacant lots

conversions occur in quasi-residential districts, neighborhood commercial districts and commercial districts (Fig. 9, right side). Of those that undergo conversion, vacant lots in quasi-residential districts most often see conversions to combined-use single dwellings, and vacant lots in neighborhood commercial districts are usually converted to combined-use apartment buildings or commercial establishments.

5.2.3 Analysis of Conversions of Land Adjoining Main Roadways

The effect of street widths in metropolitan plans was also investigated (Fig. 10). Conversions to single dwellings were suppressed but conversions to commercial use and combined-use apartment houses were increased by the widening of adjoining streets (Fig. 10, right side). Therefore, it is necessary to control the potential for development of lots in districts with streets that are marked for widening, so that development proceeds in an orderly manner in adjoining districts during the widening process.

5.2.4 Analysis of Conversions of Land Use with Respect to Zoning

The influence of zoning revisions on land use conversions was observed, paying particular attention to low-rise residential districts adjoining neighborhood commercial districts and commercial districts near railway stations (Fig. 11). If zoning does not change (present zoning), the number of single dwellings increases, but this number decreases if the land is converted to any of the following: middle-to-high-rise residential; residential; quasi-residential/neighborhood commercial; or commercial (Fig. 11, upper left). Following zoning changes, however, there were predicted to be increased conversions to combined-use apartment houses and to commercial establishments (Fig. 11, upper right, lower left); the increase in commercial establishments was particularly marked.

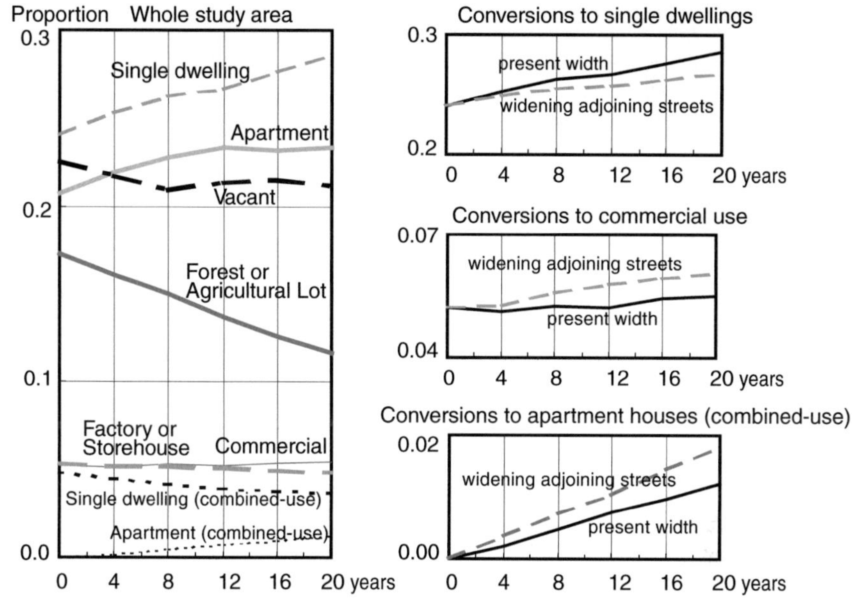

Fig. 10 Conversions of land adjoining main roadways

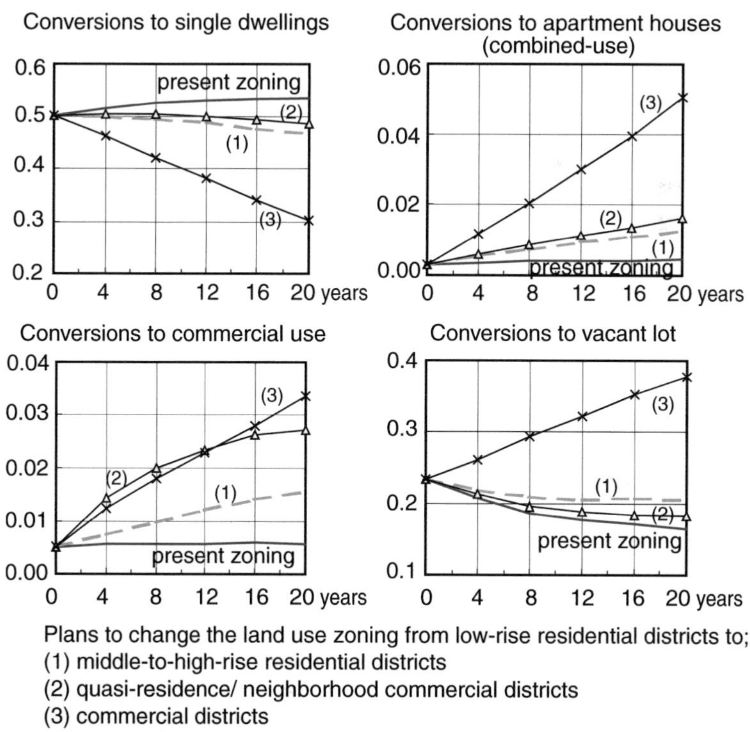

Fig. 11 Effects of policy plans on land use conversions in low-rise residential districts

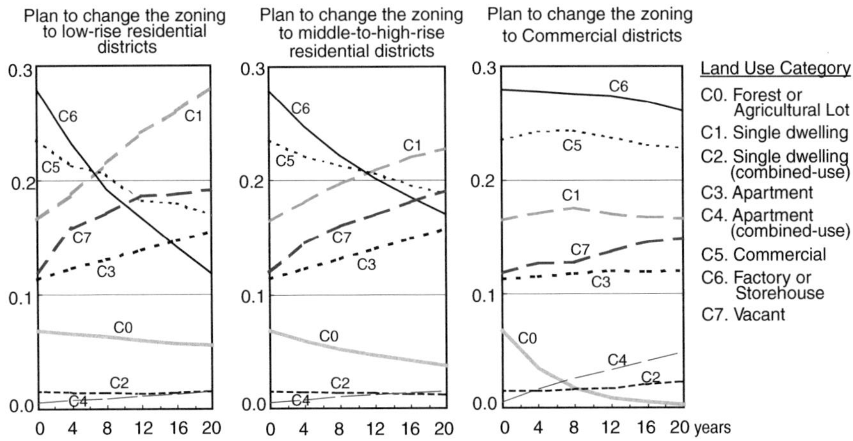

Fig. 12 Land use change in quasi-industrial or industrial districts

In sum, when zoning regulations are relaxed in especially convenient locations such as low-rise residential districts near railway stations, conversion to commercial establishments was accelerated.

It is necessary to debate revision of zoning designations in the face of the problems related to the use of empty land left after factories in quasi-industrial or industrial districts have been moved or abandoned (Fig. 12). The conversion of these areas to low-rise residential districts promotes the expansion of living spaces, including single dwellings. On the other hand, the conversion of these areas to middle-to-high-rise residential districts or residential districts promotes mixed use of nearly 20 % of the area, combining single dwellings, apartments, commercial establishments, factories, storage and still-vacant spaces. If these abandoned plots are re-zoned for commercial use, changes occur at a more sedate pace, and only forest and agriculture lots are quickly converted to residential land. Therefore, planners must note that the direction of land use conversions varies greatly from the usual direction when the initial use is industrial. Planners must pay careful need to the probable direction of conversions as they prepare to re-designate the permitted usage of lots.

6 Summary and Conclusions

For this study, a method for predicting the future designated use of single lot units was devised from a previous area-based estimation method for the probability in conversions in land use. A land use model incorporating this method was proposed. Submodels were created and validated that described essential processes in land use—changes in land category, lot division, demolition of buildings, and reconstruction of buildings. These sub-models were combined to create the final model. The accuracy and time stability of the model were validated and the range of region sizes over which the model provided acceptable estimates were investigated, and good results were obtained.

Simulations of changes in land use were then carried out and predictions of the effects of changes in zoning and street width on future land use were observed, and we demonstrated that (1) planners need to control the potential for development of lots in districts with streets that are marked for widening, so that development proceeds in an orderly manner in adjoining districts, (2) conversion to commercial establishments was accelerated, when zoning regulations are relaxed in especially convenient locations such as low-rise residential districts near railway stations, (3) it is necessary to note that the direction of land use conversions varies greatly from the usual direction when the initial use is industrial. Hence, planners must consider the probable direction of conversions as they prepare to re-designate the permitted usage of lots. Thus, the models were shown to allow investigations of specific land use policies and comparison of these with actual land use changes.

The model proposed in this research seems to be applicable only to a rather limited area, since the database used here is very precise and detailed. Sui et al. (2013) explores both the theories and applications of crowdsourcing for geographic knowledge, by situating volunteered geographic information (VGI) in the context of big-data deluge and the data-intensive inquiry. This phenomenon of VGI, which is part of a profound transformation in how geographic data, information, and knowledge are produced and circulated, will be activated in the future. Very high-detail spatiotemporal databases are, therefore, expected to be available, especially in urban area.

Moreover, the estimated parameters would vary if we choose another area for a case study. For instance, the demolition rate of buildings in a certain area would be different from that of the study area used here. However, the authors think from their experiences that the mechanism of land use conversion will be quite similar each other, at least in Japan. Further study is, therefore, needed to confirm the compatibility of the proposed models by applying them to other areas.

Acknowledgments The authors would like to acknowledge the valuable comments and useful suggestions from anonymous reviewers and Tohru Yoshikawa, Professor, Tokyo Metropolitan University, to improve the content and clarity of the chapter. A part of this chapter was presented at Journal of Architectural Planning and Engineering, AIJ, "Land use conversion model based on transition probability of lots", no. 564, pp. 203–210, 2003 (in Japanese).

References

Akaike H (1972) Information theory and an extension of the maximum likelihood principle. In: Petron BN, Csak F (eds) Proceedings of the 2nd international symposium on information theory, Akademiai kaido, Budapest, pp 267–281

Akaike H (1974) A new look at the statistical model identification. IEEE Trans Autom Control 19:716–723

Arentze T, Timmermans H (2003) A multiagent model of negotiation processes between multiple actors in urban developments: a framework for and results of numerical experiments. Environ Plan B: Plan Des 30(3):391–410

Balzter H, Braun PW, Kohler W (1998) Cellular automata models for vegetation dynamics. Ecol Model 107(2/3):113–125

Bell EJ (1974) Markov analysis of land use change: an application of stochastic processes to remotely sensed data. Socio-Econ Plan Sci 8:311–316

Bell EJ, Hinojosa RC (1977) Markov analysis of land use change: continuous time and stationary processes. Socio-Econ Plan Sci 11:13–17

Bourne LS (1971) Physical adjustment processes and land use succession: a review and central city example. Econ Geogr 47:1–15

Drewett JR (1969) A stochastic model of the land conversion process. Reg Stud 3:269–280

Gauthier M (2007) Urban planning and sustainable development. Human and social science, http://www.h-net.org/announce/show.cgi?ID=157060

Geertman S, Stillwell J (2004) Planning support systems: an inventory of current practice. Comput Environ Urban Syst 28:291–310

Harris B, Batty M (1993) Locational models, geographic information and planning support systems. J Plan Educ Res 12:184–198

Ishizaka K (1992) A consideration on analyzing method of land use transition matrix. Trans Archit Inst Jpn 436:59–69 (in Japanese)

Jahan S (1986) The determination of stability and similarity of Markovian land use change processes: a theoretical and empirical analysis. Socio-Econ Plan Sci 20:243–251

Kapelan Z, Savic DA, Walters GA (2005) Decision-support tools for sustainable urban development. Eng Sustain 158(ES3):135–142

Li H, Reynolds JF (1997) Modeling effects of spatial pattern, drought, and grazing on rates of range land degradation: a combined Markov and cellular automaton approach. In: Quattrochi DA, Goodchild MF (eds) Scale in remote sensing and GIS. Lewis Publishers, New York, pp 211–230

Li X, Yeh AG (2000) Modelling sustainable urban development by the integration of constrained cellular automata and GIS. Int J Geogr Inf Sci 14(2):131–152

McClean CJ, Watson PM (1995) Land use planning: a decision support system. J Environ Planning Manage 38(1):77–92

Miller EJ, Hunt JD, Abraham JE, Salvini PA (2004) Microsimulating urban systems. Comput Environ Urban Syst 28:9–44

Moeckel R, Schürmann C, Wegener M (2002) Microsimulation of Urban Land Use, 42nd European Congress of the Regional Science Association, Dortmund, 27-31 August 2002.

Muller RM, Middleton J (1994) A Markov model of land-use change dynamics in the Niagara region, Ontario, Canada. Landscape Ecol 9:151–157

Osaragi T (2004) Factors leading to buildings being demolished and probability of remainder. In: Marchettini N et al. (eds) The sustainable city III. WIT Press, pp 325–334

Osaragi T (2005) The life span of buildings and the conversion of cities to an incombustible state. In: Brebbia et al. (eds) Safety and security engineering IV. WIT Press, pp 495–504

Osaragi T, Aoki Y (2006) A method for estimating land use transition probability by using raster data. In: Timmermans H et al. (eds) Innovations in design & decision support systems in architecture & urban planning. Springer, pp 69–84

Owens S, Cowell R (2002) Land and limits: interpreting sustainability in the planning process (RTPI library series). Routledge, London

Parker D, Manson S, Janssen M, Hoffmann M, Deadman P (2003) Multi-agent systems for the simulation of land-use and land-cover change: a review. Ann Assoc Am Geogr 93(2):314–337

Qihao W (2002) Land use change analysis in the Zhujiang Delta of China using satellite remote sensing, GIS and stochastic modelling. J Environ Manage 64(3):273–284

Robinson VB (1978) Information theory and sequences of land use: an application. Prof Geogr 30:174–179

Saarloos D, Arentze T, Borgers A, Timmermans H (2005) A multiagent model for alternative plan generation. Environ Plan B: Plan Des 32(4):505–522

Sui D, Elwood S, Goodchild M (eds) (2013) Crowdsourcing geographic knowledge, volunteered geographic information (VGI) in theory and practice. Springer Science + Business Media, Dordrecht

Theobald DM, Hobbs NT (1998) Forecasting rural land-use change: a comparison of regression and spatial transition-based models. Geogr Environ Model 2:65–82

Torrens PM (2002) Cellular automata and multi-agent systems as planning support tools. In: Geertman SS, Stillwell J (eds) Planning support systems in practice. Springer, London, pp 205–222

Yoshikawa T (1994) A stochastic model of lot land use transition based on multinomial probability distribution. Compr Urban Stud 53:113–121 (in Japanese)

Simulating the Dynamics Between the Development of Creative Industries and Urban Spatial Structure: An Agent-Based Model

Helin Liu and Elisabete A. Silva

Abstract Creative industries have been widely adopted to promote economy growth, urban regeneration and innovation. It is expected that this strategy can produce a sustainable development model. However, in reality it is not effective enough because the implemented policy based on linear analysis is misleading. This chapter aims to fill this gap by examining the dynamics among creative industries, urban land space and urban government from a complex systems' view. It presents a general simulation framework and an agent-based model (named "CID-USST") developed in NetLogo. This is a spatially explicit model where a simplified urban space is used to represent the real urban land space. The agents involved include the creative firms, the creative workers, and the urban government. The resulting urban spatial structure is examined from two aspects: the spatial density distribution and the spatial clustering pattern of both the creative firms and the creative workers.

1 Introduction

It took more than 50 years for the concept of creative industries to evolve from "culture industry", through "cultural industries" to "creative industries" (O'connor 2007). However, it was quickly accepted and adopted world widely by urban policymakers aspiring to promote urban growth. The underlying policy rationale, growth and innovation (Foord 2008) were fuelled by the argument for "creative city" (Landry, 2000) and "creative class" (Florida 2002). As a result,

H. Liu (✉) · E. A. Silva
Department of Land Economy, University of Cambridge,
19 Silver Street, Cambridge CB4 9EP, UK
e-mail: hl362@cam.ac.uk

E. A. Silva
e-mail: es424@cam.ac.uk

S. Geertman et al. (eds.), *Planning Support Systems for Sustainable Urban Development*, Lecture Notes in Geoinformation and Cartography,
DOI: 10.1007/978-3-642-37533-0_4, © Springer-Verlag Berlin Heidelberg 2013

cities and regions around the world are trying to develop, facilitate or promote concentrations of creative, innovative and/or knowledge-intensive industries in order to become more competitive and sustainable (Girard et al. 2012).

These initiatives also provide evidence to support the argument that creative industries can contribute to urban economy growth (Piergiovanni et al. 2012; Higgs et al. 2008), raise employment rate (Mossig 2011; Higgs et al. 2008), support urban creativity (Stam et al. 2008; Scott 2006), facilitate urban regeneration (Pratt 2009; Evans 2005), and be beneficial to sustainable urban development (Kakiuchi 2012; Forum for the Future 2010).

However, much of the focus has been around the investment of cities in specific regeneration projects or flagship developments rather than addressing the nature of the infrastructure, networks and agents engaging in the city's cultural development (Comunian 2011). This directly leads to the ineffectiveness of policy in terms of urban growth and sustainable development. First, focusing on short-term attraction rather than long-term relation can lead to the waste of significant investment in cultural infrastructure and creative economies, as firms and workers tend to be 'footloose' (Florida 2002) and can move to other city or region easily. Second, creative industries are dynamic; they are not bound to certain areas within a city but can resettle themselves spatially within a very short time. Therefore, an understanding of their spatial behaviours and its relationship with the locational factors and the urban policies is critical in developing sustainable and effective urban development schemes.

This chapter introduces a system approach to explore the dynamics between the development of creative industries and urban spatial structure. First, it proposes the dynamics framework which explains the complex interactions among the involved agent classes (creative firms, creative workers and urban government). Within this framework, the government acts as advocator and regulator. To cultivate creative industries, the leverage of supportive policies are utilised by the urban government to (re)develop certain urban sites or regenerate declining urban areas. The creative firms are influenced by the government policies and constrained by their own available capital and profit rate. They have to first find office location according to location utility, and second employ/fire workers by checking their profit rate. Using a similar principle of referring to location utility, the workers have to find suitable residences to live. In addition, they also have to hunt for jobs in these creative firms. Their employment/unemployment consequently changes the profit rate and capital of the firms. Then, this chapter uses the city of Nanjing to illustrate the development of the simulation model named "CID-USST".

2 The General Framework for Simulating the Dynamics

The strategies/policies adopted by the urban/region government to attract investment in creative industries can be grouped into six categories: (1) property and premises strategies; (2) business development, advice and network building;

(3) direct grants and loans schemes to creative business/entrepreneurs; (4) fiscal initiatives; (5) physical and IT infrastructure; (6) soft infrastructure (Foord 2008).

For modelling simplification purpose, within a city realm these six categories can be generalised into two types: (1) policies that centre on financial or other soft support (i.e. as tax reduction, lower land/office and promotion of cultural diversity); (2) policies that directly relate to urban land use planning. In practice, the government sometimes offer grants and loans directly to the creative firms without requiring them to locate in certain sites. In this research, we suppose that all these policies are used as the leverage to develop certain land plots. In other words, the government prefers to allocate the policy packages to some certain land plots so as to attract firms to promote the land-use development.

The behaviour of the urban government in making supportive policies can be described as shown by the left part of section A in Fig. 1: (1) The first step is to design policy packages (in a process of combining the possible policies, say tax reduction, lower office rent for instance); (2) The government has to allocate all the policy packages spatially within the city (by changing planning permission in plots of land, which in consequence influences the firms' decision on office location and promotes land use change).

Regarding the second type of policies (land use change), strategies were generalised into three aspects: urban regeneration, new development and land-use guidance (the right part of section A in Fig. 1). This "urban regeneration" behaviour resonates with the fact that creative industries are usually integrated with urban regeneration as a strategy to renew declining urban areas. New development deals with land plots of green belt/farmland which are strictly protected. Both the creative firms and the creative workers cannot occupy these plots unless development of new offices or new housing estates is permitted by the government. The urban land-use guidance is about the strategy of defining the maximum plot ratio of each plot to control land exploitation. It is obvious these three strategies can affect the marks of the locational factors of the involved plots.

The values of each plot's locational factors are crucial for both the creative firms and the creative workers. This is because those locational factors determine the firms' final decision upon where to set their offices and the workers' final choice of where to live. There are two points which need further clarification. First, the set of locational factors for the creative firms and the creative workers should be different—their choice are based on totally different use purposes, one as working space and the other as living place. Second, depending on the case, both of the two sets of locational factors can vary accordingly, for instance, factors that determine the location of creative industries in London might be different from that in New York. Despite of these differences, the principle of deciding the final location is the same. That is, both the firms and the workers prefer the plot which generates the largest location utility among the candidate plots.

For both the firms and the workers, it is not an easy task to find the right location. In this framework, it is assumed that the tolerance of failure for both the firms and the workers is not limitless. In other words, for a firm, it will be allowed to try to find an office location, until exceeds the defined maximum time of

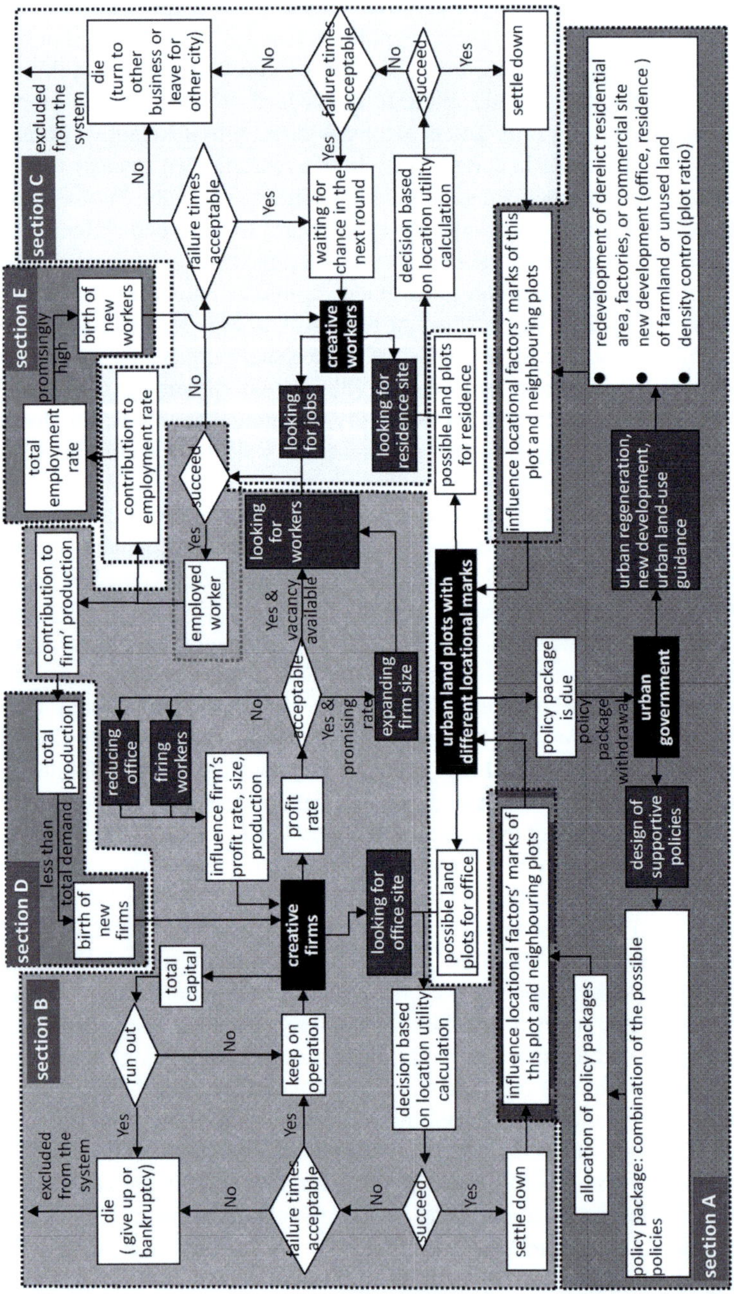

Fig. 1 The general simplified framework for dynamics simulation

attempts. Once it runs out the allowed maximum number of attempts it will give up and be permanently excluded (the lower left part of section B in Fig. 1). For a worker, failing to find a residence for too long will lead him to give up. In modelling terms, he will be excluded permanently from the system (the lower part of section C in Fig. 1).

However, what the two classes of agents need to do is not confined only to their locational behaviours. The creative firms, aiming to generate profit and avoid bankruptcy, must employ the right number of workers. The creative workers, in order to support themselves, must try to find a job in the creative firms. This is a bilateral process which can determine the fate (exclusion from the system) of each other.

To employ the right number of workers is a complex process (the upper left part and the right part of section B in Fig. 1). It is assumed that if the firm's total capital becomes negative, immediately the firm will claim bankruptcy and be excluded from the system. As a consequence, all the workers employed by this firm will lose their jobs, turning to be unemployed and waiting for the next employment opportunity. If the total capital is positive, the firm then has to decide the firm's plan in terms of size development. The basic principle is that if the profit rate is not acceptable, the firm will fire workers and reduce its size so that the costs (workers' payment and office rent) can be reduced and continue to operate. On the other hand, if the profit rate is higher enough, the firm will expand its expected size (increase available offices), expecting to employ more workers and generate more profit. Or if the firm's profit rate is positive but not high, it will keep its current size and employ new workers if more workers are needed.

The creative firms' decisions on size development influences the creative workers directly. This is because, as labour force, the creative workers have to compete with each other in the labour market so as to find a job (the upper part of section C in Fig. 1). For those unemployed, the first step to find a job is to target the potential firms which are in need of new workers. The number of candidate firms that a worker can target is not infinite but limited to a certain value. This value can be interpreted as the maximum times a worker can go for interviews in 1 month. However, those candidate firms are not randomly selected, but based on the job-hunter's expected salary. Finally, the worker will choose the firm offering the highest payment. It is possible that the job-hunter may fail in finding a position. If this happens, the job-hunter will wait for a new chance in the next round until the times of failure is beyond the defined tolerance (number of times worker can apply/be-interviewed in 1 month).

In terms of the creation of new agents, we suppose that new creative firms will be established if the total demand is greater than the total production (section D in Fig. 1). The total production is all the value the employed workers generated which should be calculated in the simulation process. It is also assumed that that the supply of new creative workers highly depends on the real-time employment rate. If this employment rate is higher than the critical rate which should be open to user's control, new creative workers will be created and ready to enter the system in the next round (section E in Fig. 1).

3 Agents and Their Action Rules

Agents and their action rules require an environment to perform their dynamic behaviour, commonly called 'the modelling environment'. The modelling environment defines the virtual space in which agents operate, serving to support their interaction with the environment and other agents. Agents within an environment may be a-spatial, spatially explicit and/or spatially implicit (Silva and Wu 2012; Wu and Silva 2010). Spatially explicit means that agents have a location in geometrical space, while spatially implicit equals to the statement that location is irrelevant (Crooks and Heppenstall 2012). Currently, there are a couple of modelling systems (software) suitable for agent-based modelling. Of these systems, in spatially explicit simulation, seven are widely used which are Swarm, MASON, Repast, NetLogo, StarLogo, AgentSheets and AnyLogic (Crooks and Castle 2012). The model to be built based on the dynamics framework proposed above is spatial explicit.

3.1 Urban Government

As discussed in the dynamics framework, the behaviour of the government is defined by two ways: design of supportive policies, land use planning schemes.

3.1.1 Design of Supportive Policies

It should be admitted that supportive policies vary across different cities. But it is possible that these policies can be classified into a number of similar groups. In this simplified policy design process the government first figures out the possible policies and then combines groups of policies to produce policy packages. In a general condition, if the number of policy groups is n, then the number of possible policy packages is $N_p = C_n^1 + C_n^2 + \cdots + C_n^n$. These policy packages then can be allocated spatially by the government according to their development schemes. It is also very important that each policy package has its own duration. once it is due it is withdrawn by the government (Table 1).

Table 1 Triggers and action rules for the urban government

Trigger	Action
There are supportive policy schemes	Designs and allocates the policy packages spatially
Duration of a policy package is due	Withdraws the policy package
The land plot can be used for office development or housing development	Sets its maximum plot ratio for development

3.1.2 Land Use Planning Schemes

In terms of land use planning, the government is supposed to act in three aspects: (1) to carry out urban regeneration which focuses on the redevelopment of declining sites; (2) to develop new housing or offices in undeveloped suburban areas (such as green belt); (3) density control, which can be defined by plot ratio (Table 1). To simplify this process, instead of specifying the ratio for every plot, we can divide the urban land space into different areas and set up the maximum plot ratio for each area.

3.2 *Creative Firms*

3.2.1 Looking for Office Site

In this model, we hold that, in terms of choosing a location, each firm has its own particular preference. In other words, if we assume that the set of locational factors that determine the firms' office location is $F_f = \{F_1, F_2, F_3, \ldots, F_n\}$ and the number of the locational factors is n (the subscript of f *refers to* "firm"). The importance that a firm attaches to each locational factor is different, which in consequence can be described as $W_f = \{w_1, w_2, w_3, \ldots, w_n\}$.

If we assume that the set of candidate plots is $CP_f = \{P_1, P_2, P_3, \ldots, P_m\}$ where m is the number of candidate plots. It is obvious that each plot has a different set of values for the location factors in F_f, which can be described as $\{m_{f1}, m_{f2}, m_{f3}, \ldots, m_{fn}\}$. Thus, the location utility each plot generates for this firm can be calculated by the formula:

$$U_f = (w_1 \times m_{f1} + w_2 \times m_{f2} + w_3 \times m_{f3} + \cdots + w_n \times m_{fn}) \times Q$$

where Q is the building quality of the plot. Then the one which generates the greatest location utility can be defined as:

$$P_{final} = P_{\max} \text{ where } P_{\max} \text{ satisfies that}$$

$$U_f(P_{\max}) = \max\{U_f(P_i), i = 1, 2, 3, \ldots, m\}$$

This plot, as explained earlier, is the site where the firm chooses to settle down. The firm's moving-into this plot will consequently raise the marks of the locational factors of this plot and its neighbouring ones. It is possible that the firm fails to find the right place. If this happens, the firm will continue to search in the next tick. But when the waiting time exceeds the maximum tolerable time, the firm is excluded from the system.

Table 2 Triggers and action rules for the creative firms

Trigger	Action
Firm's total capital <0	Dies and is permanently excluded (claims bankrupt)
Time of failing to find the right office site > the maximum times that it can tolerate	Dies and is permanently excluded (gives up and moves to another city)

3.2.2 Looking for Workers

Before deciding whether to employ or fire workers, the firm must first check its capital and profit rate (Table 2). The firm's capital can be determined by its cost for land rent, workers' salary workers' output, tax paid and the sale rate of the production. The design depends on the case being studied. If the capital runs out, the firm will collapse and lead to the unemployment of its staff. Simultaneously, its demise will also cause the decreasing of the values of the locational factors of the relevant plots.

But if the firm is successful and is generating a profit-rate higher than the critical value (open to change by the user), it is allowed to expand its office area, and employ more workers. However, considering costs, it will only employ those workers whose expected salary is approximate to the firm's wage standard (should be set by the user depending on the studied case). A firm has to decrease its expected size (and so its office area) if its profit rate is negative and lower than the critical value (open to change by the user). If the expected size is reduced to a value smaller than the real size of the firm, workers with less profit-making ability (referring to the value of worker's capability minus his real income) will be fired.

3.3 Creative Workers

3.3.1 Looking for Residence

Using the analogy of the firm's action rules in finding an office, the principle for the worker to find accommodation is to choose plot with the maximum location utility (Table 3). If we assume that:

The locational factors for the workers are $F_w = \{F_1, F_2, F_3, \ldots, F_n\}$ (w refers to "worker"), and the corresponding weights are $W_w = \{w_1, w_2, w_3, \ldots, w_n\}$;

The candidate plots are $CP_w = \{P_1, P_2, P_3, \ldots, P_m\}$, and the marks each candidate plot features for the locational factors are $\{m_{w1}, m_{w2}, m_{w3}, \ldots, m_{wn}\}$.

Then the location utility of each candidate plot for the worker can be calculated via:

$$U_w = (w_1 \times m_{w1} + w_2 \times m_{w2} + w_3 \times m_{w3} + \cdots + w_n \times m_{wn}) \times Q$$

where Q is the building quality of the plot;

Table 3 Triggers and action rules for the creative workers

Trigger	Action
Times of failing to find a job in the creative firms > the maximum times he can endure	Dies and is permanently excluded (gives up and turn to find jobs in other businesses)
Times of failing to find a place to live > the maximum times he can endure	Dies and is permanently excluded (gives up and moves to another city)

And the best location can be identified by:

$$P_{final} = P_{\max} \text{ where } P_{\max} \text{ satisfies that}$$

$$U_w(P_{\max}) = \max\{U_w(P_i),\ i = 1,\ 2,\ 3, \ldots,\ m\}.$$

Similarly, the worker's moving-into this best plot will raise the values of the locational factors of the neighbourhood. Still, the question of which factors will be affected depends on the modeller's design and purpose. However, if the worker is unable to find the right accommodation, he will continue to try until is excluded from the system (the maximum value of attempts is open to change by the user).

3.3.2 Looking for Job

The worker cannot relax even if he succeeds in finding residence. He must continue looking for a job position (Table 3). He only considers candidate firms which can provide payment higher than his expected salary. However, to reflect the two-way choice practice in employment, the payment that the candidate firms can offer is limited to two times the income expected by the worker. With this restriction, a number of firms (this number is open to change by the user) will be picked out as the candidate firms which the worker is willing to work for.

The supposition here is that the worker will choose to work in the firm which provides the best payment. But if the employment rate is worryingly low, the worker may not be able to find a job. In this situation, the worker will try two possible strategies. First, try to find a job in the next round. Second, give up directly if the time of being unemployed (open to change by the user) is too long, and as a result, he will be excluded from the system. Conversely, if the employment rate is high, greater than the critical value set by the user, new workers will be born into this dynamic system.

4 Model Implementation: The Case of Nanjing

This section uses the modelling environment of NetLogo (Wilensky 1999) in order to simulate the dynamics of the modelling framework proposed above. In NetLogo, the default agent is defined as "turtle". If the model to be built is spatially explicit, the "world" in NetLogo can act as the geographical space (Table 4).

Table 4 The concepts for agent-based modelling in NetLogo

General concepts in agent-based model	Corresponding concepts in NetLogo
Agent	"turtle" spatially implicit: set by the modeller as necessary
Environment	spatially explicit: "world"

The city used in this simulation is Nanjing, a growth pole with a population of about 7.7 million in the Yangtze River Delta. Since 2006, creative industries have been designated by the government as the pivotal industrial sectors to generate growth and facilitate urban spatial restructuring (The Nanjing Bureau of Culture, Broadcast, Television, News Media and Publication 2011; Nanjing Municipal Government 2006, 2010). As creative industries consist of a set of industrial sectors, in this study we narrow the research down to two industrial sectors: software design and cartoon design. The data (such as urban policy schemes, locational factors that determine the worker's residence choice, worker's income, locational factors that determine the firms' office location, creative firms' size and size-development plan) needed for this research are collected via a fieldwork of questionnaires and semi-structured interviews.

As this model deals with how the development of creative industries influences the urban spatial structure, it is called Creative Industries Development—Urban Spatial Structure Transformation (CID-USST). Figure 2 illustrates the user interface of the complete model.

4.1 Setting-Up Procedure

The first step in setting up the environment is to generate the simplified urban land space of Nanjing. In this case, it is based on these principles:

- It is single-centered with one CBD.
- The urban space is divided into five areas: CBD, inner city bureau, outer city bureau, inner suburb, and outer suburb. The plot ratio in each area is capped to the maximum plot ratio.
- The geographical features that influence the locational marks of each plot are generated based on the parameters in the user interface.
- The generation of facilities (such as shopping malls, libraries, underground and city roads) that influence the locational marks of the land plots obeys the principle: the closer to the city centre, the higher the probability is.
- The urban road system is a combination of ring-ways, grid, (within inner-city-bureaus) and radiant roads from the centre. The closer to the city centre, the higher the road density is.

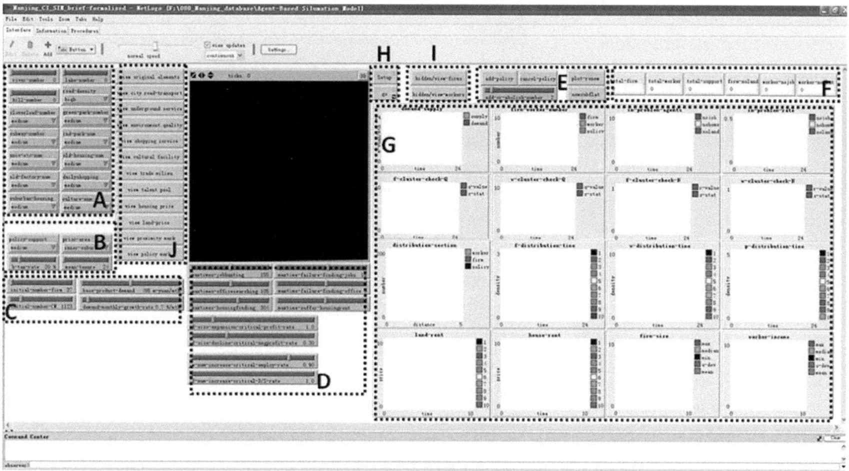

Fig. 2 The user interface of the complete CID-USST model (overview)

- The locations of the other urban elements (creative industry parks, old factories, green parks) are generated via controlling their distribution percentages in each urban area of the city.

In this case study, the urban land space is represented by the "world" (the black plain in Fig. 2). Via fieldwork, we figured out that:

1. The firms' locations are mainly determined by eight factors listed by importance (high to low) are:

   government policy guidance,
   urban road-transport (bus line),
   high-speed public transport (underground),
   cooperation and trade milieu among firms,
   geographical proximity,
   land rent,
   sharing of talent pool, and
   physical environment.

2. The factors that influence the creative workers' housing-location preference involve six determinants which, in sequence with the first as the most important, are:

   public transport (bus line and underground),
   convenience for buying daily supplies,
   housing rent/price,
   physical environment quality,
   allocation/inheritance, and
   cultural facility.

Fig. 3 Section A of Fig. 2: tools to set the geographical features

The geographical features considered (generated) in this are: river, lake, hill, urban road, regional road linkage (cloverleaf), green park, underground, industrial park, universities and research institution, declining housing site, old factory, shopping mall, cultural facilities (such as museum, library). These features can be set via the user interface, shown by section A in Fig. 2 (Fig 3).

According to the questionnaires, the most frequently adopted supportive policies are: **tax reduction, lower land/office rent, and trade/milieu promotion**. In the setting up process, these policies are repackaged and allocated to certain plots by referring to the parameters controlled by section B of Fig. 2 (Fig. 4). The parameter “b-tax-rate” means the basic tax rate that the government generally

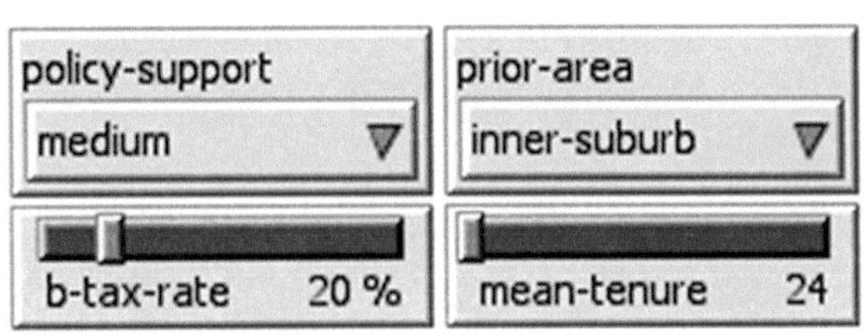

Fig. 4 Section B of Fig. 2: tools to set the policy packages

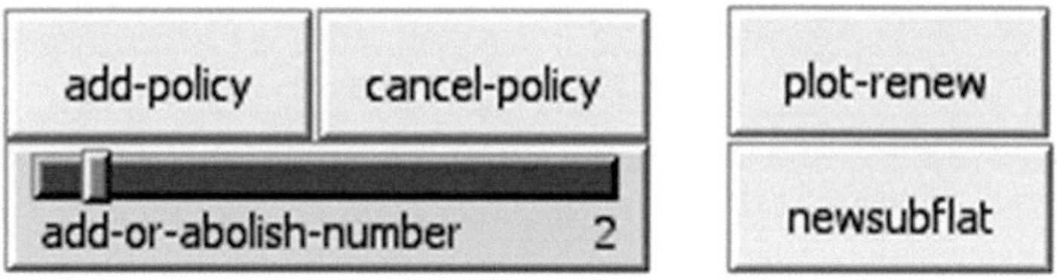

Fig. 5 Section E of Fig. 2: tools for the user to act as the urban government

applies to the firms. The "mean-tenure" is the mean duration for all the policy packages.

The government's power in urban renewal/development is simulated by section E of Fig. 2 (Fig. 5). Via it, the user can add/cancel policy packages or start new regeneration/development schemes. The button "plot-renew" points to the procedure of urban regeneration. Once it is pressed during the operation of the model, a set of plots with lower building quality (building quality <0.5) will be renewed. If the user needs, he can also approach the previously protected areas and develop new housing property on them via pressing the button of "newsubflat".

Section C of the user interface is designed for the user to set the primeval numbers of the firms and the workers (Fig. 6). It also provides the user the access (the two sliders on the right) to setting the primitive demand (in value) of the creative production (software, cartoon, animation) per month.

The critical values that closely related to the triggers of the behaviours of the agents are open to change by the user via section D (Figs. 2, 7). It consists of ten sliders. The upper six deal with the three actions of the agents: the workers to find job and to look for accommodation; the firm to look for office. The lower four is the critical values that relate to the birth of new agents (firms, workers) (Table 5).

The setting up process can be initiated by pressing the button "setup" (Fig. 8). All the system needs to execute the programme in this process is to prepare the primeval status of the system according to the details depicted in Table 5. This setting up process basically consists of 8 steps: (1) Generate a simplified urban land space; (2) Calculate the primitive locational marks (basic-mark, transport-mark, rapidtrans-mark, environment-mark, shopping-mark, culture-mark, milieu-mark, talent-mark) for each plot; (3) Calculate the basic land price (office rent, housing rent) of each plot; (4) Set the building quality of each plot; (5) Set the maximum plot ratio of each plot; (6) Set up policy packages and their locations;

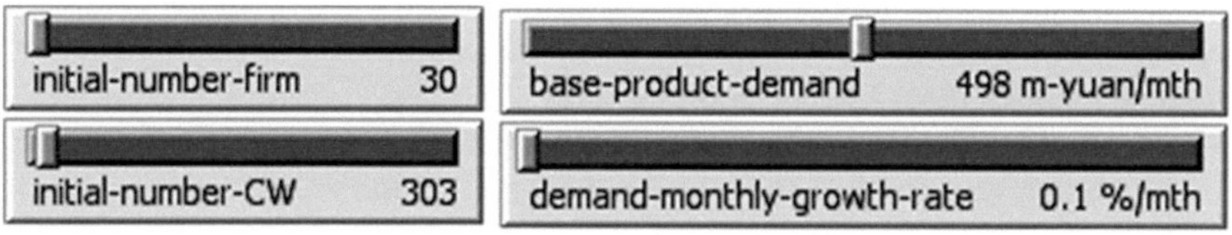

Fig. 6 Section C of Fig. 2: tools to set the primitive number of the agents and the production demand per month

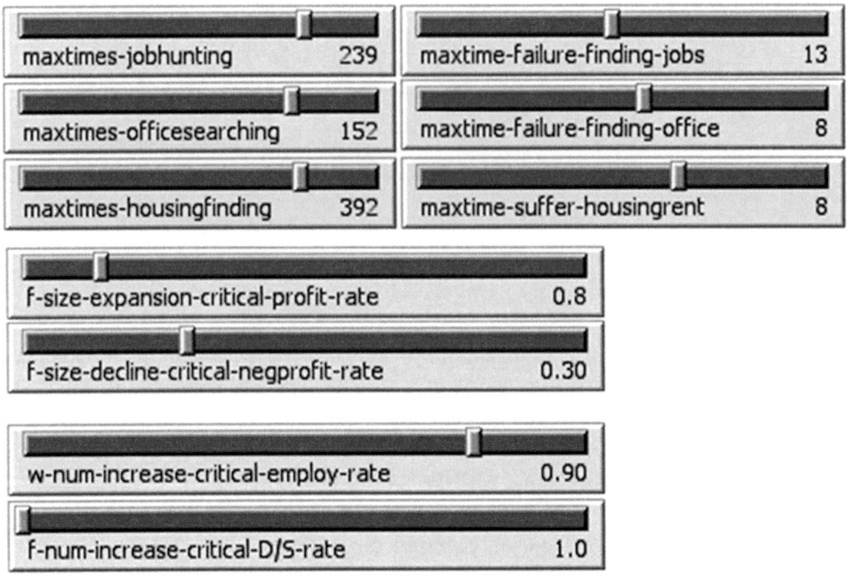

Fig. 7 Section D of Fig. 2: tools to set the critical values (and triggers)

(7) Generate the primitive firms and set their primeval status; (8) Generate the primitive workers and set their primeval status. Only when this process is complete can the model execute the dynamic procedure, which operates according to the action rules defined in the section of "agents and their action rules". The "go" button in section H directly points to the whole dynamic procedure (Fig. 8).

4.2 Running the Model in Nanjing

Once the model starts to run it is possible to display the simulation results. These results are spatially visible in the "world" by swapping views via section I and J (Figs. 2, 9, 10, 11), and also numerically readable as presented in section F and G (Figs. 12, 13).

The section of F and G are both dedicated to the demonstration of the dynamic results. The first three monitors in section F of Fig. 2 are used to record the number of total firms, total workers, and total policy packages that the government provides. The last three, in contrast, aim to show the number of agents in trouble, including firms without offices, workers with no jobs and worker with no suitable accommodations (Fig. 12).

In section G (Fig. 2), the four plots at the second row are tools to examine the spatial clustering pattern of the creative workers and the creative firms. For this purpose, two spatial statistics are used (Wong and Lee 2005). One is the variance-mean ratio (VMR) by the method of quadrat analysis, which is used by plots "f-cluster-check-Q" and "w-cluster-check-Q". The expression for VMR is:

Table 5 Setting up of the model

A	
Basic modelling settings	
Number of patches (plots)	200*200 with origin at the centre
Scale of each plot	1:10,000 (one plot represent 10,000 m^2 of land)
Time interval (tick)	One month (corresponding to real time period that the firms and the workers take action)
Urban land space settings (geographical features)	
Division of the urban land	Five areas: CBD, inner city bureau, outer city bureau, inner suburb and outer suburb
Generation of geographical features: river, lake, hill, urban road, regional road linkage (cloverleaf), green park, underground, industrial park, universities etc., old housing site, old factory, shopping mall, suburban housing, cultural facilities (museum, library, etc.)	We decide to include these features by referring to the locational factors that determines the location behaviours of the firms and the workers. These features can be set via section A in Fig. 2 (Fig. 3)
Calculate the mark of each of the geographical locational factors for each plot	It is assumed that the mark of each factor the plot has is inverse to the nearest distance from this plot to the geographical feature. In general, it is calculated by $m_i = {}^1/_{D_{i(nearest)}}$, i refers to factor i
Calculate the office rent (per month per m^2) and the housing rent (per month per m^2)	For office rent, it is calculated via $R_o = R_{o0} * \sum m_{oi}$, where {$m_o$i} refers to those geographical factors that the creative firms care about. R_{o0} is the constant for office rent. The value of R_o is identical to the real office rent For housing rent, it is calculated via $R_h = R_{h0} * \sum m_{hi}$, where {$m_{hi}$} refers to those geographical factors that the creative workers care about. R_{h0} is the constant for office rent. The value of R_h is identic! to the real housing rent
Policy marks that each plot gets (depend on the settings by section B in Fig. 2, as detailed by Fig. 4)	The policy mark that a plot can have depends on the number of policies that this plot enjoys. The possible value set is {1, 1.5, 2, 2.5}, where 1 means this plot enjoy no supportive policy, 1.5 mean it enjoys one policy, while 2 for two policies and 2.5 for three
Building quality	The mark of building quality is specified by referring to the geographical features. The value range is [0, 1], where 0 represent completely declined while 1 means brand new. Those plots of declining housing site and old factory can get marks within the range of [0, 0.5]. The suburban housing plots are regarded as brand new. The marks for the rest constructed plots will be specified randomly within the range [0, 1]

(continued)

Table 5 (continued)

B			
Settings directly related to the three agent classes			
Urban government	Design, and spatial allocation of the policy packages		Controlled by section B of Fig. 2 (Fig. 4)
	Maximum plot ratio for each plot		From the CBD to the outer suburb the plot ratio for each area is set to 2.0, 1.5, 1.25, 1 and 0.5 respectively for office development, and 1.5, 1.2, 1.0, 0.75, 0.5 for housing development
	Urban regeneration and new development		Controlled by section E of the user interface (Figs. 2, 5)
Creative firms	Primeval number: controlled by section C of Fig. 2 (Fig. 6)		
	Primeval demand and demand growth rate/month: controlled by section C of Fig. 2 (Fig. 6)		
	Looking for office	The maximum plots that the firm can have as candidate plots within a month	Controlled by section D of Fig. 2 (Fig. 7)
		Maximum time of failing to find the right office site	
	Looking for workers	Critical profit rate that determines the increase or decrease of the firm's expected size	
		Critical rate of demand to supply that determines the new births of the creative firms	
Creative workers	Primeval number: controlled by section C of Fig. 2 (Fig. 6)		
	Looking for residence	Maximum number of housing plots (number of candidate housing plots) that a worker can go for checking within a month	Controlled by section D of Fig. 2 (Fig. 7)
		Maximum time of failing to find an suitable accommodation	
	Looking for jobs	Maximum number of firms (number of candidate firms) that a worker can go for interview within a month	
		Maximum time of failing to find a job	

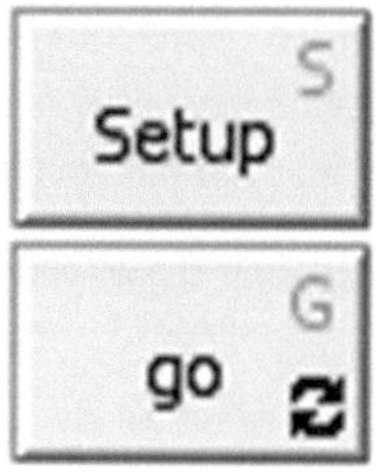

Fig. 8 Section H of Fig. 2: the "setup" button and the "go" button

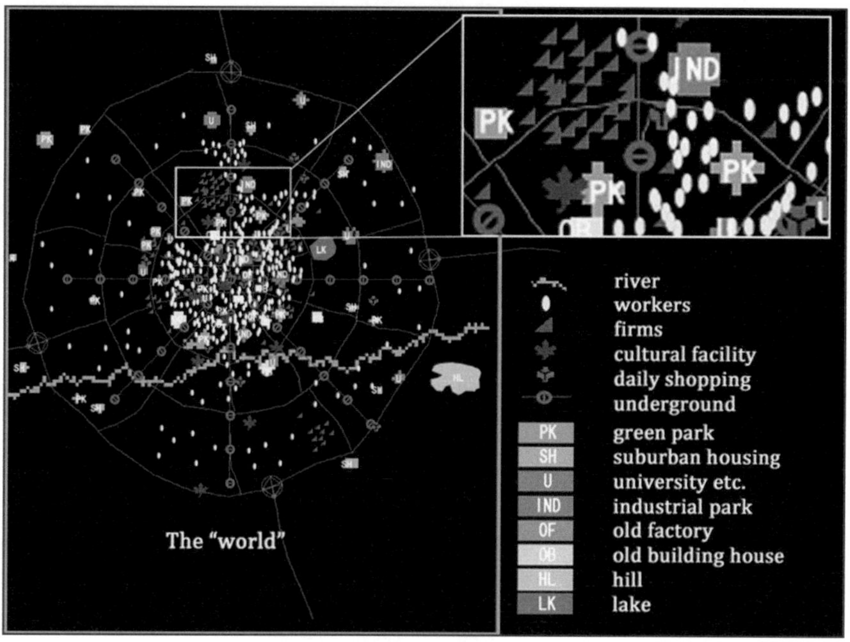

Fig. 9 The "world" as direct demonstration medium of the result

$$VMR = \frac{S^2}{\overline{X}}$$

The other one is the R statistic by nearest neighbour analysis. It is adopted by the other two plots (f-cluster-check-N and w-cluster-check-N). The expression is:

$$R = \frac{R_o}{R_e} = \frac{\sum d_i/_n}{0.5\sqrt{A/_n}}$$

where d_i is the nearest distance of each point, A is the study area and *n* is the number of point.

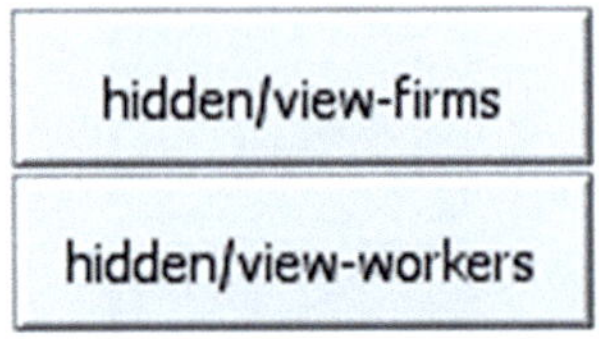

Fig. 10 Section I of Fig. 2: tools to show/hidden the two classes of agents (creative firms, creative workers)

Fig. 11 Section J of Fig. 2: tools to show the spatial distribution of the location marks

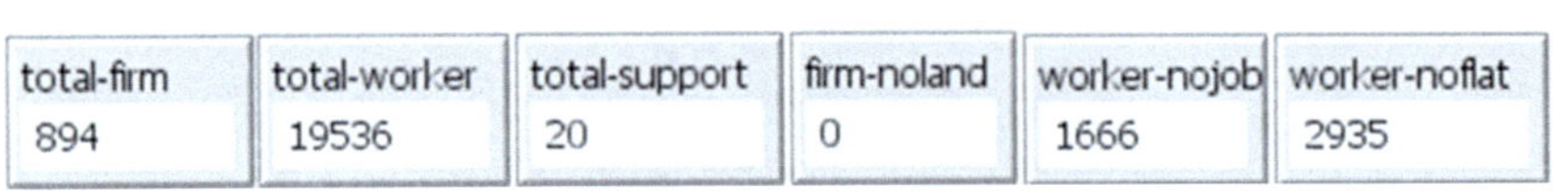

Fig. 12 Section F of Fig. 2: monitors to record the number of the agents

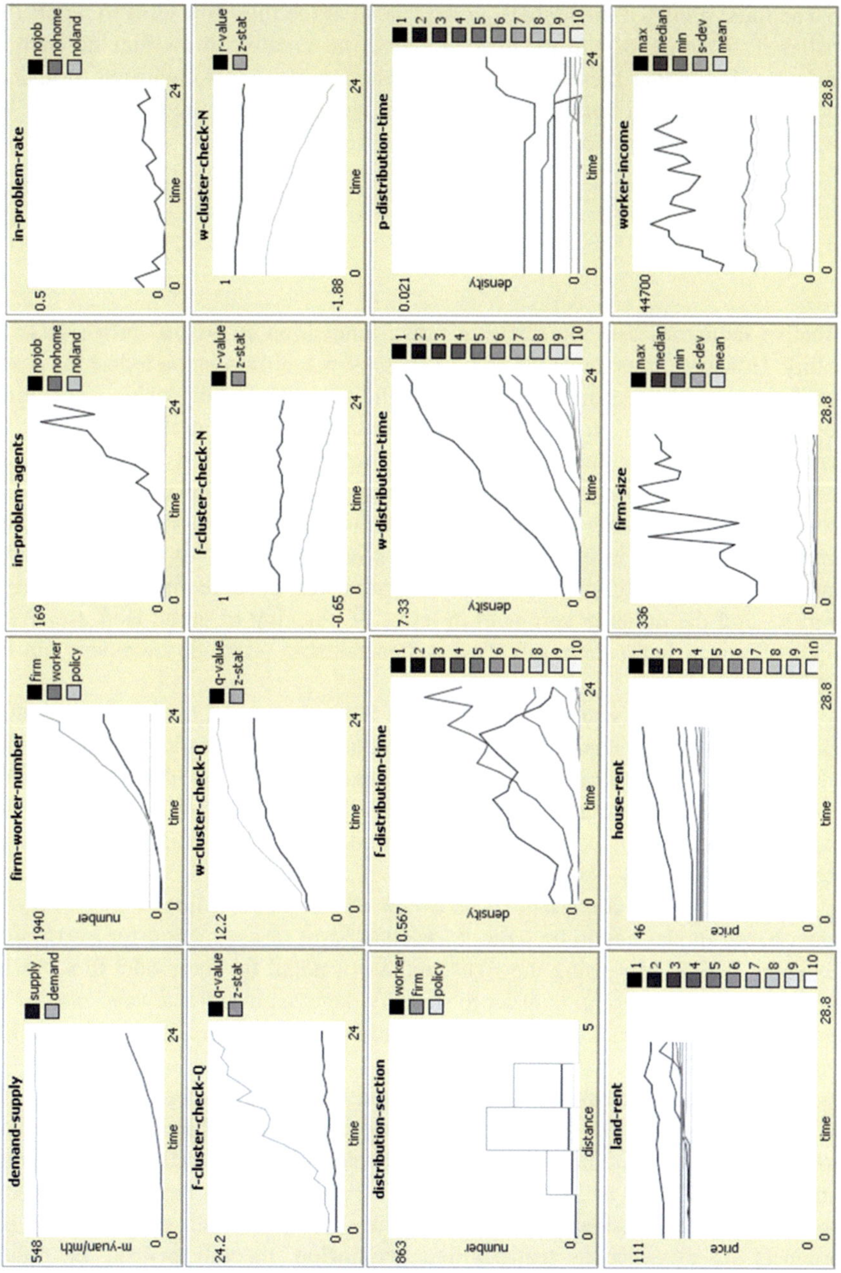

Fig. 13 Section G of Fig. 2: plots to describe the spatial distribution features and the socioeconomic features of the dynamics process

The three plots with ten labels on the legend at the third row aims to report the spatial distribution of the creative workers, the creative firms and the policy packages. Following these three plots are another two plots dedicated to understand the spatial distribution of the land rent and the housing rent of the plots by similar approach (Fig. 13).

5 Conclusion

Creative industries can contribute to urban innovation as well as urban sustainability. However, in practice the effectiveness of planning creative industries is not satisfactory and many times economically and socially unsustainable—generating negative competition, high government investment on regeneration in order to attract firms/workers but resulting in vacancy rates or short term usage and abandonment, once the government subsidy stops. One solution to this problem is to view the relationship among all the involved agents as a dynamic complex process as presented in our research, by proposing a complex system approach to understanding the complex interactions among the creative firms, the creative workers and the urban government in terms of planning of urban land use. With this we expect that a more effective and sustainable urban development can be achieved.

We present this research framework and computer model hoping that visually and quantitatively decision makers, researchers and stakeholders can see the advantages for dynamically planning land uses, firm location, demand for office and residential space and human resources—the case of Nanjing illustrates the model development process. The complete model provides an interface that allows scenarios to be easily generated.

Finally, it is also important to be aware of the model's limitations. First, an agent-based model has to be built at the right level of abstraction for every phenomenon, judiciously using the right amount of detail for the model to serve its purpose (Couclelis 2002). In order to simplify the dynamics framework, we had to omit some details (i.e. the urban residents are not regarded as an agent class in the framework).

Second, this agent-based system deals with human beings which means that it will involve agents with potentially irrational behaviour, subjective choices, and complex psychology. These factors are difficult to quantify, calibrate, and sometimes justify (Crooks and Heppenstall 2012). For instance, in the case of Nanjing, policies such as intellectual property rights protection and special outlet for products are classified as **trade/milieu promotion**. Its influence on the policy marks of the relating land plots is counted, but its further influence upon the firm's profit rate and capital cannot be quantified.

References

Comunian R (2011) Rethinking the creative city: the role of complexity, networks and interactions in the urban creative economy. Urban Stud 48:1157–1179

Couclelis H (2002) Modelling frameworks, paradigms, and approaches. In: Keith C, Brad P, Michael C (eds) Geographic information systems and environmental modelling. Prentice Hall, London, pp 36–50

Crooks AT, Castle CJE (2012) The integration of agent-based modelling and geographical information for geospatial simulation. In: Heppenstall AJ, Crooks AT, Batty LMM (eds) Agent-based models of geographical systems. Springer, Dordrecht, pp 219–251

Crooks AT, Heppenstall AJ (2012) Introduction to agent-based modelling. In: Heppenstall AJ, Crooks AT, Batty LMM (eds) Agent-based models of geographical systems. Springer, Dordrecht, pp 85–105

Evans G (2005) Measure for measure: evaluating the evidence of culture's contribution to regeneration. Urban Stud 42:1–25

Florida R (2002) The rise of the creative class: and how it's transforming work, leisure, community and everyday life. Basic Book, New York

Foord J (2008) Strategies for creative industries: an international review. Creative Ind J 1:91–113

Forum for the Future (2010) Can the creative industries lead us to a sustainable future? http://www.forumforthefuture.org/sites/default/files/project/downloads/sustainability-baseline-review-ciktn-08-10-2010.pdf Forum for the Future, and Creative Industries Knowledge Transfer Network (CITKN) Team

Girard LF, Baycan T, Nijkamp P (2012) Editorial preface. In: Girard LF, Baycan T, Nijkamp P (eds) Sustainable city and creativity: promoting creative urban initiatives. Ashgate

Landry C (2000) The creative city: a toolkit for urban innovators. London: Earthscan Publications.

Nanjing Municipal Government (2006) The outline of the eleventh five-year plan for creative cultural industries in Nanjing (Chinese)

Nanjing Municipal Government (2010) The outline of the twelfth five-year plan of Nanjing (Chinese)

Higgs P, Cunningham S, Bakhshi H (2008) Beyond the creative industries: mapping the creative economy in the United Kingdom. National Endowment for Science, Technology and the Arts (NESTA)

Kakiuchi E (2012) Sustainable cities with creativity: promoting creative urban initiatives-theory and practice in Japan. In: Girard LF et al (eds) Sustainable city and creativity: promoting creative urban initiatives. Ashgate, pp 413–440

Mossig I (2011) Regional employment growth in the cultural and creative industries in Germany 2003–2008. Eur Plann Stud 19:967–990

O'connor J (2007) The cultural and creative industries: a review of the literature. Creative Partnerships, London

Piergiovanni R, Carree MA, Santarelli E (2012) Creative industries, new business formation, and regional economic growth. Small Bus Econ 39:539–560

Pratt AC (2009) Urban regeneration: from the arts 'Feel Good' factor to the cultural economy: a case study of Hoxton, London. Urban Stud 46:1041–1061

Scott AJ (2006) Entrepreneurship, innovation and industrial development: geography and the creative field revisited. Small Bus Econ 24:1–24

Silva EA, Wu N (2012) Surveying models in urban land studies. J Plann Lit 27:139–152

Stam E, De Jong JPJ, Marlet G (2008) Creative industries in the Netherlands: structure, development, innovativeness and effects on urban growth. Geogr Ann Ser B, Hum Geogr 90:119–132

The Nanjing Bureau of Culture, Broadcast, Television, News Media and Publication (2011) The outline of the twelfth five-year plan for culture development in Nanjing (Chinese)

Wilensky U (1999) NetLogo. Center for Connected Learning and Computer-Based Modeling, Northwestern University, Evanston, IL, http://ccl.northwestern.edu/netlogo/

Wong DWS, Lee J (2005) Statistical analysis of geographic information: with ArcView GIS and ArcGIS. Wiley, Inc. pp 221–252

Wu N, Silva EA (2010) Artificial intelligence solutions for urban land dynamics: a review. J Plann Lit 24:246–265

LACONISS: A Planning Support System for Land Consolidation

Demetris Demetriou, John Stillwell and Linda See

Abstract Land consolidation is considered as the most effective land management planning approach for solving the land fragmentation problem in rural areas and more broadly to assist the implementation of regional and urban planning. This chapter focuses on a planning support system for land consolidation in rural areas called Land CONsolidation Integrated Support System for planning and decision making (LACONISS) that integrates GIS, artificial intelligence techniques and multi-criteria decision methods. The system involves four modules: the Land Fragmentation System (LandFragmentS) module measures existing land fragmentation in an agricultural context; the Land Spatial Consolidation Expert System (LandSpaCES) Design module produces alternative land redistribution plans; the LandSpaCES Evaluation module assesses these plans based on a set of criteria; and the Land Parcelling System (LandParcelS) module produces the final land partitioning plan. The final output is the land reallocation plan. The whole system has been applied to a case study area in Cyprus and this reveals that land fragmentation can be reliably measured, that alternative land reallocation plans can be produced that can successfully emulate planners' reasoning, that the most beneficial plan for various combinations of criteria and weights can be identified and that the subdivision process can be automated satisfactorily although further improvements are needed. The contribution of LACONISS is relevant to both rural and urban sustainable development since it supports and automates the land

D. Demetriou (✉) · J. Stillwell
School of Geography, University of Leeds, LeedsLS2 9JT, Leeds, UK
e-mail: demdeme@cytanet.com.cy

J. Stillwell
e-mail: j.c.h.stillwell@leeds.ac.uk

L. See
Institute for Applied Systems Analysis (IIASA), Laxenburg, Austria
e-mail: see@iiasa.ac.at

L. See
Centre for Advanced Spatial Analysis (CASA), University College London (UCL), London, UK

S. Geertman et al. (eds.), *Planning Support Systems for Sustainable Urban Development*, Lecture Notes in Geoinformation and Cartography,
DOI: 10.1007/978-3-642-37533-0_5, © Springer-Verlag Berlin Heidelberg 2013

reallocation (or land readjustment) process which is the principal component of any land consolidation project whether rural or urban. Eventually, LACONISS may constitute the foundations for developing a generic system that could be applied in any country that implements land consolidation projects.

1 Introduction

Land fragmentation is defined in the literature as the situation in which a single farm or ownership consists of numerous spatially separated parcels (Bentley 1987; Van Dijk 2003) which may be small in size and have irregular shapes. In an agricultural context, land fragmentation implies a defective land tenure structure that in many cases may inhibit agricultural development and sustainable development more generally (Demetriou et al. 2012a). In some countries, there may be additional land fragmentation factors such as the lack of road access to land parcels in certain areas and issues relating to ownership rights. For example, a parcel may have shared ownership, i.e. it may belong to more than one landowner, or there may be dual or multiple ownership, i.e. the land is owned by one person whilst the trees growing on the land are owned by someone else and a third party has ownership rights for water. All the problems associated with agricultural land fragmentation hinder mechanisation, cause inefficient production and involve large costs to alleviate adverse effects, resulting in a reduction in farmers' net incomes.

Land fragmentation is also a problem in urban areas where it may obstruct rational urban expansion or development. Legal and planning restrictions linked with individual properties may prevent an integrated planning approach in urban areas. Thus, although this chapter is specifically concerned with land fragmentation in a rural context, there is little doubt that similar land issues arise in urban areas and require solutions to ensure that urban spaces and communities become more sustainable.

Land fragmentation occurs frequently in various parts of the world. To control fragmentation through the application of certain land management measures, a land fragmentation study is usually undertaken followed by an environmental impact assessment and a feasibility study. The results from the first of these assessments can be quantified by an appropriate fragmentation index, although at present, no standard algorithm or universally accepted methodology is available for measuring land fragmentation (Van Hung et al. 2007). The most popular indices of those developed in the past include the ones by Simmons (1964) and Januszewski (1968), although as pointed out in Demetriou et al. (2013a), they both have significant drawbacks and their application can result in wrong decisions. Thus, there is clearly a need for a better methodology for measuring land fragmentation in a reliable manner.

The most favoured land management response to the land fragmentation problem is land consolidation. There are two main components: the reallocation (or readjustment) of land and agrarian spatial planning (Thomas 2006a). The first component involves finding an optimal reconfiguration of an existing land tenure structure in a given rural area based on the relevant land consolidation legislation and current practices, both of which impose various constraints on achieving the aims of a particular land consolidation project. The second involves provision of the necessary infrastructure, including roads, irrigation and drainage systems, landscaping and environmental management, settlement renewal and soil conservation.

Land reallocation consists of land redistribution and land partitioning. Land redistribution involves decision-making by combining current legislation, existing land tenure structures, rules of thumb and the experience of the planners. It involves preparing a preliminary plan to restructure ownerships and therefore the number, ownership, size, land value and approximate location of the parcels. Land partitioning, in contrast, is a design process that involves subdividing the land into smaller parcels or 'sub-spaces' in terms of parcel shape, size, land value, final location and various constraints. Land partitioning is often a trial-and-error process in practice, which uses the relevant legislation concerned, the existing land structure, design criteria, constraints and rules of thumb. The outcome is the final land consolidation plan.

Land consolidation has a number of problems such as lengthy project duration, high operational costs, conflicts of interest among landowners and lack of consensus amongst stakeholders. These latter problems are associated with the most critical component of the whole exercise, which is the land reallocation process (Sonnenberg 2002; Thomas 2006b). Thus an automated system that can transform land consolidation planning into an efficient, systematic and transparent process is clearly required (Demetriou et al., 2012a).

Proprietary Geographical Information Systems (GIS) do not currently have the capability to adequately support such complex spatial planning and decision-making problems (e.g. Geertman and Stillwell 2009). GIS are simply too generic and would need the ability to incorporate expert knowledge, produce alternative solutions or allow evaluation of these solutions without customization or knowledge of programming. In fact, planning or decision support systems for land consolidation which automate the process in a systematic and efficient manner have not yet been developed despite the considerable efforts in land reallocation research that have been undertaken since the 1960s (Rosman and Sonnenberg 1998). Instead, the focus has been on the development of isolated algorithms for land redistribution, land partitioning or the evaluation of land consolidation plans.

Some previous studies have attempted to automate the problem of land redistribution by treating it as a mathematical optimisation problem (e.g. Rosman and Sonnenberg 1998; Ayranci 2007), meaning that, although the results are sometimes optimal in terms of efficiency, they are not necessarily realistic or operationally applicable. Other studies, focusing on land partitioning (Buis and Vingerhoeds 1996; Tourino et al. 2003), have produced results that are encouraging but still far from satisfactory. Furthermore, land consolidation evaluation studies (e.g. Coelho et al. 2001; Sklenicka 2006) have also suffered from a lack of tools capable of providing detailed land reallocation inputs for *ex-ante* project evaluation. The limitations of these studies emphasise the need for new and more efficient methods and techniques to model the entire land reallocation process within an integrated planning framework.

LAnd CONsolidation Integrated Support System for planning and decision making (LACONISS) (Demetriou et al. 2012b) aims to fill this gap and is the focus of this chapter. This new system is a hybrid prototype that integrates GIS, artificial intelligence techniques (e.g. Openshaw and Openshaw 1997), namely expert systems (ES) (e.g. Giarrantano and Riley 2005), genetic algorithms (GAs) (e.g. Goldberg 1989) and multi-criteria decision methods (MCDM) (Malczewski 1999) both multi-attribute (MADM) (Sharifi et al. 2004) and multi-objective (MODM) (Deb 2001). The reader is referred to Demetriou et al. (2011, 2012a, b, c 2013a, b, c) for further conceptual and technical details. The focus of the current chapter is on the use of LACONISS as a planning support system and its application to a case study area in Cyprus (Demetriou et al. 2011). The next section summarises the land reallocation process as carried out in practice and Sect. 3 presents an explanation of each of the four modules of LACONISS. The case study is reported in Sect. 4 and the chapter concludes with a short discussion of the potential prospects for the system.

2 The Land Reallocation Process in Practice

As noted earlier, the decision to implement land consolidation measures to control fragmentation usually involves undertaking a land fragmentation study, an environmental impact assessment and a feasibility study. The latter two are beyond the scope of this chapter. The former currently involves the statistical analysis of the land tenure structure of the area under study so as to reveal potential problems and their extent to which they are associated with land fragmentation. If these analyses suggest the implementation of land consolidation, then a national organization such as the Land Consolidation Department (1993) in Cyprus promotes the project so as to receive approval by the Minister of Agriculture, Environment and Natural Resources. Thereafter, following legal acceptance, the final decision for implementation is taken by the landowners concerned. In Cyprus, a decision to go ahead is taken when 50 % plus one landowner agree with the implementation since this group should own the majority of the total land value of holdings in the area. The other landowners should then follow and the implementation stage may begin. This involves the study and construction of a main road network, an updated valuation of all land parcels, the study and construction of a complementary road network, the undertaking of carrying out preference sessions with landowners and, eventually, the final land reallocation study (Demetriou et al. 2012a).

In practice, the process of land reallocation involves the assembly of all properties belonging to different landowners in a certain area, followed by a new subdivision of land into parcels and redistribution of the land to the same landowners, based on the share (in terms of a holding's area and/or land value) of each individual's land in the whole area (Sonnenberg 2002). The process is normally carried out using a computer-aided design (CAD) system and/or a GIS. In particular, the process, as it is applied in Cyprus, involves the following nine major steps:

(1) Subdivision of the consolidated area in land blocks where each land block is enclosed by roads, streams and the external boundaries of the study area. Roads include those constructed in a previous stage of the project; the proposed new roads will provide access to all new parcels.
(2) Calculation of the total area and land value of each land block.
(3) Definition of which landowners will not be granted property in the new plan, based on the minimum area and land value limits established by the Land Consolidation Committee (LCC).
(4) Definition of which parcels will be exempted from reallocation, based on the relevant decision of the LCC.
(5) Calculation of the percentage of land (as a land value) that should be provided by landowners for public facilities which is called the contribution coefficient.
(6) Calculation of the land value that should be allocated to each landowner after the subtraction of the land value calculated by the contribution coefficient.
(7) Definition of the maximum number of parcels that can be allocated to each landowner based on the rule of a 'small-medium-large holding' as defined by the Head of the Land Consolidation Department.
(8) Calculation of the initial available land (in terms of size and value) for reallocation in each block.
(9) Reallocation of properties.

The last step is the most complex and involves land redistribution and land partitioning as noted earlier which, in practice, may be carried out as a single, iterative trial and error process. This proceeds block by block by considering how to reallocate the properties in a certain block. Firstly, the properties in the block owned by current landowners are considered, followed by the potential to transfer the properties of other landowners in that block. The aim of the planner is to produce an optimum plan based on a series of criteria categorized as parcel and landowner related, legislative/authoritative, economic, social, environmental and local (Demetriou et al. 2012a). Thus, for each land block, planners attempt to design a predefined number of parcels (as a result of land redistribution), each with as regular a shape as possible, approximating a desired land value (because the latter is the basic land reallocation criterion according to legislation) and size, subject to the limitation that the latter exceeds the minimum limits provided by legislation and that a parcel should have access to a road. The processes can be semi-structured and formulated as an algorithm as follows:

(1) begin reallocation for block i for landowner k who owns land located within block I;

(2) consider if the property of landowner *k* should be reallocated to block *i* and whether this is possible;
(3) consider other ownership attributes, i.e. parcels/shares of the landowner *k* owned in other blocks;
(4) decide how to reallocate property based on the previous steps (1–8 outlined above), using the reallocation principles provided in the legislation, government directives, advice and the criteria noted earlier. The decision refers to the number of parcels, their size, the land value, the location of each parcel and the aggregate size and land value;
(5) design the new parcel(s) in block *i*, which involves the exact definition of their shape and location. The design should take into account the existing boundaries, environmental protection, *et cetera*;
(6) calculate the remaining available land in terms of size and value for reallocation in block *i*;
(7) move to the next landowner in block *i* and execute steps (2–5) and reconsider (if necessary) the reallocation decisions that have already been made, for potential appropriate modifications;
(8) once land reallocation is completed for block *i*, go to the next block and repeat steps (1–6); and
(9) once all the landowners have been granted the property in question, then a new version of the land reallocation plan is ready.

It is obvious that the above process is laborious and hence usually takes months or even more than a year to complete depending on the size of the project. It is also clear that human capabilities are limited in terms of rapidly producing a set of alternative solutions, even using a CAD and/or a GIS. Thus, in practice a unique solution is constructed by working through each block, parcel by parcel. In addition, different planners may produce different solutions because the process is highly subjective and hence decisions may be questioned by farmers in many cases.

3 LACONISS Components

In contrast to the conventional process explained above, LACONISS transforms the whole planning approach into a systematic, efficient, transparent and automated process as outlined in the operational framework (Demetriou et al. 2012b) illustrated in Fig. 1. The whole planning approach is based on Simon's (1960) three-stage decision-making model, i.e. intelligence, design and choice that reflects the three sub-systems of LACONISS embedded within an ArcGIS platform. Land Fragmentation System (LandFragmentS) integrates GIS and multi-attribute decision making (MADM) methods which scan and measure the extent of land fragmentation (Intelligence phase). Thus, if the planner judges that land fragmentation is a problem, it becomes possible to proceed to the next sub-system called Land

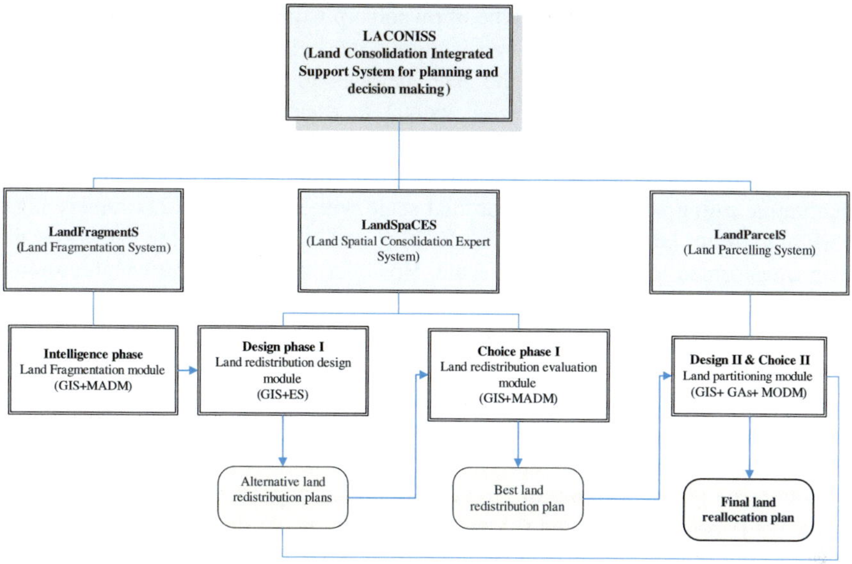

Fig. 1 The operational framework of LACONISS (Demetriou et al. 2012b)

Spatial Consolidation Expert System (LandSpaCES) which consists of two modules for design and evaluation respectively. The Design module integrates GIS with an expert system (ES) and automatically generates a set of alternative land redistribution plans which are then passed to the Evaluation module which assesses these alternative solutions based on the classical MADM ('Choice phase I') so as to identify the most beneficial plan. The optimal solution is then input to the last sub-system, that is, Land Parcelling System (LandParcelS) that integrates GIS with a genetic algorithm (GA) and multi-objective decision-making (MODM) methods and automatically generates the final land reallocation plan which provides an optimal configuration of parcels in terms of shape, size and land value ('Design and Choice phase II'). The development tools are Visual Basic for Applications (VBA) and ArcObjects.

3.1 The LandFragmentS Module

Traditionally, MADM involves selection from a number of alternative solutions which are described by attributes (or criteria) (Malczewski 1999). This method is used for measuring the performance of an existing system (i.e. land tenure) compared against an ideal situation; the measure might take the value of 0 for the most amount of fragmentation and 1 for the least amount. The planner first selects which land fragmentation factors to include in the analysis from the following six factors: the dispersion of parcels; the size of parcels; the shape of parcels; the accessibility of parcels; and the type of ownership which is a twofold criterion, i.e. dual ownership (when land and trees and/or water belong to different landowners) and shared ownership (where the land belongs to different landowners). The planner must then assign a relevant weight to each factor which represents its importance using one of two available methods: direct rating of a numerical value and qualitative rating. Direct rating involves assigning a value of less than 1 while qualitative rating is rating via an ordinal scale with seven levels (extremely high; very high; high; intermediate; moderate; low; and very low). These categories are then transformed to a quantitative scale. However, the sum of the weights should equal 1 in both methods.

Once the factors and weights are chosen, the system calculates a score for each factor and ownership and builds a land fragmentation table (with factors in columns and ownerships in rows). These values are then standardised between 0 and 1. Value functions are used for standardisation (Beinat 1997) which capture of human judgement within mathematical expressions. These value functions have been constructed by a group of land consolidation experts. A land fragmentation index (LFI) is then calculated for each ownership, using the weighted summation method, and then a global land fragmentation index (GLFI) is calculated for the whole area as the average of the LFIs. For both the ownership and global levels, the percentage contribution of each factor is also calculated. A tool for undertaking sensitivity analysis tool is also built into the system which allows the user to identify the sensitivity of the outcomes to the weights selected. The final GLFI then indicates whether land consolidation is required. If this is found to be true, the next step involves the design of alternative land redistribution plans using the Design module.

3.2 The LandSpaCES Design Module

The LandSpaCES Design module (Demetriou et al. 2011) integrates GIS with an ES, the most traditional artificial intelligence technique that attempts to emulate human reasoning for solving complex decision-making problems such as land redistribution. A knowledge base and an inference engine are the main components of a typical ES. For this module, the knowledge base consists of 74 IF–THEN rules. These have been extracted from legislation, the experience of the planners, rules of thumb and other related documents. A cadastral map, databases regarding land parcels, landowners and ownerships and facts, which are decision variables defined by the planner representing different scenarios, are provided as inputs to the system. The inference engine examines which rules have been satisfied by the facts, which are then returned as decisions (in the form of a land redistribution map) to the planner. New land redistribution solutions are produced when the facts change. Therefore the system can automatically generate a set of alternative solutions. These are then passed onto the next module for evaluation.

3.3 The LandSpaCES Evaluation Module

The Evaluation module integrates GIS with the classical form of MADM. This enables planners to evaluate a set of alternative solutions in a flexible manner. A set of land redistributions for evaluation and a set of criteria to assess these alternatives are selected by the planner. The available criteria are: the mean size of the new parcels; the mean parcel concentration coefficient (PCC); the change in the number of landowners; the percentage of ownerships 'completed' that had less than the minimum size limit according to the legislation to which extra land is added to reach the minimum limit; and the mean landowner satisfaction rate (LSR). The PCC and LSR are new concepts which have been developed by Demetriou et al. (2013c).

An effect table is then produced which contains the alternatives as columns and the criteria as rows. The performance of each alternative for each criterion is represented by a score in the appropriate cell of the table. Appropriate values functions are used to standardise the scores between 0 and 1, representing the worst and best performances of alternatives, respectively Weights are then chosen for each criterion using one of the two available methods mentioned previously in LandFragmentS. A decision rule (or aggregation model) then calculates the overall performance of each alternative followed by a ranking of the alternatives. In the final stage, a sensitivity analysis is undertaken are carried out on both the weights of the criteria and the performance scores. The method of Triantaphyllou (1997) is used, which assesses the robustness of the ranking and produces a final recommendation (the most beneficial solution) which is passed to the next module for automated land partitioning.

3.4 The LandParcelS Module

The LandParcelS module (Demetriou et al. 2013b) integrates GIS, genetic algorithms (GAs) and multi-objective decision making (MODM) methods. Based on the theory of evolution, GAs are stochastic optimisation techniques that are used for solving complex non-linear optimisation problems. MODM is a design process that searches for the best solution among an infinite or a very large number of feasible alternatives within the feasible region.

The first step is select the optimisation parameters that represent the objectives for the new parcels, i.e. shape (F1), size (F2) and land value (F3) so as to compose an appropriate fitness function (Demetriou et al. 2013b). Each objective term is calculated through an appropriate formula which will equal zero when the result represents an optimum. Thus, the further the outcome of each term is from zero, the worse it is. In the situation where more than one input parameter is optimised, a weight is required for each parameter. In the optimisation process, this weight represents the importance of the objective. It is also possible to add a penalty function (R) which can be used to penalize infeasible solutions, e.g. solutions that do not provide parcels with access to roads. The fitness function will have a value of zero if all parcels of a block have: the optimum shape (F1 = 0), which reflects a rectangle with a length: breadth ratio of 2:1; the predefined size (F2 = 0); the predefined land value (F3 = 0) and accessibility to a road (R = 0). After these parameters are defined, a random population of solutions is then generated by defining which land block will be partitioned and the size of the population. In this research Thiessen polygons were used to generate the random or new solutions.

To run the GA, the number of generations (or iterations) must be chosen and the elitist factor, which is the percentage of the best solutions from each generation that will be directly transferred to the next one, must be specified. A selection method must be then be chosen to fill the mating pool with the same number of individuals as found in the initial population based on their fitness value in an iterative process. New individuals (offspring) are then created by applying the genetic operators to parent individuals. The crossover operator combines two randomly selected parents from the mating pool and then introduces changes through mutation. The new offspring generated are then evaluated using the fitness measure chosen. The iterative process then ends if the termination criterion is met, returning the best solution generated at that point. Otherwise, the iterative process continues. The result consists of two database tables with information about the process of evolution for each generation. Visualisation of the resulting subdivision of each block is then possible using the GIS environment. By running the system for all land blocks in the consolidation area, the final land consolidation plan is produced.

4 System Implementation

LACONISS has been implemented using a case study area in Cyprus covering the land consolidation of the village of Chlorakas located in the District of Paphos (Demetriou et al. 2011). The outputs for each module are presented and discussed. The case study area showing the situation before and after land consolidation is illustrated in Fig. 2.

4.1 Measuring Fragmentation in the Study Area

LandFragmentS was run for the case study area using equal weights for all factors. Figures 3, 4, 5 present the resulting distributions for the Simmons and Januszewski indices and the new land fragmentation index (LFI), respectively. The two existing indices present very similar patterns although the Januszewski index gives relatively high values with an average of 0.841 and a minimum of 0.364. In contrast, the Simmons index gives lower values with an average of 0.785 and minimum of

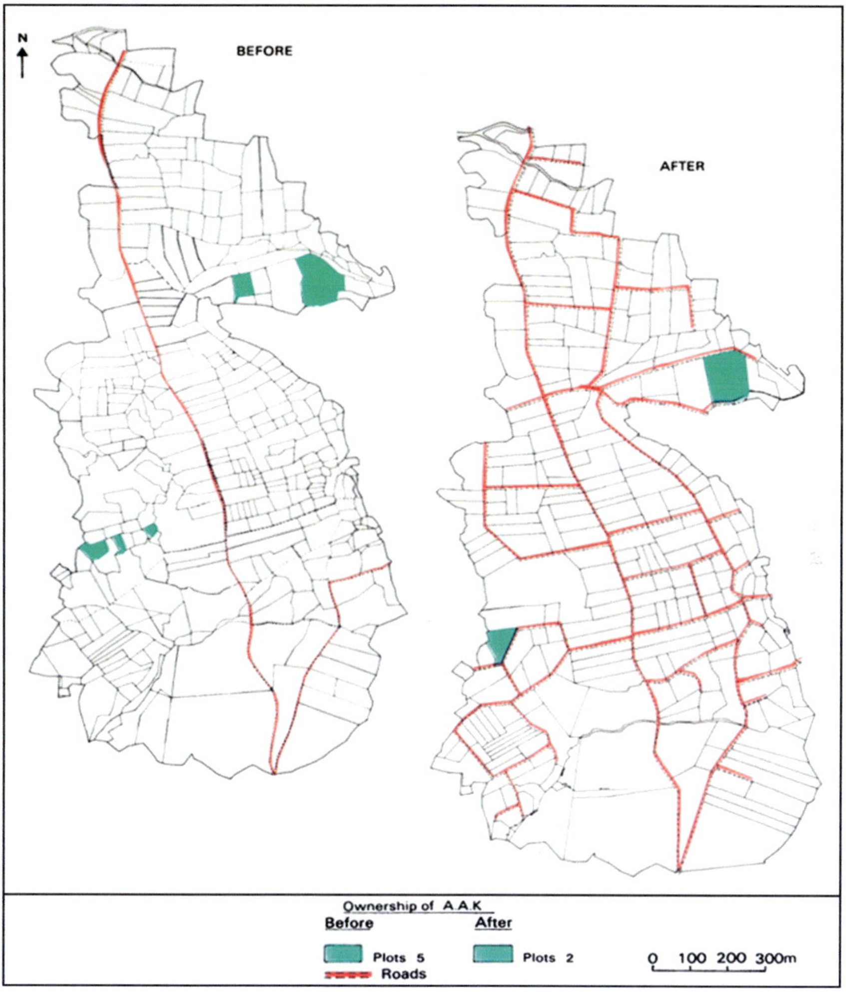

Fig. 2 The case study area before and after land consolidation (LCD 1993)

0.160. In contrast, the new index (LFI) results in considerably lower values compared to both existing indices, i.e. an average of 0.512 with a minimum of 0.216 and a maximum of 0.839; In addition, no holding achieves an LFI of 1 in contrast to both existing indices in which around 50 % of ownerships have resulted in the highest value of 1.

The results show that both existing indices underestimate the problem of land fragmentation (with an average around 0.8) compared to the new index where the latter takes six critical factors into account instead of only two. If the two existing indices had been used, it is clear that policy decisions made from their use would have been wrong. In contrast, the GLFI result of approximately 0.5 suggests that

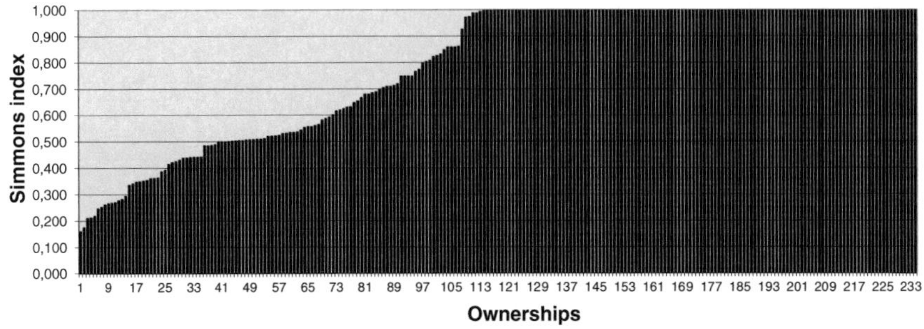

Fig. 3 Distribution of the Simmons index across holdings

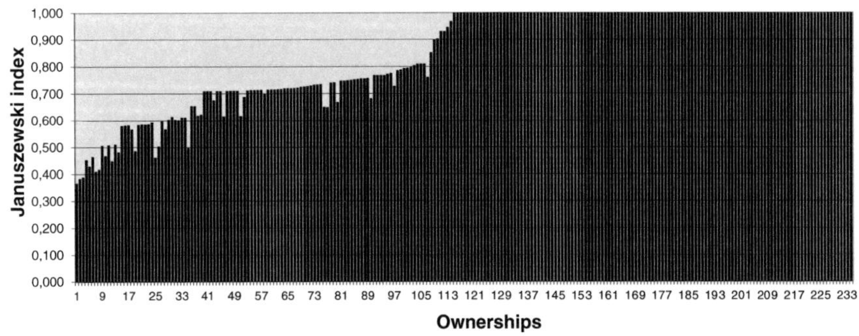

Fig. 4 Distribution of the Januszewski index across holdings

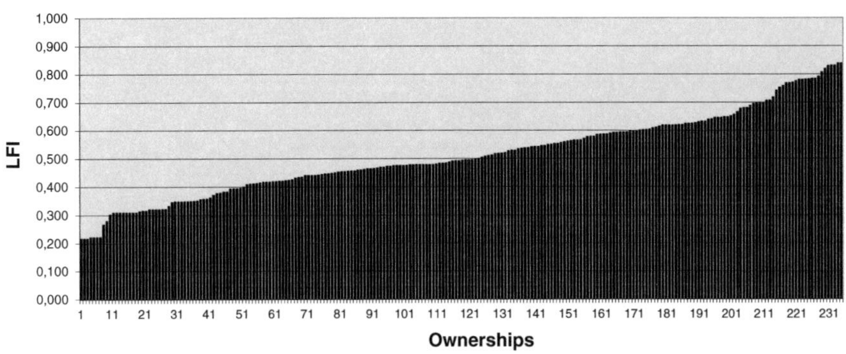

Fig. 5 Distribution of the new LFI index across holdings

there is a significant land fragmentation problem in the area. It is comforting to note that land consolidation was actually carried out in this study area. The results have therefore shown that the GLFI is more reliable and overcomes the deficiencies associated with the existing indices (Demetriou et al. 2013a). More flexibility could be added to the system, however, by allowing users to add more factors or an interactive facility could be provided for constructing the value functions.

The results have also shown that MADM can be used for spatial systems analysis in contrast to the more conventional assessment of a discrete number of alternative solutions. In LandFragmentS, MADM has been used to explore and measure the performance of the land tenure system compared to an ideal system and in evaluating the shape of a parcel compared to an optimum standard. These approaches can be applied to other disciplines that focus on system or object evaluation.

4.2 Designing Alternative Solutions

The system performance of the 'Design module' can be measured by comparing the agreement between the solution generated by the system and the solution of the human experts for the case study area. A land redistribution map and three associated database tables containing information regarding the new parcels, landowners and ownerships are the outputs of this module. The evaluation of the system has shown that it is capable of replicating the decisions of the planners between 63 and 100 % across nine criteria. Moreover, the system considerably outperforms the planners by reducing the time required for a solution from days to minutes. The system also contributes towards the generation of more objective solutions since the process is standardised, systematic and transparent (Demetriou et al. 2011).

Although the results were shown to be successful, a number of improvements could be made. For example, it would be useful to provide an editable knowledge base and more rules could be added to enhance performance. The addition of extra data such as the preferences of the actual landowners needs to be considered in the future. The development of an explanation facility regarding the decisions made would also be a very useful feature to facilitate communication between the planners and the landowners. Finally, the system could be tested with more case studies in order to provide more robust conclusions regarding the performance. The system was run with ten different sets of facts to generate ten alternative land redistributions representing different planning scenarios which are evaluated in the next section.

4.3 Evaluating the Alternative Solutions

Four different weighting scenarios illustrate the ranking of the ten alternatives generated in the 'Design module': equal weights (scenario 1); descending order of weights (scenario 2); ascending order of weights (scenario 3); and expert judgment (scenario 4). Figure 6 provides the results of these rankings.

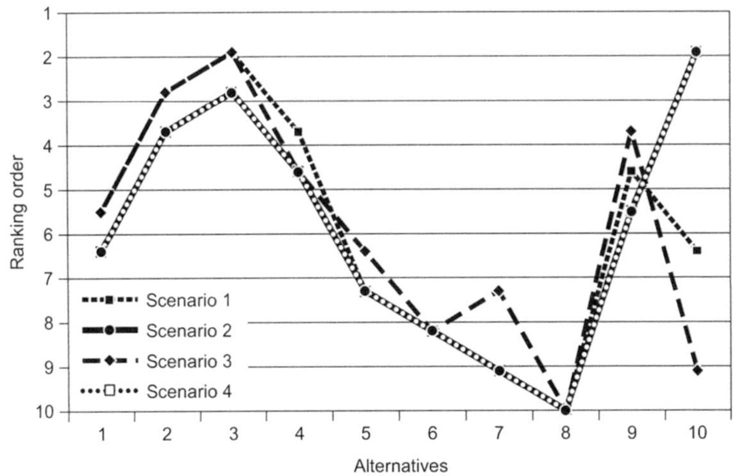

Fig. 6 Ranking of alternatives for four different criteria weighting scenarios (Demetriou et al. 2012c)

The results show some interesting outcomes. For example, alternative 1 (i.e. the solution from the experts) is outperformed by alternatives 2, 3, 4 and 9, so the system can produce solutions that are superior to those of the experts. There is also no single alternative that produces the best performance across all scenarios. For example, alternative 3 is ranked the highest in scenarios 1 and 3 while alternative 10 is the best in scenarios 2 and 4. Alternative 10 ranks third and seventh in scenarios 1 and 3 while alternative 3 is ranked second in scenarios 2 and 4. The results show that alterative 3 has a better balance across performance scores, i.e. there is a trade-off between all of the criteria. Thus, alternative 3 is the best as it achieved the most stable behaviour by rank order across all scenarios and thus is passed as an input to the next module.

4.4 Generating the Final Parcels

Two typical land blocks were chosen to demonstrate the LandParcelS module which differed in the degree of complexity. Block I with a size around 3 ha contains six parcels whilst Block II with a size of around 5 ha contains 10 parcels. As noted earlier, land partitioning is handled as a multi-objective optimization problem based on a fitness function which was utilized for a combination of objectives so as to optimise: shape and size (F1 and F2); shape and land value (F1 and F3); and shape, size and land value (F1, F2 and F3). In contrast to single-objective optimization which yields a unique optimum solution, multi-objective problems with conflicting objectives will produce a different optimum solution for each objective. The final result is a set of equally optimal solutions but that contain trade-offs between the objectives (Deb 2001). This set of optimal solutions lie on the Pareto-optimal front.

These encouraging results indicate that advancements have been made towards solving this complex spatial problem. However, the results differ depending upon the degree of complexity. For example, solutions that are close to the optimum are generated for the block with the lower complexity while the results were further from the optimum when optimizing two objectives, e.g. size and land value even though they were close to optimum for shape. Similar results were found when optimizing all three objectives at the same time.

The final subdivision of block I is shown in Fig. 7 where shape and size are optimized (case a in Fig. 7), shape and land value (case b in Fig. 7) and shape, size and land value (case c in Fig. 7). For instance, case a has a fitness of 0.112, F1 of 0.181 and a F2 of 0.094 meaning that F1 has been improved by 31.44 % and F2 by 85.17 % compared to the initial subdivision. The parcel shape (F1) and size (F2) are on average distant from the optimum by 18.1 and 9.4 %, respectively. In accordance with this, Fig. 8 shows the Pareto-optimal front (dashed line) for case a. The optimum solution corresponds on the origin of the two axes F1 and F2 i.e. 0,0. The best solution, which is the solution that dominates all the others based on a certain weighting scheme, is marked with a triangle and falls on the Pareto-optimal front. All the other solutions belong to the non-Pareto-optimal set.

The results generated here are encouraging but there is clearly a need to further improve the performance of the algorithm in terms of the final outcome. This could involve the development of a new algorithm for space partitioning in which the parameters are directly involved in the process of optimization, which is not the case in the current process. Another solution would be to introduce a so-called guidance (or learning or local) optimiser which would be embedded within the current optimisation process. There is also room for improvement in terms of the

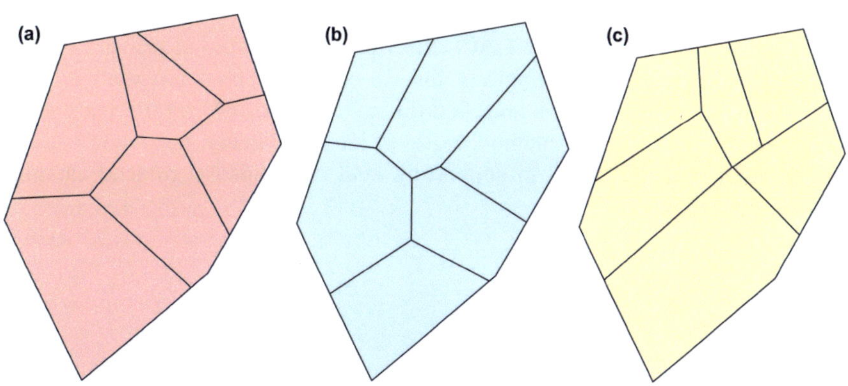

Fig. 7 The final outcome for block I for three optimisation cases

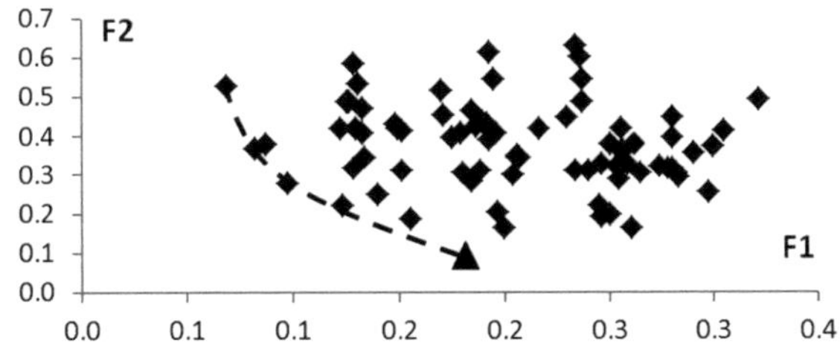

Fig. 8 The Pareto-optimal front for case a

running time of the algorithm, which currently takes approximately five to ten hours for each land block. The use of parallel computing might be a solution or the algorithm could be redeveloped with a more powerful programming language.

5 Conclusions

A prototype planning support system (PSS) for land consolidation has been outlined in this chapter with a focus on rural sustainable development. The application has been demonstrated in a land consolidation area of Cyprus. The PSS demonstrated in the chapter integrates planning and decision-making within a framework for land reallocation where the process has been transformed by integrating GIS, artificial intelligence techniques and multi-decision making methods into a systematic, automated, transparent and efficient process. This new system is, therefore, capable of solving some of basic problems associated with the process.

Two suggestions for further research and development include: a) making improvements to each module of LACONISS in order to overcome the limitations noted earlier; and b) move the system from a prototype demonstrated in a single country system to a more generic, commercialised or open source software where the user could customise the application based on land consolidation legislation and practices of a particular country. Many researchers have attempted to construct such a system since the 1960s. LACONISS has the potential to provide the cornerstone for reaching this ambitious aim. In addition, planners within different disciplines and specialisations may find the system rationale, several parts of the design and the development methodologies of interest. Finally, one might consider the potential of LACONISS in supporting land consolidation projects in urban areas in the future.

References

Ayranci Y (2007) Re-allocation aspects in land consolidation: a new model and its application. J Agron 6(2):270–277

Beinat E (1997) Value functions for environmental management. Kluwer Academic Publishers, Dordrecht

Bentley W (1987) Economic and ecological approaches to land fragmentation: in defense of a much-maligned phenomenon. Annu Rev Anthropol 16:31–67
Buis A, Vingerhoeds R (1996) Knowledge-based systems in the design of a new parcelling. Knowl-Based Syst 9:307–317
Coelho C, Pinto PA, Silva M (2001) A systems approach for the estimation of the effects of land consolidation projects (LCPs): a module and its application. Agric Syst 68:179–195
Deb K (2001) Multi-objective optimization using evolutionary algorithms. Wiley, West Sussex
Demetriou D, Stillwell J, See L (2011) LandSpaCES: a spatial expert system for land consolidation. In: Geertman S, Reinhardt W, Toppen F (eds) Advancing geo-information science for a changing world. Springer, Berlin, pp 249–274
Demetriou D, Stillwell J, See L (2012a) Land consolidation in Cyprus: Why is an integrated planning and decision support system required? Land Use Policy 29(1):131–142
Demetriou D, Stillwell J, See L (2012b) An integrated planning and decision support system (ipdss) for land consolidation: theoretical framework and application of the land redistribution modules. Environ Plan B: Plan Des 39(4):609–628
Demetriou D, See L, Stillwell J (2012c) A new methodology for the evaluation of land reallocation plans. ISPRS Int J Geo-Inf 1(3):272–293
Demetriou D, Stillwell J, See L (2013a) A new methodology for measuring land fragmentation. Computers, Environment and Urban Systems 39:71–80
Demetriou D, See L, Stillwell J, (2013b) A parcel shape index for use in land consolidation planning. Transactions in GIS
Demetriou D, Stillwell J, See L (2013c) A spatial genetic algorithm for automated land partitioning. Int J Geogr Inf Science (under review)
Geertman S, Stillwell J (2009) Planning support systems: best practice and new methods. Springer, Dordrecht
Giarrantano J, Riley G (2005) Expert systems: principles and programming. Course technology. Canada
Goldberg D (1989) Genetic algorithms: in search optimisation and machine learning. Addison Wesley, Crawfordsville
Januszewski J (1968) Index of land consolidation as a criterion of the degree of concentration. Geographia Polonica 14:291–296
Land Consolidation Department (LCD) (1993) Land consolidation in Cyprus. Ministry of Agriculture and Natural Resources of Cyprus, Nicosia
Malczewski J (1999) GIS and multi-criteria decision analysis. Wiley, New York
Openshaw S, Openshaw C (1997) Artificial intelligence in geography. Wiley, West Sussex
Rosman F, Sonnenberg J (1998) New method for the design of the reallocation plan in land consolidation projects. FIG conference. In: Proceedings of the XXI FIG Congress Brighton, Developing the Profession in a Developing Word, 19–25 Jul, Brighton, UK
Sharifi A, Herwijnen M, Toorn W (2004) Spatial decision support systems. Lecture Notes. ITC, International Institute for Geo-Information Science and Earth Observation, The Netherlands
Simmons AJ (1964) An index of farm structure, with a Nottinghamshire example. East Midlands Geogr 3:255–261
Simon A (1960) The new science of management decision. Harper and Row, New York
Sklenicka P (2006) Applying evaluation criteria for the land consolidation effect to three contrasting study areas in the Czech Republic. Land Use Policy 23(4):502–510
Sonnenberg J (2002) Fundamentals of land consolidation as an instrument to abolish fragmentation of agricultural holdings. In: Proceedings of the XXII International FIG Congress, 19–26 Apr, Washington, D.C., USA. http://www.fig.net/pub/fig_2002/Ts7-4/TS7_4_sonnenberg.pdf
Thomas J (2006a) What's on regarding land consolidation in Europe? In: Proceedings of the XXIII FIG Congress, Shaping the Change, 8–13 Oct, Munich, Germany. http://www.fig.net/pub/fig2006/papers/ts80/ts80_03_thomas_0311.pdf

Thomas J (2006b) Property rights, and fragmentation and the emerging structure of agriculture in Central and Eastern European countries. J Agric Dev Econo [online] 3(2). ftp://ftp.fao.org/docrep/fao/009/ah757e/ah757e00.pdf
Tourino J, Boullon M, Gonzalez X (2003) A GIS-embedded system to support land consolidation plans in Galicia. Int J Geogr Inf Sci 17(4):377–396
Triantaphyllou E (1997) A sensitivity analysis approach for some deterministic multi-criteria decision making methods. Decis Sci 28(1):151–194
Van Dijk T (2003) Dealing with central European land fragmentation. Eburon, Delft
Van Hung P, MacAulay G, Marsh S (2007) The economics of land fragmentation in the North Vietnam. Aust J Agric Resour Econ 51:195–211

Qualitative and Quantitative Comparisons of Agent-Based and Cell-Based Synthesis Estimation Methods of Base-Year Data for Land-Use Microsimulations

Kazuaki Miyamoto, Nao Sugiki, Noriko Otani and Varameth Vichiensan

Abstract Land-use microsimulation is becoming an indispensable function in a planning support system for sustainable urban development because it provides the detailed information necessary for decision making on emerging issues at the household or firm level. In land-use microsimulations, there are two approaches for estimating base-year micro-data: cell-based population synthesis, which generally uses the iterative proportional fitting method, and agent-based methods. This chapter compares these two methods qualitatively and quantitatively. The qualitative comparison shows that neither one is superior in every aspect. The cell-based method is preferred when the microsimulation deals with data sufficiently simple, while the agent-based method is preferred when accurate and/or numerous micro-data attributes are demanded. Similarly, the quantitative comparison based on a goodness-of-fit evaluation does not show a single superior method for all applications. These findings suggest a way for selecting a better method based on the conditions of the microsimulation model and the purpose of its application.

K. Miyamoto (✉)
Faculty of Urban Life Studies, Tokyo City University, 8-9-18 Todoroki,
Setagaya-ku, Tokyo 158-8586, Japan
e-mail: miyamoto@tcu.ac.jp

N. Sugiki
Transportation Department, Docon Co., Ltd, 4-1, 5-chome, 1-jo,
Atsubetsu-cho, Atsubetsu-ku, Sapporo 004-8585, Japan
e-mail: ns1491@docon.jp

N. Otani
Faculty of Informatics, Tokyo City University, 3-3-1 Ushikubo-nishi, Tuzuki-ku,
Yokohama 224-8551, Japan
e-mail: otani@tcu.ac.jp

V. Vichiensan
Faculty of Engineering, Kasetsart University, 50 Phaholyothin Rd., Lardyao, Jatujak,
Bangkok 10900, Thailand
e-mail: fengvmv@ku.ac.th

S. Geertman et al. (eds.), *Planning Support Systems for Sustainable Urban Development*, Lecture Notes in Geoinformation and Cartography,
DOI: 10.1007/978-3-642-37533-0_6,

1 Introduction

In the field of urban modelling, microsimulation is emerging as a popular way to describe detailed changes in land-use and transportation in metropolises. It provides the detailed information necessary for decision making on emerging issues, many of which should be studied at the level of individual households or firms. Therefore, urban microsimulation is becoming an indispensable function in a planning support system for sustainable urban development.

Land-use microsimulation requires the micro-level modelling of a dataset of individuals, such as populations, households, or firms. Although this chapter focuses on micro-population data, intrinsically used for trip generation in travel demand models, the concept is also applicable to households and firms in land-use models. An individual is characterized through many attributes, such as age and type of household. In a microsimulation, the necessary attribute of every individual in a study area must be prepared for the simulation's base year. However, such information is not available because the retrieval of individual micro-data from the administrative registration system is neither allowed nor desirable for privacy reasons. Therefore, most microsimulation models use a synthetic population created from readily accessible aggregate data such as national censuses and information obtained from sample surveys. Micro-data estimation may proceed in one of two ways: cell-based or agent-based approaches. The popular cell-based population synthesis is built mostly by applying the iterative proportional fitting (IPF) method (Pritchard and Miller 2009; Müller and Axhausen 2011; Lee and Fu 2011), while the agent-based approach is emerging (Moeckel et al. 2003; Miyamoto et al. 2009, 2010a, b; Sugiki et al. 2013). Therefore, it is reasonable to compare these two approaches.

The aim of this chapter is to compare these two approaches from various viewpoints for selecting the appropriate one based on the type of microsimulation model and purpose of its application.

The presented comparisons are made qualitatively and quantitatively. The qualitative comparison discusses and descriptively compares the advantages and disadvantages of each approach. The quantitative comparison presents an empirical examination using data from the Sapporo metropolitan area of Japan. A set of complete information data obtained from the person trip survey is used as a pseudo-observed dataset. Stochastic samplings from this pseudo-observed dataset produce several sampling datasets, which are used to estimate base-year micro-data by employing and comparing both the cell-based and the agent-based approaches. The goodness-of-fit values between the pseudo-observed dataset and the estimated datasets produced by the cell-and agent-based methods are then calculated for each sampling case. Although the goodness-of-fit between the cell-based datasets is conventional, that for the agent-based datasets is novel in this field. Here, the one proposed by Otani et al. (2010, 2012) is employed. The findings drawn from these comparisons finally suggest a technique for selecting the appropriate method.

The remainder of this chapter is organized as follows. Section 2 provides the relevant backgrounds of the cell- and agent-based population synthesis methods. Sections 3 and 4 present the qualitative and quantitative comparisons of the two methods, respectively. Finally, the chapter concludes with suggestions for method selection in Sect. 5.

2 Population Synthesis Estimation Methods

2.1 Cell-Based Methods

2.1.1 IPF Procedure

The IPF method, proposed by Deming and Stephan (1940), is a popular technique for population synthesis. While Beckman et al. (1996) were the first to apply IPF to the population synthesis problem (see Guo and Bhat 2007), the same approach was also proposed in Miyamoto et al. (1986) to synthesize households in a metropolis using the simultaneous probability maximization principle with margins as constraints, basically equivalent to IPF in terms of the solving procedure.

In principle, IPF produces the number (Z_{ijk}) of agents, individuals, or households, in the cells of a multidimensional (i, j, k) table, regarded as a limited number of individual types defined by a set of attributes. The IPF method uses a sample dataset to consider the contingencies between the dimensions or attributes, with the condition that the summations of Z_{ijk} into lower dimensions should fit the margins given by the census data. The general procedure of IPF is shown in Fig. 1 for a two-dimensional table for simplicity. Although the data obtained by using this approach are useful for microsimulation models, they are not micro-data on individual agents but rather the number of agents by type.

Fortunately, Guo and Bhat (2007) improved IPF by overcoming the zero-cell-value problem and the inability to control the statistical distributions of both household- and individual-level attributes. Pritchard and Miller (2009) also improved the IPF procedure by adding a function to allow many more attributes per agent using a Monte Carlo simulation based on a sparse list-based data structure. Both improvements are useful developments that remain within the scope of cell-based approaches, which is inevitable as long as the IPF procedure is used.

2.1.2 Goodness-of-Fit

In discussing the goodness-of-fit of micro-data, Pritchard and Miller (2009) argued that a synthetic population could be tested for its goodness-of-fit against the true population's characteristics if the complete "ground truth" is known, but they did not indicate a method for carrying out this goodness-of-fit test. In the absence of

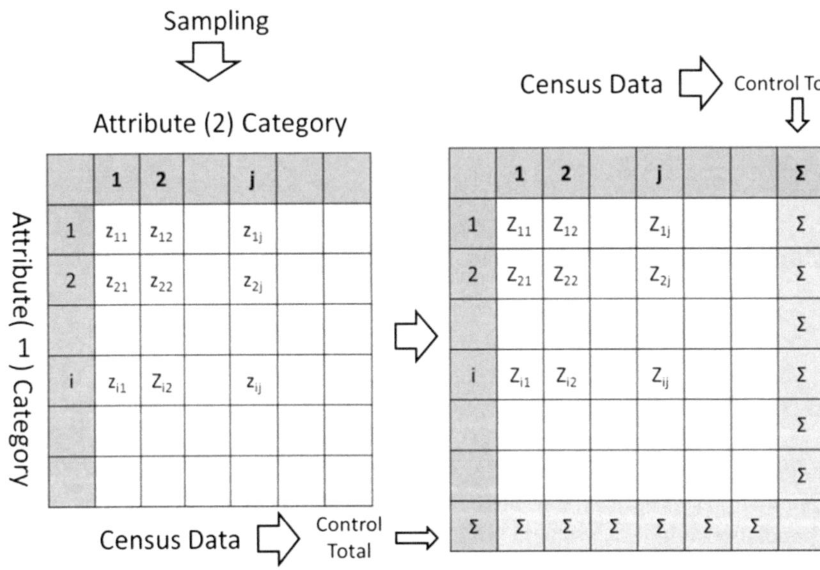

Fig. 1 IPF procedure for cell-based estimations

such ground truth, they referred to a method of cross-classifying the synthetic population in order to form a table using IPF-based procedures. If the individual is classified using two categories (i, j) for simplicity, the synthetic population can be compared with a validation table using the distance-based standardized root mean square error (SRMSE) statistic (Knudsen and Fotheringham 1986):

$$SRMSE = \frac{\sqrt{\frac{1}{IJ}\sum_{i,j}\left(\hat{Z}_{ij} - Z_{ij}\right)^2}}{\frac{1}{IJ}\sum_{i,j} Z_{ij}} \tag{1}$$

A smaller value indicates a better fit. This statistic is calculated for each of the validation tables in turn and the goodness-of-fit statistics in each type are then averaged to provide an overall goodness-of-fit for the type, overcoming the manipulation problem in the calculation.

2.2 Agent-Based Methods

2.2.1 Monte Carlo simulation

Figure 2 shows the agent-based approach used by Moeckel et al. (2003). To carry out a microsimulation, they used Monte Carlo sampling procedures with an agent in order to select as many features as necessary. In this case, the number of features

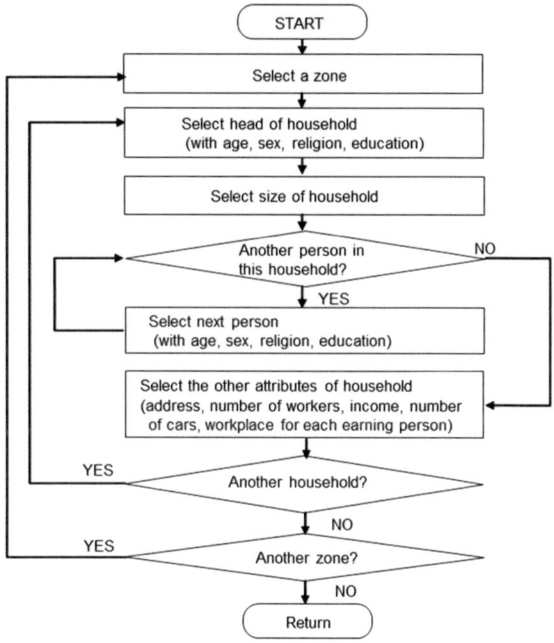

Fig. 2 Agent-based synthetic population generator proposed by Moeckel et al. (2003)

was limited only by the possibility of determining reasonable relationships between the selected attributes.

Following the basic idea proposed by Moeckel et al. (2003), Miyamoto et al. (2009) built an agent-based system using a Monte Carlo approach. This system originally dealt with the ages of four-member households only, but it was later extended to cope with both continuous- and discrete-attribute variables for a household as an agent (Miyamoto et al. 2010a, b; Sugiki et al. 2013), as shown in Fig. 3. The system uses sample data that contain all information from the micro-data to reflect the contingencies between attributes, along with the existing statistical data as the control total in each zone. To reproduce the correlation between continuous-attribute variables, independent variables that can be obtained through the principal component analysis of the original variables are introduced into the system and employed as intervening variables (Sugiki et al. 2013). In addition, the system is composed of several iteration processes so that data production satisfies the available control total by attribute. An indicator to evaluate the goodness-of-fit between the two sets of micro-data, detail of which is described in next section, is also proposed in order to ensure that system development remains rational and objective. The present chapter uses a simplified version of this system as an agent-based estimation method.

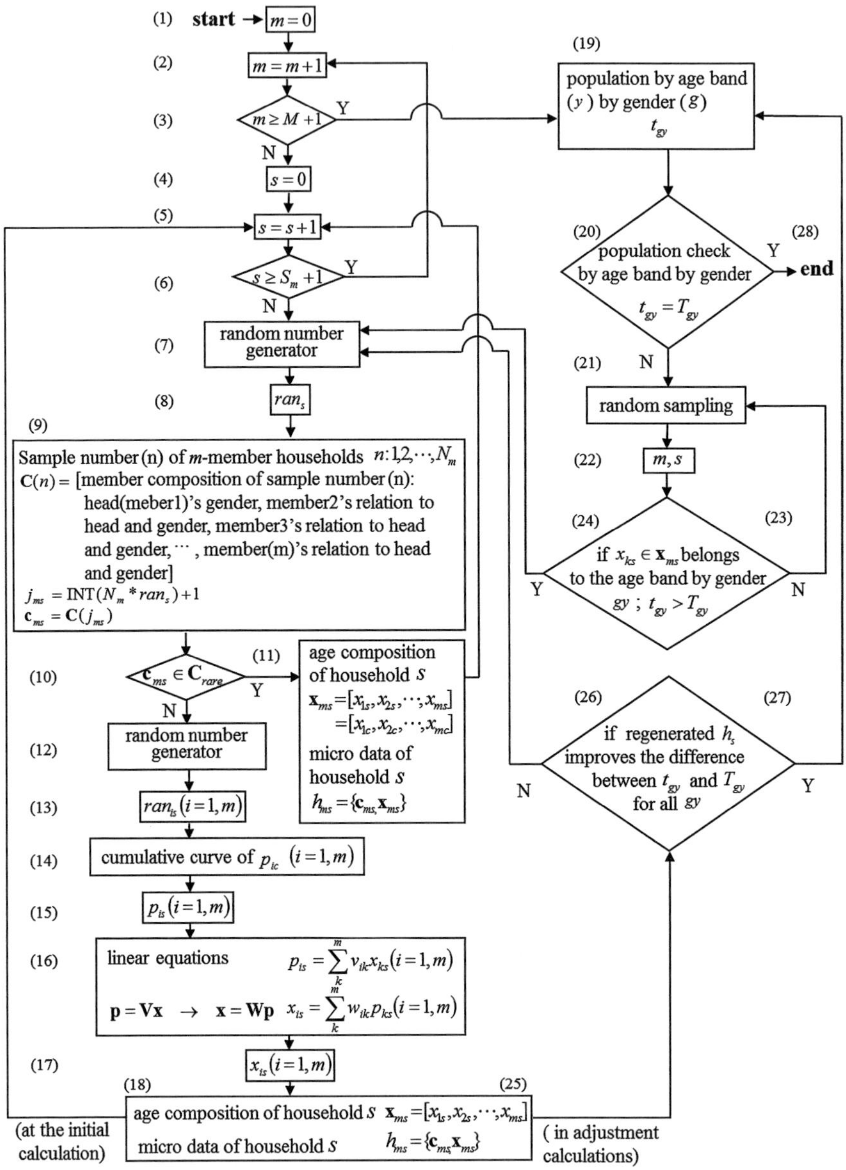

Fig. 3 Agent-based population synthesis flowchart of Miyamoto et al. (2010a)

2.2.2 Goodness-of-Fit

The micro-dataset is denoted as a set whose elements are vectors. The observed dataset A and j-th estimated dataset E_j are denoted as follows:

$$A = \{\mathbf{a}_i = (a_{i1}, a_{i2}, \ldots, a_{iZ}) | 1 \leq i \leq N\} \tag{2}$$

$$E_j = \{\mathbf{e}_i^j = (e_{i1}^j, e_{i2}^j, \ldots, e_{iZ}^j) | 1 \leq i \leq N\} \tag{3}$$

where Z is the number of attributes and N is the number of agents. The a_{ik} and e_{ik}^j, values in A and E_j, respectively, denote the values of the k-th attribute in the i-th agent.

The average and variance statistics are usually used to compare the similarities between the two sets. However, the problem is not testing whether the distributions match. As no existing indicator was available, the authors propose the minimum value of the summation of the squared distance, $Dis(\mathbf{a}_i, \mathbf{e}_k^j)$, to determine the fitness between particular elements in the two sets. To avoid both memory overflow and the excessive influence of vacant values (999), the ceiling value of the squared difference between elements, *DiffMax*, is used as follows:

$$Dis(\mathbf{a}_i, \mathbf{e}_k^j) = \sum_{l=1}^{Z} \min((a_{il} - e_{kl}^j)^2, DiffMax) \tag{4}$$

Then, the goodness-of-fit, $Fit(E_j)$, of the estimated dataset, E_j, is defined as follows:

$$Fit(E_j) = \min_{\sigma \in R_n} \sum_{i=1}^{N} Dis(\mathbf{a}_i, \mathbf{e}_{\sigma(i)}^j) \tag{5}$$

where R_n means the set of all bijections from the set $\{1, 2, \ldots, N\}$ onto itself. As R_n has $N!$ elements, it is necessary to calculate the squared distance summation $N!$ times for one estimated dataset.

Although calculating the squared distance summation is straightforward, as N grows, the number of elements in R_n becomes larger and the calculating time increases abruptly, a problem known as the "combinatorial explosion problem." To calculate the goodness-of-fit, the minimum value is thus selected after calculating all the squared distance summations. In other words, calculating the goodness-of-fit is the same as searching for the minimum value in numerous squared distance summations; this is the combinatorial optimization where the genetic algorithm works efficiently. Of the many types of genetic algorithm methods, the authors adopt symbiotic evolution to find the bijection that minimizes the squared distance summation. Symbiotic evolution maintains two separate populations: a partial solution population whose individuals represent partial solutions and a whole solution population whose individuals are combinations of individuals in the partial solution population and represent whole solutions. This approach results in a fast and efficient genetic search and prevents convergence to suboptimal solutions (Otani et al. 2012).

3 Qualitative Comparison

This section presents a qualitative comparison based on the following perspectives: the characteristics of agents, the number of attributes and their categories, the estimation norm, uniqueness, complexity, data availability, and the compatibility between the agent-and cell-based estimates.

3.1 Characteristics of Agents

Because cell-based methods are used to estimate the number of agents that belong to a cell in a multidimensional table, regarded as a limited number of individual types defined by a set of attributes, they are not used to estimate agents directly. Thus, agents are characterized by the representative attributes by which the cell is defined. In other words, many agents have the same attributes.

By contrast, agent-based methods directly estimate each of the agents individually. Therefore, each estimated agent is unique in the dataset, making the values of the attributes continuous and rendering any attribute combination possible.

3.2 Number of Attributes and Their Categories

When generating complex sets of micro-data, the IPF procedure used in cell-based methods has its limitations. First, it implies very high standards in terms of the reliability of the initial input data for the cells of the matrix. Second, the IPF method must set zero cells to 0.1 or 0.01, which influences the probabilities and is hardly theoretically founded (Moeckel et al. 2003). In addition, cell-based approaches (i.e., IPF and its extensions) present the following challenges:

- Neither attributes nor their categories can be set universally.
- The discrete setting of categories causes a problem similar to the modifiable area unit problem in zoning. For example, certain attributes, such as age and income, are originally continuous.
- Additional attributes generate more zero-cell problems and reduce reliability.

The IPF procedure may have been devised to reduce the computation and memory burden of computers by setting a limited number of agent types. However, this approach does not necessarily overcome these issues, because the combinations of attribute categories are apt to reach a huge number, while most cells have a zero value. Meanwhile, agent-based methods have essentially no limit to the number of attributes and categories they can use.

3.3 Estimation Norm

Cell-based methods use the IPF procedure, which has a robust norm based on the entropy maximization principle, thus establishing a set of solving algorithms. However, agent-based methods have a weaker norm in the current formulation because the estimation is carried out through multiple Monte Carlo simulations. These simulations stochastically estimate each agent based on the probabilistic distributions of the attribute values obtained from a sample set. Although they may be based on likelihood criteria, the function cannot be formulated.

3.4 Uniqueness

Because cell-based methods use the IPF procedure, the estimated cell-based data are unique as long as the IPF algorithm is fixed. However, because agent-based methods use the Monte Carlo simulation, the resulting estimation will be different for each simulation trial. Thus, it is necessary to select a set of micro-data for the base year from among the estimated datasets.

3.5 Complexity

Although there are variations in IPF procedures (Müller and Axhausen 2011), the general algorithm is straightforward. The estimation framework for agent-based methods is rather ad hoc and complicated, such as that shown in Fig. 3. They are not an optimization problem but rather a solving method using simulations.

3.6 Data Availability

Both methods are built using the same data, namely generally accessible aggregate data provided by a national census with additional information obtained from sample surveys. Thus, they are virtually identical.

3.7 Compatibility Between Agent-and Cell-Based Estimates

Because cell-based methods are not used to estimate agents directly, agents can be characterized only by using the corresponding attributes by which the cell is defined. In other words, many agents must have the same attributes. Pritchard and

Miller (2009) relaxed this stipulation by adding a function to IPF in order to allow many more attributes per agent using a Monte Carlo simulation based on a sparse list-based data structure. Because agent-based methods estimate a set of agents directly, agent data can be easily converted into cell-based data.

4 Quantitative Comparison

The quantitative comparison presented in this section is carried out using real micro-datasets of cells and agents. The former dataset is used to evaluate the robustness of the two methods, while the latter is applied to validate the effectiveness of the micro-data synthesis. Target data are individuals with two attributes: age and number of household members. The original categories for these attributes are set as follows: age has 18 categories (0–4, 5–9, 10–14, 15–19,..., 80–84, and ≥85) and the number of household members has seven categories (1, 2, 3, 4, 5, 6, and ≥7).

4.1 Data

The data used in the quantitative comparison are obtained from the current person trip survey for the Sapporo metropolitan area of Japan. Complete information is available for 19,394 households; 10,000 households are randomly sampled from the original dataset to create a pseudo-observed dataset that consists of 24,115 individuals. Eight sample datasets of 1,000 households, which consist of approximately 2,400 individuals, are then obtained by randomly sampling the pseudo-observed dataset.

4.2 Estimation

4.2.1 Cell-Based Method

A cell-based estimation using the IPF method, as shown in Fig. 1, is conducted using the following procedure.

Step 1 Set the control totals for each marginal column and row in the census data.
Step 2 Set the sample data to cells as seed values in the estimation table.
Step 3 Set the zero cells to very small values (e.g., 0.01) in the estimation table.
Step 4 Calculate the column and row totals.

Step 5 Calculate the evaluation value as the sum of the absolute differences between the marginal column and the row totals in the estimation table and the control totals.
Step 6 Calculate the adjustment ratio from the marginal column and row totals in the estimation table and the control totals.
Step 7 Renew each cell value by multiplying the present value by the adjustment ratio in the estimation table.
Step 8 Repeat steps 4–7 until the evaluation falls below a pre-determined convergence value.

4.2.2 Agent-Based Method

The agent-based estimation using a Monte Carlo simulation is conducted using the population synthesis system shown in Fig. 3. This system provides a household-based agent dataset containing the age of each household member and the number of household members as attributes. It can then easily convert individual-based micro-data using these two attributes.

4.3 Comparison Framework

The quantitative comparison between the cell- and agent-based methods is presented in Fig. 4. It is assumed that a pseudo-observed dataset for simulations that have complete information is available. Some sample datasets are obtained from the pseudo-observed dataset using a specific sampling rate. The control totals are given by the total number of individuals by age and the number of household members for the pseudo-observed dataset.

This study estimates eight sets of base-year data from a sample dataset based on the two methods: cell data according to the cell-based method, agent data according to the cell-based method, cell data according to the agent-based method, and agent data according to the agent-based method. First, cell data are estimated according to the cell-based method using the IPF procedure. Second, agent data are obtained by assuming that all agents that belong to a cell have the same corresponding attributes. Third, the agent data obtained from the agent-based method are estimated using the Monte Carlo simulation system. Lastly, the cell data derived from the agent-based method are obtained by aggregating them using the categories of the attributes. Two hierarchical categories, called "original" and "aggregated" cells, are then set for each attribute in order to compare the effect of cell size, as shown in Fig. 5.

The goodness-of-fit is calculated for both cell and agent data between the estimated and pseudo-observed datasets to quantitatively compare the two methods of estimation for the two cell cases. The SRMSE in Eq. (1) is used to compare

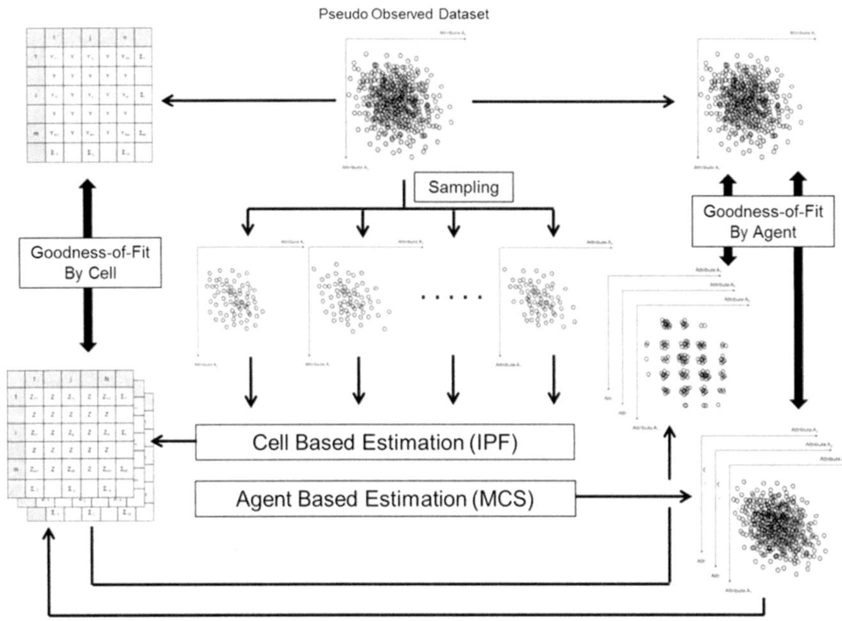

Fig. 4 Quantitative comparison framework

		Aggregated Categories (2)		1		2		3		
		Original Categories (2)		1	2	3	4	5	6	7
Aggregated Categories (1)	Original Categories (1)			Number of Household Members						
				1	2	3	4	5	6	7
1	1	Age	0–4							
	2		5–9							
	3		10–14							
2	4		15–19							
	5		20–24							
	6		25–29							
3	7		30–34							
	8		35–39							
	9		40–44							
4	10		45–49							
	11		50–54							
	12		55–59							
5	13		60–64							
	14		65–69							
	15		70–74							
6	16		75–79							
	17		80–84							
	18		85+							

Fig. 5 Two hierarchical cells for quantitative comparison

the two datasets in the cell-based method, while the goodness-of-fit index in Eq. (5) is applied to compare the two datasets in the agent-based method.

4.4 Comparison by Cell

The cell-based goodness-of-fit values of the cell- and agent-based methods for the eight sample datasets for the original and aggregated cells are shown in Tables 1 and 2, respectively. The goodness-of-fit of the agent-based estimation is slightly better than that of the cell-based one, on average, for the original cell case. The difference in the goodness-of-fit values between the cell- and agent-based methods is not significant, however. In samples 2, 5, and 7, no significant difference is shown, while a better fit is exhibited in samples 1, 3, 4, 7, and 8. On the contrary, the cell-based method generally shows better performance for the aggregate cells, likely because no zero-cell is included in the aggregated cell case, whereas there are some in the original cell case.

4.5 Comparison by Agent

The agent-based goodness-of-fit values of the cell- and agent-based methods are calculated six times for each sample dataset. The averages and standard deviations of the agent-based goodness-of-fit values for the original and aggregated cells are shown in Tables 3 and 4, respectively. The goodness-of-fit of the agent-based estimation is significantly better than that of the cell-based estimation, on average, except for sample 5. Moreover, the standard deviation results show that the agent-based estimation is more stable than the cell-based one. In the aggregated cell case, it is natural that the agent-based method shows significantly better performance than the cell-based method.

Table 1 Cell-based goodness-of-fit values of the cell- and agent-based estimations (original cells)

Case	Population of sample	Cell-based goodness-of-fit	
		Cell-based estimation (IPF)	Agent-based estimation (MCS)
Sample 1	2,346	0.184	0.170
Sample 2	2,357	0.183	0.192
Sample 3	2,434	0.208	0.171
Sample 4	2,415	0.184	0.175
Sample 5	2,416	0.213	0.219
Sample 6	2,452	0.163	0.162
Sample 7	2,443	0.220	0.182
Sample 8	2,467	0.223	0.167
Average		0.197	0.180

Table 2 Cell-based goodness-of-fit values of the cell-and agent-based estimations (aggregated cells)

Case	Population of sample	Cell-based goodness-of-fit	
		Cell-based estimation (IPF)	Agent-based estimation (MCS)
Sample 1	2,346	0.046	0.050
Sample 2	2,357	0.055	0.079
Sample 3	2,434	0.044	0.070
Sample 4	2,415	0.063	0.058
Sample 5	2,416	0.071	0.082
Sample 6	2,452	0.043	0.045
Sample 7	2,443	0.096	0.088
Sample 8	2,467	0.082	0.062
Average		0.063	0.067

Table 3 Agent-based goodness-of-fit values of the cell-based and agent-based estimations (original cells)

Case	Population of sample	Agent-based goodness-of-fit			
		Cell-based estimation (IPF)		Agent-based estimation (MCS)	
		Average	Standard deviation	Average	Standard deviation
Sample 1	2,346	3,997	27.3	2,539	15.5
Sample 2	2,357	4,189	38.1	2,932	17.5
Sample 3	2,434	2,976	34.5	2,369	16.8
Sample 4	2,415	3,517	51.8	3,028	13.6
Sample 5	2,416	2,969	12.1	3,336	16.8
Sample 6	2,452	2,486	50.8	2,444	22.6
Sample 7	2,443	3,116	22.9	2,594	9.3
Sample 8	2,467	3,782	40.5	2,863	13.1

Table 4 Agent-based goodness-of-fit values of the cell-based and agent-based estimations (aggregated cells)

Case	Population of sample	Agent-based goodness-of-fit			
		Cell-based estimation (IPF)		Agent-based estimation (MCS)	
		Average	Standard deviation	Average	Standard deviation
Sample 1	2,346	19,100	12.0	2,539	15.5
Sample 2	2,357	18,824	18.5	2,932	17.5
Sample 3	2,434	19,397	12.6	2,369	16.8
Sample 4	2,415	19,217	13.0	3,028	13.6
Sample 5	2,416	19,495	10.9	3,336	16.8
Sample 6	2,452	19,790	43.9	2,444	22.6
Sample 7	2,443	18,462	11.5	2,594	9.3
Sample 8	2,467	19,474	23.2	2,863	13.1

5 Concluding Remarks

This chapter compared cell- and agent-based population synthesis methods qualitatively and quantitatively. The findings are summarized as below.

First, the qualitative comparison showed that both methods have advantages and disadvantages. In particular, agents should be defined by attributes in accordance with the capacity of the microsimulation model into which the estimated base-year dataset is input. If the model deals with data sufficiently simple for the cell-based method to estimate, the computation costs of the agent-based method increase. By contrast, if the applicable model is sensitive to the accuracy of the attributes and/or is able to deal with numerous attributes, the cell-based method will face difficulties estimating detailed base-year data.

Second, the quantitative comparison failed to draw a clear conclusion about which method is superior, although the numerical results showed that the agent-based method has a slightly better goodness-of-fit than the cell-based one. This unclear difference is probably caused by the size of the pseudo-observed dataset and thus that of the sampled set. However, the agent-based method clearly showed significantly better agent-based estimation performance.

These findings suggest a way for selecting a better method based on the conditions of the microsimulation model and the purpose of its application.

Acknowledgments This study was supported by a Grant-in-Aid for Scientific Research (23360228) from the Japan Society for the Promotion of Science.

References

Beckman RJ, Baggerly KA, McKay MD (1996) Creating synthetic baseline populations. Transp Res A-Pol 30:415–429

Deming WE, Stephan FF (1940) On a least squares adjustment of a sampled frequency table when the expected marginal totals are known. Ann Math Stat 11:427–444

Guo JY, Bhat CR (2007) Population synthesis for microsimulating travel behavior. Transp Res Record 2014:92–101

Knudsen DC, Fotheringham AS (1986) Matrix comparison, goodness-of-fit, and spatial interaction modelling. Int Reg Sci Rev 10:127–147

Lee DH, Fu YF (2011) Cross-entropy optimization model for population synthesis used in activity-based microsimulation models. In: TRB 90th Annual Meeting Compendium of Papers

Miyamoto K, Ando J, Shimizu E (1986) A housing demand model based on disaggregate behavioral analysis. J Jpn Soc Civ Eng 365:79–88

Miyamoto K, Sugiki N, Kitazume K, Suzuki A, Vichiensan V (2009) An estimation method of household micro-data for the base year in land-use micro simulation. In: Proceedings of CUPUM'09

Miyamoto K, Sugiki N, Otani N, Vichiensan V (2010a) An agent based estimation method of household micro-data for the base year in land-use microsimulation. In: TRB 89th Annual Meeting Compendium of Papers

Miyamoto K, Sugiki N, Otani N, Vichiensan V (2010b) An agent based estimation method of household micro-data including housing information for the base year in land-use microsimulation. In: Selected Proceedings of 12th WCTR

Moeckel R, Spiekermann K, Wegener M (2003) Creating a synthetic population. In: Proceedings of CUPUM'03

Müller K, Axhausen KW (2011) Population synthesis for microsimulation: state of the art. In: TRB 90th Annual Meeting Compendium of Papers

Otani N, Sugiki N, and Miyamoto K (2010) Goodness-of-fit evaluation method between two sets of household micro-data for land-use microsimulation model. In: Selected Proceedings of 12th WCTR

Otani N, Sugiki N, Miyamoto K (2012) Goodness-of-fit evaluation method for agent-based household microdata sets composed of generalized attributes: Transp Res. Record 2254:97–103

Pritchard DR, Miller EJ (2009) Advances in agent population synthesis and application in an integrated land use/transportation model. In: TRB 88th Annual Meeting Compendium of Papers

Sugiki N, Otani N, Vichiensan V, Miyamoto K (2013) Agent-based household micro-datasets: an estimation method composed of generalized attributes with probabilistic distributions from sample data and available control totals by attribute. Asian Transp Stud 2(1):3–18

Application of Land Use Model Combined with GIS and RS Technology in Supporting Urban Spatial Planning

Rui Zhou, Hailong Su, Xinjun Wang, Yuanman Hu and Fenge Zhang

Abstract This chapter illustrates an application of the Conversion of Land Use and Its Effects at Small regional extent (CLUE-S) model, combined with Geographic Information System (GIS) and Remote Sensing (RS) technology, in simulating future land use and landscape pattern change under different land use scenarios in Xinzhuang town, Jiangsu province, China. It is based on multi-temporal high-resolution remotely sensed images. Three different scenarios are designed to simulate the future patterns of the study area. Kappa coefficients are applied to evaluate the feasibility of CLUE-S model for supporting spatial planning. The results showed that the increase of construction land and decrease of paddy fields would be the dominant trend of future land use change. Plenty of farmlands and ecological land will be encroached by construction land in the next 20 years. The landscape pattern will be more fragmented, disaggregated and disconnected, and the landscape will become more diversified and homogenous. The prediction accuracy of CLUE-S is satisfactory; it means that this model can provide scientific support for land use planning and policy making.

R. Zhou · H. Su · X. Wang (✉) · F. Zhang
Research Center for Urban Planning and Development, Fudan University, Shanghai 200433, China
e-mail: fudanwxj@163.com

R. Zhou
e-mail: zhourui1215@163.com

H. Su
e-mail: fdsuhailong@126.com

F. Zhang
e-mail: zhangfe@126.com

R. Zhou · Y. Hu
State Key Laboratory of Forest and Soil Ecology, Institute of Applied Ecology, Chinese Academy of Sciences, Shenyang 110164, China
e-mail: huym@iae.ac.cn

S. Geertman et al. (eds.), *Planning Support Systems for Sustainable Urban Development*, Lecture Notes in Geoinformation and Cartography,
DOI: 10.1007/978-3-642-37533-0_7,

1 Introduction

Land use changes are driven by the spatial–temporal interactions between biophysical and human dimensions at different scales (Veldkamp and Lambin 2001; Verburg et al. 2004). They affect the ecological, physical, and socioeconomic processes of a region in various ways (Forman 1995; Brookes 2001). Thus, increased efforts have been made to understand the processes, trends and driving forces of land use change and its ecological consequences (Verburg et al. 1999; Parker et al. 2003; Turner et al. 2007). Identifying the primary causes, processes and trends of land use change are crucial for urban planning, utilization of regional resources, and environmental management (Ojima et al. 2002; Zhao et al. 2013). Land use change model is a useful tool for analyzing driving forces and processes, understanding the causes and consequences, predicting the possible futures of land use change, and assessing ecological impacts and decision-making for land use planning (Luo et al. 2010). Scenarios analysis with land use modeling can provide support for land use planning (Verburg et al. 2004) and help inform policymakers of possible future patterns under different policy restraint conditions (Koomen and Stillwell 2007). Land use change models can support an examination of future land use change under different scenarios (Liu et al. 2011). They are helpful tools in providing reproducible data to supplement our capabilities to analyze land-use change and make better-informed decisions (Costanza and Ruth 1998). Consequently, spatially-dependent land use models are indispensable for sustainable land use planning (Guan et al. 2011).

Land use planning has attracted increasing attention over the last decade since the growing negative impacts of urban sprawl, such as consumption of prime agricultural land and open space have been realized (Batisani and Yarnal 2009). The conflicts between urban development and farmland preservation or ecological protection have been becoming increasingly contentious all over the world, especially in China (Zheng et al. 2012). Whether it can be handled properly will have implications for food security, ecological security and social security. It is related to important issues such as sustainable development of the economy, society, and ecology (Koomen and Ding 2006; Clover and Eriksen 2009). The topic has caught the attention of land-use and urban planners (Carsjens and van der Knaap 2002; Deal and Schunk 2004; Lee et al. 2009). Planners worldwide thus seek to steer land-use developments through a wide range of interventions that either constrain certain developments or favor them. For the formulation of adequate spatial policies, the involved parties normally make use of models that simulate possible spatial developments (Koomen et al. 2008).

Land use change modeling and simulation has increasingly become a popular tool in land use planning and policy formulation. Veldkamp and Lambin (2001) emphasized the importance of land use change modeling as a planning tool for projecting alternative land use pathways into the future. Some land use models have been successfully applied in supporting region planning. However, considering the complexity of model application and actual conditions in China, land use change models have rarely been applied in assisting spatial planning. Therefore, most spatial planning maps, especially urban planning maps, were schematically made without scientific support. The planning of future urban land use is designed by the need of economic and population growth, and the planners' experiences and topographical and geologic characteristics. Such planning processes lack scientifically quantitative approaches and models.

In the past 10 years, different land use change models have been developed with various objectives and backgrounds (Verburg et al. 2004). As a typical spatial-explicit and empirical-statistical model, the CLUE-S model is well-known. It can better understand the processes that determine changes in the spatial pattern of land use and explore possible future changes in land use at the various spatial scales (Verburg et al. 2008). It can also specify the scenario conditions for future land use in detail (Verburg et al. 2002; Verburg and Veldkamp 2004). Compared to the relatively subjective land use models based on decision-making behavior of locators (Parker et al. 2003; Berger et al. 2006), the CLUE-S model is based on land use change processes and its simulation result is more objective and persuasive. However, it lacks multi-temporal historical land use data. The advantage of this model is the explicit attention for the functioning of the land-use system as a whole, the capability to simulate different land-use types at the same time and the ability to simulate different scenarios. CLUE-S model has been successfully applied in simulating land-use changes based on different spatial and non-spatial policies (Verburg et al. 2006; Overmars et al. 2007).

In this study, three scenarios were designed to systematically describe possible alternative land use views of the future. These scenarios considered land use planning, farmland and ecological environment protection over the next 20 years. The objective of this study is to simulate a broad range of future spatial developments and offer a full overview of the potential land use change trends under different scenarios based on CLUE-S model. First, we provide a detailed account and the implementation issues of the methodology. Second, spatio-temporal changes in land use and landscape patterns under different scenarios were analyzed. Third, the predicted results of urban construction area under the three scenarios in 2020 were compared with a prospective plan map of 2020 to evaluate the feasibility of a land use change model for supporting spatial planning. Finally, we discuss the evaluation of CLUE-S model and present the conclusion. This study will be helpful for urban planners and decision-makers to better understand the complexities of land use change and make scientifically sound decisions for future land use planning and management.

2 Materials and Methods

2.1 Study Area

As a typically area of rapid urbanization and industrialization in the south of Jiangsu province, Xinzhuang town (120°32′–120°44′E, 31°29′–31°37′N) is located in the Changshu city (Fig. 1). It is one of the two significant development central towns in the latest master plan of Suzhou city, due to its prominent location and convenient location for water and land transportation. It is in the east of Shajiabang resort district, in the west of Wuxi city, approximately 50 km from Suzhou city and Wuxi city, and 190 km from Nanjing city and Hangzhou city. The study area is about 104.26 km^2, including 2 district agencies, 20 villages, 3 neighborhood committees and a farm named South Lake. In recent years, the economy of Xinzhuang town has been growing rapidly. In 2008, the GDP of the study area was 6.058 billion CNY, and the population was about 127,500, including more than 52,000 adventitious workers. Urban and rural industrial and residential land has continuously expanded with economic and industrial growth in Xinzhuang, especially after 2000 (Zhou et al. 2010); Urban sprawl has occupied plenty of farmland and caused substantial change to the area's landscape and environment in the past decades (Zhou et al. 2011).

2.2 Data Preparation and Processing

In this study, many data sets, including 1991 aerial photographs at 1:10000, 2001 IKONOS image and 2009 QUICKBIRD image, were used to acquire land use maps. The 1:10000 topographic maps and 1:10000 digital elevation model (DEM) of the study area were collected from Geographic information center of Jiangsu province. A total of 405 evenly distributed field-survey points for land use information were sampled via field surveys in 2009 with the help of a global positioning system (GPS) with ±1 m error for ground-truthing.

Other data used in this study includes: (1) Statistical yearbooks of 1991, 2001 and 2009 of Changshu city obtained from Changshu Statistical Bureau; (2) Statistical yearbooks from 2007 to 2010 of Xinzhuang town obtained from Xinzhuang Statistical Bureau; (3) The urban planning and land use maps and textbooks of Xinzhuang town obtained from Xinzhuang Planning Bureau of Land and Resources.

Firstly, the 2009 QUICKBIRD image was geometrically corrected and geocoded to the Transverse coordinate system, using the 1:10000 topographic maps with ERDAS IMAGE 9.1 software. Secondly, the image-to-image method was applied for the geo-referenced registration of images in 1991 and 2001 with the total Root Mean Squared (RMS) error of less than 0.5 pixels. Thirdly, an image enhancement of intensifying visual distinction among features was performed to

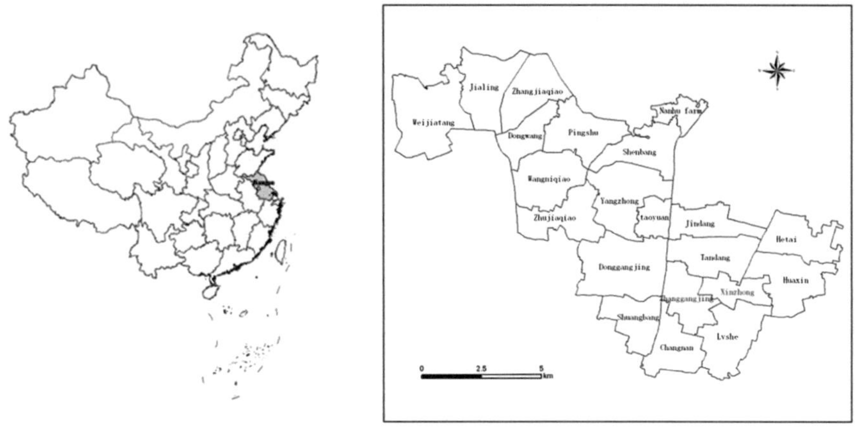

Fig. 1 Location of Xinzhuang town in Jiangsu province, China

increase the amount of information. In succession, image interpretation symbols of different image elements were added accompanying by field investigations, which could be consulted in the process of artificial visual operations. Finally, visual interpretation was carried out on images of 1991, 2001 and 2009, and land use maps were acquired with the help of ancillary data including the topographic map and ground survey information. Land use types were divided into 9 classes: paddy field, dry land, forestland, water area, urban and rural construction land, aquaculture land, grassland, vegetable field, and orchard land). Because of the restriction of area percentage of each land use type in CLUE-S model, the 9 land use types were integrated into 6 categories for simulating: paddy field, dry land (including vegetable field), forestland (including grassland and orchard land), water area, aquaculture land, urban and rural construction land. The 405 field-survey points were used to examine the accuracy of the image classification. The Kappa coefficient was 95.2 % in 2009, 93.5 % in 2001, and 92.1 % in 1991. Both ERDAS Imagine 9.1 and ARCGIS 9.0 were applied to integrate the data using standard GIS features. Due to the different resolutions of remotely-sensed images, all the results of classification were re-sampled at 20 × 20 m for further analysis.

2.3 The CLUE-S Model

The CLUE-S is a new version based on early CLUE model (Veldkamp and Fresco 1996; Verburg et al. 1999). It was based on an empirical analysis of location suitability combined with the dynamic simulation of competition and interactions between the spatio-temporal dynamics of land use systems, and specifically developed for the spatially explicit simulation of land use change (Verburg et al. 2002). This version has been applied in case studies with a local to regional extent

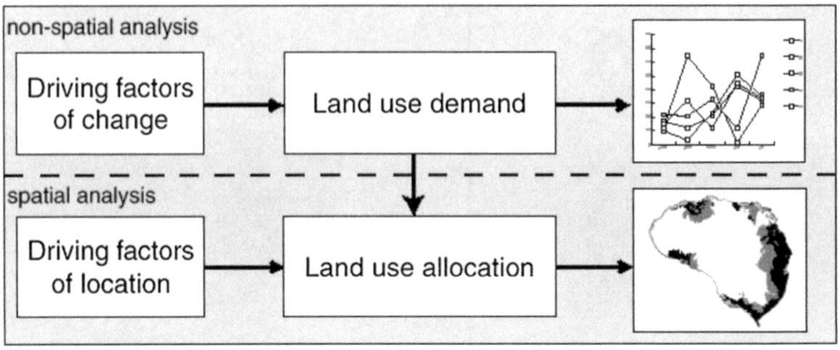

Fig. 2 Modules within the CLUE-S model (Verburg et al. 2002)

and the spatial resolution ranging from 20 to 1,000 m (Verburg and Veldkamp 2004; Overmars et al. 2007).

The model is sub-divided into two distinct modules: non-spatial demand module and spatially explicit allocation procedure. The non-spatial module calculates the area change for all land use types at the aggregate level. Within the second part of the model these demands are translated into land use changes at different locations within the study region using a raster-based system (Fig. 2). Allocation of each land use type is based on a combination of empirical analysis, spatial analysis, and dynamic modeling. Empirical analysis is applied to determine the relationships between spatial distribution of land use and a number of proximate factors that drive or constrain land use change. Based on the competitive advantage of each land use at a location, the competition among land uses for a particular location is simulated. The schematic representation of the procedure to allocate change in land use in CLUE-S model is in Fig. 3 (Verburg et al. 2002). Verburg et al. (2006) provided its basic structure. The actual allocation process depends on the constraints and preferences defined by the user based on the characteristics of the land use type or the assumed processes and constraints relevant to the scenario.

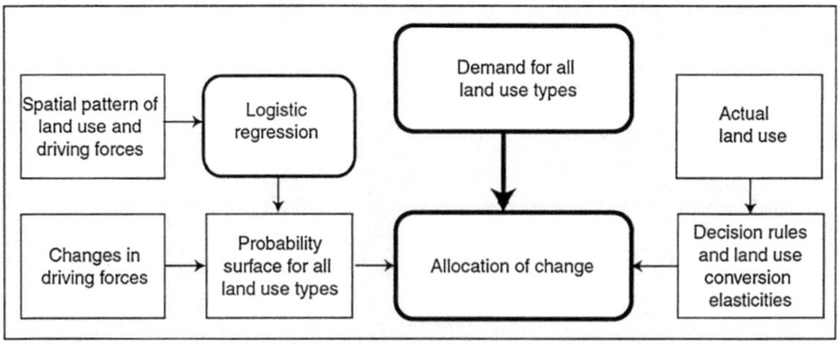

Fig. 3 Overview of the procedure to allocate changes in land use to a raster based map (Verburg et al. 2002)

2.3.1 Scenarios Design and Assumptions

In this study, three scenarios were designed to represent different implementations of the spatial policies and restrictions: (1) The Historical Trend (HT) scenario was formulated based on historical land use change from 1991 to 2009. Land use area demand was forecasted via the autoregressive integrated moving average (ARIMA) approach, which employs a time series analysis based on the historical land use area changes. Besides 1991, 2001, and 2009, the land use data in this study area for the years from 1991 to 2009 were obtained from statistical yearbooks of Changshu city. (2) The Urban Planning (UP) scenario was designed based on the urban planning and land use planning schemes of Xinzhuang town, which emphasized compact urban development and basic farmland preservation. The areas which demand different land use types in the UP scenario were adjusted from the results of the HT scenario. (3) The Ecological and environmental protection (EE) scenario was designed based on related ecological and environmental protection policies in the study area over the next 20 years, the expectation of this scenario was to maintain the ecological land and improve the increasingly worsening ecological problems. The land use area demand in the EP scenario was adjusted based on the results from the UP scenario.

2.3.2 Logistic Regression

The demand for land by the different land use types determines the overall competitive capacity of the different land use types, but the location suitability is also a major determinant of the competitive capacity of the different land use types at a specific location. Generally, conversions of land use are expected to take place at locations which have the highest 'preference' for a specific type of land use at a given moment. It can be calculated as a probability of a certain grid cell by logistic regression as follows (Bucini and Lambin 2002):

$$\log\left(\frac{P_i}{1-P_i}\right) = \beta_0 + \beta_1 X_{1,i} + \beta_2 X_{2,i} + \cdots + \beta_n X_{n,i} \tag{1}$$

where P_i is the probability of a grid cell for the occurrence of the considered land use type i and the X are the driving factors; β_0, β_1, …, β_n are the beta values of logistic regression for driving factors. The value of Relative Operating Characteristics (ROC) proposed by Pontius and Schneider (2001) was used to evaluate the fit of the regression model. A completely random model gives ROC a value of 0.5 while a perfect fit results with ROC value of 1.0. If the value of ROC is below 0.7, the accuracy of the model is low; the accuracy will be preferable if an ROC value is above 0.7 (Pontius 2000).

The driving factors of land use change taken into account as potential determinants were selected based on literatures and fieldwork in the study area, including 11 factors: distance to major road, distance to minor road, distance to

Table 1 Results of logistic regression for different land use types in 2009

Driving factor	Exp(β) Construction land	Dry land	Paddy field	Aquaculture land	Forestland	Water area
A	0.9987	0.9998	1.0002	1.0007	0.9991	1.0000
B	0.9995	1.0000	1.0001	1.0002	1.0001	1.0001
C	0.9989	0.9999	0.9955	1.0037	0.9983	1.0017
D	1.0000	1.0000	1.0000	1.0000	1.0000	0.9999
E	1.0000	1.0000	1.0000	1.0000	1.0000	1.0000
F	0.9998	1.0003	1.0003	0.9996	0.9996	1.0000
G	1.0003	0.9998	0.9997	1.0001	0.9998	0.9999
H	0.9999	0.9998	1.0000	1.0002	0.9999	1.0000
I	1.0004	1.0003	0.9993	0.9994	1.0008	1.0002
J	1.0048	0.9908	1.008	1.0063	1.0037	0.7906
K	0.9777	0.998	1.0015	1.0061	1.0029	1.0015
Constant	0.0279	0.4873	10.3249	0.0157	0.9809	8.2511
ROC	0.851	0.759	0.792	0.812	0.815	0.988

A distance to major road; *B* distance to village government; *C* distance to minor road; *D* GDP; *E* gross industrial product; *F* grain output; *G* per capita income of resident; *H* gross agricultural product; *I* population density; *J* distance to river; *K* distance to rural settlement; *ROC* relative operating characteristics

river, distance to village government, distance to rural settlement, GDP, gross industrial product, gross agricultural product, grain output, population density, per capita income. The logistic regression models, based on the GIS dataset, were constructed to determine the relations between land use changes and a set of potential driving factors. The logistic regression results got through SPSS software were shown in Table 1. The spatial distribution of all land use types could be well explained by the selected driving variables as indicated by the high ROC test statistics (>0.7). The derived regression models were used to calculate suitability maps for different land use types.

2.3.3 Simulation Spatial and Temporal Resolution Setting

A series of test scenarios were set using 10–50 m resolution with 10 m steps. The results showed that the highest spatial resolution in the CLUE-S model was 20 m in Xinzhuang town. Therefore, the simulation spatial resolution was set as 20 × 20 m grid in this study, including 946 rows and 685 columns.

The land use in 2009 was simulated based on that in 1991 and 2001 with CLUE-S model to validate its applicability in the study area. The forecast periods were 18 and 8 years, respectively. The predicted land use map in 2009 was compared with the actual land use in 2009 by utilizing the Kappa coefficient. It has been applied to any model that predicts a homogenous category in each grid cell. The Kappa result was 0.75 and 0.80 for 1991 and 2001, which indicated that the two maps show a relatively high consistency with 18 years. CLUE-S was then used to predict land use change for the 18-year period beginning in 2009 with 1-year steps in the study area.

2.3.4 Land Use Conversion Matrix and Elasticity Setting

There are two sets of parameters in the CLUE-S model (land use conversion matrix and elasticity) that could influence the pattern of land use change. The land use conversions restricted by land use policies, restrictions and land tenure could be reflected in a land use conversion matrix. The rows of the matrix indicate the land use type at time step t and the columns indicate the land use type at time step (t + 1). Moreover, land use type specific conversion settings were defined and implemented by the relative elasticity for change (ELAS) in the model (Verburg et al. 2002). The relative elasticity ranges between 0 (easy conversion) and 1 (irreversible conversion). The value of this factor is set based on expert knowledge and can be adjusted during the calibration stage.

In this study, we assumed that the construction land would not be converted to other land use types. Based on the reference data during 1991–2009 and expert knowledge, the values of conversion elasticity for different land use types were tuned so that they were suitable for the calibration of the model. The final conversion elasticity values of paddy field, dry land, forestland, water area, aquaculture land, urban and rural construction land in the model during 2009–2027 were 0.4, 0.5, 0.6, 0.7, 0.4, and 0.9, respectively.

2.4 Landscape Metrics

In order to evaluate and compare the simulated results based on CLUE-S model under the three scenarios of the strategy in the study area, four landscape-level spatial metrics were selected to reflect future landscape pattern changes based on their ecological meanings, including number of patches (NP), landscape shape index (LSI), Shannon's evenness index (SHEI), and Contagion index (CONTAG), these metrics were calculated using Fragstats 3.3 software at the landscape level.

3 Results and Analysis

3.1 Spatio-Temporal Change of Land Use in Future Under Different Scenarios

Different land use types with regard to future area changes are shown in Fig. 4, which displayed different change trajectories from 2010 to 2027. The main land

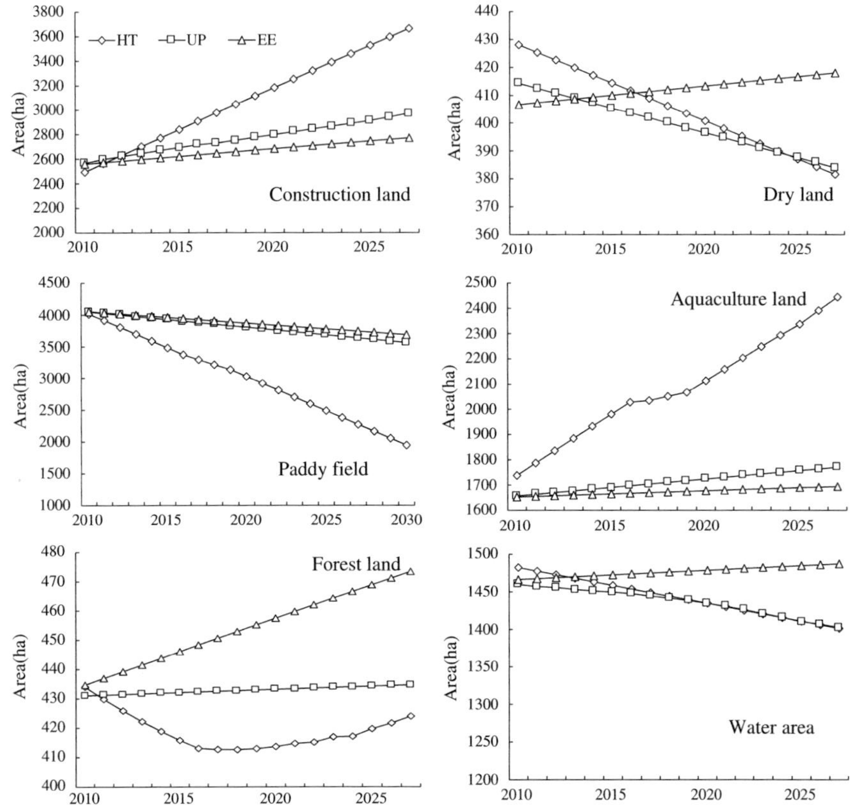

Fig. 4 Area change of different land use types in the study area under three scenarios

use types of construction land and aquaculture land showed the similar increasing trend under the three scenarios. The average annual increase rate of construction land and aquaculture land under HT scenario were 68.97 and 41.57 ha/a, which were obviously higher than the other land use types. The transfer matrix analysis showed that the increased area of construction land was mainly converted from paddy field. Unlike the other two scenarios, under the EE scenario, the forestland and water area would increase, for their higher ecological service values. In addition, the dry land under EE scenario was well protected, and its area increased slightly. The area of paddy field would decrease under all scenarios. This is particularly true for the HT scenario, which showed the area of paddy field decreased by 2,088.50 ha from 2009 to 2027, while the area under UP and EE scenario decreased by 472.25 and 346.81 ha, respectively.

In summary, the expansion of construction land and the decrease of paddy field were the dominant changes of land use in the study area under the three scenarios (Fig. 5). Most of the increasing construction lands were due to the sprawl from the existing construction area through occupation of the surrounding paddy field, dry

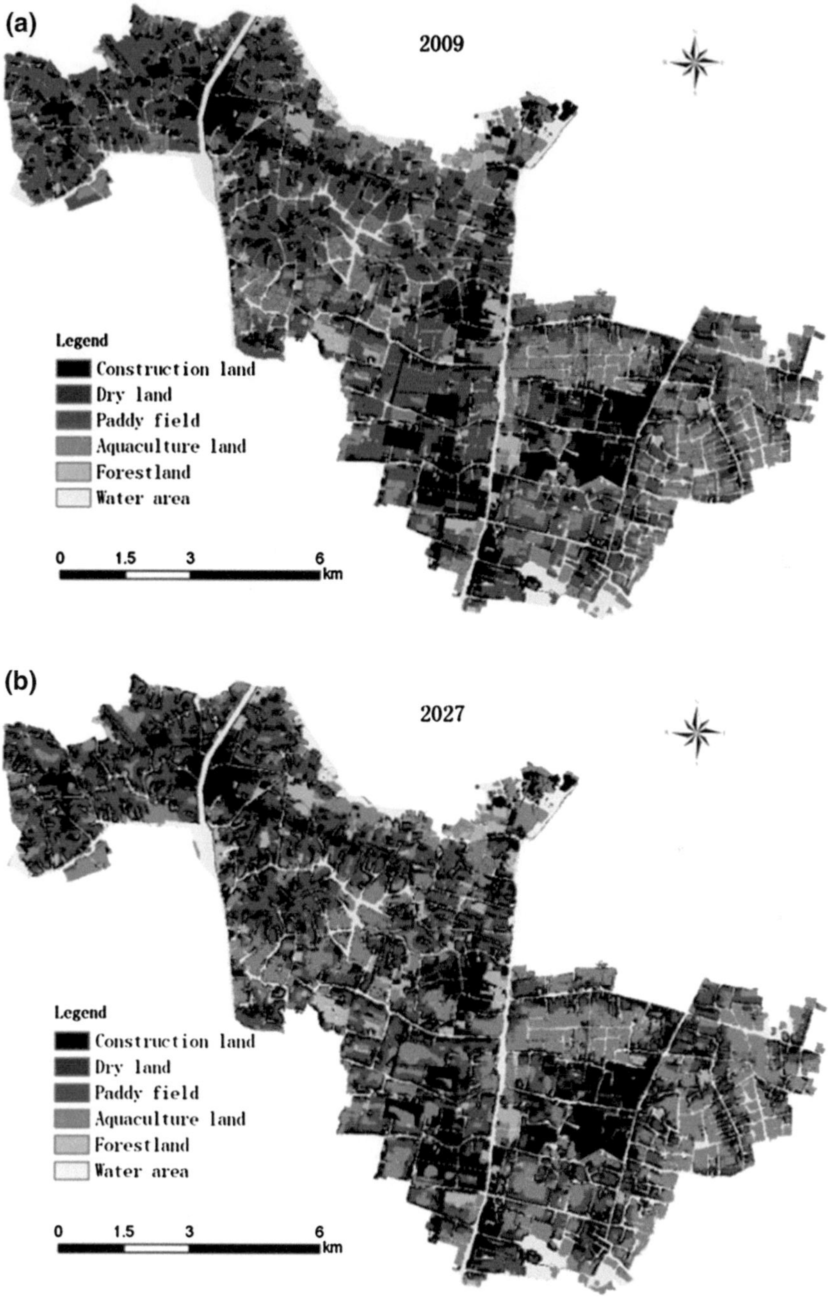

Fig. 5 The land use map in 2009 and simulated results of land use in 2027 under different scenarios. **a** Land use of Xinzhuang town in 2009. **b** Simulated results under HT scenario. **c** Simulated results under UP scenario. **d** Simulated results under EE scenario

Fig. 5 continued

land, forestland, and so on. It occurs mainly at the urban–rural fringe, nearby existing construction land and around major roads and water area. The decreased paddy fields were mainly located near existing construction land, and along water area and major roads. As elsewhere, the farmland in this area was also more likely to be encroached by construction land, which means the conflicts between urban and rural development and farmland preservation will gradually increase. Consequently, in the process of plan-making, the urban planners and decision-makers in Xinzhuang town need to highlight the issues and strengthen the control of its land use by confirming the amount of urban and rural construction land scientifically, and reasonably prohibiting the unplanned sprawl of construction land, in order to realize the sustainable development of land use.

3.2 Landscape Pattern Change Under Different Scenarios

The number of patches at the landscape level over the study area increased under UP and EE scenarios from 2010 to 2027, which means that the landscape would be more fragmented, especially under UP scenario. The number of patches would increase rapidly before 2018, and then decline quickly under HT scenario. Such change may be due to the discrete construction land increasingly linked together after 2018, for the rapid growth of urban land and rural settlement area. The increasing trend of landscape shape index under each scenario from 2010 to 2027 indicates that the landscape patches would become increasingly disaggregated. Shannon's evenness index increased under UP and EE scenarios, which means the landscape pattern would be toward more diversified and homogenous. Under HT scenario, the SHEI showed a parabolic trend. Namely, the value increased first, then decreased for the construction land; it would become the dominant landscape and make the land use type patches more uneven in future. Contagion index showed a decreasing change trend under UP and EE scenarios, which means that the connectivity of future landscape would decline and the landscape would be more fragmented (Fig. 6).

3.3 Comparison Analysis of Simulated Results with Urban Planning Maps

To evaluate the feasibility of land use change model in supporting spatial planning, a comparison analysis was made between simulated urban construction land of 2020 and perspective plan map in 2020 from Xinzhuang town urban planning (2006–2020). The simulated urban construction land in 2020 under different scenarios and the perspective plan map in 2020 are shown in Fig. 7. It is observed that the urban construction land in the planning map was regular and compact with

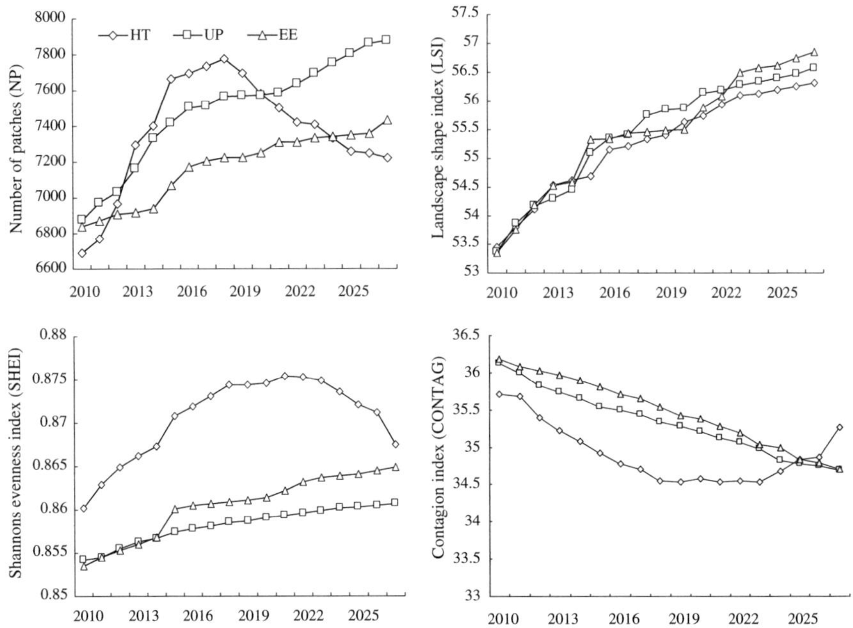

Fig. 6 Landscape metrics at landscape level under the three scenarios

clear boundaries, while the simulated urban construction land in 2020 under different scenarios was rather spatially dispersed with complicated boundaries. However, the simulated map may more closely reflect the real situation of urban growth pattern, while the planning map is more like subjective blueprint with few objectively and scientifically technological and methodological supports. It is well known that the Kappa coefficient can be used for measuring the consistency between two maps based on a contingency table and for accuracy assessment as a whole. The simulated results and planning map were compared by analysis of Kappa coefficient. Table 2 shows that the consistency between simulated maps and planning map was more than 0.5, especially for simulated location and quantity precision, which was more than 0.8 and 0.6, respectively. This means that the three scenarios were designed reasonably and the prediction accuracy of CLUE-S model was excellent.

4 Discussion and Conclusions

Based on three periods of land use data extracted from high-resolution remote sensing images, natural and socioeconomic data, the CLUE-S land use change model was used. This was then combined with GIS and RS technology to successfully simulate future land use change trend under three scenarios. The land use

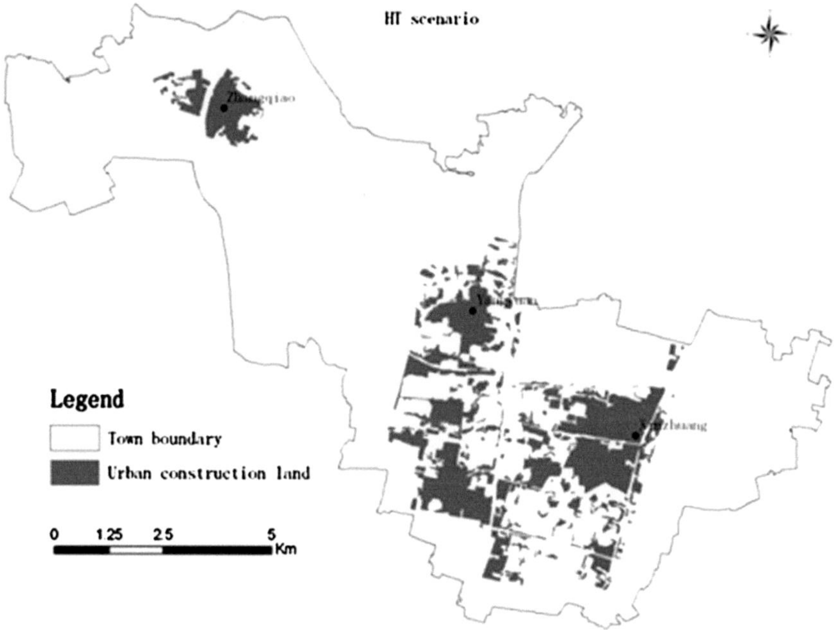

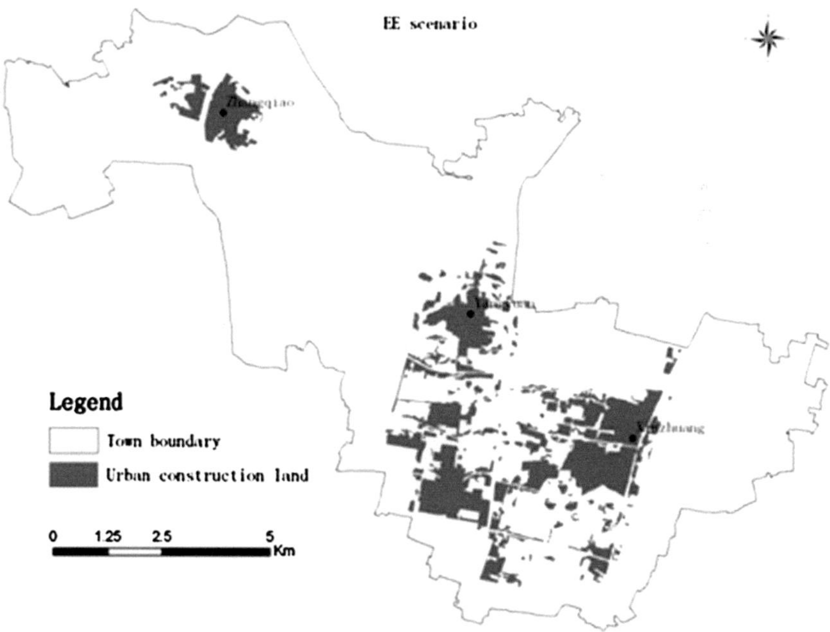

Fig. 7 The simulated urban construction land under three scenarios and prospective plan map in 2020

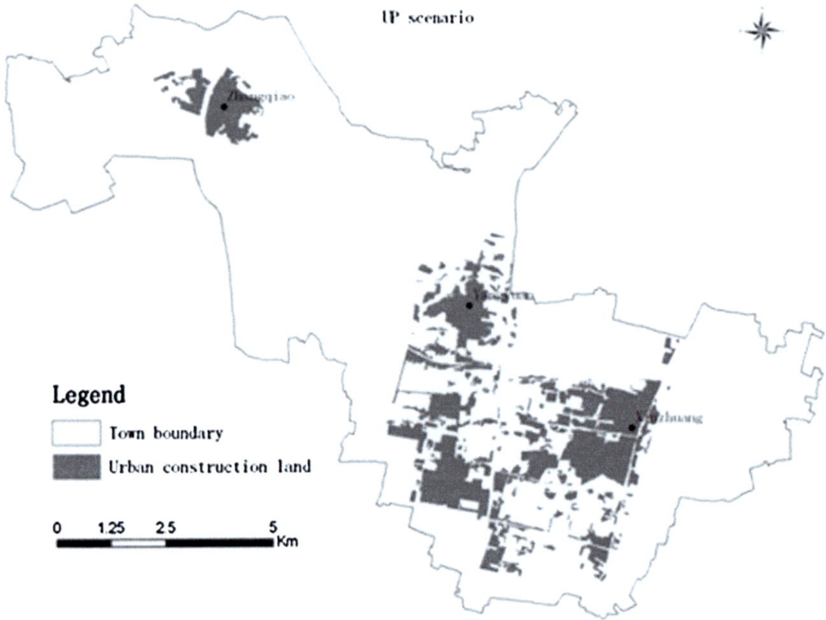

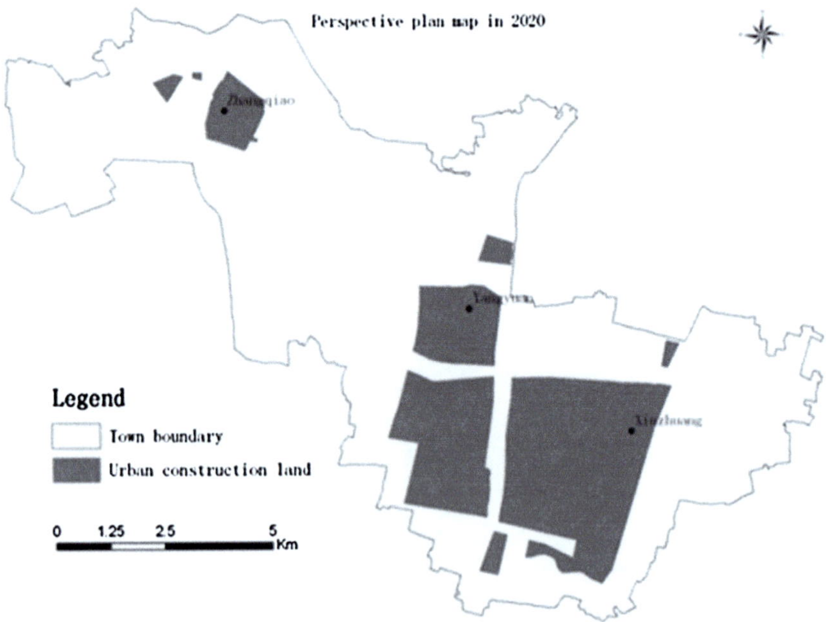

Fig. 7 continued

Table 2 Results of Kappa coefficients of simulated map with plan map in 2020

Comparison	K standard	K location	K quantity
UP scenario	0.551	0.831	0.663
EE scenario	0.533	0.824	0.647
HT scenario	0.564	0.815	0.691

modeling and prediction results showed that the increase of construction land and decrease of paddy field would still be the main trend of land use change in the study area. A good deal of farmlands and ecological land, especially around existing construction land, major roads and water areas, would be mainly transformed to urban land and rural settlement in the next 20 years. Results of landscape pattern metrics analysis showed that the landscape would be more fragmented, disaggregated and disconnected, and the landscape pattern was toward more diversified and homogenous. Therefore, a reasonable constraint and control policy should be made for urban expansion and land use change. Such a policy, or policies, would improve the current land use trends and establish an ecological safety pattern for urban and rural development. This is an effective way towards smart protection and smart growth, promoting regionally sustainable development. The prediction accuracy of CLUE-S was satisfactory, no matter the location, quantity or overall accuracy of land use change, which suggests the feasibility of land use change model for land use planning supports.

In summary, our study presented an important contribution to land use modeling for supporting spatial planning, and successfully simulated and predicted the spatio-temporal changes of future land use. The simulated future land use maps under different scenarios could serve as an early warning system for understanding the future effects of land use changes. Furthermore, the simulation results can also be considered as a strategic guide for urban land use planning. It would help local authorities better understand the complex land use system and develop the improved urban development and land use management, which can better balance urban expansion, basic farmland and ecological environment conservation. Above all, the land use change model can be a helpful and scientific tool for supporting urban land use planning and policy making. Planners and decision-makers should pay more attention to the potential consequences of land use change in the process of policy-making. Our findings and discussions will not only give decision-support for Xinzhuang town, but also for other similar areas with rapid rural urbanization in China.

Acknowledgments The research work presented in this chapter is supported by the National Natural Science Foundation of China (Project No. 50808048) and the National Science and Technology supporting Program of China (Project No. 2006BAJ10B05-03).

References

Batisani N, Yarnal B (2009) Uncertainty awareness in urban sprawl simulations: lessons from a small US metropolitan region. Land Use Policy 26:178–185

Berger T, Schreinemachers P, Woelcke J (2006) Multi-agent simulation for the targeting of development policies in less-favored areas. Agric Syst 88:28–43

Brookes CJ (2001) A genetic algorithm for designing optimal patch configurations in GIS. Int J Geogr Inf Sci 6:539–559

Bucini G, Lambin EF (2002) Fire impacts on vegetation in Central Africa: a remote-sensing-based statistical analysis. Appl Geogr 22:27–48

Carsjens GJ, van der Knaap W (2002) Strategic land-use allocation: dealing with spatial relationships and fragmentation of agriculture. Landscape Urban Plan 58:171–179

Clover J, Eriksen S (2009) The effects of land tenure change on sustainability: human security and environmental change in southern African savannas. Environ Sci Policy 12:53–70

Costanza R, Ruth M (1998) Using dynamic modeling to scope environmental problems and build consensus. Environ Manage 2:183–195

Forman RTT (1995) Land mosaic: the Ecology of landscape and region. Cambridge University Press, New York

Guan DJ, Li HF, Takuro I, Su WC, Tadashi N, Kazunori H (2011) Modeling urban land use change by the integration of cellular automaton and Markov model. Ecol Model 9:3761–3772

Koomen E, Ding CR (2006) Land use efficiency, food security, and farmland preservation in China. Land Lines 18:2–7

Koomen E, Stillwell J (2007) Modeling land-use change. In: Koomen E (ed) Modeling land-use change: progress and applications. Springer, Dordrecht

Koomen E, Piet R, de Ton N (2008) Modeling land-use change for spatial planning support. Ann Reg Sci 42:1–10

Lee CL, Huang SL, Chan SL (2009) Synthesis and spatial dynamics of socio-economic metabolism and land use change of Taipei metropolitan region. Ecol Model 220:2940–2959

Liu M, Hu YM, Zhang W, Zhu JJ, Chen HW, Xi FM (2011) Application of land-use change model in guiding regional planning: a case study in Hun-Taizi River Watershed, Northeast China. Chin Geogr Sci 5:609–618

Luo GP, Yin CY, Chen X, Xu WQ, Lu L (2010) Combining system dynamic model and CLUE-S model to improve land use scenario analyses at regional scale: a case study of Sangong watershed in Xinjiang, China. Ecol Complex 7:198–207

Ojima D, Lavorel S, Graumlich L, Moran E (2002) Terrestrial human-environment systems: the future of land research in IGBP II. IGBP Glob Change Newslett 50:31–34

Overmars KP, Verburg PH, Veldkamp T (2007) Comparison of a deductive and an inductive approach to specify land suitability in a spatially explicit land use model. Land Use Policy 3:584–599

Parker DC, Manson SM, Janssen M, Hoffmann MJ, Deadman PJ (2003) Multi-agent systems for the simulation of land use and land cover change: a review. Ann Assoc Am Geogr 93:316–340

Pontius J (2000) Quantification error versus location error in comparison of categorical maps. Photogram Eng Remote Sens 8:1011–1016

Pontius RG, Schneider LC (2001) Land-cover change model validation by an ROC method for the Ipswich watershed, Massachusetts, USA. Agric Ecosyst Environ 85: 239–248

Turner BL, Lambin EF, Reenberg A (2007) The emergence of land change science for global environmental change and sustainability. PNAS 104:20666–20671

Veldkamp A, Fresco LO (1996) CLUE: a conceptual model to study the conversion of land use and its effects. Ecol Model 2:253–270

Veldkamp A, Lambin EF (2001) Editorial: predicting land-use change. Agric Ecosyst Environ 85:1–6

Verburg PH, Veldkamp A (2004) Projecting land use transitions at forest fringes in the Philippines at two spatial scales. Landscape Ecol 19(1):77–98
Verburg PH, Veldkamp A, Bouma J (1999) Land-use change under conditions of high population pressure: the case of Java. Glob Environ Change 9:303–312
Verburg PH, Soepboer W, Limpiada R (2002) Land use change modeling at the regional scale: the CLUE-S model. Environ Manage 3:391–405
Verburg PH, Schot P, Dijst M, Veldkamp A (2004) Land use change modeling: current practice and research priorities. GeoJournal 4:309–324
Verburg PH, Overmars KP, Huigen MGA (2006) Analysis of the effects of land use change on protected areas in the Philippines. Appl Geogr 2:153–173
Verburg PH, Eickhout B, van Meijl H (2008) A multi-scale, multi-model approach for analyzing the future dynamics of European land use. Ann Reg Sci 42:57–77
Zhao RF, Chen YN, Shi PJ, Zhang LH, Pan JH, Zhao HL (2013) Land use and land cover change and driving mechanism in the arid inland river basin: a case study of Tarim River, Xinjiang, China. Environ Earth Sci 68:591–604
Zheng XQ, Zhao L, Xiang WN, Li N, Lv LN, Yang X (2012) A coupled model for simulating spatio-temporal dynamics of land-use change: a case study in Changqing, Jinan, China. Landscape Urban Plan 3:51–61
Zhou R, Li YH, Hu YM, He HS, Wu ZF (2010) Dynamic changes of landscape in Xinzhuang town based on high-resolution remote image and GIS. Chin J Ecol 5:1000–1007 (in Chinese)
Zhou R, Hu YM, Su HL, Li YH (2011) Changes of paddy field landscape and its influence factors in a typical town of south Jiangsu Province. Acta Ecologica Sinica 20:5937–5945 (in Chinese)

Planning Support Systems for Fiscally Sustainable Planning

Scott N. Lieske, Roger H. Coupal, Jeffrey D. Hamerlinck, Donald M. McLeod and Anna M. Scofield

Abstract Local government's need for accurate assessments and projections of the fiscal consequences of development is well established and persistent. This analysis demonstrates the use of a geographic information science-based planning support system to project residential growth and the fiscal consequences of development. The cornerstone of the analysis is a spatial index of urban form which captures clustering and dispersion of the built environment. A regression model indicates the spatial index to be a statistically significant determinant of expenditures on policing services in the study area. Modelled future growth was spatially and temporally disaggregated to indicate future residential growth at different planning horizons. Spatial indices were calculated for these planning horizons and incorporated into the econometric model for *ceteris paribus* evaluation of the effect of change in urban form on public service expenditures. Results demonstrate planning informed by PSS modelling has the potential to realize savings on public service expenditures.

S. N. Lieske (✉)
Sustainability Research Centre, University of the Sunshine Coast,
Maroochydore DC, Queensland 4558, Australia
e-mail: slieske@usc.edu.au

R. H. Coupal · D. M. McLeod · A. M. Scofield
Agricultural and Applied Economics, University of Wyoming,
Laramie, Wyoming 82071, USA
e-mail: Coupal@uwyo.edu

D. M. McLeod
e-mail: DMcLeod@uwyo.edu

A. M. Scofield
e-mail: ascofiel@uwyo.edu

J. D. Hamerlinck
Wyoming Geographic Information Science Center, University of Wyoming,
Laramie, Wyoming 82071, USA
e-mail: jhamerl2@uwyo.edu

S. Geertman et al. (eds.), *Planning Support Systems for Sustainable Urban Development*, Lecture Notes in Geoinformation and Cartography,
DOI: 10.1007/978-3-642-37533-0_8, © Springer-Verlag Berlin Heidelberg 2013

1 Introduction

Sustainability, from the perspective of local government finance, is in part a situation where the costs of providing a public service are minimized rather than exacerbated by development decisions. Emerging research in local government fiscal modeling demonstrates a statistically significant relationship between increasing dispersion of the built environment and increasing costs of public service provision. The goal of this research is to inform public service decision making and support sustainable planning and urban development by making clear the fiscal implications of a local government land-use planning and regulatory framework. The approach taken is to use a geographic information science-based planning support system (PSS) to link a public service expenditure model with a growth model. The growth modeling technique employed here is a straightforward build-out analysis based on future land use and density requirements specified in a developing comprehensive plan. Build-out rather than a more sophisticated growth model is preferred as it ties clearly and directly to the existing planning and regulatory framework in the study area. Zoning, along with subdivision control law and to a lesser extent growth control laws and building codes are the primary regulatory tools associated with land development in the study area. In order to compare and contrast the impacts of differing land-use patterns a base scenario emerging from the comprehensive planning process is contrasted with an alternative land-use scenario. The alternative land-use scenario has decreased densities in outlying areas and a larger urbanized area. The hypothesis is the alternative land-use scenario, with a more concentrated built environment, will offer greater efficiency than the base scenario. Given plausible results (indicating the econometric modeling, growth modeling, and the linking of the two are valid), the degree of economic efficiency realized, the magnitude of fiscal change and change in efficient area across scenarios may be used as a cornerstone for proactive fiscally sustainable planning.

PSS operationalize various tools needed to develop this analysis including scenario planning, build-out analysis, spatial and temporal growth modeling, and impact assessment. By using PSS to critically evaluate land-use and density specifications within the context of future growth it may be possible to more clearly demonstrate the future impacts of current policies and regulations including whether mechanisms in place help to realize or conflict with a community's vision. Van Eck and Koomen (2008) describe this approach as "ex-ante evaluation of the spatial effects of planning measures" (p. 124) and note the importance and need for tools to do this. By enabling ex-ante evaluation of the potential impacts of extant planning policies and regulations it is hoped to be able to realize broader impacts from this research including conclusive impact and benefit from PSS, spatial modeling and spatial planning.

2 Background

This research used the CommunityViz® PSS to evaluate the long-term (25 year) influence of changing urban form on expenditures for policing services in Laramie County, Wyoming, USA. Developed by the Orton Family Foundation (Rutland, VT), CommunityViz is a modular system built on the ArcGIS platform (ESRI, Inc.; Redlands, CA). CommunityViz includes two integrated components: Scenario 360 and Scenario 3D. Scenario 3D allows display of three-dimensional landscape and structure information with real-time movement and object manipulation in a semi-realistic setting. Scenario 360 provides functionality for assessing the potential impacts of specific, proposed land use actions. CommunityViz Scenario 360 is used here to develop a build-out analysis, distribute modeled dwelling units spatially and temporally across the study area, and evaluate the fiscal and related impacts of alternative land uses using a scenario analysis and series of indicators.

2.1 Local Government Expenditure Modeling

The fiscal implications of differing urban forms are important considerations for local government financial health and as evaluation criteria in planning processes. Burchell and Listokin (1978) articulate the need for fiscal analysis stating that municipalities, "…must be aware of the public costs associated with private development, major rezonings, annexations, or alternative land use plans" (p. 1). Fiscal analysis may be used to predict budget deficits, allow local governments to quantitatively consider land-use policy decisions, levels of service, capital improvement plans and long term financing needs including current and future revenue streams. Fiscal analyses are helpful in short and long-term land-use policy and financial planning (Tischler 2002a). For comprehensive planning, Tischler recommends evaluation of different plan alternatives early in the planning process, prior to making the plan. This allows planners, "…to determine if land use … and location assumptions generate net revenues or net costs to the jurisdiction" (Tischler 2002b, p. 4). Goetz (2007) further emphasizes the importance of the issue when he juxtaposes the need for the fiscal evaluation of planning alternatives with the idea of planning as a non-repeatable experiment. Once constructed, residential land-use patterns are "largely irreversible" (p. 20).

Important research in local government expenditure modeling includes Davenport (1926), Fabricant (1952), Hirsch (1970), Borcherding and Deacon (1972) and Heikkila (2000). These works demonstrate the use of econometric techniques to model local government expenditures on public services as a function of determinants including factor inputs and neighborhood characteristics. Examples of the latter include population, income, density and urbanization. Often, but not always, low density is associated with higher per capita costs of public service provision and higher density is associated with lower per capita costs of service provision. Lieske et al. (2012) add to this body of literature by developing a spatial index of urban form that was a statistically significant determinant of local government expenditures. The econometric model put forward here is based on the tradition of local government expenditure modeling from the 20th Century augmented with the methods of Lieske et al. (2012) for quantifying urban form.

2.2 Scenario Planning

In planning contexts, scenarios may be viewed as different plausible development alternatives. Scenarios can serve as thinking tools, communication tools and evaluative tools. As thinking tools, the construction of scenarios "...requires people to uncover and cope with forces that are driving change in their environment" (Avin and Dembner 2001, p. 22). Observing different patterns allows consideration of the driving forces behind those patterns (Veeneklaas et al. 1995). Implicitly and explicitly tying driving forces to scenarios allows consideration of different assumptions and stimulates critical thinking (Xiang and Clarke 2003). Scenarios change and improve thinking because the discussion moves from "What do you think might happen?" to "What else might happen?", "Why might this happen?" and "How might this happen?" (Avin and Dembner 2001, p. 22). As a communication tool, scenarios serve a bridging function, allowing an exchange of information between scenario creators and scenario users (Xiang and Clarke 2003). This exchange of information means scenarios are useful as a communication and education tools between planners and their audiences. Finally, as evaluative tools, scenarios allow the answering of "What if?" questions quantitatively, qualitatively and visually. Evaluation of scenarios brings about the key benefit from scenario planning, "...the knowledge that, having explored many alternatives well, the selected course of action is defensible and prudent" (Avin and Dembner 2001, p. 20).

2.3 Build-Out

Build-out analysis is a calculation of the development capacity of land based on current or proposed zoning regulations and associated density values. Walker and Daniels (2011) call build-out analysis "...the most common and basic type of growth projection" (p. 76). Build-out analysis enables a community to illustrate and evaluate the impacts of their zoning regulations including lot size, floor area ratio, lot coverage, building height, number of stories, setbacks and density. On one hand the analysis may appear somewhat hypothetical because it specifies the very upper limit of legally allowable growth. On the other hand, it is critical for decision makers and citizens to know the long-term impacts of their land-use regulatory framework and if that framework helps realize or conflicts with the vision of the community. For example:

Often, a build-out analysis presentation shocks audiences. By showing a map or 3-D model of what the community would look like if it were built out, you can draw the public's attention to the negative effects of poorly planned growth and weak zoning (Walker and Daniels 2011, p. 77).

2.4 *Spatial and Temporal Growth Modeling*

CommunityViz® TimeScope™ functionality allows mapping and analyzing temporal change by enabling the storage and presentation of historical data as well as modeling future growth by specifying build rate details including growth rate and growth rate type (e.g. linear, exponential etc.). TimeScope also allows specification of the order in which features are built over time, for example based on development criteria such as suitability. "TimeScope calculates where and when new development will occur based on assumptions that you provide about growth rates and growth patterns" (Walker and Daniels 2011, p. 80).

2.5 *Impact Assessment*

Impact analysis involves quantifying outcomes from different spatial patterns and their interaction with any related non-spatial drivers. Impact analysis is premised upon and directly supports the rational planning process where data are transformed into conclusions and recommendations (Hanna and Culpepper 1998). As a tool that is both evaluative and communicative, impact analysis helps people understand the implications of discrete design and policy choices on the larger community (Snyder 2003).

The quantification of impacts is accomplished by creating indicators. Indicators are calculated values which measure impact or performance (Walker and Daniels 2011). Indicators can be used to represent nearly anything quantifiable within a planning process. They can be as simple as counting features or measuring the size of an area or as sophisticated as the user cares to develop. Indicators facilitate direct comparison of criteria before and after events and/or among scenarios. Van Eck and Koomen (2008) recommend indicators be understandable to policy makers, accurately present modeling results, and allow one to discern differences between scenarios.

3 Methods

Laramie County, Wyoming was selected for study because county planners were recently using CommunityViz to develop a build-out analysis as part of a comprehensive plan update. The county is an example of a fast-growing community with considerable development occurring beyond the urban fringe. The county seat, Cheyenne, is the state capital of Wyoming and primary node and source of service provision in the area. Growth in Cheyenne and Laramie County has been spurred by the northern expansion of the front range of Colorado, the proximity of Cheyenne to the junction of two major interstate highways, and oil and gas development.

3.1 Scenarios

Two scenarios are developed, a base scenario and an alternative land-use (Alt. LU) scenario. The base scenario captures anticipated future land uses and associated densities that are part of the current planning effort. The alternate land-use scenario attempts to ascertain if changed land use and density values can lead to lower costs of public service provision. The alternate land-use scenario represents a slight expansion of the urbanized area of the county and the establishment of agricultural (Ag) zoning where development is restricted to one dwelling unit per 640 acres.

3.2 Build-Out

Preliminary econometric modeling did not indicate the pattern of commercial or industrial development to be a statistically significant determinant of policing services expenditures. The emphasis of the build-out analysis is therefore residential rather than commercial or industrial development. Commercial development is included only to the extent needed to have a reasonable certainty that there is not an over estimation of residential development due to the complete exclusion of commercial development.

Steps in developing a build-out analysis are to (1) chose a land-use analysis layer, (2) specify density rules and (3) specify any additional development parameters based on both the regulations of the study area and the desired accuracy and level of detail of modeled output. The analysis layer used here is a grid with cells one square kilometer in size that covers the study area. CommunityViz® build-out functionality allows one to enter settings for densities and additional parameters through an interactive wizard-type series of input windows. Build-out parameters used for the two scenarios are identical and presented in Appendix I. CommunityViz offers both numeric and spatial build-out functionality. Numeric build-out calculates building capacity based on area and density rules. Spatial build-out generates data by creating a GIS layer of building points. The land-use layers used to develop the build-out analyses for the two scenarios are presented in Fig. 1.

One of the additional parameters that may be specified in a build-out model is development constraints. Constraints incorporated here include public lands

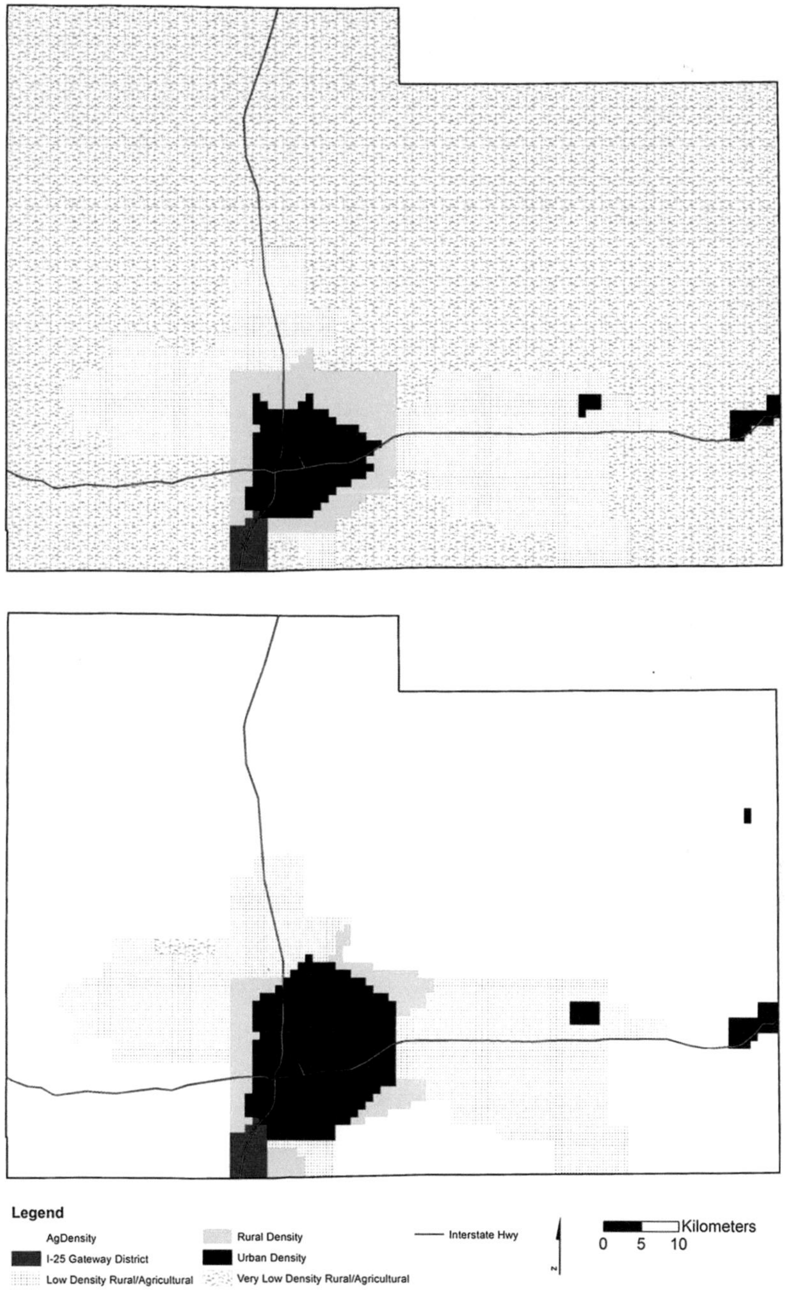

Fig. 1 Land uses for the base scenario (*above*) and alternate land-use scenario (*below*)

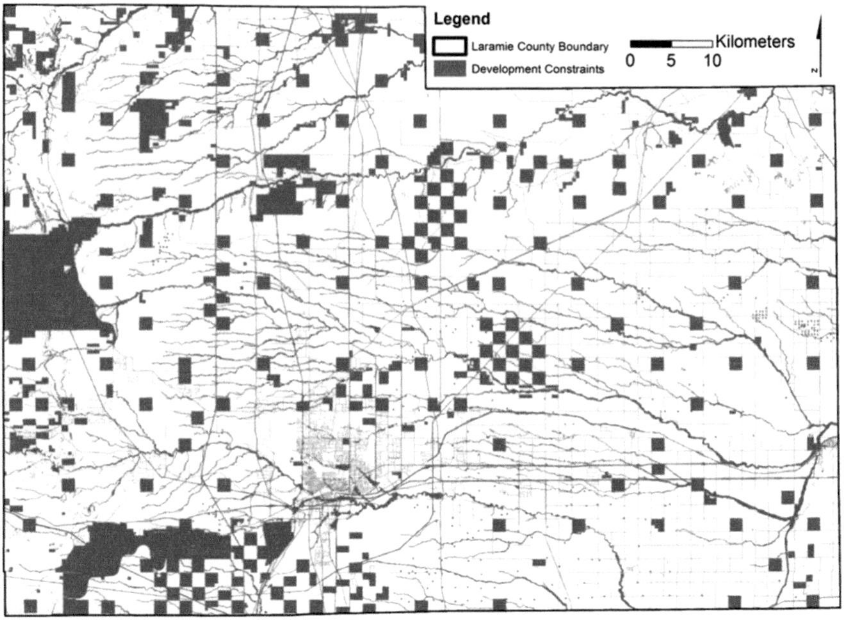

Fig. 2 Laramie County, Wyoming development constraints

managed by the U.S. Bureau of Land Management, State of Wyoming owned lands, lands under conservation easement, oil and gas wells indicated as permitted by the Wyoming Oil and Gas Conservation Commission as of April 2012 with 350 foot buffers, Western Area Power Authority power transmission lines buffered to 120 feet, county road rights of way, distributed U.S. Department of Defense sites, areas with slope greater than ten per cent, 100 year floodplains, and petroleum pipelines from the Wyoming Pipeline Authority buffered to 120 feet on each side.

As seen in Fig. 2 development constraints in the county are extensive and widely distributed. Identical constraints were incorporated in each scenario.

3.3 Spatial and Temporal Growth Modeling

CommunityViz® Timescope™ functionality was used to temporally and spatially disaggregate build-out results. Critical Timescope parameters are start and end date, growth rate and build order. Build order specifies where features are more and less likely to be built over time. The Timescope start date was set to 2010, the year of the latest data incorporated into the fiscal modeling. The Timescope end date was set to 2037 establishing a 25 year planning horizon from the 2012 update to the Laramie County comprehensive plan. Growth rate is indicative of development pressure (Theobald and Hobbs 1998) and set at two percent annually.

While census data indicate a 1.2 per cent annual increase in population and 1.6 per cent annual increase in housing units for the time period 1990–2010, Timescope will only accept integer values for growth rate. Two percent was selected rather than one percent as two percent more closely tracks the pace of residential development in the study area. Preliminary efforts at developing a spatial regression-based development suitability model were inconclusive therefore build order was specified based solely on proximity to existing residential development. Statistically, proximity to existing buildings is the best indicator of desirability (Walker and Daniels 2011), and presumably, the location of future growth. Using Timescope, build-out results were temporally disaggregated to indicate modeled future growth in 2012, 2022, 2032 and 2037. These years correspond to planning horizons relevant to the Laramie County comprehensive plan update where 2012 represents current conditions, 2022 a ten year planning horizon, 2032 a twenty year planning horizon and 2037 a twenty-five year planning horizon.

3.4 The Fiscal Model

Based on Hirsch (1970), Borcherding and Deacon (1972), Heikkila (2000) and Lieske et al. (2012) a per service aggregate expenditure function is presented as follows:

$$E = f(\mathrm{N_x},\ \mathrm{M},\ \mathrm{I}) \qquad (1)$$

where,

$\boldsymbol{E}$ represents total annual per service local government expenditures on inputs in dollars,

$\boldsymbol{f}$ (.) represents an allocation of resources for service provision,

$\mathbf{N_x}$ represents a combination of neighborhood characteristics relevant to expenditures,

M is a spatial index that captures urban form by land use; and,

I is a vector of essential inputs related to $\boldsymbol{E}$.

The dependent variable in the model is operating expenditures of the Laramie County Sheriff's department. This and the explanatory variables incorporated in the econometric model are summarized in Table 1. All data are time series covering 21 years from 1990–2010. In all cases data represent 21 observations (n = 21).

Following Lieske et al. (2012) results of the regression model may be used to calculate a spatially defined service-based impact for each grid cell in the analysis. A property tax revenue model may also be developed. By contrasting the spatially defined service-based impact with revenue for each cell, fiscal efficiency (whether a cell contributes more in tax payments than it demands in services) and inefficiency (where the modeled demand for service is greater than revenue) may be mapped in a spatially explicit fashion.

Table 1 Definition of variables

Included variables	Definition	Units
LE_{EXP}	Operating expenditures of the Laramie County Sheriff's Dept.	Millions of $USD 2010
Res	Spatial index representing residential urban form	Spatial index
Rural Pop	Population of unincorporated Laramie County	Individuals
LEO	Officers in the Albany County Sheriff's Dept.	Individuals
T	Time	Years

Extension of the base (2010) fiscal model to incorporate the results of the spatial and temporal growth modeling relied heavily on CommunityViz® dynamic attribute functionality to calculate new spatial indices for the planning horizons of interest. Dynamic attributes are formula-driven attributes of a GIS layer, much like formula-based cells in a spreadsheet, which automatically re-calculate when inputs change. After running the build-out and Timescope models, the first step in extending the expenditure model was to use a dynamic attribute to count the number of buildings in each grid cell indicated by Timescope as having been built. The sum of the building value for each cell was determined by a dynamic attribute that calculates the mean residential building value in the current or nearest cell where mean residential building value is greater than zero, multiplying this value by the count of new dwelling units in the grid cell, then adding this modeled building value to any building value from existing residences located within the cell. The next step was to calculate local Moran's *I* values for each cell based on the sum of building value within that cell. As detailed in Lieske et al. (2012), the spatial index is calculated as the average of these local Moran's *I* values for cells where the sum of building value is greater than zero. This process was repeated in order to develop a spatial index for each planning horizon. Following the governmental expenditure projection methods of Hirsch (1961), calculation of a spatial index for each planning horizon allowed *ceteris paribus* evaluation of the change in urban form in each planning horizon on expenditures.

The revenue model required similar adjustments in order to incorporate the outputs of the build-out and Timescope spatial and temporal growth modeling. The first step in expanding the revenue model was to count the number of buildings in each grid cell indicated by Timescope as having been built. Dynamic attributes where used to determine lot size and value per acre. Lot size was assigned based on zone. Value per acre was calculated similarly to mean building value where the value per acre per cell was taken from that cell if it was greater than zero or from the nearest neighboring cell if the original cell was undeveloped. Given the count of residential buildings, lot size, land value per acre and mean building value for each cell it was a simple matter of following the Laramie County assessor's formula to model assessed value. Property tax for modeled buildings was determined by multiplying assessed value per cell by the appropriate tax rate. In order to determine the property tax paid for each cell the property tax for modeled residential buildings was simply added to the property tax paid by existing residences. Spatially explicit revenue models where developed for each planning horizon.

Property tax may be contrasted with the revenue model on a cell by cell basis to estimate fiscal efficiency in a spatially explicit fashion. The annual share of overall property tax dedicated to the provision of policing services is approximately 8 % over the time period of the study. Evaluation of fiscal efficiency is accomplished by multiplying property taxes by 8 % to indicate the portion of property taxes applicable to policing services. An if/then dynamic attribute formula is used to calculate a Boolean value indicating whether or not a grid cell contributes more in revenue than the value of the service-based impact of the cell, or vice versa.

3.5 Impact Analysis

Fiscal impact evaluation is both map and indicator based. Maps may be used to indicate areas of efficient and inefficient residential development. Indicators may be used to summarize the impacts of different development patterns on revenues, expenditures, dwelling units, area impacted by efficient and inefficient development, and the spatial index of urban form. Like dynamic attributes, CommunityViz® indicators are formula driven and update automatically when inputs change.

4 Results

The mapped build-out results for 2037 shown in Fig. 3 illustrate the differing future land uses presented in Fig. 1 yield substantially different patterns of residential growth. Both scenarios result in an increase in residential development in and around the core urban area of the county, the City of Cheyenne. The alternate land-use scenario sees substantially increased residential development in the area northeast of the present municipal boundary of Cheyenne and substantially less residential development scattered throughout the remaining areas of the county. It is important to note the number of dwelling units accommodated in the alternate land-use scenario is slightly more than the base scenario in each planning horizon. In 2037, the year of maximum difference, the alternate land-use scenario allows 3,000 more dwelling units than the base scenario.

Mapped and indicator-based results for this analysis are presented for 2012, 2022, 2032 and 2037. The presentation of results across several years both helps one understand and communicate the continuous nature of temporal change and, according to Walker and Daniels (2011) generates more interest and a greater call to action than the theoretical maximum that is the more usual output of a build-out analysis.

Fig. 3 Build-out results through 2037, base scenario (*top*), alternate density (*bottom*)

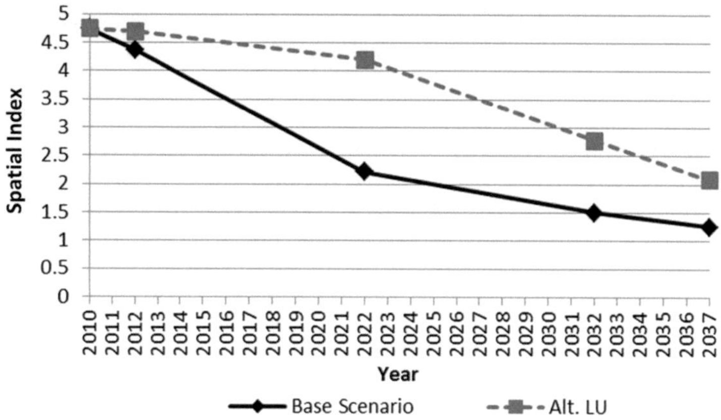

Fig. 4 Change over time of the spatial index of urban form

Urban form quantified as a spatial index allows one to quickly ascertain the changing character of the built environment across the time period of the study. Because the spatial index is based on the Moran's *I* measure of spatial autocorrelation a larger spatial index indicates greater clustering in the built environment and a lower spatial index indicates decreased clustering of the built environment. As shown in Fig. 4 changes in the spatial index over the time period of the study indicate de-clustering of the built environment over the time period of the study for both scenarios, but greater dispersion for the base scenario and relatively greater clustering for the alternate land-use scenario.

Results of the econometric modeling (Table 2) show a statistically significant relationship between urban form as quantified using the spatial index and law enforcement expenditures. Urban form is specified with standard and the quadratic form of the spatial index (Res and Res^2). The sign on the standard form of the spatial index is negative, indicating that as the built environment becomes more dispersed expenditures on policing services increase. The positive relationship between expenditures and the square of the spatial index demonstrates that as clustering decreases expenditures increase at an increasing rate. The positive sign on rural population indicates expenditures increase commensurately with population.

Table 2 Regression model coefficients

Parameter	Coefficient	Standard error	t	P	Lower 95 % CI	Upper 95 % CI
Constant	17.256	5.49	3.14	0.007	5.48	29.03
Res	−6.802	2.097	−3.24	0.006	−11.3	−2.3
Res^2	0.522	0.126	4.15	0.001	0.25	0.79
Rural population	0.000	0.000	1.06	0.307	−0.000	0.000
Officers	0.063	0.29	2.17	0.047	0.001	0.124
$Officers^2$	−0.000	0.000	−2.17	0.048	−0.001	−0.000
Time	−0.135	0.213	−0.62	0.543	−0.599	0.33

The positive sign on officers and negative sign on officers squared indicates that increasing the number of law enforcement officers increases expenditures at a decreasing rate.

Maps indicating residential areas of Laramie County that are efficient and inefficient for the provision of policing services for the planning horizons of the study are presented in Fig. 5. Looking at the maps in Fig. 5 and the number of efficient (Fig. 6a) grid cells, it is apparent both scenarios result in increases in fiscally efficient cells around the large urban center of the county, Cheyenne. Although, there are more grid cells that are fiscally efficient in the alternate land-use scenario than in the base scenario across the time period of the study. The effect of the Agricultural District zoning can be seen in the number of inefficient

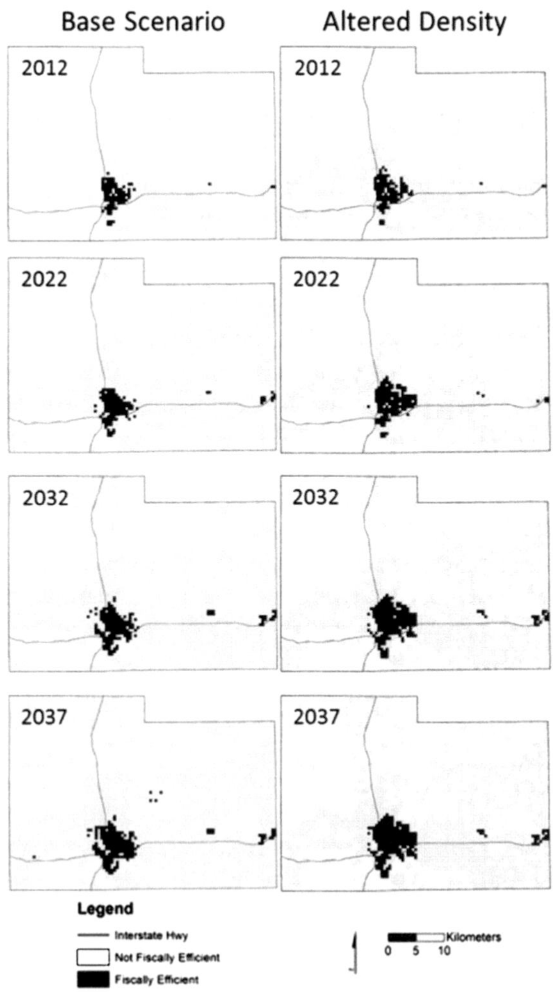

Fig. 5 Modeled fiscal efficiency for the provision of county policing services by year: 2012–2037

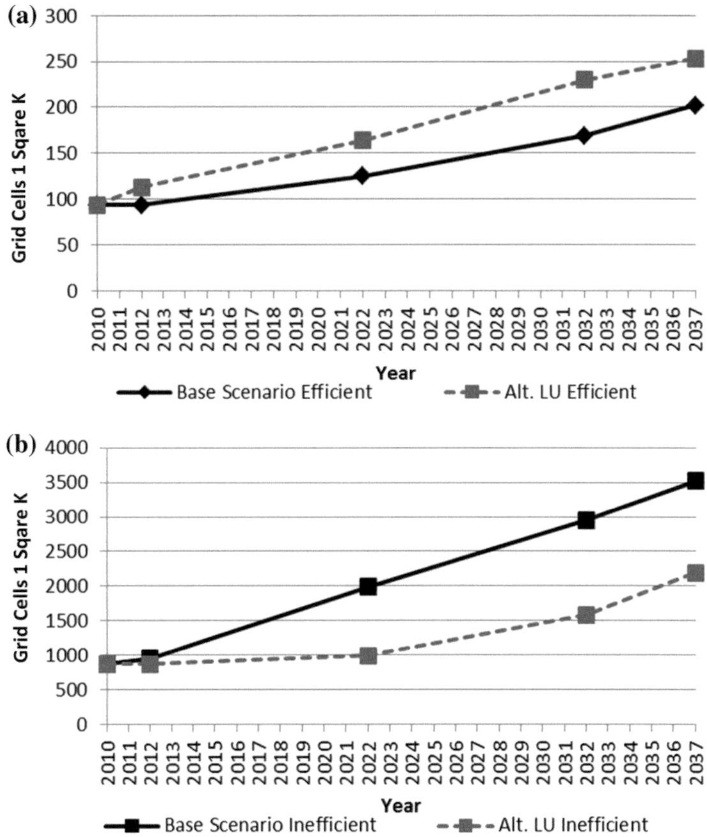

Fig. 6 **a** Numbers of efficient one square kilometer grid cells in each scenario. **b** Numbers of inefficient one square kilometer grid cells in each scenario

cells (Fig. 6b) which is considerably less in the alternate land-use scenario than the base scenario.

In addition to the mapped outputs and counts of efficient and inefficient grid cells, fiscal impacts may be gauged by looking at the revenues and expenditures associated with each scenario. As shown in Fig. 7 expenditures for the base scenario increase both more rapidly and to an overall higher annual level than for the alternate land-use scenario. Revenues for both scenarios remain approximately equal throughout the time period. While both scenarios indicate residential property taxes allocated to policing services are insufficient to cover the cost of policing services, the alternate land-use scenario requires smaller expenditures and therefore performs better over the course of the modeling than the base scenario.

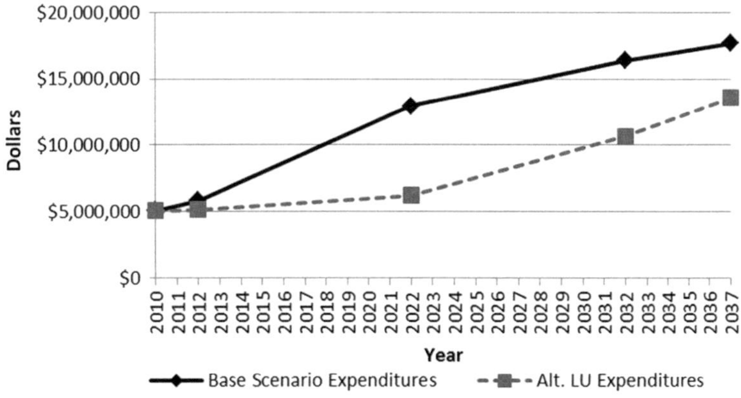

Fig. 7 Modeled annual expenditures associated with the provision of county policing services

5 Discussion and Conclusion

This chapter demonstrates an econometric model of public service expenditure that incorporates urban form can be linked with a growth model in such a way that produces plausible results for public service expenditures. Fiscal modeling and growth modeling are linked via land use and density specifications that are key components of the regulatory and policy framework of the study area. These links make clear the direct connection between those frameworks and potential future expenditures on public services. Results, both mapped and numeric, are in turn useful for local government planning and decision making.

In a discussion of land resource economics, van Kooten (1993) posits that demonstrating inefficiency is insufficient justification for planning; planning must lead to efficient land use. Given that benchmark, neither the base or alternate land-use scenario meets the standard needed for government intervention to bring about improved outcomes. The fact that neither scenario closes the gap between revenues and expenditures leads to the question if there is a way to further optimize results. What the scenarios suggest is further research on the relationship between urban form and public service expenditures to see if urban form can be manipulated to bring about substantially greater efficiencies in the costs of public service provision.

From the benchmark provided by van Kooten it would be desirable to see, for example, if one could realistically model land use and density in a such a way that results in fiscal sustainability, where the gap between revenues and expenditures closes rather than widens as time passes. The econometric model results suggest it is necessary to alter urban form in such a way that it increases rather than decreases the spatial index. This would likely require further expanding the urban area, increasing residential densities in the urban core and the areas surrounding the urban core in the county, and decreasing residential densities in the more outlying areas. Theobald and Hobbs (1998) note that land-use change models are noticeably affected if "...even small clusters of moderate density are at some distance from an urban center" (p. 68). This is likely the case here. In addition to there being trade-offs between development in the urban core of the county and outlying areas there may be tradeoffs associated with levels of development within higher density clusters of development, specifically the two developing clusters of higher density indicated as fiscally efficient that lie east of Cheyenne as shown on the maps in Fig. 5.

The modeling highlights a number of features of PSS including scenario planning, build-out, spatial and temporal data disaggregation and the quantitative assessment of planning alternatives with indicators. Given the largely irreversible nature of changes to the built environment, PSS enable an inexpensive and low-risk opportunity for ex-ante evaluation of planning regulations and policies. Another potential benefit of this modeling approach is that it could be incorporated into the day to day permitting work of a planning office. Having defined the relationship between trends of change in urban form and change in county revenues and expenditures for policing services over time it is a small leap to consider site specific additions to model inputs and resultant changes in model output. Furthermore, while the focus of this discussion is development of optimal solution(s) to a planning problem, it is noted there are no restrictions on integrating model outputs with a public engagement planning process in order to incorporate public values in fiscally sustainable planning and decision making.

Among the broader impacts of this work are support for spatial planning, spatial modeling and PSS adoption generally through promotion of the benefits of PSS using demonstration projects (Vonk et al. 2005) that meet the demands of planners and successfully address standard and desirable planning tasks (Vonk and Geertman 2008). By embedding the approach within an existing planning and regulatory framework and producing plausible results, this research may help address PSS bottlenecks including lack of awareness of the potential applications of PSS (Geertman and Stillwell 2009; Vonk et al. 2005). The forward looking nature of the project also is well aligned with Klosterman's (2009) mention of the normative effort to move the planning profession toward planning and away from administration. Looking forward, in addition to further refining data inputs and modeling procedures, there are opportunities for research in demonstrating the positive communication, group cognitive, process impact and business case benefits of PSS and scenario planning for local government fiscal sustainability.

Acknowledgments This work was supported by the United States Department of Agriculture National Institute of Food and Agriculture.

Appendix 1: Build-out Settings for Both Scenarios

Polygon layer containing land-use information			LaramieGrid1 K		
Attribute specifying land-use designation			Zone		
Unique identifyer			OBJECTID		
Density rules					
	Dwelling units			FAR	
	Quantity	Measurement	Equivalence	Quantity	Measurement
AgDensity	0.011563	DU per acre	1 DU/640 acres	0	FAR
I-25 gateway district	1	DU per acre	1 DU/1 acre	0.25	FAR
Low density/agricultural	0.02857	DU per acre	1 DU/35 acres	0.1	FAR
Rural density	0.1	DU per acre	1 DU/10 acres	0.1	FAR
Urban density	1	DU per acre	1 DU/1 acre	0,1	FAR
Very low density rural/agricultural	0.02857	DU per acre	1 DU/35 acres	0.1	FAR
Mixed-use land are percentages	Dwelling units		Floor area		
Designation	Percent	Measurement	Percent	Measurement	
I-25 gateway district	25	1 DU per acre	75	0.25 FAR	
Low density/agricultural	97	0.02857 DU peracr	3	0.1 FAR	
Rural density	98	0.1 DU per acre	2	0.1 FAR	
Urban density	93	1 DU per acre	7	0.1 FAR	
Very low density rural/agricultural	93	0.02857 DU per acre	7	0.1 FAR	
Efficiency					
Designation	Efficiency percent				
AgDensity	95				
I-25 gateway district	77				
Low density/agricultural	100				
Rural density	90				
Urban density	77				

(continued)

(continued)

Very low density rural/agricultural	100						
Building information designation	DU per Building	Building footprints area sq feet		Floors			
AgDensity	1	0		1			
1-25 gateway district	1	39,204		1			
Low density/agricultural	1	47,916		1			
Rural density	1	47,916		1			
Urban density	1	43,560		1			
Very low density rural/agricultural	1	47,916		1			
Constraints	constraints11_floodplains_final_prj						
	Layer			Value or attribute specifying DU/bldg			
Existing buildings	Address_modified	1					
	Value or attribute specifying floor area 0						
Spatial layout			Minimum Separation Distance				Setback
Designation			feet		Layout pattern	Road or line layer	feet
AgDensity			60		Random	Streets	30
1-25 gateway district			60		Random	Streets	30
Low density/agricultural			60		Random	Streets	30
Rural density			60		Random	Streets	30

(continued)

(continued)

Urban density	60	Random	Streets	30
Very low density rural/agricultural	60	Random	Streets	30
Spatial buildings	Building type		Footprint size	
Designation				
AgDensity	Points		0	
I-25 gateway district	Points		39,204	
Low density/agricultural	Points		47,916	
Rural density	Points		47,916	
Urban density	Points		43,560	
Very low density rural/agricultural	Points		47,916	
End of spatial phase				
Set the number of times buildout will attempt to place a random building				18

References

Avin U, Dembner JL (2001) Getting scenario-building right. Planning 67(11):22–27

Borcherding TE, Deacon RT (1972) The demand for services of non-federal governments. Am Econ Rev 62:891–901

Burchell RW, Listokin D (1978) The fiscal impact handbook: estimating local costs and revenues of land development. Center for Urban Policy Research, Brunswick, New Jersey

Davenport D (1926) An analysis of the cost of municipal and state government and the relation of population to the cost of government, net taxable income and full value of real property in New York State. J.B. Lyon Company, Albany, New York

Fabricant S (1952) The trend of government activity in the United States since 1900. National Bureau of Economic Research

Geertman S, Stillwell J (2009) Planning support systems: content, issues and trends. In: Geertman S, Stillwell J (eds) Planning support systems: best practices and new methods. Springer Science and Business Media B.V, Netherland, pp 1–26

Goetz SJ (2007) The economic case for state-level land use decision-making. J Reg Anal Policy 37(1):20–24

Hanna KC, Culpepper BR (1998) GIS in site design. Wiley, New York

Heikkila E (2000) The economics of planning. Center for Urban Policy Research, New Brunswick, New Jersey

Hirsch WZ (1961) Projections: governmental expenditures, economic activity and population. In: Bollens JC (ed) Exploring the metropolitan community. University of California, Berkeley, pp 368–388

Hirsch WZ (1970) The economics of state and local government. McGraw-Hill, Inc., New York

Klosterman R (2009) Planning support systems: retrospect and prospect. In: S Geertman, J Stillwell (eds) Planning support systems: best practices and new methods. Springer Science and Business Media B.V, Berlin, Germany, pp v–vii

Lieske SN, McLeod DM, Coupal R, Srivastava S (2012) Determining the relationship between urban form and the costs of public services. Environ Plan B 39(1):155–173

Snyder K (2003) Tools for community design and decision-making planning support systems in practice. In: Geertman S, Stillwell J (eds) Planning support systems in practice. Springer, Heidelberg, pp 99–120

Theobald DM, Hobbs NT (1998) Forecasting rural land-use change: a comparison of regression- and spatial transition-based models. Geogr Environ Model 2(1):65–82

Tischler PS (2002a) Analyzing the fiscal impact of development. TischlerBise, Bethesda, Maryland

Tischler PS (2002b) Fiscal impact analysis, reader beware: some Caveats. TischlerBise, Bethesda, Maryland

Van Eck JR, Koomen E (2008) Characterising urban concentration and land-use diversity in simulations of future land use. Ann Reg Sci 42:123–140

van Kooten GC (1993) Land resource economics and sustainable development. University of British Columbia Press, Vancouver

Veeneklaas FR, van den Berg LM, Schoonenboom IJ, van de Klundert AF (1995) Scenarios as a tool. In: Schoute J, Finke PA, Veeneklaas FR, Wolfert HP (eds) Scenario studies for the rural environment. Kluwer Academic Publishers, Dordrecht

Vonk G, Geertman S (2008) Improving the adoption and use of planning support systems in practice. Appl Spat Anal 1:153–173

Vonk G, Geertman S, Schot P (2005) Bottlenecks blocking widespread usage of planning support systems. Environ Plan A 37(5):909–924

Walker D, Daniels T (2011) The planners guide to communityViz: the essential tool for a new generation of planning. The Orton Family Foundation, Middlebury, Vermont

Xiang W, Clarke K (2003) The use of scenarios in land-use planning. Environ Plan B 30:885–909

Part II
Environmental Planning and Modelling

Generalisation of Planning Data as a Contribution to Strategic Environmental Assessments (SEA): The Example of a City-Wide Biotope-Type Assessment for Berlin

Michael Förster, Antje Köppen, Johanna Ferretti, Johann Köppel and Birgit Kleinschmit

Abstract GIS-based Strategic Environmental Assessments (SEA) generally require the processing of a variety of environmental and planning data. Therefore, it is necessary to combine thematic datasets in different spatial resolutions. This chapter presents a method for the generalization of a city-wide biotope-type assessment and an urban structure and inventory of open and green spaces map for the city of Berlin. The method includes the aggregation of an ecological assessment of a biotope-type map to a more generalized block-based inventory of open and green spaces map by areal weighting. The results show that the combination of the two datasets can be very sufficiently utilized to detect areas of high ecological significance in early stages of environmental planning. The application of a 1 ha minimum mapping unit leads to a better integration of biotope-type information into the block structure than the utilization of a 3 ha-size criterion.

M. Förster (✉) · A. Köppen · B. Kleinschmit
Geoinformation in Environmental Planning Lab, Technische Universität Berlin,
Str. d. 17. Juni 145 10623 Berlin, Germany
e-mail: michael.foerster@tu-berlin.de

A. Köppen
e-mail: antje.koeppen@tu-berlin.de

B. Kleinschmit
e-mail: birgit.kleinschmit@tu-berlin.de

J. Ferretti · J. Köppel
Environmental Assessment and Planning Research Group, Technische Universität Berlin,
Str. d. 17. Juni 145 10623 Berlin, Germany
e-mail: jferretti@zedat.fu-berlin.de

J. Köppel
e-mail: johann.koeppel@tu-berlin.de

S. Geertman et al. (eds.), *Planning Support Systems for Sustainable Urban Development*, Lecture Notes in Geoinformation and Cartography,
DOI: 10.1007/978-3-642-37533-0_9,

1 Introduction

The Strategic Environmental Assessment (SEA) Directive was introduced in the European Union by the European Commission in 2001 (EC 2001). Since then, the SEA has developed into a key instrument for sustainable development (e.g., Dalay-Clayton and Sadler 2005; Jones et al. 2005). This European Directive requires an environmental assessment of the effects of formal plans and programs which set a framework for subsequent planning levels (Therivel 2004). In terms of the practical implementation of the SEA tiering is an important aspect. Tiering can be defined as avoiding duplication of thematic information in succeeding assessments of policies, plans, programmes and projects. Once an issue has been assessed at a higher planning level it is not required to be considered at a lower level, other than perhaps to provide essential detail not provided in the prior assessment (Jones et al. 2005). SEA studies involve the consideration of many different environmental variables presenting complex interrelationships, which vary in time and space. Therefore methods and techniques in such strategic assessments should aim at simplifying the frequently complex issues under consideration at strategic decision-making levels (Fischer 2007). In practice the wide acceptance of Geographic Information Systems (GIS) applications in environmental studies has permitted the development of spatially explicit approaches (Gonzales Del Campo 2008; Marull et al. 2007; Vanderhaegen and Muro 2005).

Particularly for the SEA as a cross-sectional environmental assessment instrument, normally a huge amount of environmental data is needed to develop a GIS-based spatially explicit method for analyses. For supporting informed decisions, data quality and scale are key considerations. Although the utilization of GIS data is extensive, in practice several barriers limit the widespread use of spatial data. A main obstacle is the difficulty to integrate data from different sources because of their heterogeneities (Gonzales Del Campo 2008; Kieler et al. 2007; Vanderhaegen and Muro 2005), because of:

- Syntactic heterogeneities—caused by different data models (relational or object oriented) or geometric models (grid or vector),
- Schematic heterogeneities—caused by different conceptual data models (for example, different generalization hierarchies in regard to faunistic mappings such as grid or point based mappings),
- Semantic heterogeneities—caused by differences in the meaning, interpretation and usage of identical or similar information (chemical water quality as an indicator for the environmental assets fauna and water etc.).

To overcome those heterogeneities the INSPIRE (INfrastructure for Spatial InfoRmation) Directive (2007/2/EG) aspires to establish an infrastructure for spatial information in the European Community for the purposes of community environmental policies or activities which may have an impact on the environment (EC 2007). It aims at allowing users to identify and access spatial or geographical information from a wide range of sources in an interoperable way for a variety of uses (Gonzales Del Campo 2008; Vanderhaegen and Muro 2005). The INSPIRE Directive advantageously affects the integration of environmental data for the further development of the SEA.

For developing a city-wide strategic planning and decision support instrument it is necessary to combine various sources of planning data of different scales and attributes, a challenge usually addressed by disciplines such as spatial statistics, where spatial aggregation methods are necessary to create meaningful units in (geo)statistical analyses (Flowerdew and Green 1992; Goodchild and Lam 1980; Gotway and Young 2002). Therefore, different methods were developed to combine data collected within differing zonal systems. If data for a specific variable is available for one zonal system but is required for use in a different one, it can be described as a source unit which needs to be transposed to a target unit by areal interpolation (Flowerdew and Green 1989). Through those areal interpolation methods different data can be combined within a single zonal system and jointly analyzed. A typical form of areal interpolation is based on weighting by area. The averaging method assigns the average value of the original units to the newly aggregated unit. The weights are derived from the proportion of the target unit overlapping each source unit. The implicit assumption in this method is that the variable of interest is evenly distributed in the source unit, which is often not the case in reality (Flowerdew and Green 1989, 1994; Goodchild and Lam 1980).

The aim of this chapter is to develop an aggregation method for the generalization of detailed planning data for enhancing the use of this data in large-scale analyses like SEAs using the example of the city of Berlin (see Fig. 1). To that purpose, an aggregation method was developed for a "biotope-type assessment" (BTA), which is based on the city-wide "biotope-type map" (BTM, see SenStadt 2009), shown in Fig. 2. The BTM focuses on the classification of habitat types containing biodiversity-based information. In contrast, the "urban structure and inventory of open and green spaces map" (USM) focuses on the classification of different land use types in general. Both maps complement one another and can contribute useful information for a SEA. Therefore, this chapter presents a method that aggregates both information types and potentially enhances their usability for strategic planning in Berlin.

2 Data and Methods

2.1 Data

To evaluate the effects of planning data generalization, two different data sets representing the current land use for the whole city of Berlin were selected. The biotope-type mapping (BTM) with a scale of 1:5,000 represents the "habitats in which certain plants and animals form a long-term association" (SenStadt 2009). It is based on primary data gathered by detailed field mapping and aerial image

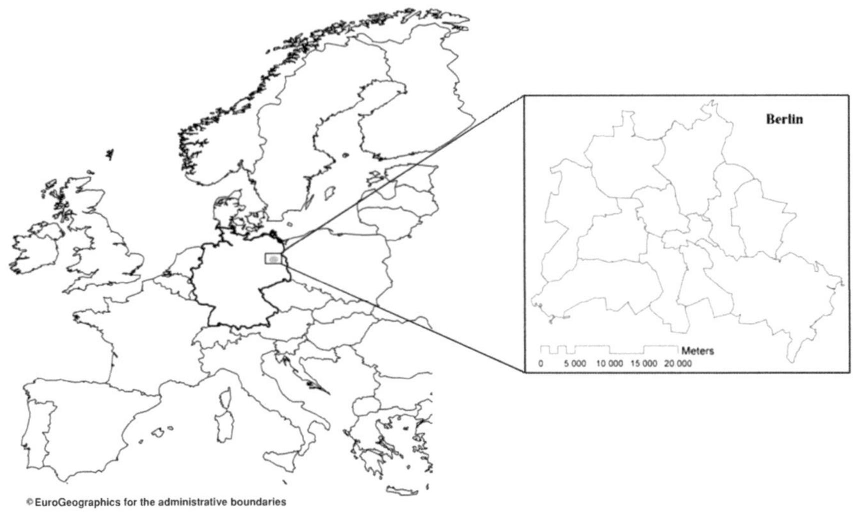

Fig. 1 Study area Berlin

Fig. 2 Assessment matrix for the ecological significance of biotope-types (extraction) with assessment map (BTA—*left*) and block structure of the urban structure and inventory of green and open spaces map (USM—*right*)

interpretation as well as on pre-existing geodata such as urban structure, green and open space. The reference units for the primary data are habitat boundaries, and are predominantly statistical block structures for the secondary data. The biotope-type map consists of around 59,000 polygons (SenStadt 2009).

For the subsequent integration of the BTM into the USM, it was necessary to simplify the complex BTM information. Thus, a SEA Assessment Framework was used. Within this framework, the BTM was reclassified to create the BTA, which is restricted to assessment classes relevant to a SEA. The BTA distinguishes a restriction area, indicating areas with legally binding criteria or restrictions (e.g., protection areas), and four precautionary areas (I–IV, see Fig. 2), representing the environmental quality for each environmental asset according to Directive 2001/42/EC (Ferretti et al. 2009; Koeppen et al. 2010).

The second data-set is the "urban structure and inventory of open and green spaces map" (USM). The USM is a combination of the urban structure for residential areas and an inventory of green and open spaces from 2008, and has a scale of 1:50,000 (SenStadt 2008a, b). The reference unit is the statistical block structure, and the data set consists of more than 25,000 statistical blocks and block segments for the whole city (Schneider et al. 2007; SenStadt 2010; Welsch 2008).

3 Areal Weighting with Aggregation Rules

The first step for the aggregation of the BTA into the USM was areal weighting (see Fig. 3). Areal weighting is a method assigning the average value of the original zones (in this case the BTA) to new zones (in this case the block structure of the USM). Hence, the areal weighting aggregates the assessments of intersecting biotope-types within a block to a mean value. By that, a summed-up value of the ecological significance for each single block is calculated.

By just including the areal weighting as described above, the aggregation method considers only the block structure of the USM as the target zone system. This is appropriate for the intended strategic scale level of 1:50,000. However, in some parts the method is too coarse for an assessment of ecological significance. Especially in the less-developed boundary area of Berlin some large blocks show a size of over 1,000 ha. These blocks are often characterized by large agricultural areas with few structures. Nevertheless, the BTA can have a high variance of polygons of different ecological significance within the coarser block polygons of the USM. In these cases of over-generalization, aggregation rules were developed to sub-divide USM polygons in order to integrate additional BTA information into the USM statistical blocks.

The first criterion to identify statistical blocks where the original biotope-type geometry should be integrated is in cases where an absolute deviation of two or more categorical assessment steps between a single biotope-type assessment and the block-based mean value of the assessment occurs. For example, if the main part of a block is characterized by habitats with a low ecological significance assessed as precaution area III (with a mean of 2.3 for the total block) and the same block also contains one smaller polygon with a biotope-type of high ecological significance assessed as restriction area (5.0), this information will be integrated into the USM (thematically and geometrically). Without this additional step, the information about the existence of that exceptional habitat within that strategic level block would only be considered in more detailed planning stages. Therefore, a comparison of the calculated block-based mean values with the original BTA was accomplished by an overlay analysis.

The second criterion for the inclusion of polygons from the BTA was the biotope size. It is not suitable to integrate a very small habitat with a size of only a

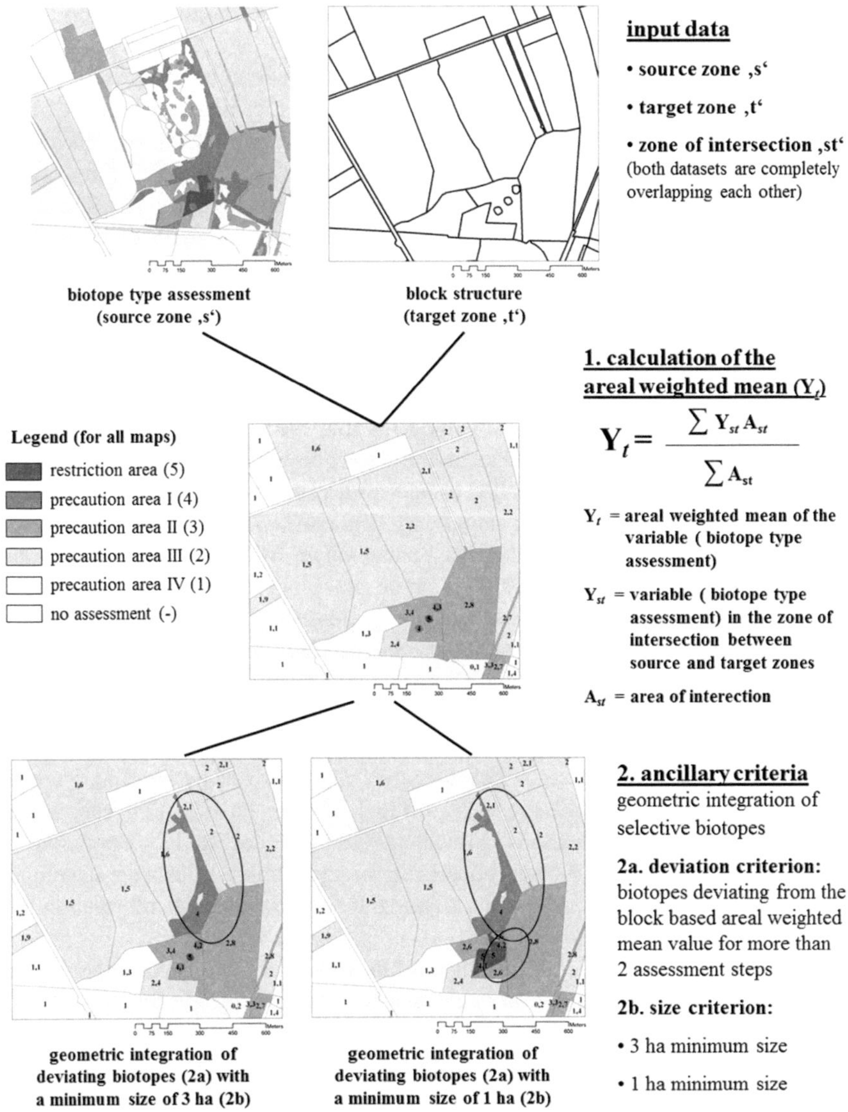

Fig. 3 Flow chart of the developed aggregation method (calculation formula for the areal weighted mean according to Flowerdew and Green 1994)

few square meters into the block structure for a city-wide analysis even if a sufficient deviation exists. Therefore, two different minimum sizes were determined and the planning applicability for each size was evaluated. First, a minimum size of 3 ha was considered due to the fact that this is a minimum size for the illustration of relevant city-wide plans in Berlin, such as the Land Use Plan or the Landscape Program (SenStadt 2007). Next, a minimum size of 1 ha was tested in

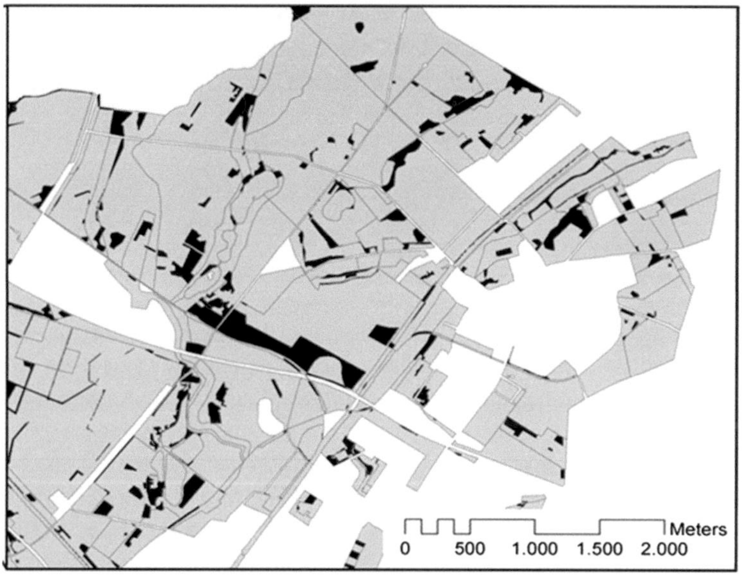

Fig. 4 Allocation of detected 4,500 blocks (*grey*) subdivided by deviating biotopes (*black*)—example from the north-east of Berlin

recognition of the fine granularity of the BTA as well as to increase the amount of useful biotope-type information for integration into the USM. For both results the biotope-types deviating from the block-based mean by two or more assessment steps (see Fig. 4) were analyzed according to their size.

By including these aggregation rules, relevant biotopes were able to be integrated into the block structure of the USM. However, the biotope-type geometry is not perfectly aligned with the statistical block geometry of the USM, which can cause sliver polygons. Therefore, resultant sliver polygons were assigned to neighbouring biotopes with the largest shared border.

4 Results

After implementing the described aggregation method, the results were analyzed for the whole city of Berlin. Altogether about 9,000 biotopes were detected which deviate from the block-based mean by at least two assessment steps. These biotopes are located in approximately 4,500 USM statistical blocks. These blocks are predominantly situated in the border area of the city of Berlin, where the average polygon size of the USM is higher and the BTA in contrast can be very detailed (see Fig. 4 for an example). The detected USM polygons have an average size of 6.9 ha. The deviating biotopes within these blocks on the contrary have an average size of only 0.1 ha. Analyzing the range of ecological significance in five

Table 1 Range value and ancillary statistic values of detected blocks with deviating ecological significance of occurring biotopes according to Fig. 2

Range value	Number of blocks	In percentage of total	Area in ha	Average area in ha
2	1382	31	5838	4
3	1552	34	7950	5
4	1123	25	9687	9
5	382	9	6146	16
6	45	1	1580	35
	$\sum$ 4484	$\sum$ 100	$\sum$ 31200	Ø 6.9

assessment steps plus the category of no assessment (all in all six different assessment categories) of the biotopes occurring in these blocks, 90 % of the detected blocks contain biotopes with a differing ecological significance of two through four assessment steps (see Table 1). Analyzing the average size of these blocks in relation to the range values it is obvious that the average size is increasing together with the range of ecological significance. Blocks with a range value of five show an average size of 16 ha, and blocks with a range value of six have up to 35 ha.

Due to the small size of the biotope-types deviating from the block-based mean for two or more assessment steps, the 4,500 generated blocks were filtered by the two size criteria. For the minimum size of 3 ha only 29 biotopes were detected, which were integrated into 24 USM blocks. For the minimum size of 1 ha in contrast, 277 polygons were identified, which were integrated into 201 blocks (see Fig. 5a, b). The minimum size evaluations show that most of the biotopes deviating from the block-based mean value for more than two assessment steps are too small for a meaningful geometric integration on a city-wide scale.

For the verification of the different integration results it is also important to analyze the semantic dimension of the integration process. Therefore, the thematic

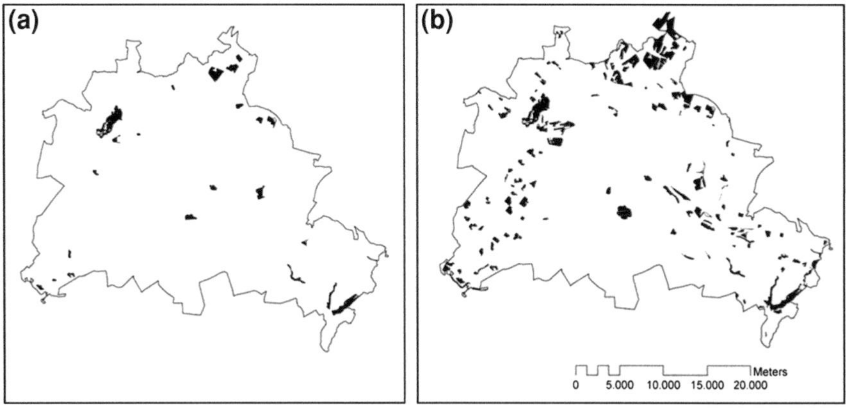

Fig. 5 **a** Blocks with an integration of detected biotope-types with 3 ha, **b** Blocks with an integration of detected biotope-types with 1 ha

attributes from both the BTA and USM data sets within the detected polygons were evaluated. Table 2 shows the classification of the identified 24 blocks where BTA polygons were integrated into the USM block zonation under the 3 ha minimum criterion. It can be stated that most of the subdivided blocks are categorized as water, farmland and vacant areas. The average size of more than 50 ha of the divided blocks indicates that predominantly very large blocks are affected. Analyzing the biotopes which subdivided the USM blocks, it can be asserted that mostly forests, standing waters and fields were integrated into the USM map. Table 4 shows the similar evaluation of subdivided polygons for the 1 ha minimum criterion. It can be seen, that blocks with the USM category "forest" are especially affected using this more detailed evaluation. Analyzing the deviating biotopes with a size of more than 1 ha (see Table 5), it can be stated that almost 150 biotopes of the total 277 integrated biotopes belong to the categories forests and green spaces. This classes show an average size of about 1.7 ha and a total size of 260 ha. This is about 50 % of the total area of the integrated biotopes. Generally it can be observed that the variety in both the USM and BTA classifications of the subdivided blocks ascends in relation to the reduction of the minimum size criterion. Noting the average assessment categories of the integrated biotopes according to Fig. 2 in relation to the frequency in Tables 3, 4, 5 it can be seen that there is no correlation. It can be assumed that biotopes with both high (e.g., forests) as well as low (e.g., built-up areas) ecological significance are likely to be integrated into the block structure. This can be explained by the more quantitative and assessing character of the criteria applied, especially regarding the BTA.

In Tables 2, 4 the calculated average mean BTA values of the affected blocks before and after the geometric integration of the deviating biotopes are illustrated as well as the difference between both values. For both size criteria (1 and 3 ha) a marginal decrease is visible. Thus, the effect of the geometric integration of the biotopes on the block-based mean value of the ecological significance is not very significant. In total, biotopes with a higher ecological significance were integrated into these blocks due to the general decrease of the block-based mean value. No change in assessment category can be detected on average for any of the affected

Table 2 Blocks subdivided by biotopes deviating from the block-based mean value for more than two assessment steps with a size of more than 3 ha classified by the USM

USM classes	Number of blocks	Average area in ha	Sum area in ha	Mean value before integration	Mean value after integration	Average mean difference
Water	6	128.30	769.82	3.66	3.88	0.21
Farmland	5	78.15	390.75	1.86	1.86	−0.01
Vacant areas	5	32.11	165.49	2.56	2.56	0.00
Forest	3	18.22	50.49	3.40	3.30	−0.10
Public facilities	2	38.22	76.44	1.97	1.81	−0.15
Parks/Green spaces	2	54.35	108.70	2.59	2.33	−0.25
Sports Facilities	1	23.54	23.54	2.56	2.36	−0.21
Total	∑ 24	Ø 53.27	∑ 1585.24	Ø 2.66	Ø 2.58	Ø −0.08

Table 3 Main biotope-type categories of the BTA deviating from the block-based mean value for more than two assessment steps with a size of more than 3 ha

Main categorization of biotope-types	Number of biotopes	Average assessment category	Average area in ha	Sum area in ha
Forests	7	4.9	3.77	26.37
Standing waters	6	3.3	7.53	45.17
Fields	5	1.8	4.84	24.18
Green spaces, herb fringe fields and grassland	4	2.5	3.49	13.98
Bushes, tree rows and groves	3	4.0	6.96	20.88
Built-up areas, traffic facilities	2	1.0	4.31	8.61
Flowing waters	1	1.0	8.94	8.94
Anthropogenic regosol sites and ruderal fields	1	3.0	3.09	3.09
Total	$\sum$ 29	Ø 3.07	Ø 6.13	$\sum$ 151.21

Table 4 Blocks subdivided by biotopes deviating from the block-based mean value for more than two assessment steps with a size of more than 1 ha classified by the USM

USM class	Number of blocks	Average area in ha	Sum area in ha	Average mean value before integration	Average mean value after integration	Average mean difference
Forest	76	20.31	1590.8	2.84	2.66	−0.18
Vacant areas	35	19.21	690.88	2.52	2.55	0.03
Farmland	22	43.93	966.47	1.91	1.92	0.00
Public facilities	16	22.23	355.70	2.34	2.26	−0.07
Water	16	65.58	1049.3	3.31	3.46	0.15
Parks/Green spaces	12	35.36	424.27	2.90	2.77	−0.13
Low development with predominantly small business and industrial use	8	15.25	121.98	1.96	1.60	−0.35
Meadows and pastures	4	19.85	79.39	2.84	2.77	−0.07
Sports facilities	3	13.72	41.17	2.79	2.19	−0.60
Cemeteries	2	16.66	33.32	2.83	3.00	0.17
Low buildings with yards	2	15.55	31.11	2.37	1.98	−0.38
Allotment gardens	1	6.96	6.96	3.33	2.87	−0.46
Dense development with predominantly small business and industrial use	1	10.09	10.09	1.24	1.04	−0.20
Little or non-built-up green and open spaces	1	60.57	60.57	1.79	1.56	−0.23
Postwar high-rise development	1	12.91	12.91	2.00	1.53	−0.47
Horticulture	1	10.75	10.75	1.57	1.27	−0.30
Total	$\sum$ 201	Ø 24.3	$\sum$ 5485.7	Ø 2.4	Ø 2.2	Ø −0.2

Table 5 Main biotope-type categories of the biotope-type mapping deviating from the block-based mean value for more than two assessment steps with a size of more than 1 ha

Main categorization of biotope-types	Number of biotopes	Average assessment category	Average area in ha	Sum area in ha
Forests	91	4.57	1.74	158.14
Green spaces, herb fringe fields and grassland	61	2.30	1.68	102.66
Built-up areas and traffic facilities	28	1.07	1.70	47.53
Anthropogenic regosol sites and ruderal fields	26	1.69	1.53	39.73
Bushes, tree rows and groves	22	3.82	2.36	51.96
Standing waters	17	3.88	3.77	64.11
Fields	13	1.46	2.78	36.20
Rivers and streams	11	1.36	2.36	25.93
Bogs and marshes	3	5.00	1.32	3.95
Green and open spaces	3	1.00	1.22	3.67
Dwarf shrub heaths	1	5.00	1.01	1.01
Special biotopes	1	1.00	1.49	1.49
Total	$\sum$ 277	Ø 2.68	Ø 1.91	$\sum$ 536.37

classes in the USM, apart from the type "sports facilities" for the one-hectare-size criterion (see Table 4). Here, a decrease of the average mean value can be detected. Before integrating deviating biotopes, the value was 2.79, which corresponds to the assessment category precaution area II, and after the integration the value changed to 2.19, which corresponds to the assessment category precaution area III (see Fig. 2).

Analyzing the amount of legally protected biotopes which were detected for both sizes, it can be stated that for the 3 ha minimum mapping unit a total of 13 biotopes were protected according to the Federal Nature Conservation Act and 8 biotopes under the Habitats Directive. For the 1 ha minimum mapping unit 118 biotopes were protected according to German law, and 75 areas were protected under the Habitats Directive.

The comparison of the city-wide areal distribution of all assessment categories for the three aggregation steps in Fig. 6 shows an almost equal distribution. The highest deviation can be noticed for precaution area IV. This is caused by the fact that with over 30,000 ha, most habitats in the original biotope-type assessment were categorized as precaution area IV. Thus, due to the calculation method of the block-based mean value, the approximation of this assessment category in relation to precaution area III is very high.

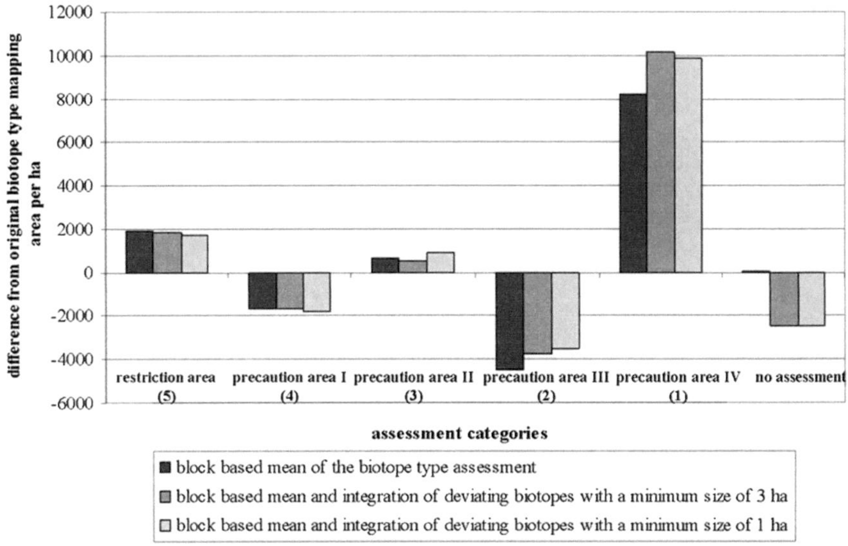

Fig. 6 Difference of the area (in ha) for each assessment step from the original biotope-type mapping

5 Discussion

In this study an aggregation method for the generalization of over-detailed planning data, in this case a biotope-type assessment, is described. Thereby the method includes an aggregation of a biotope-type map-based assessment of ecological significance to the coarser block-based inventory of the open and green spaces map by areal weighting. Additionally, the ancillary criteria assessment deviation and two different minimum sizes were applied for the integration of certain biotopes into block structure as a target zoning system.

The application of the presented aggregation method for planning data is fairly new in planning processes. It is more established in disciplines such as geostatistics, where overcoming different zoning systems between various data sets has always been a common challenge. However, the recently increasing adoption of GIS in complex environmental planning questions and the required early integration of environmental issues in planning processes makes the incorporation of a large variety of planning and environmental data at different scales necessary. This is especially relevant for decision-making at an early, more strategic planning stage. Therefore, the presented method shows a way to adjust a detailed planning data-set to small-scale map by a spatial aggregation. The result shows that the areal weighting of the detailed biotope-type assessment on the block structure of USM includes significant parts of ecologically valuable areas. Hence, this method can help to present block-based ecological significance for early small-scale environmental analysis in an appropriate way. It shows a possibility to selectively integrate important detailed information and to bring them up to a small-scale for a first early inspection. The intended loss of information normally associated with an aggregation process (Gotway and Young 2002) was restricted to an acceptable degree. The application of the 1 ha-size criterion leads to a higher amount of integration of biotope-type information into the small-scale block structure of the USM than the 3 ha-size criterion (see Fig. 5). Moreover, a higher amount of legally protected biotopes were included, which can have a high relevance in later planning decisions. Therefore, the 1 ha-size criterion should be preferred to approach the fine resolution of the biotope-type assessment.

It should be stated that the application of the deviation criterion as presented in this study considers absolute deviation. Deviating biotopes with a lower ecological significance and with a higher ecological significance as the block-based mean value got integrated likewise. In future developments it would be interesting to limit the deviation criterion to biotopes with a higher ecological significance, or to additionally integrate further technical criteria, such as the exclusive preference of specific main biotope-type categories such as forests or the unconditioned integration of legally protected biotopes.

6 Recommendations

The presented and discussed method can be used for enhancing the quality of data for large-scale analyses such as Strategic Environmental Assessments. The selective integration of specific biotope-assessment information leads to an enhancement of the data set quality for the use at a strategic planning level, because important detailed information such as legally protected biotopes were not considered in previous SEAs.

Therefore, one recommendation derived from this study is to include the presented method as a part of a Planning Support System (PSS) of a SEA. The detected 8 (respectively 75) sites which are protected under the Habitats Directive within the city of Berlin could be permitted for further urban development at an early planning stage. The procedure can be implemented very easily in already available Planning Support Systems by selecting sites with a significant share of protected areas. This can prevent time-consuming and costly protection measurements at a later planning stage, such as the Environmental Impact Assessments.

Acknowledgments This research project has been funded by the Deutsche Bundesstiftung Umwelt (DBU). The authors thank Mr. H. Brandl and Mr. D. Bangert from the Senate Department for Urban Development Berlin for their committed technical support.

References

Dalai-Clayton B, Sadler B (2005) Strategic environmental assessment—a sourcebook and reference guide to international experience. Earthscan, London

European Commission (EC) (2001) Directive 2001/42/EC of the European parliament and of the council of 27 June 2001 on the assessment of certain plans and programmes on the environment (SEA Directive). Official J European Communities L197:30–37

European Commission (EC) (2007) Directive 2007/2/EC of the European parliament and of the council of 14 March 2007 establishing an infrastructure for spatial information in the European community (INSPIRE). Official J European Communities L108:1–14

Ferretti J, Koeppel J, Koeppen A (2009) Strategic environmental scenario tool for urban planning processes in Berlin. In: Proceedings of the 3rd international congress on environmental planning and management, pp 251–255

Fischer T (2007) Theory and practice of strategic environmental assessment towards a more systematic approach. Earthscan, London

Flowerdew R, Green M (1989) Statistical methods for inference between incompatible zonal system. In: Goodchild M, Gopa S (eds) The accuracy of spatial data bases. Taylor & Francis, London, pp 239–247

Flowerdew R, Green M (1992) Developments in areal interpolating methods and GIS. In: Annals of regional science, vol 26. Springer, Berlin, pp 67–78

Flowerdew R, Green M (1994) Areal interpolation and types of data. In: Fotheringham S, Rogerson P (eds) Spatial analysis and GIS. Taylor & Francis, London, pp 121–145

Gonzales Del Campo A (2008) Incorporating spatial data and GIS to improve SEA of land use plans: opportunities and limitations: case studies in the Republic of Ireland. PhD dissertation, Institute of Technology, Dublin

Goodchild M, Lam N (1980) Areal interpolation: a variant of the traditional spatial problem. Geo-Processing 1:297–312

Gotway C, Young L (2002) Combining incompatible spatial data. J Am Stat Assoc 97:632–648

Jones C, Baker M, Carter J, Jay S, Short M, Wood C (2005) Strategic environmental assessment and land use planning. Earthscan, London

Kieler B, Sester M, Wang H, Jiang J (2007) Semantic data integration: data of similar and different scales. Photogrammetrie Fernerkundung Geoinf (PFG) 6:447–457

Koeppen A, Foerster M, Kleinschmit B, Koeppel J, Ferretti J (2010) SUPPORT—a strategic planning and decision support instrument for environmental urban planning in Berlin. In: Schrenk M, Popovich V, Zeile P (eds) Proceedings of REAL CORP 2010, Vienna, 17–20 May

Marull J, Pino J, Mallarach J, Cordobilla M (2007) A land suitability index for strategic environmental assessment in metropolitan areas. Landscape Urban Plan 81:200–212

Schneider T, Goedecke M, Lakes T (2007) Urban and environmental information system: application of remote sensing for planning and governance—potentials and problems. In: Netzband M, Stefanow WL, Redman C (eds) Applied remote sensing for urban planning, governance and sustainability, Springer, Berlin, pp 199–219

Senate Department for Urban Development (SenStadt) (2007) Berlin land use plan—contents and conventional signs http://stadtentwicklung.berlin.de/planen/fnp/pix/erlaeuterungen_fnp/Contents_and_Conventional_Signs_col.pdf, p 16

Senate Department for Urban Development (SenStadt) (2008a) 06.07 Urban structure. Map description. http://www.stadtentwicklung.berlin.de/umwelt/umweltatlas/e_text/ekc607.pdf, p 16

Senate Department for Urban Development (SenStadt) (2008b) 06.01 Actual use of built-up areas 06.02 inventory of green and open spaces. Map description. http://www.stadtentwicklung.berlin.de/umwelt/umweltatlas/e_text/ekc601.pdf, p 13

Senate Department for Urban Development (SenStadt) (2009) 05.08 Biotope types. Map description. http://www.stadtentwicklung.berlin.de/umwelt/umweltatlas/e_text/ek508.pdf, p 9

Senate Department for Urban Development (SenStadt) (2010) Berlin digital environmental atlas http://stadtentwicklung.berlin.de/umwelt/umweltatlas/edua_index
Therivel R (2004) Strategic environmental assessment in action. Earthscan, London
Vanderhaegen M, Muro E (2005) Contribution of a European spatial data infrastructure to the effectiveness of EIA and SEA studies. Environ Impact Assess Rev 25:123–142
Welsch J (2008) Stadtklima im kommunalen planungsprozess. UVP-Report, vol 22, pp 221–224 (in German)

Using MapTable® to Learn About Sustainable Urban Development

Peter Pelzer, Gustavo Arciniegas, Stan Geertman and Jaap de Kroes

Abstract Sustainable urban development is a critical issue in the Netherlands. The country is densely populated, which causes conflicts between environmental concerns and spatial development. Environmental policy integration is proposed as a way to improve the integration of environmental values into spatial planning with the help of learning processes. This chapter evaluates the extent to which the combination of a map-based touch table and an area-specific environmental profile are of added value to environmental policy integration. The case study is the application of the map-based touch table, called MapTable® for the development of a sustainable neighborhood in the region of Utrecht, the Netherlands. It was found that MapTable® facilitates learning processes by providing a platform for communication among stakeholders from different backgrounds. Nonetheless, it must be ensured that all stakeholders are equally included, and that the process suits the application of a map-based touch table in combination with an area-specific environmental profile.

P. Pelzer (✉) · S. Geertman
Urban and Regional Research Centre Utrecht (URU), Utrecht University, Heidelberglaan 2 3584 CS, Utrecht, The Netherlands
e-mail: p.pelzer@uu.nl

S. Geertman
e-mail: s.c.m.geertman@uu.nl

G. Arciniegas · J. de Kroes
Mapsup, Tolboomweg 2a 3784 XC, Terschuur, The Netherlands
e-mail: gustavo@mapsup.nl

J. de Kroes
e-mail: jaap@mapsup.nl

S. Geertman et al. (eds.), *Planning Support Systems for Sustainable Urban Development*, Lecture Notes in Geoinformation and Cartography,
DOI: 10.1007/978-3-642-37533-0_10,

1 Introduction

Environmentally sustainable urban development is a critical issue in the Netherlands. The country is densely populated, which causes conflicts between environmental concerns and spatial developments. Moreover, climate change poses challenges to several Dutch cities with regards to water management. Attempts have been made to integrate environmental concerns into urban and regional planning, particularly since the publication of the Brundtland Report in 1987 (e.g., Brundtland 1987; De Roo et al. 2012). The inclusion of environmental values in urban development is part of the concept of environmental policy integration (EPI). The main premise of this approach is the integration of environmental values into specific sectoral policies—in this regard, urban planning (e.g., Lafferty and Hovden 2003; Nilsson and Persson 2003; Nilsson 2005). However, this integration has proved a far from easy task, as spatial planning "tends not to pay enough attention to sustainability and environmental concerns" (Simeonova and Van der Valk 2009, p. 244). De Roo et al. (2012, p. 203) cite an evocative example from a practitioner:

> It is shocking to see staff from different departments [within the same municipality] shaking hands and introducing themselves for the first time. Environmental and Spatial Planning departments still operate very much in parallel, and are not – or not sufficiently – aware of each other's work.

Hence, the question arising from this issue is how environmental values can be better integrated into the urban planning process. This chapter argues that this process of integration should be perceived as a learning process in which actors from varying backgrounds gain a better understanding of the role of environmental values in a planning problem. The importance of learning has been emphasized in a range of studies focusing on spatial planning and public administration (e.g., Schön 1983; Te Brömmelstroet 2010; Zanon 2010). Moreover, learning is argued to be viable solution for EPI (Nilsson and Persson 2003).

The aim of this study was to improve the learning process related to EPI through the combination of two tools: area-specific environmental profiles and planning support systems (PSSs). Area-specific environmental profiles describe the extent to which specific environmental indicators are relevant to a specific area. The underlying rationale of such profiles is that the relevance of environmental values differs among spatial areas (De Roo et al. 2012; Weber and Driessen 2010). For instance, in an industrialized area, noise pollution is usually not a big issue, whereas in a residential neighborhood it can cause a lot of harm. It is also assumed that by applying area-specific profiles, higher sustainability goals can be achieved, since environmental values do not lead to the imposition of juridical restrictions on urban development, and in this case form an integral part of a vision that relates to local characteristics and functionalities. As area-based environmental profiles have an explicit spatial component, they can be depicted on a map or incorporated in a geographic information system (GIS). They can therefore be translated into a PSS, which Geertman (2008, p. 217) defines as: "geoinformation technology-based instruments that incorporate a suite of components that collectively support some specific parts of a unique professional planning task". Recent studies (e.g., van der Hoeven 2009; Te Brömmelstroet 2010) emphasize that a PSS can be helpful in supporting learning processes. In this, discussions and interaction among the involved stakeholders are crucial. For this purpose, it is assumed that a PSS like a map-based touch table has the potential to encourage and facilitate discussion (Arciniegas 2012).

This chapter presents an evaluation of the extent to which the application of a map-based touch table is useful to support learning processes in which the prime purpose is to integrate area-specific environmental profiles into spatial planning practices. The chapter is structured as follows. Section 2 elaborates on the debate about EPI, area-specific environmental profiles, and learning processes. Section 3 describes the potential of PSSs, and in particular a map-based touch table, to facilitate learning processes, as well as their limitations. Section 4 introduces MapTable®,—a map-based touch table developed by the geo-communication company MAPSUP. It makes use of an area-specific environmental profile called the "Sustainability Profile of the Location" developed by IVAM. Section 5 presents the application of MapTable® in the sustainable area development of Rijnenburg in the Netherlands. Section 6 relates the findings to the theoretical framework, and draws lessons for the future. Section 7 presents the main conclusions.

2 Learning About Sustainability in Urban Planning

In recent years, sustainability has gained attention in the urban planning literature (e.g., Campbell 1996; Rydin 2010). In particular, climate change is seen as a topic that is critical for the discipline of planning (e.g., Uittenbroek et al. 2013). Whereas sustainability encompasses several dimensions such as people, planet and profit, the focus here is on the "planet" dimension, namely the environmental sustainability.

2.1 Environmental Integration as Learning

Despite good intentions, it has proven very difficult to integrate environmental values into the planning process (De Roo et al. 2012; Simeonova and van der Valk 2009). The challenge has been described more explicitly by researchers in the EPI field, for whom EPI is:

> … the incorporation of environmental objectives into all stages of policymaking in non-environmental policy sectors, with a specific recognition of this goal as a guiding principle for the planning and execution of policy; accompanied by an attempt to aggregate presumed environmental consequences into an overall evaluation of policy, and a commitment to minimise contradictions between environmental and sectoral policies by *giving principled priority to the former over the latter* (Lafferty and Hovden 2003, p. 9 – emphasis added).

Note that in this definition, the central element of integration is that environmental values are prioritized over other values (such as profitability and aesthetics). The question thus becomes *how* environmental values could be given more priority in urban planning processes.

Policymaking is far too complex, ambiguous, and fuzzy for simple and straightforward solutions. For instance, stakeholders in policymaking frame problems in terms of environmental dimensions in different ways. Urban designers, for example, put aesthetics central, whereas a transport planner focuses on traffic flows. To incorporate environmental values effectively in a policy process, it is necessary to think beyond traditional values. For this purpose, EPI is addressed as a learning process in which stakeholders learn about environmental values (Nilsson and Persson 2003; Nilsson 2005). While there are many types of "learning" in the field of policymaking and planning, the focus here is on the learning processes of individual stakeholders. The notion of "frames" is central to understanding learning. Based on the work of Erving Goffman (1974), Benford and Snow (2000, p. 614) define frames as "schemata of interpretation that enable individuals to locate, perceive, identify, and label occurrences within their life space and the world at large." In planning and policymaking, the work of Rein and Schön (1994) has shown that a change of frame, or the incorporation of other frames, leads to learning processes (Nilsson 2005). Put differently, learning occurs when a stakeholder in the planning process perceives a planning issue through a different lens.

It is not an easy task to bring about a frame change, either radically or incrementally, since frames tend to be rooted in the different rationalities of the involved stakeholders (Carton 2007). Moreover, the knowledge base of individuals in the planning process also differs. Or, as stated by Healey (2007, p. 245): "What we know exists in many forms, from systematized accounts and analyses, and practical manuals, to stories exchanged in the flow of life, and skills exercised in doing practical work." Transport planners, for instance, tend to rely strongly on systematized knowledge, whereas for urban designers experiential knowledge is of more importance. Nonetheless, there is one aspect that unites all the stakeholders in urban planning: space.

2.2 A Spatial Discussion

For urban planning, EPI is an essentially spatial endeavor. The critical question is always where developments occur, and what the local environmental impacts and restrictions are. Another important aspect of space in EPI is the characteristics of the local context. Environmental policy is not a one-size-fits-all policy, but must deal with the particularities of local circumstances. For instance, the siting of a windmill might conflict with a residential neighborhood, but not with a business park. Area-specific environmental profiles are thus developed according to the context (see De Roo et al. 2012; Weber and Driessen 2010). Two of these tools are

LOGO (a Dutch acronym for "Local Area Typology and Environmental Quality") and MILO (a Dutch acronym for "Environmental Quality in the Physical Environment"), which were developed in the early 2000s. The input of these tools is based on the local circumstances and reveals the environmental ambitions for a specific locality. MILO is particularly useful in areas where the environmental burdens are not too strict.

> MILO aims at improving the environmental quality of certain parts of the city by integrating area-specific environmental ambitions in the spatial planning process. This often leads to higher environmental quality than legally required. Yet reviews indicate that MILO's success can be assured only in areas with limited environmental constraints (Weber and Driessen 2010, p. 1130).

By making the environmental ambitions explicitly spatial, it is easy to incorporate EPI in spatial planning disciplines like urban design, spatial planning, and landscape architecture. However, area-based environmental profiles such as MILO and LOGO are relatively new tools; extensive experience of their application in practice is therefore lacking (De Roo et al. 2012).

Whereas area-specific environmental profiles make EPI spatially specific, there is underused potential to make environmental values also spatially visible. In other words, as De Roo et al. (2012, p. 113) state: "area qualities can also be visualized in the form of maps, which can reinforce the related discussion." In fact, maps are a logical platform, since they tend to be part of the working practice of the involved actors. For environmental analysts, maps are very useful to depict noise contours or air pollution, whereas for an urban designer they are an instrument to sketch a vision for an area. Put differently, planning actors have different frames about the function of maps. Carton (2007) argues that three dominant frames represent the role of maps in spatial planning: analysis, design, and negotiation.

The analysis frame addresses maps as a way to systematically understand what the problem is. Stakeholders within this frame rely strongly on quantitative methods and logical analysis; the map is used as a research model. Typical examples of stakeholders in policy settings are environmental analysts and transportation planners. In the design frame, maps are a way to support an intuitive and creative process. It is a way to communicate ideas—for instance, about urban form—that are not easily translated through text. Typical stakeholders relating to this frame are urban designers and landscape architects, whose knowledge base is to an important extent experiential. In the negotiation frame, the content of a map is of minor importance. A map is envisioned as an instrument to be used strategically in a planning process. Typical stakeholders related to this frame are policy advisors, process-oriented spatial planners, and stakeholders with political commitments. Table 1 shows the three frames, the typical stakeholders, the knowledge bases, the functions of a map, and the roles of area-based environmental profiles.

Central to this chapter is the question how the analytical frame can be adopted, or at least be understood, by stakeholders with a design or negotiation frame. This is based on the premise that the depiction of environmental values (noise, air quality, etc.) tends to be based on the knowledge of environmental analysts with an

Table 1 Frames about maps in planning

Frame	Typical stakeholders	Knowledge base	Function of a map	Role of area-based environmental profiles
Analytical	Environmental analysts, transport planners, financial analysts, GIS specialists	Systematized	Research model	To integrate environmental values into spatial planning
Design	Urban designers, architects/ landscape architects	Experiential	Design language	To show environmental restrictions and opportunities of visual designs
Negotiation	Policy advisors, some spatial planners, politicians	Pragmatic	Decision agenda	As a strategic instrument in the planning process

Inspired on Carton (2007)

analytical frame. Hence, EPI can be achieved in a learning process in which reframing is the central aim. Guidelines on this process are provided by Simeonova and van der Valk (2009, pp. 257–58): "it demands a decision-making process based on strong social interaction and learning between actors (…)." This statement is in line with planning scholars who emphasize the communicative and interactive nature of planning, based on Jürgen Habermas's notion of communicative rationality (Healey 1992; Innes and Booher 2010). Discussion, collaboration, and communication are seen as the central elements of a planning process. However, the discussion of environmental values requires the use of quantitative models and large amounts of data. Hence, a tool to facilitate the process must be sensitive to the communicative process with stakeholders from different backgrounds, and should be able to provide them with quantitative information and analysis tools. Based on recent studies that address the use of PSSs in collaborative planning settings (e.g., Arciniegas 2012; Te Brömmelstroet 2010), we hypothesized that a user-friendly PSS is suited for this task.

3 Planning Support Systems

A broad range of instruments can support integrated planning and environmental policy, varying from paper maps to complex environmental models. PSSs are designed to combine the user friendliness of the former with the information richness of the latter. In a recent review of PSS applications by Geertman and Stillwell (2009), about one third of the studies deal explicitly with environmental issues. Moreover, in the PSS literature a shift can be observed from a direct instrumental application of a PSS, towards a much more interactive and communicative use, which mirrors the shift from rational planning to collaborative planning (Geertman 2006; Pelzer 2012). In the case of, for instance, the land-use scanner (a large-scale model about future land uses in the Netherlands), learning plays a key role:

> The system is developed to support the *discussion* on the long-term adaptability of the Netherlands to flood risk. It aims to facilitate the learning of the user on the subject, instead of giving unambiguous answers on what management strategy is preferable. This is a significant difference with the more traditional *decision* support systems. (Van der Hoeven et al. 2009, p. 162, emphasis in original).

In a similar vein, Te Brömmelstroet (2010) argues that the key role of PSSs is to facilitate learning processes. In a recent study by Arciniegas (2012), a PSS in the form of a map-based touch table was used to support learning processes. Touch tables comprise software that is user friendly, and hardware in the form of an interactive table that can function as a platform for discussion. Such touch tables provide the opportunity to combine intuitive sketching with quantitative analyses. Until now, however, only a few studies have described the practical application of touch tables in planning (Alexander et al. 2012; Arciniegas 2012; Dieber 2003; Hofstra et al. 2008), and none of these studies explicitly relates to EPI.

4 The MapTable®

The map-based touch table that is empirically analyzed in this chapter is MapTable®. Besides the hardware (the actual table), the system comprises GIS software (ArcGIS and CommunityViz) and an area-based environmental profile called the "Sustainability Profile of the Location". These elements are explained in more detail below.[1]

---

[1] Since the use of specific jargon, brand names, and abbreviations might be somewhat confusing, the following table presents the central terms.

Generic term	In the Rijnenburg case study
Map-based touch table	MapTable®
GIS software	ArcGIS® and CommunityViz®
Area-specific environmental profile	Sustainability profile of the location (SPL)

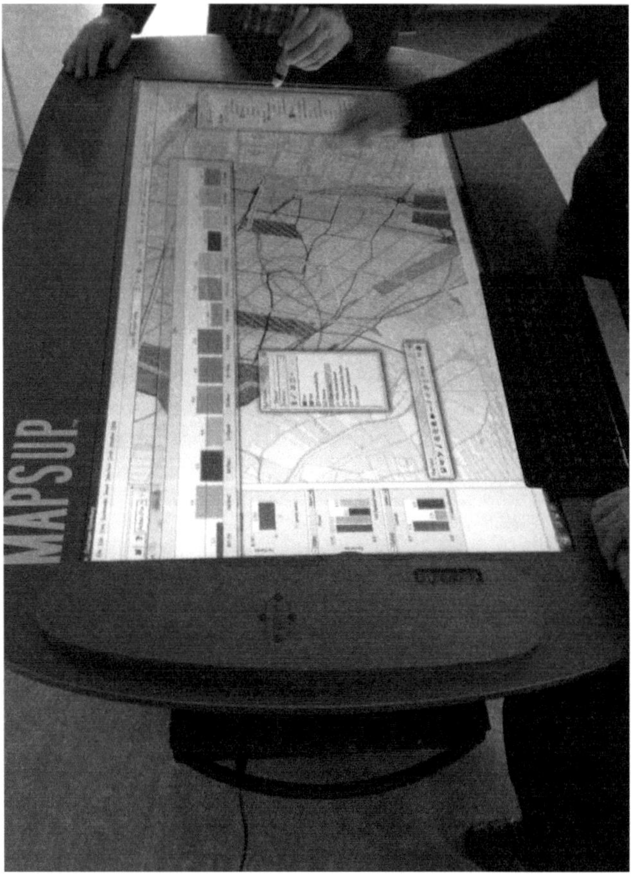

Fig. 1 The MapTable®

4.1 The Tool

4.1.1 Hardware: MapTable®

MapTable® (see Fig. 1) is a large (116 cm in diagonal) touch table, developed and brought to the market by MAPSUP (www.mapsup.nl). The table stands on four multidirectional wheels and can easily be moved around. Its screen can be rotated, tilted, and adjusted in height. MapTable® can be operated with a digital pen (or stylus), finger multi-touch gestures, or a wireless mouse and keyboard. It supports a wide range of group tasks to which maps are central, including visioning and idea sketching, the structuration of local knowledge, and real-time geographical calculations.

4.1.2 Software: ArcGIS and CommunityViz

MapTable® works as the interface between users and a large variety of planning tools (which are typically run within the ESRI ArcGIS® environment), including the Sustainability Profile of the Location. The Sustainability Profile of the Location tools described in this chapter were developed with CommunityViz Scenario 360 (http://www.communityviz.com/), which is a set of extensions for ArcGIS, specifically designed to support land-use planning.

4.1.3 The Area-Based Environmental Profile: The Sustainability Profile of the Location

The Sustainability Profile of the Location (SPL) for Rijnenburg was based on fourteen indicators, which mainly concern such environmental issues as noise, energy, water, and ecology (www.toolboxrijnenburg.nl). The indicators were selected in close cooperation with the Municipality of Utrecht. To make the indicators accessible for users from different backgrounds, it was chosen to have 1–10 scores (the widely accepted Dutch school grading system) rather than the original indicators. Although the SPL for Rijnenburg was unique, the logic and methodology of the SPL had already been developed and applied by IVAM (www.ivam.uva.nl). The SPL was implemented in a GIS in order to visualize it on MapTable.

4.2 Using MapTable®

Three main tasks can be carried out with MapTable. It can be used to modify existing land-use functions, to implement and sketch sustainability measures, and to calculate and portray the SPL indicators. As users design a district's land-use plan, the SPL scores are calculated or updated in real time. The tool allows its users to modify the land-use functions of a region by painting new land uses directly on the MapTable map. It can also be used to draw additional sustainable themes, such as windmills, sound barriers, cycle tracks, insulation, bus stops, biogas generators, etc.

MapTable serves as the interface between users and the GIS-based SPL. The SPL tool is interactive: Users can provide input, see the results in real time, and then modify the input and recalculate new output. The SPL comprises three major components, namely a spatial component, an interface with interactive sliders or parameter space, and a chart space.

The spatial component, or map space, is essentially a layer-based GIS that shows relevant map layers that can be combined and turned on and off, and on which new designs can be made by drawing on MapTable. Amongst these maps, two dynamic layers are part of the SPL, namely a land-use layer—which contains existing functions that can be modified—and a layer with additional sustainable

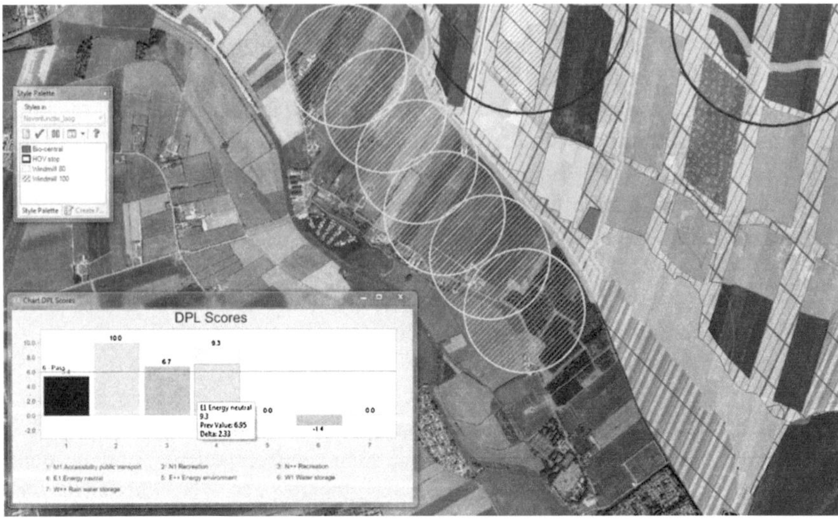

Fig. 2 Drawing additional sustainability functions on MapTable. Note that on the dynamic chart, the sustainability profile of the location score for "energy neutral" increases as a result of inserting this six-windmill array

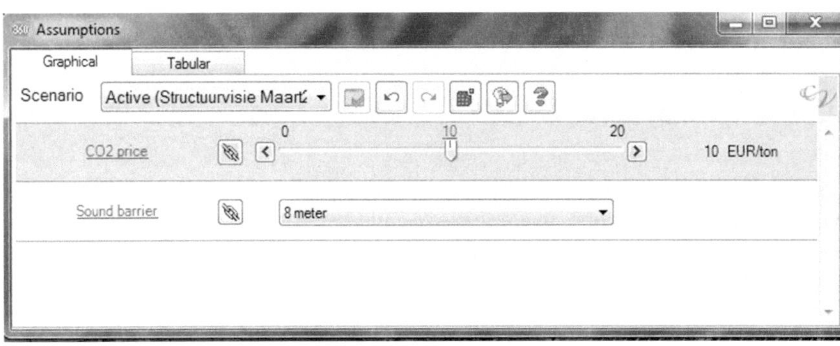

Fig. 3 Interface with interactive sliders and drop-down menus

functions, such as windmills, that can be added to the plan. Existing land-use functions can be modified by painting new land uses using a painter tool. Users touch on a land-use palette to choose a land-use type and "paint" this new class on existing parcels on the map. New or additional sustainable functions can be added to or drawn on the map. Examples are lines representing cycle tracks, or circles representing windmills and their coverage. These shapes can be selected from a list and inserted into the map (see Fig. 2). Changes to existing land-use functions as well as additional functions result in changes to the SPL scores portrayed in the chart space.

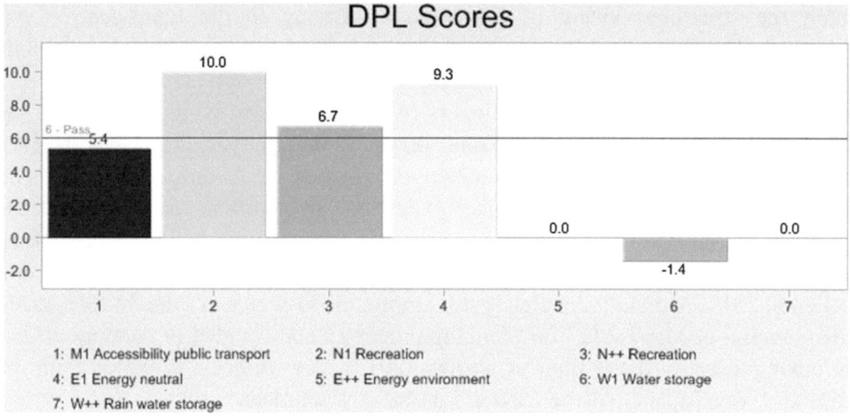

Fig. 4 Dynamic chart showing the sustainability profile of the location (here, "DPL") scores for various sustainability themes

The parameter space comprises an interface that has a number of interactive sliders and drop-down menus, with which users can specify values for parameters that are relevant to each sustainability theme. Values are set by touching and dragging a slider to a desired position, or by selecting a new value from a list of choices (see Fig. 3). For example, the interface can be used to modify the height of existing sound barriers or to increase the price of CO_2 per ton. Changes to these input values result in changes to the SPL scores in the dynamic chart space.

The chart space contains the main output of the SPL, which is presented as a bar chart containing the SPL score for each theme of sustainability deemed relevant to Rijnenburg (see Fig. 4). The values on this chart are updated as soon as changes are made on the map (by painting new land uses or drawing additional functions) or to input values (by changing parameter values on the interface).

5 Experiences with MapTable® in Utrecht: Rijnenburg

5.1 Rijnenburg: A Sustainable Neighborhood

The Municipality of Utrecht has strong ambitions with regards to sustainable urban development. For example, its plans to develop a new neighborhood—Rijnenburg—southwest of the city and build 7,000 dwellings there, can be seen as an exemplary case, as the intention is to incorporate environmental values that are significantly above average, and for the neighborhood to produce energy rather than consume it.

In 2008, the municipality brought together an urban design firm, several environmental research agencies (which provided the SPL), and a geo-communication firm (which provided MapTable®) to concretize its vision of the new neighborhood. The Province of Utrecht actively supported the project. Two years later, the structured vision of Rijnenburg—"Living in the Landscape"—was finalized. The sessions with MapTable® had helped the participants to include sustainability aspects in the plan. See Fig.5 for practical application.

However, the commercial developers (who owned most of the land) were not convinced of the feasibility of the plans, partly because they had not been involved in the MapTable® sessions. They challenged many of the assumptions in the plan, which led to a pause in the project execution. In addition, as a result of the financial crisis in 2010, developers are much more cautious about undertaking large-scale developments. As a consequence, the MapTable® project was cancelled in 2011. Although the area is still empty, the Province of Utrecht recognized the potential of MapTable® for planning processes and decided to continue its use in other projects. At the time of writing (2013), new projects in which both the SPL and MapTable® will be used are under development.

5.2 *MapTable® in Rijnenburg*

There are two reasons why the municipality decided to use the combination of MapTable® and the SPL. First, when a support tool had to be chosen in 2008, the SPL was one of the few methods that could provide a quantitative and integral analysis of sustainability in an area—one of the explicit aims in the case of Rijnenburg. Second, there was a need for a tool that would facilitate not only quantitative analysis but also drawing, and MapTable® does this. The assumption was that because of the possibility to draw, urban designers would actively think and work together.

Since 14 indicators had to be included in MapTable®, the preparation process was very lengthy and time intensive. In total, six workshops were organized. The workshops focused both on developing a plan for Rijnenburg and on getting the SPL indicators right in MapTable®. They were led by an environmental analyst from the municipality and facilitated by a GIS specialist from MAPSUP. Most of the participants were environmental analysts with a specific expertise. The group comprised external consultants (water, peat meadow) and specialists from the municipality and the Province of Utrecht (air, noise, traffic).

The urban designers from the external firm were less actively involved. They did not actually draw on MapTable®, as was the intention beforehand, but provided three sketches of scenarios, which were entered into the system. Then, with the help of the municipality's urban designers (not to be confused with the leading designers from the commercial firm), the environmental analysts started to make drawings on MapTable. Attempts were made to include the municipality's financial experts, but they did not participate in the sessions because of time pressure and a lack of interest. The commercial developers were also not included in the workshops. Consequently, three types of actors had intense experience with MapTable®: GIS specialists, environmental analysts, and urban designers. Their experiences are described below.

Fig. 5 MapTable® depicting the Rijnenburg area during a demonstration at a conference. *Photo* Toolbox Rijnenburg (www.toolboxrijnenburg.nl)

5.3 Experiences of the Planning Actors

5.3.1 GIS Specialists

The GIS firm MAPSUP seconded a GIS specialist to provide technical facilitation during the sessions with MapTable®. Although the specialist had no input in the process in terms of content, such a specialist is a critical actor for the usage of MapTable. His or her main role is to translate the suggestions and remarks made by the planning actors into the right map images by using the ArcGIS software on MapTable®. Or as the GIS specialist who was most closely involved in Rijnenburg remarked: "*When an environmental analysts talked about noise, I had to make sure I directly depicted the noise map, otherwise the momentum would have been lost*". One of the difficulties the GIS specialist had to cope with, was the continuous addition of new factors by planning actors, which made the underlying model of the SPL very complex, leading to sensitivity to technical errors and lengthy calculation times.

5.3.2 Environmental Analysts

One of the starting points in Rijnenburg was that environmental analysts should actively provide planning solutions for the area. This is in contrast to the traditional role of Dutch environmental analysts, who tend to participate in later stages of the planning process, when legal requirements have to be met. Environmental analysts comprised the majority of the people who were involved in the MapTable sessions, and their expertise covered such subfields as air quality, noise pollution, and water management. In general, the environmental analysts evaluated the use of MapTable® positively, in the sense that it helped them to envision the diverse topics of a planning problem in an interconnected way. One challenge was the handling of the SPL scores: While the scores help the participants to think integrally and to connect different actors, some of the environmental analysts found it difficult to think in terms of scores of 1–10, rather than the well-known environmental indicators (e.g., decibels for noise pollution).

5.3.3 Urban Designers

An even greater challenge than the participation of environmental analysts was the inclusion of urban designers. Someone closely involved said: "*For me the process would be successful if the urban designers would stay around MapTable® until the end*". Although this was the case, the application of MapTable® disrupted the urban designers' working practice, to which sketching and visualization are central. In particular, two issues became clear during the sessions in Rijnenburg. First, the working practices of urban designers and landscape architects, to which rough and intuitive sketches are central, are difficult to handle on MapTable®. This is because, on the one hand, vague images are difficult to translate into a GIS, and on the other hand, the drawing itself is part of the designer's creative process. Second, the integral approach that was applied in Rijnenburg was a barrier to creativity. For example, the urban designer who attended most of the sessions remarked, "*If you emphasize integrality too much, it hampers focus and creativity*".

5.4 Communicating with MapTable®

All of the involved actors recognized that the central function of MapTable® is to improve communication. The role of MapTable® has two dimensions regarding communication. First, the inclusion of the table as such (the hardware) leads to a group dynamics in which it is easier to come up with ideas and to think beyond professional roles. Second, the use of the SPL indicators leads to intensive discussions about the incorporation of environmental issues in the spatial plan. The indicators force the actors to be very explicit about their proposed interventions and the expected effects, since the results are directly depicted by MapTable®. Or as one of the involved actors who led most of the workshops remarked: "*They are forced to show the spatial impact of their discipline*". Besides the fact that issues become much more explicit, the development of such a "spatial language" encourages the understanding of the issues of relevance to other disciplines and the interrelatedness of problems. A case in point is windmills (see also Fig. 2), which lead to a significant increase in energy production, but also to new noise contours where it is not possible to construct dwellings.

However, there were also problems. Technical errors and long waiting times in some instances hampered the dynamics of the workshop. Even the tiniest hiccough could have a significant impact. As someone involved in the organization of the workshops remarked: "*When MapTable had to calculate for a couple of minutes, or measures couldn't be presented right away, people lost their concentration*". A more fundamental issue is the difference in working practices between urban designers and environmental analysts. MapTable® was in many instances a burden on, rather than a catalyst for, the design process. It forced the designers to be very explicit at an early stage and to continuously think integrally. Whereas this was a very convenient approach for the environmental analysts, for urban designers it interrupted their working practices too much. Illustrative of that is a remark made by the main urban designer involved in MapTable®: He called the session a "side project", rather than a fundamental part of the development of the plan for Rijnenburg.

6 Discussion: A Map-Based Touch Table for the Inclusion of Environmental Values

This section places the experiences with MapTable® in Rijnenburg in a wider context. It evaluates the extent to which a map-based touch table is useful in supporting EPI, and draws lessons for future research and practice.

6.1 Learning with a Map-Based Touch Table

6.1.1 Learning as Reframing

In this chapter, reframing is considered the outcome of a learning process. This seems to make sense from a theoretical perspective: The strong differences between environmental analysts and urban designers can be explained by the fundamentally different frame they have of maps. Providing a solution through MapTable® was more difficult, however. It turned out to be relatively easy to connect with stakeholders with a similar frame (i.e., environmental analysts). However, connecting the work of urban designers (design frame) and environmental analysts (analytic frame) proved much more of a challenge. Although MapTable® encouraged interaction among actors with frames that are far apart, this interaction did not necessarily lead to learning processes. Consequently, while MapTable® is clearly capable of improving the learning processes of actors who have similar working practices and knowledge bases, more is needed to improve the communication among actors with frames that vary more significantly.

6.1.2 Area-Specific Environmental Profiles

The area-specific environmental SPL has positive influences on EPI as a learning process. First, the sustainability scores ranging from 1–10 are easy to understand by actors with different frames. Environmental analysts are encouraged to discuss their expertise in a language that is understandable to other planning actors. Moreover, since an area-specific profile focuses on the characteristics of a neighborhood or a region, the discussion about sustainability remains concrete, rather than becoming abstract. However, some challenges were observed too. In Rijnenburg, there was hardly any unsolvable conflict between values: High environmental standards could relatively easily be included in the plan. However, this is not the case in dense urban settings, where it is not always possible to meet the demands of the various disciplines. For instance, although a new road might harm the air quality in a neighborhood, it might be a vital new connection for the city. However, area-specific environmental profiles also have to cope with a conflict of values that is more generic. Whereas environmental values can be expressed in a quantitative score, this is much more difficult, if not impossible, for aesthetic values. Such values are central to the work of urban designers, which explains the hesitance of urban designers in Rijnenburg to focus too much on environmental integration.

6.1.3 A Map-Based Touch Table

In short, a map-based touch table supports a learning process in two ways. First, the table's hardware leads to an open and constructive group dynamics. Planning actors stand around the table and feel free to make possible interventions visible on the map and to exchange arguments. The second element is a greater challenge, however. The purpose of the SPL indicators in combination with the CommunityViz software is to develop a spatial language, which leads to the reframing of the involved planning actors. As argued, it is a challenge to bridge the design and the analytic frame, and it can be questioned whether a GIS-based PSS like MapTable® suffices for this purpose. In the following subsection, some lessons are drawn that could improve the application of a map-based touch table in learning processes.

6.2 Lessons for the Future

Four lessons can be drawn for the future application of map-based touch tables to facilitate learning processes. Although these lessons are based on the experiences with MapTable® in Rijnenburg, where the objective was to improve learning for EPI, they can be applied more widely.

First, it is important to pay explicit attention to the organization of the process. In the Rijnenburg case study, the application of MapTable® and the area-based environmental profiles required changing the conventional planning and design process. Environmental concerns had to be included at an earlier stage in the planning process, which meant that environmental analysts had to get involved much earlier and become much more active. Hence, when considering the application of a map-based touch table, it is important to ensure that both the process and the sessions are carefully prepared. Based on experiences in a range of cases with MapTable®, it is proposed to have four types of session in a sequential order: exploration phase (getting to understand the area and the data/models that are needed), problem definition phase (defining the issues that have to be solved in the area), solution phase (brainstorming, during which possible interventions are proposed), and design phase (drawing and calculating interventions on the map-based touch table). This approach would have two advantages: There would be ample time in the process for the analysis of the problem, which encourages learning processes, and the exploration session would allow stakeholders time to get acquainted with the map-based touch table.

Second, the selection of stakeholders in each stage should be made with sufficient care. In Rijnenburg, for instance, the developers were involved too late and, partly as a consequence, did not endorse the results of the MapTable® sessions. Moreover, in sessions that focus explicitly on sub-dimensions of environmental concerns (noise, water, etc.), it is wise to include the relevant experts. However, in more design-oriented sessions, one professional from an environmental background (the Municipality of Utrecht has "Environmental Coordinators") might be sufficient, along with, for instance, an urban designer, a transport planner, and a commercial developer. However, this does not guarantee a successful learning process, as the next lesson reveals.

Third, in this chapter, reframing is considered the outcome of a learning process. For a map-based touch table to facilitate this, it has to bridge the demands of an analytical frame, a design frame, and a negotiation frame. In Rijnenburg, the focus was on the analytic frame, and less on the design frame. Nonetheless, map-based touch tables could in the future play a more significant role, since urban designers also use maps, albeit in a different, more intuitive way. Depending on the characteristics of the process and the demands of the designers, two directions are possible for future application. First, a map-based touch table is very suitable for design software like SketchUp (www.sketchup.com), which now tends to have ArcGIS plugins, making exchange very easy. Depending on the stage of the planning process, either more design or more analytical software could be used. Second, however, it can be questioned whether it is desirable to eliminate the paper-based drawing by urban designers from the process, since this is considered a very valuable endeavor for a range of stakeholders. The challenge then becomes how to translate these drawings into a GIS environment while preserving the original aims and relating them to data and analysis.

7 Conclusion

The purpose of this study was to assess the extent to which the application of a map-based touch table can support learning processes in which the prime purpose is to integrate environmental values into spatial planning practices. As a general conclusion, a map based touch table supports such learning processes. In the Rijnenburg case in-depth discussion about a range of sustainability aspects in relation to a planning issue, which gives the involved planning actors new insights. Moreover, the combination of the Sustainability Profile of the Location with the MapTable® ensured that environmental values played an important role throughout the various stages of the planning and design process. Nonetheless, some aspects stand out and require more attention in future research and practice.

It is quite common practice to apply a planning support system in relative isolation from the political and private sector arena, and this was the case in Rijnenburg. The advantage of this approach is that it provides a safe and comfortable setting for learning processes. The disadvantage is that the results of the sessions are not connected to the political dynamics. A way to make such a connection is to organize map-based touch table sessions with all the relevant stakeholders (e.g., city council members, commercial developers) and to communicate the results of the sessions in a timely manner. A wide consensus is also critical for environmental policy integration, since although reframing the processes of a set of stakeholders is a step forward, it is far from a guarantee that environmental values will be included in the actual development.

Moreover, the Rijnenburg case revealed that the use of an area-specific profile on a map-based touch table is very fruitful to improve the analyses, but steps should be taken regarding the communication among stakeholders from different backgrounds. A case in point is the role of urban designers, for whom it is not always easy to connect to the structured and systematized nature of an area-specific environmental profile. However, this problem shows that Carton's (2007) ideas about frames in relation to maps are a very relevant heuristic tool. The difference between the analytic frame and the design frame became very clear during the application of MapTable®, whereas actors with a negotiation frame (e.g., developers, politicians) were absent from the sessions.

This chapter provides solutions for the application of a map-based touch table to facilitate learning processes related to sustainable development, but has raised several questions. How could stakeholders with a political or commercial interest be appropriately involved in the application of a map-based touch table? And could the burgeoning field of geodesign (e.g., McElvaney 2012) further specify the requirements for the application of a map-based touch table to support actors with different frames, in particular urban designers? Finally, how can the planning process be adjusted in such a way that environmental values are included much earlier in the planning process, while ensuring that the change is not too radical for the involved institutions and stakeholders?

At the time of writing (2013), a new pilot study is being developed in which the Sustainability Profile of the Location and MapTable® will be applied again. The lessons drawn from the Rijnenburg case are being taken into account in the development of this study, which we hope will provide answers to the questions raised above. After all, the endeavors with a map-based touch table are a learning process not only for planning actors, but also for software developers and university researchers.

Acknowledgments The authors would like to thank Dymph Hoffmans from the Municipality of Utrecht, Joop Machielse from the Province of Utrecht, Mattijs Gevers from the urban design company Juurling & Geluk, and former MAPSUP owner Sjoerd Verhagen for sharing their valuable insights about the use of MapTable® in Rijnenburg and the two anonymous reviewers for their valuable and constructive comments. This study has partly been made possible by the knowledge initiative Connecting Sustainable Cities of the Netherlands Organisation for Scientific Research (NWO).

References

Alexander KA, Janssen R, Arciniegas GA, O'Higgins TG, Eikelboom T, Wilding TA (2012) Interactive marine spatial planning: siting tidal energy arrays around the mull of Kintyre. PLoS ONE 7(1):e30031. doi:10.1371/journal.pone.0030031

Arciniegas G (2012) Map-based decision support tools for collaborative land use planning. Free University Amsterdam PhD thesis

Benford R, Snow D (2000) Framing processes and social movements: an overview and assessment. Ann Rev Sociol 26:611–639

Brundtland GH (1987) Our common future. World Commission on Environment and Development, Brussels

Campbell S (1996) Green cities, growing cities, just cities? Urban planning and the contradictions of sustainable development. J Am Plan Assoc 62(3):296–312

Carton L (2007) Map making and map use in a multi-actor context: spatial visualizations and frame conflicts in regional policymaking in the Netherlands. PhD thesis, TU Delft

De Roo G, Visser J, Zuidema C (2012) Smart Methods for Environmental Externalities. Ashgate, Farnham

Dieber M (2003) Paint the town: lessons learned. 2nd annual conference on PPGIS, urban and regional information systems association. In: Proceedings of the conference on public participation GIS (PPGIS), Portland, 20–22 July 2003

Geertman S (2006) Potentials for planning support: a planning-conceptual approach. Environ Plan B: Plan Des 33:863–880

Geertman S (2008) Planning support systems: a planner's perspective. In: Brail (2008, ed) Planning support systems for cities and regions. Lincoln Institute of Land Use Policy, Cambridge, pp 213–30

Geertman S, Stillwell J (2009) Planning support systems: best practices and new methods. Springer/GeoJournal Library, Berlin, p 95

Goffman E (1974) Frame analysis: an essay on the organization of experience. Harper Colophon, New York

Healey P (1992) Planning through debate. The communicative turn in planning theory. Town Plan Rev 63(2):143–162

Healey P (2007) Urban complexity and spatial strategies. Towards a relational planning for our times. Routledge, Oxon/New York

Hoeven der EJ, van der Aarts JH, van der Klis, Koomen E (2009) An integrated discussion support system for new Dutch flood risk management strategies, in planning support systems: best practices and new methods. In: Geertman S, Stillwell J (eds) (Springer/GeoJournal Library 95, Berlin/Heidelberg/New York), pp 74–159
Hofstra H, Scholten H, Zlatanova S, Scotta A (2008) Multi-user tangible interfaces for effective decision-making in disaster management. In: Nayak S, Zlatanova S (eds) Remote sensing and GIS technologies for monitoring and prediction of disasters. Springer, Heidelberg/Berlin, pp 243–266
Innes J, Booher D (2010) Planning with complexity. An introduction to collaborative rationality for public policy. Routledge, Oxon
Lafferty W, Hovden E (2003) Environmental policy integration: towards an analytical framework. Environ Polit 12(1):1–22
McElvaney S (2012) Geodesign: case studies in regional and urban planning. ESRI Press, Redlands
Nilsson M, Persson A (2003) Framework for analyzing environmental policy integration. J Environ Policy Plan 5(4):333–359
Nilsson M (2005) Learning, frames and environmental policy integration: the case of Swedish energy policy. Environ Plan C: Gov Policy 23(2):207–226
Pelzer P (2012) Planning knowledge through technology: potentials for PSS from the knowledge debate. Paper presented at the 12th international conference on design and decision support systems in architecture and urban planning, Eindhoven, The Netherlands 27–29 Aug 2012
Rein M, Schön D (1994) Frame reflection. Toward the resolution on intractable policy controversies. Basic Books, New York
Rydin Y (2010) Governing for sustainable urban development. Earthscan, London
Schön DA (1983) The reflective practitioner. How Professionals Think in Action. Temple Smith, London
Simeonova V, Van der Valk A (2009) The need for a communicative approach to improve environmental policy integration in urban land use. J Plan Lit 23(2):241
Te Brömmelstroet M (2010) Making planning support systems matter: improving the use of planning support systems for integrated land use and transport strategy-making. PhD thesis, University of Amsterdam
Uittenbroek C Runhaar H, Janssen-Jansen L (2013) Mainstreaming climate adaptation into urban planning: overcoming barriers, seizing opportunities and evaluating the results in two Dutch case studies. Reg Environ Change 13:399–411. doi:10.1007/s10113-012-0348-8
Weber M, Driessen P (2010) Environmental policy integration: the role of policy windows in the integration of noise and spatial planning. Environ Plan C Gov Policy 28:1120–1134
Zanon B (2010) Planning small regions in a larger Europe: spatial planning as a learning process for sustainable local development. Eur Plan Stud 18(2):2049–2072

Websites

www.communityviz.com
www.ivam.uva.nl
www.mapsup.nl
www.toolboxrijnenburg.nl

Ecosystem Services, Green Infrastructure and the Role of Planning Support Systems

Brian Deal, Varkki Pallathucheril and Tom Heavisides

Abstract Traditional efforts to improve the quality of life in our communities have, at times, had detrimental effects on both the environment and the very issues the investments were designed to address. Sustainable development decisions must take into account–in social, economic and environmental terms–the long-term impacts of planning and investment decisions. In this chapter we argue that advances in computation techniques and network infrastructure enable the next generation of planning support systems to support such an accounting. We describe our experiences and lessons learned from application of the LEAM planning support system and a Web-based GeoPortal in helping to sustain critical green infrastructure resources in the state of Illinois.

1 Introduction

Nussbaum (1993) has argued that improving the quality of life of our communities is one of the least costly and most viable economic development strategies available. Efforts to foster an improved quality of life have traditionally focused on a narrow definition of economic development. As a consequence a large portion of

B. Deal
Department of Urban and Regional Planning, University of Illinois at Urbana-Champaign, Illinois, USA
e-mail: deal@illinois.edu

V. Pallathucheril
School of Architecture and Design, American University of Sharjah, Sharjah, UAE
e-mail: vpallathucheril@aus.edu

T. Heavisides (✉)
Division of Ecosystems and Environment, Illinois Department of Natural Resources, Illinois, USA
e-mail: Tom.Heavisides@illinois.gov

S. Geertman et al. (eds.), *Planning Support Systems for Sustainable Urban Development*, Lecture Notes in Geoinformation and Cartography,
DOI: 10.1007/978-3-642-37533-0_11, © Springer-Verlag Berlin Heidelberg 2013

the investments made have focused on built, physical infrastructure projects. This traditional approach has had detrimental effects not only on the environment and climate, but on the very things the investments were intended to improve—the characteristics that define a better quality of life. For example, infrastructure investments intended to increase mobility and reduce travel times by increasing road capacity have instead lead to sprawling patterns of development, longer commutes, increases in vehicle miles travelled, and increased total carbon emissions. Likewise, the root causes behind concerns like climate change, urban water run-off and loss of open-space have been traced to conventional infrastructure investments and related urban planning decisions.

Achieving a more sustainable future for our communities necessitates a departure from the traditional approach to planning and decision making to take into account the long-term impacts of planning and infrastructure investment decisions in social, economic and environmental terms. Assessing important planning decisions in these terms however, requires the ability to: (1) prudently identify existing ecosystem resources and the potential services they provide, (2) forecast future changes to our cities, and (3) assess how these changes will affect the identified resources and services.

The term green infrastructure is increasingly being used to describe important ecosystem resources as a backbone for the delivery of necessary ecosystem services. According to Leigh Ann MacDonald of the Conservation Fund of North Carolina, green infrastructure is defined as:

> … the interconnected network of waterways, wetlands, woodlands, wildlife habitats, and other natural areas; greenways, parks, and other conservation lands; working farms, ranches and forests; and wilderness and other open spaces that support native species, maintain natural ecological processes, sustain air and water resources and contribute to health and quality of life. (McDonald et al. 2005)

Although this and related definitions provide a conceptual basis for understanding green infrastructure, there is still a gap in the planning realm, i.e. integrating the concept with local land-use decisions. Long-term sustainable development planning must take into account existing green infrastructure and the effect that development decisions may have on its ability to provide services into the future. For example, the protection and development of green infrastructure offers a two-way path toward mitigating climate change by restoring natural greenhouse gas mitigation systems to a landscape while simultaneously reducing potential emissions by encouraging a conscious and adaptive growth management structure.

Future changes to cities and regional urbanized areas are increasingly being forecasted using large-scale urban simulation models. The Land-use Evolution and impact Assessment Model (LEAM) planning support system (PSS), developed at and deployed by the LEAM laboratory at the University of Illinois, incorporates one such simulation model. LEAM uses state of the art dynamic-spatial modeling technologies to develop simulations of probable future development patterns given varying planning policy or investment scenarios and the likely impacts on regional

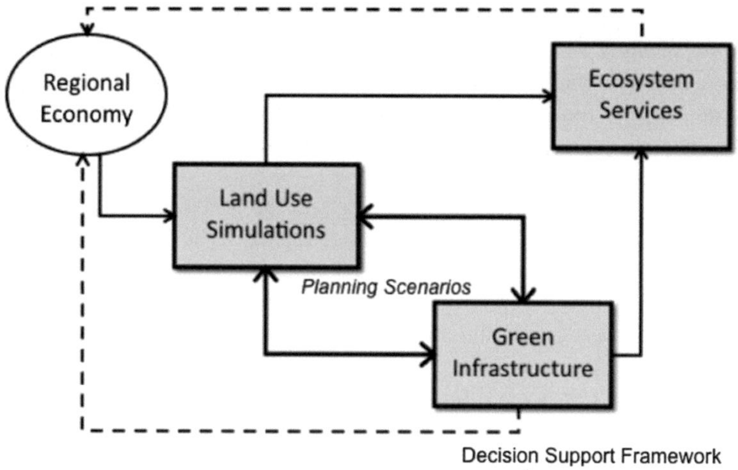

Fig. 1 A decision support framework linking land use modeling, green infrastructure resources, and ecosystem valuation

infrastructure and resources. These simulations have been used in regional planning and decision making across the United States.

This chapter describes a pragmatic first step in an effort to link land-use modeling, green infrastructure resources, and ecosystem valuation in a planning support system (Fig. 1). The PSS loosely couples a dynamic-spatial model to a beneficiaries-based ecosystem valuation model and presents the resulting information in a Web-based portal. The long term vision for this work is to provide relevant information about ecosystem services and green infrastructure at varying scales to leverage land-use decision-making at different political levels across jurisdictions.

We first review the current state of thinking on ecosystem services, their valuation, and green infrastructure. Following this, we briefly describe the LEAM PSS and the value it adds to regional planning processes. We sketch an implementation of how this PSS has been applied to support green infrastructure planning in the state of Illinois in the United States. We conclude with some thoughts about the issues that emerge from this implementation.

2 Ecosystem Services and Green Infrastructure

The urgency of economic conditions, nationally and internationally, has resulted in monumental efforts to revive the economy by increasing public spending, including investments in infrastructure (Editorial New York Times, "Mr. Obama's Energy Future" Published: February 25, 2009). While the debate about this intervention continues, there is increasing recognition that this view of the economy does not adequately include the contributions of ecosystem services (Daly 1997; Ruhl 2006; Farley 2008). "Ecosystem services are the conditions and processes through which natural ecosystems, and the species that make them up, sustain and fulfill human life" (Daly 1997). Further, ecosystems and the services they provide are often argued to be critical natural capital essential to environmental sustainability (Ekins 2003).

Simply put, ecosystems, their functions, and the services they provide are all interrelated and constitute infrastructure fundamental to human health, safety, and welfare. The literature confirms what appears intuitively apparent (the theoretical and practical importance of ecosystem services) and makes a strong argument for including these services in planning and decision making (Wackernagel 1999). However, given the rate at which ecosystems continue to be degraded and converted to other uses, it is clear the importance of green infrastructure is not being recognized quickly enough (Costanza 1997). Further, as major efforts to fix the economy focus on investment in built infrastructure, an opportunistic next step would be to develop real world examples to demonstrate the benefits of also investing in green infrastructure.

2.1 The Role of Valuation

Land-use decisions at different spatial scales can leverage valuations of relevant ecosystem services, and information about green infrastructure at varying scales. Proper valuation analysis can demonstrate the long-term implications of development in socio-economic and environmental terms across a region. The information presented in economic terms, can better guide management decisions about economic development relevant to land acquisition, conservation and restoration efforts especially as they implicate green infrastructure resources. As information moves from the state to regional hubs and further to local jurisdictions, it facilitates dialogue and proactive decision-making across all planning jurisdictions or levels (Fig. 2).

Landscape-scaled ecological models for analyzing and assessing ecosystem services have been developed and applied in various studies in urban and other intensively used landscapes (Opdam et al. 2006; Mörtberg et al. 2007). These models address the relationships between landscape patterns, ecosystem dynamics and biodiversity (Alberti 2010; Johst et al. 2011), including the role of habitat amount and connectivity. In planning applications, such models often result in the outlining of green infra-structure or ecological networks, consisting of core areas with high nature values or habitat patches, connected by corridors for maintaining connectivity (Jongman et al. 2004; Opdam et al. 2006). In these applications, green infrastructure areas can become spatially static and dependent on the legal protection of the appointed areas, while surrounding areas take on the brunt of the development pressures, at times reducing the effectiveness of the green infrastructure systems. Some have argued that the static nature of these solutions inhibit alternative solutions (Hostetler et al. 2011).

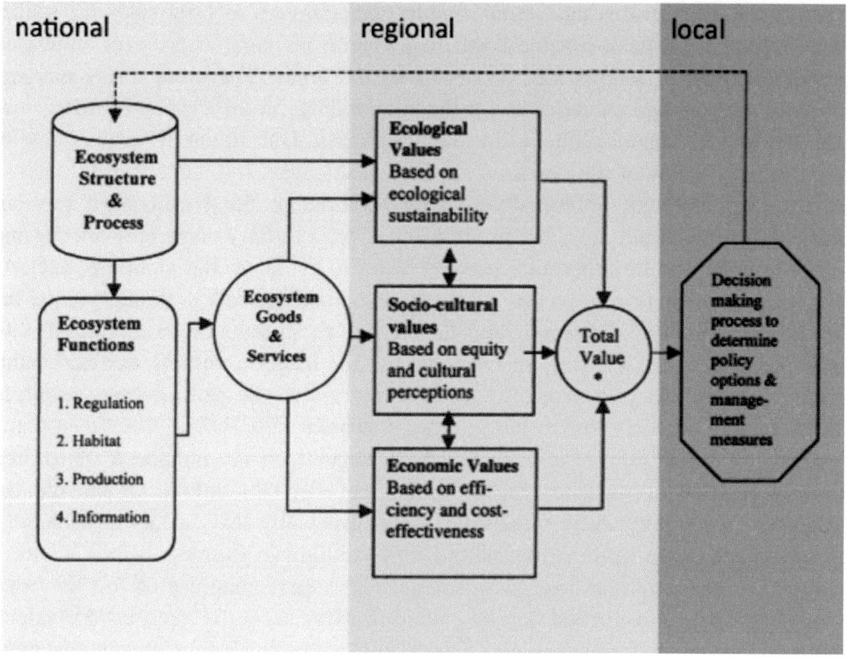

Fig. 2 Hierarchical green infrastructure identification and planning process across scales (from De Groot et al. 2002)

The above and related literature (see De Groot et al. 2002) provides an avenue for discerning the relationship between green infrastructure and planning decisions. One approach uses socio-economic valuation techniques to define the role of ecosystem services in dollar/unit terms. However, due to the difficulty of incorporating its complex valuation structures in public decision-making processes, ecosystems valuation discussions are often left out of critical policy debates.

How a society values its green infrastructure and the services it delivers has been widely discussed in the literature (Vemuri 2006). Many researchers argue that ecosystem services are significantly undervalued (Costanza et. al. 1998; Daly 1996) and that the global economy is but a subsystem of these global services. Many others however, including many economists in the US, consider green infrastructure and ecosystem services to be something outside of what is considered 'the economy' (Daly 1996). Ecosystem services are seen as just another input to industrial, value added, and economically driven processes. This approach misses the economic benefits of standing trees, healthy wetlands, abundant fisheries, and other biota, not to mention the human sense of biophilia that E. O. Wilson contends we as humans need and crave (Wilson 1984).

Natural processes add value (value in an anthropomorphic sense) in a multitude of ways. In economic terms, we derive significant amounts of utility from ecosystems and their green infrastructure services. This utility is both direct and indirect. Direct services have both visible markers such as boot sales for hiking and revenue from park use fees; and less visible markers such as the increased property values of houses located near forested areas. The more direct markers allow us to complete an estimate for the economic value of a green infrastructure resource (a Use Value) without too much difficulty. Unfortunately, we are able to measure only a few of the potential services in this way. Indirect services, such as energy from the sun, nutrient cycling by bacteria, or flood mitigation services performed by wetland areas are more indirect, with tertiary connections to us and as a consequence, are inherently more difficult to evaluate. For example, bacteria turn wastes into food-producing nutrients in the soil, insects pollinate plants, the sun provides energy and water, and plants take these valuable commodities and turn them into food. We pay for the food, but the bacteria, insects, sun and water come naturally—as part of the basic infrastructure of the earth. Nature provides these services without cost in the traditional sense.

Many of these useful services however, depend on the existence of healthy systems (forests or wetlands for example) in order to continue to provide the amenities that we value. For example, trees (especially leafy ones) in urbanized areas can provide a buffer to buildings from sunlight in summer, winds in storm events, etc. These mitigation activities reduce the required engineered solutions that a building owner must invest in. The green infrastructure—the trees, have in effect, replaced (complemented) the grey infrastructure—the structural elements that must be constructed and engineered to resolve the same issues. The ways in which green infrastructure benefits are identified and incorporated into local decisions about land use is one critical cog missing from typical land use discussions.

How does the cost of these services translate into policy decisions? Hanemann (1994) notes that humans are 'cognitive misers' who 'satisfies' or do the minimum when it comes to dealing with complex choices. Due to the difficulty of incorporating indirect values in decision making processes, and the complexity of considering green infrastructure effects, it has been easier for decision makers to avoid the difficult questions on how to handle them. In order to improve policy on these important services, we must first work on ways to reduce the complexity and opacity inherent in assessing their relative value. Equally important is our ability to make information on these services accessible and readily understandable, i.e. reducing the complexity of green infrastructure information. In other words, a municipality must be able assess and compare the relative costs, benefits, and implications of various land use policy scenarios on critical green infrastructure resources (Cropper 2000). Most local governments however, are not properly equipped to discern the value of local green infrastructure, and as such these important assets can be missing from local decision making processes.

The current literature on ecosystem services valuation has primarily focused on the "supply" or the "source" side of the equation—i.e. how much of a critical resource is available and at what cost. A key link to integrating the concepts of green infrastructure and ecosystem services in practice is the identification of the "demand" side. It answers questions of not only how much, but how quickly will it will be used. Disaggregated valuation information that is localized and spatially explicit also addresses questions of where the resources are and where the demand originates. When linked to spatially explicit green infrastructure data, the relationship between the values and the local costs of physical replacement becomes clear and these costs appear avoidable.

Arguments against ecosystem-service valuation notwithstanding [i.e. nature cannot be appropriately assigned a dollar value (Cobb 1994)], it is clear that valuation helps bridge the gap between economic and ecological perspectives for sustainable development. The point is not to precisely determine the value of nature, but highlight the finite nature of these resources.

2.2 The Case of New York's Water Supply

A widely circulated example of the hidden values of ecosystem services can be found in the case of New York City and its nearby Catskills Mountains. The city historically piped its municipal drinking water from a watershed in the Catskills. The mostly wooded watershed produced clean, clear, good tasting water that required little treatment. Development in the Catskills began booming in the 1990s however, enough to cause the water quality of the watershed to drop below EPA minimum standards. Estimates for mechanical treatment of the water were has high as 8 billion dollars. The EPA waived its treatment requirements in exchange for an agreement that New York City take measures to protect the natural filtration systems of the Catskills watershed. To date, the city has purchased over 300,000 acres of natural lands, put around $25 million into agricultural best practice programs, subsidized sewage treatment, bought agreement from Catskills communities with economic development funding, and implemented numerous other watershed restriction measures. The plan has seen success; the city has the largest unfiltered system in the nation, and produces higher quality water than it did in 1993. The watershed protection actions have cost, in total, around $1.5 billion (Chichilnisky 1998).

In this case, the choice was clear; implement an alternative plan fast, or start building an extremely expensive filtration plant. Unfortunately, most local government decisions are far more nuanced, and it is not always so readily apparent which policy is more cost-effective. As the New York example shows, providing local decision-makers with better information about the possible impacts of development on ecosystem services and potential alternatives can lead to a more enlightened decision-making process that can save the municipal government money and improve the sustainability of local development decisions.

2.3 The Case of Johnson Creek in Portland

The Lents area of southeast Portland is one of the oldest and most diverse neighbourhoods in the city. It is also an area that experiences a high risk that Johnson Creek will flood each winter. The creek's past history includes: 37 flood events since 1941 with 28 of these resulting in direct property damage and the remainder 'nuisance events' (10-year flood event or less) (from Evans 2004).

Storage of the nuisance events was the intended target of the 2002 Johnson Creek Restoration Plan. It was intended to manage flood storage, water quality, and fish and wildlife habitat, while promoting natural floodplain function (Evans 2004). In its 2004 follow up study titled: Comparative Valuation of Ecosystem Services: Lents Project Case Study, the city attempted to analyze the tradeoffs in approaches between flood control measures that that provide ecosystem function benefits and those that do not. Questions considered for analysis include: What is the return on investment for ecosystem service-oriented projects and in different locations?

The results of their comparative valuation analysis found that using natural or ecosystems based approaches for the flood abatement project could provide services of more than $30 million in economic value to the city over a 100-year time period (~ $300,000/year). It also found that 5 ecosystem services would see an increase in productivity at quantifiable levels as a result of floodplain improvements that included riparian restoration measures.

These examples demonstrate how the identification of the causal relationship between ecosystems and the benefits that humans derive from them would help in significant cost-savings and preservation efforts within highly urbanized areas. The methods used also recognized the long-term implications for their efforts and found cost-effective ways to resolve the issues. Similar fore-sight in planning for urban areas that could done more quickly would help in sustainable development by incorporating the ecological and economic perspectives in more of the day-to-day or routine decision making tasks. The following sections show the potential of using the existing technologies described above in land-use modeling and information dissemination for assessing the relative costs of various policy and investment choices.

3 Planning Support Systems

During the last two decades, spatial analysis tools, geographic information systems (GIS), and remote-sensing (RS) technologies have been widely implemented to monitor, analyze and visualize urban growth phenomena. Although GIS-based tools provide useful analysis and have been widely used to assist urban planners, the static representations on which they are built are not sufficient to study the dynamics of urban systems (Klosterman 1999; Hopkins 1999). Maps and satellite images are limited to static displays of past and current data sets. These images merely portray the current state of a system, offering neither the reasons why things appear as they do nor projections of possible future states. The causal mechanisms associated with land use change are still not well understood, in part due to the complexity of urban systems.

Computer-based urban system simulation models are now being employed to forecast and evaluate land use changes (Brail 2008; and Deal 2008). Simulation models enable planners to view and analyze future outcomes of decisions and policies before they are put into action. These models have the ability to help improve our fundamental understanding of land use transformation dynamics and the complex interactions between urban change and sustainable systems (Brail 2001; Deal 2001, 2009).

3.1 Land Use Evolution and Impact Assessment Model

LEAM is one such PSS and seeks to address the need in planning and policy-making for answers to what-if? and so-what? questions (Deal 2003, 2008; Deal and Pallathucheril 2007). Fundamentally, LEAM consists of two major organizational parts: (1) a land-use change model (LUC)—defined by a dynamic set of sub-model drivers that describe the local causality of change and enable easy addition and removal of variables and the ability to play out 'what-if' scenarios; and (2) impact assessment models that facilitate interpretation and analysis of land-use change depending on local interest and applicability—these help to answer 'so-what' questions and explicate the consequences of a land-use scenario.

The LEAM LUC model utilizes a structured lattice surface with state-change conditions that evolve over time. The lattice surface, however, is not flat but shaped by both biophysical and socioeconomic factors. The land-use transformation potential of individual cells is evaluated by explicitly quantifying the forces (drivers) that contribute to change. Understanding the causal mechanisms of change provide local decision makers an opportunity for testing policy and investment choices and are a critical component for completing scenario-planning exercises. Driver sub-models are locally dependent and derived through both analysis and local stakeholder interaction. An open architecture and modular design facilitates incorporation of additional local drivers needed to improve the explanatory power of the model.

A regional econometric, input–output model determines the regional demand for residential, commercial and open space land (Sarraf et al. 2005). Households and jobs are forecast and converted into land demand using sector-based economic and demographic analysis instead of sub-regional constraints on demand used in other approaches (see Wu and Martin 2002). The estimated demand for residential and commercial development serves as a target for regional land allocation. Market variables increase or decrease development rates based on how well the regional demand targets were met or not met. The model then generates likely development pattern simulations that help in visualizing the projected change. Figure 3 is the probability surface for the Chicago metropolitan region, and Fig. 4 the associated land-use change map.

Once model simulations are established, scenario descriptions of alternative land-use policies, investments decisions, growth trends and unexpected events

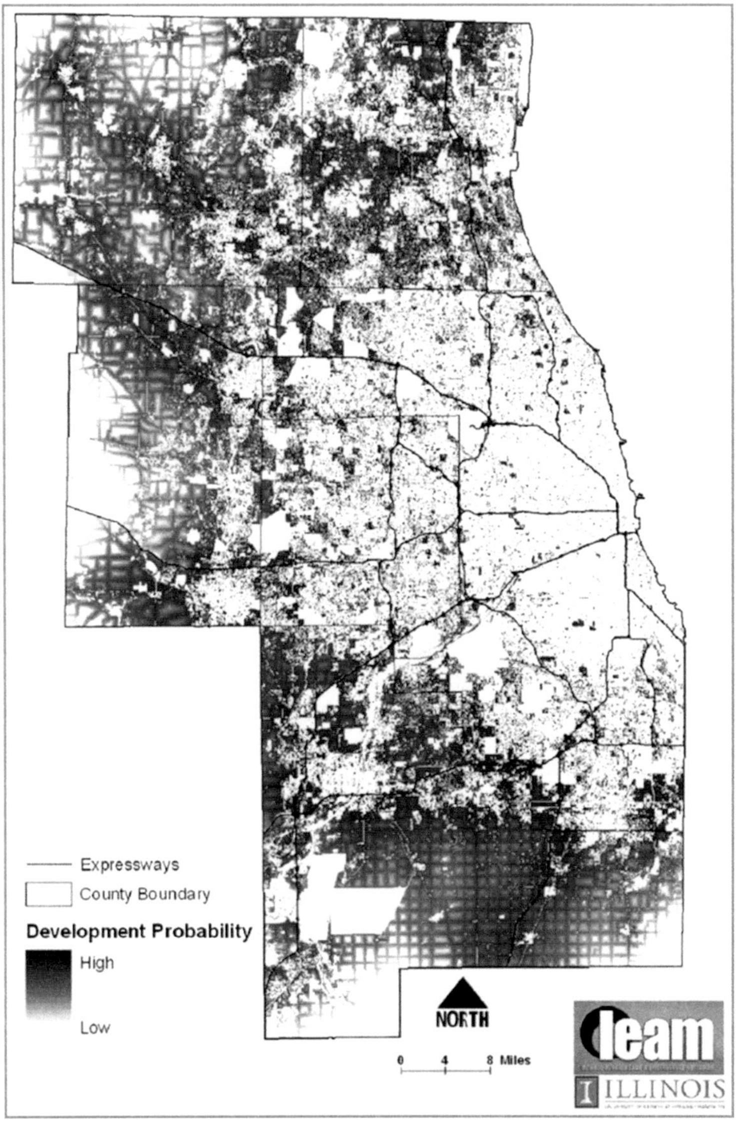

Fig. 3 A LEAM probability map for the metropolitan Chicago region. *Darker areas* are areas with higher probabilities for development

(among others) can be tested, analyzed and compared for regional importance. Simulated outcomes are described in graphs, charts, text and map form and are used in engaging local dialogue and in analyzing the potential implications of the changes described (Deal and Schunk 2004). The assessment of probable impacts is important for understanding the 'so what' of scenario simulations. A visual and quantitative representation of each scenario's outcome provides both an intuitive

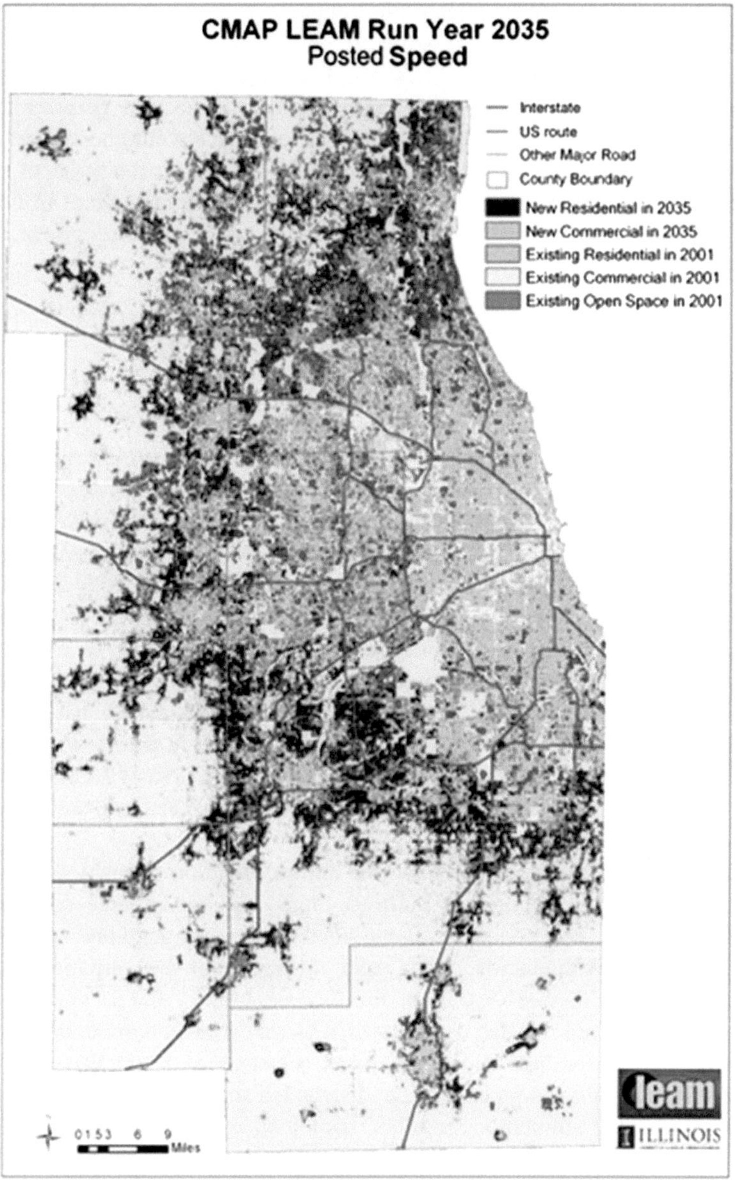

Fig. 4 A LEAM change map for the metropolitan Chicago region. *Dark areas* are areas that developed over the 30 year modeled timespan

means of understanding and a basis for analyzing the implications of potential decisions. These representations act as a catalyst for discussion and communal decision-making.

LEAM scenarios can explicate the impact of diverse public policy and investment choices on green infrastructure and its critical component ecosystem resources (Deal and Pallathucheril 2009). This approach reveals areas for mitigation and conservation, and also highlights the connection between natural resources and a region's economic vitality. The approach would be most effective if it is scaled up beyond regional jurisdictional borders to the scale of natural boundaries. Highlighting green infrastructure and ecosystem services in this way could hasten the departure from the traditional practice of development from a strictly economic perspective.

3.2 Visualization and Access to Information

The transformation from data into information is a vital part of any PSS. The importance of effective visualization techniques for dealing with complex data sets has been long recognized, especially in the field of natural resource research (e.g. Ekstrom 1984; Rosenfeld and Kak 1982). GIS has provided a much needed technological boost in quickly analyzing large amounts of spatial data. On the other hand, these systems impose an extremely steep learning curve and are relatively inaccessible to a vast majority of people.

Visualization facilitates extraction of information from complex data sets by providing salient information in clear and concise terms; they can provide a 'snapshot' of information that the user might find useful in decision making. But what happens when there are thousands of snapshots to view? LEAM produces vast quantities of data on a myriad of planning scenarios. How do we make them accessible and useful?

In LEAM, access to complex data sets and information is enabled via a GeoPortal. The LEAM GeoPortal is built on Plone, the open-source content management system (www.plone.org). Plone objects were designed and implemented for storing LEAM scenarios, plans, and other spatial and aspatial data. An experimental MapViewer (based on the Google Maps API) is embedded and visualizes spatial data sets derived from LEAM, valuation and green infrastructure analysis. Map exploration allows the usual selection of raster and vector map overlays, the ability to zoom and pan, and an added ability to dial back and forth in time within LEAM simulations results. This allows users to viscerally experience land-use changes associated with complex policy and investment scenarios (and associated changes in valuation) in a systematic manner.

The GeoPortal allows content of different types—from text to images to GIS layers—to be managed in a consistent and uniform manner, effectively simplifying the reporting and analysis process. Full text searches can be combined with spatial queries to help locate information and data of interest. A built-in workflow capability supports a create-review-publish cycle for new content. This vastly simplifies participation by numerous individuals each with a different role to play and with differential access to the content of the GeoPortal.

The advantages of this approach are numerous. The GeoPortal provides users with a simplified view into highly complex data sets. Users are today familiar with using and interacting with Web-based map services. Users can view the data in satellite mode and relate information to on-the-ground realities. Because of the ability to pan, zoom and move around the data, users get a real sense of how land use changes might affect their community. They can locate themselves in space and identify issues of primary importance to them relatively quickly and easily.

4 Support for Planning Green Infrastructure in Illinois

Working with the Illinois Department of Natural Resources, the authors are delivering information (green infrastructure identification and analysis, valuation, modeling results) for the state of Illinois. This section discusses the concepts and conceptual framework for undertaking such a task along with the potential for such practices to succeed in other areas and at different scales.

4.1 The Illinois Green Infrastructure GeoPortal

To support planning of green infrastructure in Illinois, an instance of the GeoPortal was created and populated with content from different parts of the state (www.leam.illinois.edu/ilgreeninfrastructure). Figures 5 and 6 provide two views of the GeoPortal. The Illinois Green Infrastructure (ILGI) GeoPortal integrates activities and data that lead to plan formulation and policymaking by facilitating access to content. This creates real opportunities to engage decision makers, stakeholders and the general public in the process of making both plans and planning decisions especially when paired with information on green infrastructure resources and the stresses that urbanization places on them.

The GeoPortal facilitates exploring potential development scenarios within a river basin while tracking their implications to the regional green infrastructure system and the related costs to that system can be done easily and succinctly by non-technical decision-makers. The ability to manage content at higher levels of aggregation means that green infrastructure plans can be collections of one or more digital documents. In the case of policy scenarios, this could be time series data of varying scenario attributes. Data can be aggregated and disaggregated at a relevant scale to initiate dialogue. Information is available to the most-relevant scale of planning, allowing for both top-down and bottom-up discussions.

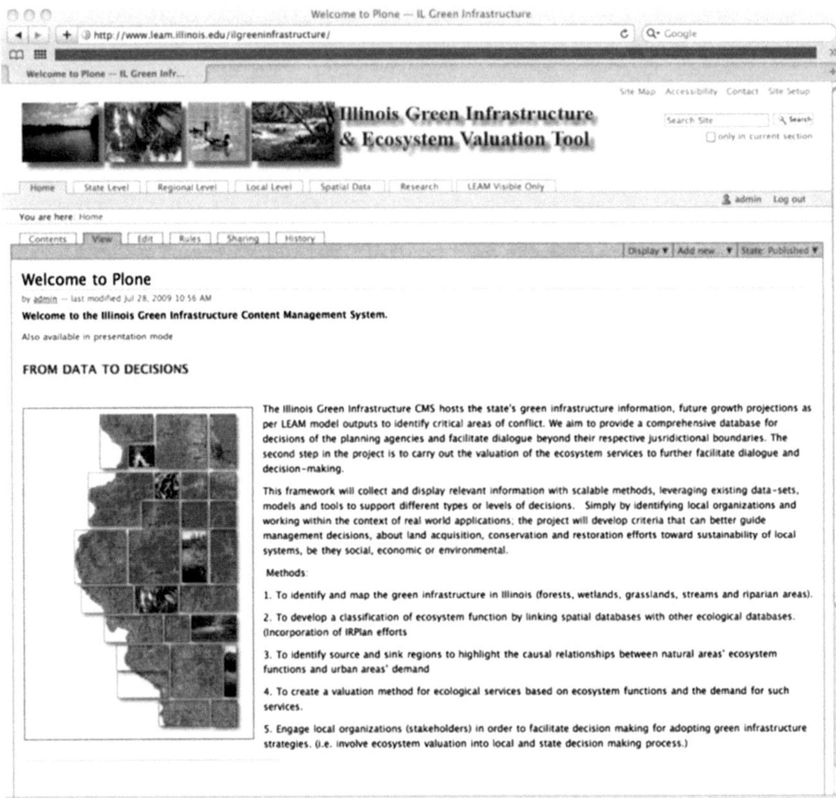

Fig. 5 The Illinois Green Infrastructure GeoPortal (ILGI GeoPortal) a web based planning and decision support system

4.2 Implementation

Various parts of the ILGI GeoPortal have been implemented and partially tested. This work represents a first step in an effort to (1) link land-use modeling, green infrastructure resources, and ecosystem valuation techniques and (2) present the outcomes from various analyses in a Web-based information portal. Various aspects of this functionality are described below and include (1) Mapping green infrastructure and modeling future land-use, (2) Classifying services and estimating demand, (3) Accounting for regional variations in demand and use, (4) Valuing ecosystem services, and (5) Incorporating valuations in resource protection scenarios. These are further discussed below:

1. *Mapping green infrastructure and modeling future land-use*. These are two fundamental analytical steps both of which raise issues of spatial scope and resolution. Statewide green infrastructure can be mapped from land-cover data

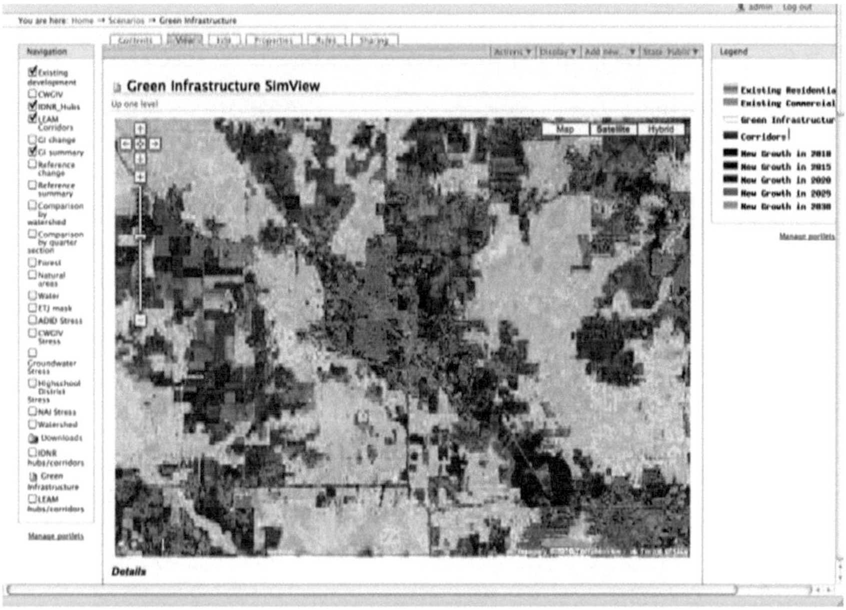

Fig. 6 Detail of a green infrastructure map of McHenry County in satellite mode. *Light areas* are identified green infrastructure areas. *Dark areas* are areas of potential future growth in the county

(Fig. 7). The extent and configuration of green infrastructure is a function of spatial and functional contiguity. The process of modeling land-use change is described above in Sect. 3.1 and produces a number of land-use scenarios (a future state of land-use and the factors that outcome).

The spatial scope and resolution of these analyses are an issue. While data are gathered at state and even national levels, the precision and relevance of these data at the local level are often questionable. Considerable ground truthing and evaluation are required at the local level, which provides an opportunity for also eliciting buy-in from local stakeholders. Statewide land-use models also pose challenges. While relevant land use decisions take place at fine scales, statewide dynamic models and their output require hundreds of millions of computations (LEAM uses a 30 $\times$ 30 m grid). While computationally feasible, doing so would require access to extraordinary computing resources. The approach we have taken is to disaggregate the process into regional areas, then aggregate the information back statewide through a networked series of GeoPortals. An example of regional green infrastructure resources (for McHenry County, IL) accessed via a GeoPortal can be seen in Fig. 8. Knowledge aggregated from these portals are key to effective deliberation and policy-making at the state level.

2. *Classifying ecosystem services and estimating demand.* A number of approaches can be taken to classifying ecosystem services, and some of these are

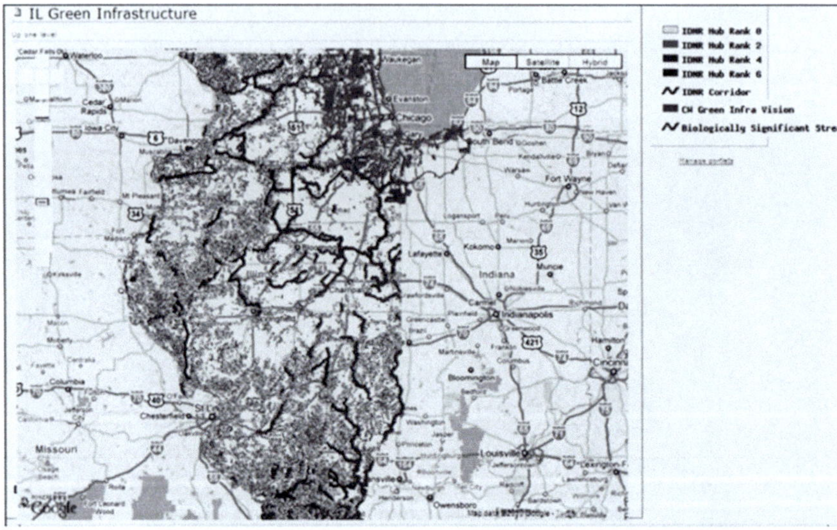

Fig. 7 A map of green infrastructure resources in the state of Illinois as seen through the GeoPortal map viewer. Hubs are *grey* and corridors are *dark*

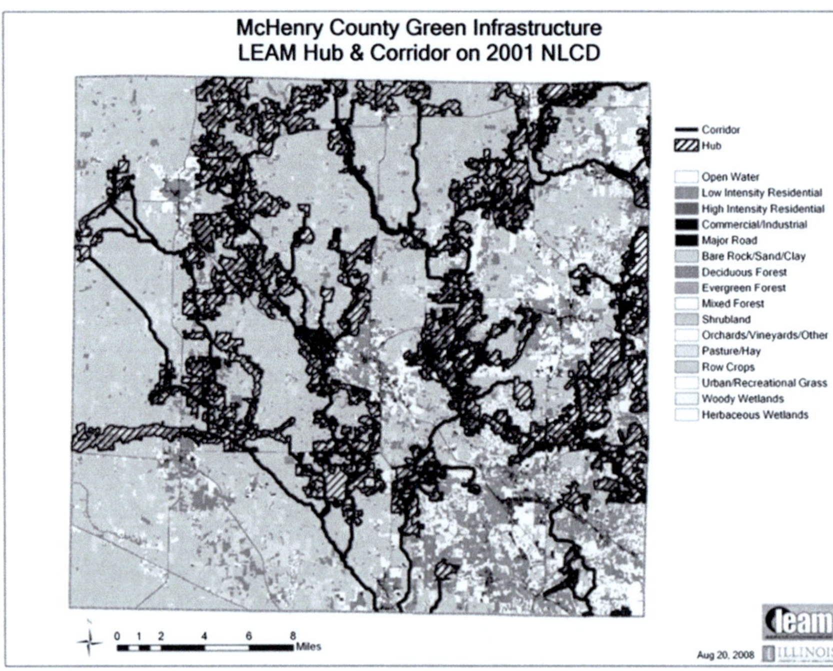

Fig. 8 A map of green infrastructure resources for McHenry County Il in the Chicago metro area. *Dark areas* are identified green infrastructure resources in the county

discussed above. In our case, we took a beneficiaries-based approach. We did this in order to improve development decisions by enhancing the understanding of ecosystem services among planners and developers. A beneficiaries-based approach takes into account the socio-cultural and economic values of different regions to arrive at a total, community-oriented value of any given ecosystem service. Representing services in direct economic terms makes the process more transparent and allows clearer communication between stakeholders and decision makers.

3. *Accounting for regional variations in demand and use of ecosystem services.* A particular ecosystem service, say storm water retention, may be more valued in the more developed Chicago metro region as compared to smaller metro regions downstate. Such variations must be reflected in the statewide picture that is being assembled rather than applying a one-size-fits-all value. Similarly, ecosystem degradation and its implications would be more pronounced in regions, such as the Chicago metro region, where the quantity and intensity of urban development is greater. As a result, the need for action may be more urgent in some places rather than others. To further complicate the picture, the source and the sink for an ecosystem may reside in two separate jurisdictions. This will necessitate inter-governmental and inter-agency dialog to arrive at a consensus on the demand and use of such services.
4. *Valuing ecosystem services within identified service areas.* Generalized Source-Sink Models (GSSMs) and ARtificial Intelligence for Ecosystem Services (ARIES) could help in identifying causal relationships between different parts of the landscape. For instance, regions of high impervious surface likely indicate source areas and dense canopy cover are indicators of sink areas. This kind of knowledge about existing natural resources is recorded in each one of 30×30 m gridcells within the region. These data provide a basis for identifying the interconnections between the source and sink areas and estimating the total value of the ecosystem services provided.
5. *Incorporating ecosystem valuation into resource protection and management scenarios.* At this point, the demand for various ecosystem services has been established and the location of beneficiaries identified. The direct implications of ecosystem degradation is then assessed in a particular region within the state. LEAM simulations of different regional futures are typically associated with different probabilities of change in land-use. The intensity and type of development in each cell would be the basis for estimating future resource demand in each scenario. The resulting change in ecosystem services and cost to the beneficiaries is used to evaluate different growth scenarios (Figs. 9 and 10). Alternative scenarios can be tested to reveal the effectiveness of conservation and restoration policies, to understand the long-term implications, and to arrive at the most sustainable alternative. Comparison maps of policy choices are then generated to show differences. We note that values of ecosystem service may not always decrease. The value of an ecosystem service could increase in the beneficiaries-based approach if a higher value is assigned to a fixed service supply as demand increases.

Ecosystem Service (Existing)

5.00	7.00	8.00	10.00
13.00	15.00	10.00	12.00
12.00	10.00	14.00	15.00
6.00	9.00	7.00	10.00

X

Probability as/ LEAM

0.05	0.01	0.03	0.05
0.03	0.05	0.10	0.07
0.04	0.06	0.15	0.08
0.06	0.09	0.20	0.10

Change is ES

0.25	0.07	0.24	0.50
0.39	0.75	1.00	0.84
0.48	0.60	2.10	1.20
0.36	0.81	1.40	1.00

ES (Reference Scenario)

4.75	6.93	7.76	9.50
12.61	14.25	9.00	11.16
11.52	9.40	11.90	13.80
5.64	8.19	5.60	9.00

Ecosystem Service (Existing)

5.00	7.00	8.00	10.00
13.00	15.00	10.00	12.00
12.00	10.00	14.00	15.00
6.00	9.00	7.00	10.00

X

Probability as/ LEAM

0.07	0.05	0.09	0.08
0.00	0.00	0.00	0.00
0.00	0.00	0.00	0.00
0.10	0.20	0.20	0.15

Change is ES

0.35	0.35	0.72	0.80
0.00	0.00	0.00	0.00
0.00	0.00	0.00	0.00
0.60	1.80	1.40	1.50

ES (Conservation Easement Scenario)

4.65	6.65	7.28	9.20
17.00	**20.00**	**15.00**	**14.00**
17.00	**15.00**	**20.00**	**17.00**
5.40	7.20	5.60	8.50

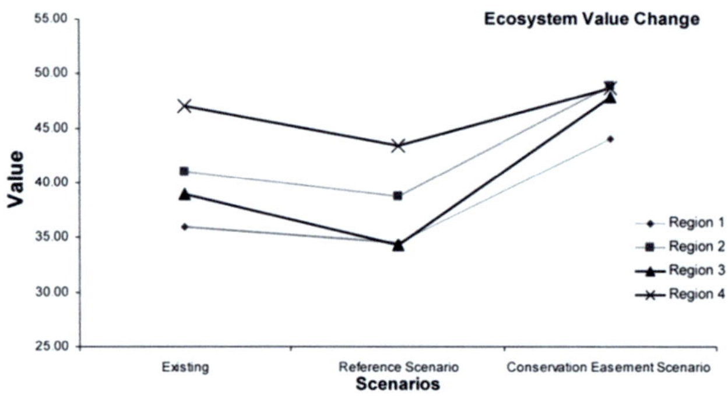

Fig. 9 Quantifiable ecosystem service 'value' scenario analysis using LEAM. For each cell, LEAM probabilities are multiplied by ES valuation for a final score—that can be mapped over space and time

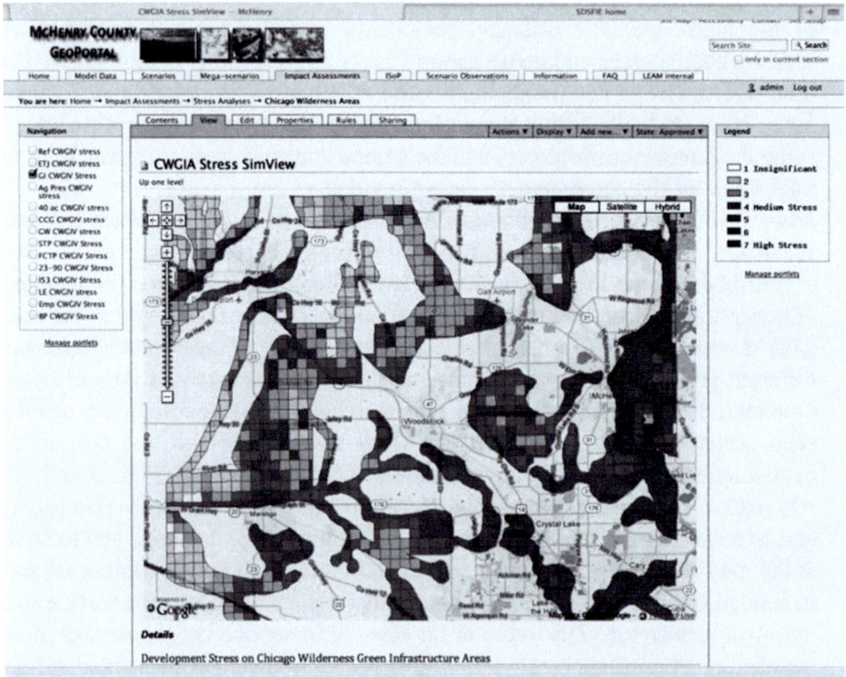

Fig. 10 An ecosystem 'score map' for McHenry County, IL. *Darker areas* have higher valuation scores

5 Conclusion

A pro-active approach to sustainable development makes use of modeling, simulation, and information management technologies to identify critical resources, green infrastructure networks, their value to the community, and probable future areas of future urban transformation within an accessible planning support system framework. This framework would help in both identifying areas for mitigation and conservation, but also highlight the connection between natural resources and a region's economic vitality. The issue would be most effectively addressed if it is analyzed regionally but scaled up beyond regional jurisdictional borders to appropriate scale of natural boundaries. The incorporation of green infrastructure and ecosystem services in the decision-making dialog would aid in a departure from the traditional practice of development from a strictly economic perspective.

The ILGI GeoPortal represents such a system. The GeoPortal attempts to provide transparency (all data used in the portal is downloadable), accessibility (Web-based), and manageability (using a content management system). The system enables access to data and analysis on green infrastructure resources, their value and the potential for urbanization to degrade them. Its ultimate purpose however, is to allow users to engage with planning decisions based on better comprehension. In terms of planning process, it also can provide a place for interactive forums for discussion, collective work, and commentary. Different levels of access to the GeoPortal can be given to different participants, while maintaining a common working area. Also, information can be made open to public to make the reasoning behind the planning decisions clear. The GeoPortal can also help remove overlaps in information and facilitate an open and collaborative process.

Easy access and direct comparative analysis of different policy or investment scenarios for areas beyond the region (that include their implications to regional ecosystem services) can help facilitate inter-regional dialogue. This is important because changes in one area/jurisdiction may affect the ecosystem services of another region. The impact on the beneficiaries of services is also affected by investments not made in their region. This is critical as the tradeoffs among the direct commercial benefits of economic development can be balanced with conservation and restoration actions that are coordinated between the affected areas within the region. This could foster a cooperative community of regions instead of the traditional regional basis for analysis where outcomes are in effect, a zero-sum game. The decisions regarding the management and conservation of resources would continue to be made at the local level, with information about the long-term implications of the actions made available at a statewide scale for inclusion in larger scale policy deliberations.

Our use of the phrase traditional economic development in describing our work does not mean we disregard the economic perspective. We in no way want to diminish the importance of traditional economic development in terms of jobs, income and other typical economic measure of growth and progress. This work does suggest however, that sustainability would be better served by a progressive economic transformation that included both traditional and non-traditional measures. This transformation would include the integration of these modern planning tools, better accounting for and valuing natural resources and the services they provide into a more efficient and effective planning and decision making process.

Future work lies in a more methodical and systematic pilot application and refinement. Testing of alternative valuation approaches, and ways to infiltrate local planning processes using the technologies described along with a better understanding of the traditional economic implications of the processes.

Acknowledgments We are grateful for the fine work of others that precede this work. We are especially grateful to Yong Wook Kim and Meghna Dutta for their support and dedication.

References

Alberti M (2010) Maintaining ecological integrity and sustaining ecosystem function in urban areas. Curr Opin Environ Sustainability 2:178–184

Brail RK, Klosterman RE (2001) Planning support systems: integrating geographic information systems, models and visualization tools. ESRI Press Redlands, California

Budthimedhee K, Li J, George RV (2002) ePlanning: a snapshot of the literature on using the World Wide Web in Urban Planning. J Plan Lit 17(2):227–246

Chichilnisky G, Heal G (1998) Economic returns from the biosphere. Nature 391(6668):629

Cleveland WS (1985) The elements of graphing data. Wadsworth Advanced Books and Software, Monterey

Costanza RV, Alexey (2004) Landscape simulation modeling:a spatially explicit, dynamic approach series. Springer, New York

Costanza R, d'Arge R, de Groot R, Farber S, Grasso M, Hannon B, et al (1998/4) The value of the world's ecosystem services and natural capital. Ecol Econ 25(1): 3–15

Cox DJ (1990) The art of scientific visualization. Acad Comput 46:20–56

Cropper ML (2000) Has economic research answered the needs of environmental policy? J Environ Econ Managt 39(3):328–350

De Groot RS, Wilson MA, Boumans RMJ (2002) A typology for the classification, description and valuation of ecosystem functions, goods and services. Ecol Econ 41(3):393–408

Deal B (2001) Ecological urban dynamics: the convergence of spatial modeling and sustainability. J Building Res Inf 29(5):381–393

Deal B (2008) Sustainable land-use planning: the integration of process and technology. Academic Publishing, VDM Verlag, Saarbrücken

Deal B, Pallathucheril V (2003) The land evolution and impact assessment model (LEAM): will it play in Peoria? In: proceedings of the 8th international conference on computers in urban planning and urban management, Sendai, May 27–29

Deal B, Pallathucheril V (2007) Developing and using scenarios. In: Hopkins LD, Zapata MA (eds) Engaging the future: forecasts, scenarios, plans, and projects. Lincoln Institute for Land Policy, Cambridge Massachusetts, pp 221–242

Deal B, Pallathucheril V (2009) Sustainability and urban dynamics: assessing future impacts on ecosystem services. Sustainability 1(3): 346–362. (http://www.mdpi.com/2071-1050/1/3/346)

Deal B, Schunk D (2004) Spatial dynamic modeling and urban land use transformation: a simulation approach to assessing the costs of urban sprawl. J Ecol Econ 51(1–2):79–95

Deal B, Sun Z (2006) A spatially explicit urban simulation model: landuse evolution and impact assessment model (LEAM). In: Ruth M (ed) Smart growth and climate change: regional development, infrastructure and adaptation. Edward Elgar, Inc., Northampton, pp 181–203

Ekstrom MP (1984) Digital image processing. Academic Press, San Diego
Evans D (2004) A comparative valuation of ecosystem services: lents project case study. Prepared for city of Portland watershed management program. Eco Northwest
Farber SC, Costanza R, Wilson MA (2002/6) Economic and ecological concepts for valuing ecosystem services. Ecol Econ 41(3): 375–392
Geertman S, Stillwell J (2003) Planning support systems in practice. Springer, New York
Hanemann WM (1994) Valuing the environment through contingent valuation. J Econ Perspect 8(4):19–43
Hopkins LD (1999) Structure of a planning support system for urban development. Environ Plan B 26:333–343
Hopkins LD, Zapata MA (eds) (2007) Engaging the future: forecasts, scenarios, plans, and projects. Lincoln Institute, Cambridge
Hostetler M, Allen W, Meurk C (2011) Conserving urban biodiversity? Creating green infrastructure is only the first step. Landscape Urban Plan 100:369–371
Johst K, Drechsler M, van Teeffelen AJA, Hartig F, Vos CC, Wissel S, Wätzold F, Opdam P (2011) Biodiversity conservation in dynamic landscapes: trade-offs between number, connectivity and turnover of habitat patches. J Appl Ecol 48:1227–1235
Jongman RHG, Külvik M, Kristiansen I (2004) European ecological networks and greenways. Landscape Urban Plan 68:305–319
Kammeier HD (1999) New tools for spatial analysis and planning as components of an incremental planning-support system. Environ Plan B: Plan Des 26:365–380
Klosterman RE (1999) The what if? Collaborative planning support system. Environ Plan B 26:393–408
McDonald L, Benedict M, O'Conner K (2005) Green infrastructure plan evaluation frameworks, J Conserv Plan 1:12–43. http://www.journalconsplanning.org/2005/volume1/issue1/allen/manuscript.pdf
Mörtberg UM, Balfors B, Knol WC (2007) Landscape ecological assessment: a tool for integrating biodiversity issues in strategic environmental assessment and planning. J Environ Manage 82:457–470
Nussbaum MC, Sen A (1993) The quality of life. Oxford University Press, New York
Opdam P, Steingröver E, Rooij SV (2006) Ecological networks: a spatial concept for multi-actor planning of sustainable landscapes. Landscape Urban Plan 75:322–332
Rozenfield A, Kak AC (1982) Digital picture processing. Academic Press, San Diego
Ruth M (2006) Smart growth and climate change: regional development, infrastructure and adaptation. Northampton. Edward Elgar, Inc, Massachusetts
Sarraf S, Pallathucheril VG, Donaghy K, Deal B (2005) Modeling the regional economy to drive land-use change models. Paper presented at the 46th annual conference of the association of collegiate schools of planning, Kansas City, Missouri
Sipes JL (2003) Visualizing community form—new digital tools help communities shape their own futures (CommunityViz, a program for landscape architects involved with community planning, land use planning, and natural resource management). Landscape Archit 93: 56–61
Waddell P (2002) UrbanSim—modeling urban development for land use, transportation, and environmental planning. J Am Plan Assoc 68:297–314
Wu FL, Martin D (2002) Urban expansion simulation of Southeast England using population surface modelling and cellular automata. Environ Plan A 34:1855–1876
Wu FL, Webster CJ (2000) Simulating artificial cities in a GIS environment: urban growth under alternative regulation regimes. Int J Geog Inf Sci 14:625–648

Urban CO_2 Planning: A Decision Support System

Ivan Blecic, Arnaldo Cecchini, Matthias Falk, Serena Marras, David R. Pyles, Donatella Spano and Giuseppe A. Trunfio

Abstract Patterns of urban development affect flows of material and energy within urban settlements and exchanges with their surroundings. In recent years the quantitative estimation of the components of the so-called *urban metabolism* has increasingly attracted the attention of researchers from different fields. To contribute to this effort we developed a modelling framework for estimating carbon exchanges in relation to alternative land-use scenarios. The framework bundles three components: (1) a Cellular Automata model for the simulation of the urban land-use dynamics; (2) a transportation model for estimating the variation of the transportation network load and (3) the Advanced Canopy-Atmosphere-Soil

I. Blecic (✉) · A. Cecchini · G. A. Trunfio
DADU—Department of Architecture, Design and Urban Planning,
University of Sassari, Alghero, Italy
e-mail: ivan@uniss.it

A. Cecchini
e-mail: cecchini@uniss.it

G. A. Trunfio
e-mail: trunfio@uniss.it

M. Falk · D. R. Pyles
LAWR—Land, Air and Water Resources, University of California, Davis, CA, USA
e-mail: mfalk@ucdavis.edu

D. R. Pyles
e-mail: rdpyles@ucdavis.edu

S. Marras · D. Spano
DipNeT—Department of Science for Nature and Environmental Resources,
University of Sassari, Alghero, Italy
e-mail: serenam@uniss.it

D. Spano
e-mail: spano@uniss.it

S. Marras · D. Spano
CMCC—Euro-Mediterranean Centre for Climatic Changes, IAFENT-Sassari, Sassari, Italy

S. Geertman et al. (eds.), *Planning Support Systems for Sustainable Urban Development*, Lecture Notes in Geoinformation and Cartography,
DOI: 10.1007/978-3-642-37533-0_12,

Algorithm (ACASA) model tightly coupled with the mesoscale weather forecasting model WRF. We present and discuss the results of an example application on the City of Florence.

1 Introduction

What is the connection between urbanisation and climate change?

It comes, of course, as a tautology to say that cities and human settlements in general are a primary source of emissions responsible for human-induced climate change. The question is then not *if*, but *how* different patterns of urbanisation and land use influence flows of energy and materials responsible for climate change. For example, sprawled, low-density cities usually have higher per capita energy consumption for transportation than compact cities and, thus, higher carbon emissions. Human activities in the city, such as heating, and the characteristics of buildings play an important role in the dynamics of carbon and heat fluxes, as do the extension and distribution of urban green areas, city's overall infrastructure and the degree of industrial development.

But this relation also runs in the opposite direction: the climate itself influences urban matter and energy fluxes. On the most obvious account, a city with continental climate consumes more energy for heating and cooling than does a city with a more temperate climate. This is to say that, in general, all the urban fluxes of matter and energy interact in a complex way with the local climate and weather conditions.

Our initial question cannot then but turn to multidisciplinary efforts for answers. One such ongoing effort is to measure, estimate and model the so-called urban metabolism (Wolman 1965; Newman 1999), that is, the fluxes of energy and matter between urban fabric, landscape and the atmosphere. Several advanced models, operating at different spatial and temporal scales, have been developed for this purpose. Related to that, scholars have developed tools and quantitative indicators to support urban planning and management for a more sustainable metabolism in future cities.

We here present a contribution to this ongoing effort: a software framework for modelling and estimating carbon fluxes for alternative future scenarios of urban development.

In Sect. 2 we outline the architecture and illustrate some characteristics of the framework's main components. In Sect. 3 we present an application on the City of Florence. In Sect. 4 we offer few conclusive remarks.

2 The Components of the Framework

The modelling framework we present bundles three main components: (1) a Cellular Automata model of urban land-use dynamics (White et al. 1997; White and Engelen 2000; Blecic et al. 2009); (2) a transportation model for estimating the impact of different land-use scenarios on the transportation network load (Tsekeris and Stathopoulos 2003, 2006); (3) a Soil–Vegetation–Atmosphere Transfer model (SVAT) (Pyles et al. 2000, 2003; Staudt et al. 2010; Marras et al. 2010, 2011; Falk et al. 2010), tightly coupled with the mesoscale weather forecasting model WRF (Skamarock et al. 2008) for simulating interactions related to carbon fluxes between the city, environment and local weather.

Figure 1 illustrates the basic mechanics of the modelling framework: the first step in the workflow is the land-use dynamics CA module which generates future land-use scenarios, that is, maps of possible future land uses. The input for the CA module are the maps of current land uses and street network, the zoning regulations by the planning authority, the physical suitability of the cells to develop into specific land uses, and a set of alternative projections of the aggregate demand for different land uses (that demand may be derived from independent studies or external (off-line) models).

The maps of future land uses generated by the CA module, together with the street network, are next fed into the transportation module to estimate future road traffic. The current road traffic data are used for its calibration.

Lastly, both future land uses and road traffic are used by the coupled WRF-ACASA model which simulates the carbon exchanges between soil, vegetation and atmosphere, and their interaction with local weather. WRF-ACASA's output are the estimated future maps of CO_2 fluxes in the urban area under consideration.

In the following three subsections we present the specifics of the three main components.

2.1 The CA-Based Land-Use Model

For the simulation of land-use dynamics we adopted a well known Constrained Cellular Automata (CCA) modelling approach (White et al. 1997; White and Engelen 2000). We set three design objectives before our CA module: (1) the ability to operate at a reasonably high spatial resolution; (2) the possibility to

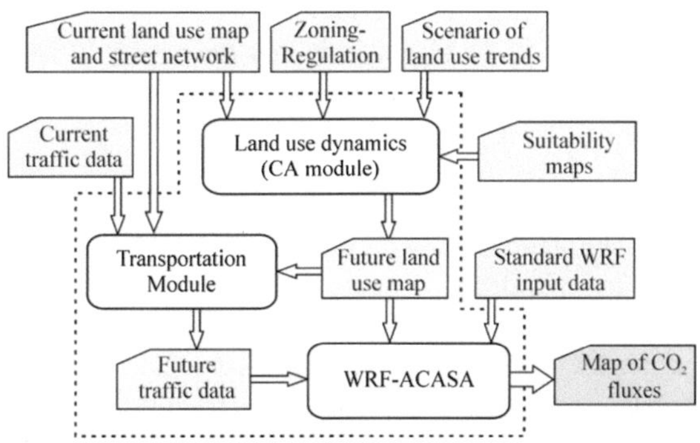

Fig. 1 Outline of the modelling framework with the most relevant data exchanges between the components

adequately simulate spatial processes which determine land-use patterns; and (3) the capability to process suitable representations of relevant landscape features as well as legal and planning restrictions on land uses.

Given an aggregate demand for different land uses, the CCA model allocates land uses in space based on local CA transition rules and on cells' characteristics.

The aggregate demand is assumed to depend on large-scale processes (demography, development of specific economic sectors, and so on) and can be determined by a-spatial demographic and economic dynamic models or by projections based on historical data (White et al. 1997; White and Engelen 2000).

The CCA can, hence, be viewed as a procedure to estimate a spatial distribution of an aggregate land-use demand by taking into account local interactions between different land uses as well as the physical, environmental and institutional characteristics of each cell.

In a CCA model, each cell is defined by a set of properties representing relevant physical, environmental, social and economical characteristics, its distance from the transportation network, and the regulatory constraints imposed upon it (like land-use zoning). This underlying data structure can both conceptually and practically be represented in a GIS (White et al. 1997; Batty and Xie 1994; Clarke and Gaydos 1998), and the cellular space on which the CA model operates can indeed easily be obtained from the layers of a raster GIS.

The cellular space consists of a rectangular grid of square cells, corresponding to the resolution of the data used as the source of land cover. Each cell is defined by:

- a *suitability factor* $S_j \in [0, 1]$ for each of the j dynamic land uses; it represents the suitability of the cell to uphold different land uses (it can, for example, be computed as a normalised weighted sum or product of relevant physical and environmental factors of each cell); suitability factors can either be pre-calculated in the GIS environment (in this case they remain constant during the simulation) or dynamically computed by the CCA model itself during the simulation;
- an *accessibility factor* $A_j \in [0, 1]$ for each land use, reflecting the importance of vicinity of the transportation network for different land uses (e.g. commercial uses generally require better accessibility than residential). These quantities are computed by the CCA module itself before the start of the simulation using the street network map as input;
- a value $Z_j \in [0, 1]$ which defines the degree of legal or planning permissibility of the j-th land use (for example due to zoning regulations);
- the cell's *current land use*.

The CA model includes both dynamic (which change during the simulation) and static land uses (which do not change but still influence other dynamic cells in their neighbourhoods in terms of attractive or repulsive effects). We derived the current land uses from the CORINE land-cover data. More detailed data can in principle be used, if available. Static land uses include road and rail networks, subways, airports, water bodies, agricultural and natural areas not available for development, and so on. The actively modelled dynamic land uses in the present application include: *continuous urban fabric, discontinuous urban fabric, industrial areas and commercial areas*. As mentioned before, their aggregate demand is given as exogenous.

Pivotal for the transition of cells from one to another land use is the so-called *transition potential* P_j (for each land use j), which expresses the cell's propensity to acquire j-th use. In fact, the first phase of the transition function, executed at each step by all cells, is to compute the transition potentials (one for each dynamic land use) on the basis of cell's suitability, accessibility, degree of legal land-use permissibility, as well as on the land uses of the neighbouring cells. In our model, the neighbourhood is defined as the circular region around the cell with the radius of 1 km, which we consider sufficient to capture local spatial processes by the CA transition rules. The transition potentials are calculated as:

$$P_j = A_j S_j Z_j N_j \tag{1}$$

where N_j is the so-called *neighbourhood effect*—the sum of all the attractive and repulsive effects of land uses and land covers within the neighbourhood—on the j-th land use which the current cell may assume. Since, in general, more distant cells in the neighbourhood exercise smaller influence, in our version of the model the factor N_j is computed as:

$$N_j = I_k + \sum_{c \in V} f_{ij}(d_c) \tag{2}$$

where the summation is extended to all the cells in the neighbourhood V (excluding the owner cell itself), i denotes the current land use of the cell $c \in V$, d_c is the distance between the owner cell and the neighbouring cell c, and $f_{ij}(d)$ is a parameterised function expressing the influence of the i-th land use at the distance d on the potential land use j. In addition, the positive term I_k, where k denotes the current land use of the cell, accounts for the effect of the cell on itself (zero-distance effect) and represents the inertial effect due to the transformation costs from one land use to another.

The second phase of the transition function takes place on a non-local basis and consists of transforming each cell into its highest-potential land use, under the global constraint of the overall number of cells of each land use imposed by the exogenous demand for land uses at that time step. The details of this procedure can be found in White et al. (1997) and White and Engelen (2000).

Since the transition function depends on several parameters, the model needs first to be calibrated, which can be done using the available historical spatial data of the area under study (Straatman et al. 2004; Engelen and White 2007; Blecic et al. 2010).

The final output of the CCA model are maps of the predicted evolution of land uses in the area of interest over a predefined period of time. By varying the inputs of the CCA model (zoning status, transport networks, distribution of facilities and services), the model can be used to explore future urban development scenarios of the area under alternative planning policies.

The CCA model, together with the transportation model described in the next subsection, was implemented using the MAGI C++ class library (Blecic et al. 2009).

2.2 The Transportation Model

The aim of the transportation module is to capture the long-term average vehicle load on the road network. In particular, out model provides an estimate of the load variation related to the future scenarios of land uses. For this purpose we adapted a dynamic formulation of the gravity model of trips distribution (Tsekeris and Stathopoulos 2003, 2006; Erlander and Stewart 1990). It uses an origin-destination (OD) trip matrix, expressing the distribution of trip demand, which is computed on the basis of relevant land uses (that is, the ones modelled by the CA module described above) considered as trip sources and attractors.

In the model, the street network is composed of a set $\mathcal{N}$ of nodes and a set $\mathcal{L}$ of directed links. Also, the entire urban area is partitioned in a set K of zones, each including nodes $n \in \mathcal{N}$ which represent sources and/or destinations for vehicular trips (see Fig. 2). In particular, for each couple of zones $\left(k_i, k_j\right) \in \mathcal{K} \times \mathcal{K}$, an OD matrix $\boldsymbol{X}$ provides the number of trips x_{ij} for the time period under consideration. In the model, the matrix $\boldsymbol{X}$ is estimated on the basis of the mix of land uses in both source and destination zones.

At the end the iterative procedure, the total amount of trips $\sum_i \sum_j x_{ij}$ are allocated and the final estimate of the load is available for every link l_k in the network.

Running the model for both the current land-use map and future land-use scenarios provides thus estimates about possible future street network load variations.

Thanks to its simplicity, this approach allows for the simulation of real road networks using reasonable computational resources. In addition, it requires a relatively small amount of parameters to be calibrated.

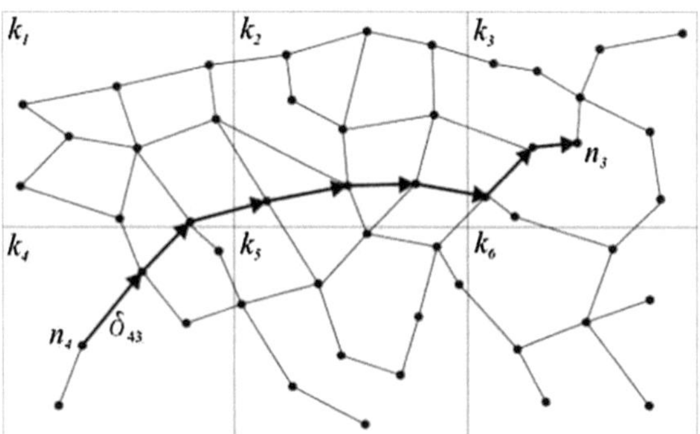

Fig. 2 Partition of the urban area into sectors and allocation of a packet of trips between two sectors during a step of the algorithm

On the negative side, our modelling takes into account only car traffic and in the present version does not account for other means of transportation—such as railway and subway—which may have impact on carbon emissions. A further issue is the lack to adequately account for public bus transportation. Considering how we estimate the travel demand based on land uses and how we allocate it on car trips, this would require *ad hoc* tweaking of the algorithm which we haven't contemplated in this version of the model. While both of these observations provide material for further refinement of the transportation module, we hold that, given the scale and level of abstraction at which the land-use modelling is done, the results provided by the transportation model are still meaningful and, under certain *coeteris paribus* assumptions, are acceptable approximations for the purpose of assessing variations of the carbon emissions due to the road traffic.

2.3 The WRF-ACASA Coupled Model

SVAT (Soil–Vegetation–Atmosphere Transfer) models simulate interactions between the Earth surface and the atmosphere above it. They describe how atmospheric processes (transfer of heat, moisture, and momentum due to turbulent motion) affect the exchanges of energy (heat), moisture, and trace gases between soil, vegetation, water bodies and the atmosphere itself. Biophysicists and ecologists use SVAT models to determine how plants and plant communities respond to environmental conditions.

In our framework, the purpose of the SVAT model is to simulate the processes of CO_2 exchanges and fluxes shaped by future scenarios of urbanisation generated by the land-use and transportation models. It therefore provides the key link in the chain noosing the environment to the patterns of urban development.

The specific SVAT model we adopted is the so called Advanced Canopy-Atmosphere-Soil Algorithm (ACASA), a multilayer model originally developed by the University of California Davis. It simulates the exchange of heat, water vapour and CO_2 within and above the canopy of aboveground portion of the plant community of the urbanised area (Pyles et al. 2000, 2003). ACASA has been widely applied on natural and agricultural ecosystems (Staudt et al. 2010a, b, 2011; Marras et al. 2011). To work properly in urban environments, the model has recently been adapted to account for the anthropogenic contribution to heat exchange and carbon production (Marras et al. 2010).

To further account for local meteorology, ACASA has been coupled with the Weather Research and Forecasting model (WRF) (Falk et al. 2010). WRF (Skamarock et al. 2008) is a mesoscale numerical weather prediction system which, thanks to its flexibility and computational efficiency, is widely used for both operational forecasting and atmospheric research. According to recent validated results (Falk et al. 2010), ACASA-WRF produces reliable patterns of heat and CO_2 fluxes at high spatial resolutions over urban areas.

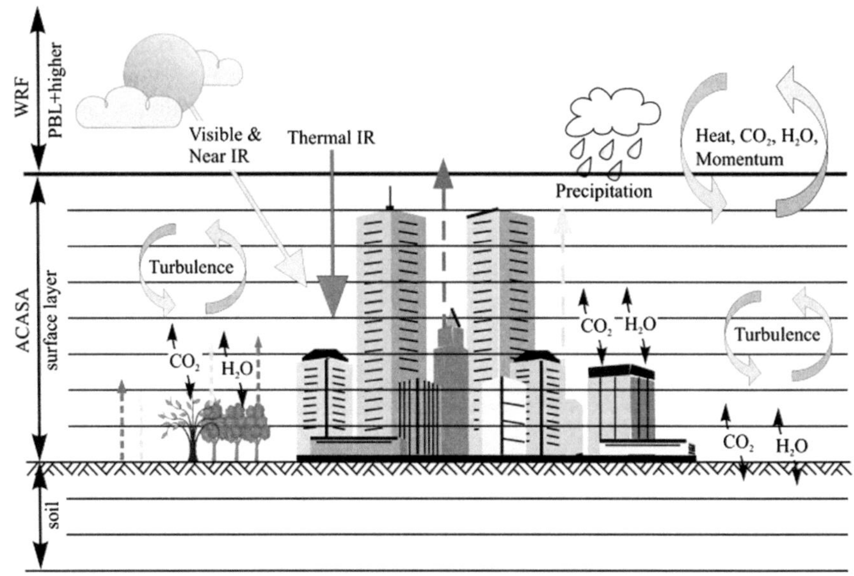

Fig. 3 Scheme of the processes modelled by the coupled model WRF-ACASA

In Fig. 3 we illustrate the processes, interactions and exchange fluxes modelled by the coupled ACASA-WRF model.

Given the space limitations and the primary intended audience of this publication, we relegate further details and background references on ACASA-WRF model to the Appendix of this chapter.

In order to simulate carbon fluxes in urban environment under alternative future scenarios, the coupled WRF-ACASA model uses the land-use and traffic scenarios produced respectively by the CA and transportation simulation modules. For this purpose, the CA module is able to export future land-use projections into the binary format accepted by the WRF Preprocessing System (WPS). The coupling was based on a suitable table of conversion between CORINE codes, which are used by the CA module, and the USGS+UCM codes used by WRF.

3 The Example Application

The key objective of this work is to evaluate the impacts of future planning alternatives on carbon emissions. In this context, we here present an example application of the modelling framework on the City of Florence, for which the WRF-ACASA model has been calibrated through measured fluxes collected in the city centre.

The first phase of the setup was dedicated to the selection of the proper input data sources for the CCA module. In particular, after a preliminary analysis of the available spatial data, we decided to adopt the 100 m-resolution CORINE land cover (CLC) as the input land-use layer (see Fig. 4). The choice was supported by the fact that, although not very accurate in urban areas, CLC data are widely available in Europe. But as we already mentioned, the CCA model is designed to also use more accurate CORINE-like raster data, when and where available.

In order to effectively incorporate the zoning regulation data into the CCA module, we carried out a semiautomatic pre-processing of the Florence urban masterplan. In Fig. 5 we show the masterplan prescriptions reduced to land-uses permissions and prohibitions relevant for the CA simulation, both in terms of land-use types simulated by the CA (continuous urban fabric, discontinuous urban fabric, industrial, commercial, agriculture), and in terms of its spatial scale of operation (pixel lattice representing 100 $\times$ 100 m^2). This zoning data layer was subsequently imported into the CCA model, assigning to each cell a *permissibility factor* of urban development for each land use, which is an important component of the CA transition rules (see Sect. 2.1).

Another essential input used by both the CCA and the transportation module is the road network of the area. In order to improve the reusability of the modelling

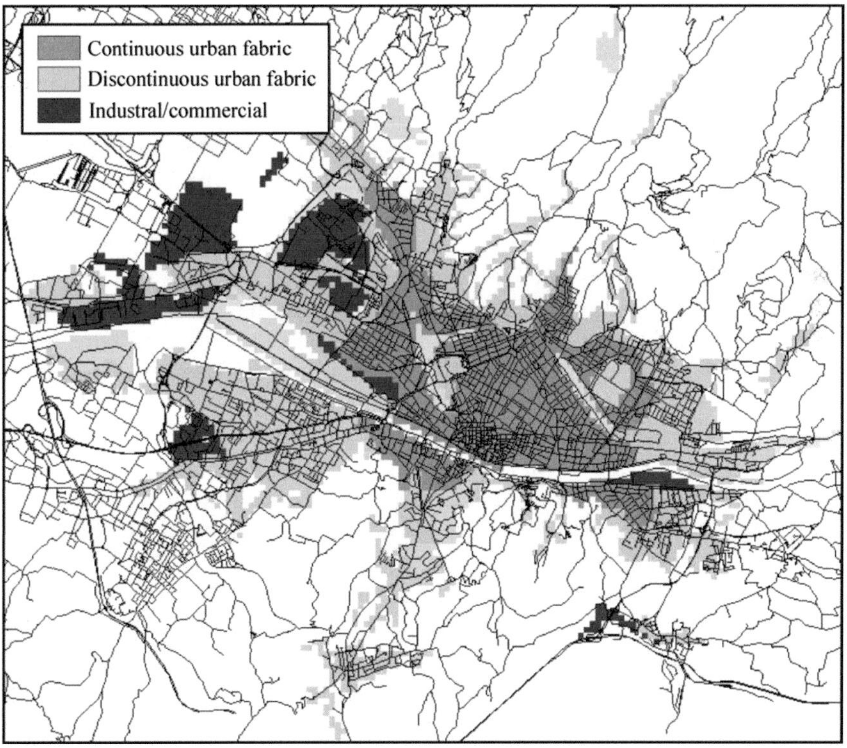

Fig. 4 Florence current land cover from CORINE (CLC2000) inventory, and street network imported from the OpenStreetMap project

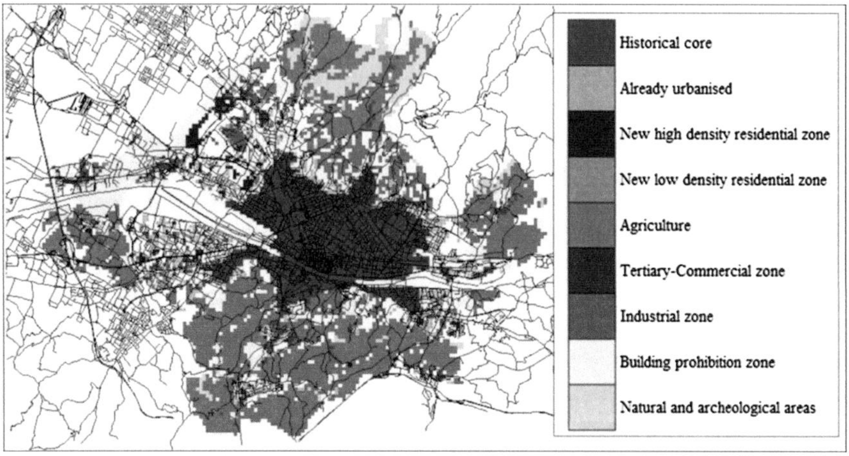

Fig. 5 Schematic zoning regulation for Florence derived from the urban masterplan

framework, the transportation module was designed to import XML data from the *OpenStreetMap* collaborative project, which provides free street network data with notable level of detail and geographic coverage.

In the example application on the City of Florence, we carried out a pre-processing phase for the calculation of the cells accessibilities (see Sect. 2.1). This calculations were done through the CCA module using the current street network represented in Fig. 4. During the same phase, we used a simple formula to derive the suitability factors from a Digital Elevation Model of the urban area under consideration. In addition, we calibrated CCA parameters using the CLC1990 dataset, the map of Florence provided by the Urban ATLAS EU project, and additional spatial data provided by the Municipality of Florence.

Then, using the land cover (Fig. 4) and the planning regulation map (Fig. 5) as inputs for the CCA model, we were able to generate several future land-use scenarios. For example, the future land use projection represented in Fig. 6 corresponds to a 20 year evolution of the urban area.

Based on the land-use scenarios, and after a preliminary calibration with the available traffic data, the transportation model was able to estimate the variation of the load of vehicles on the road network. For example, driven by the future land-use scenario in Fig. 6, the transportation model generated the map of load variations for each road segment on the network, as depicted in Fig. 7.

Finally, the WRF-ACASA model, using the land-use scenario and traffic load information generated maps of CO_2 fluxes. The model was run on the area within the grey rectangle in Fig. 6.

In Fig. 8 we compare average CO_2 fluxes for March and April for the current land uses (based on Corine 2000) and the 2020 future land-use scenario, which showed a 10 % increase continuous urban fabric, 20 % increase in discontinuous urban fabric, and 10 % increase in commercial/industrial land uses (see Fig. 6). As

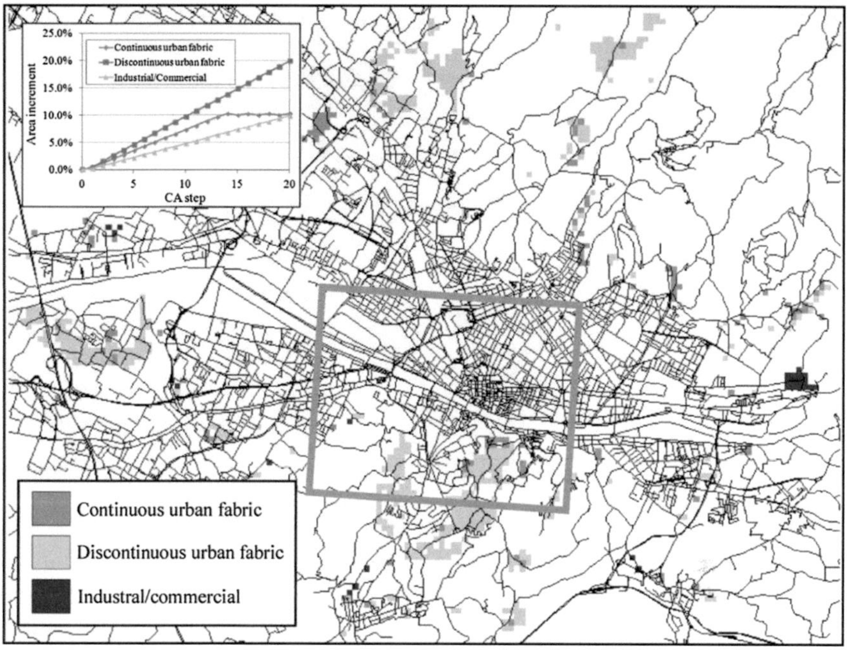

Fig. 6 A future land-use scenario obtained by the CCA module in 20 simulation steps. Only the cells where the land use has changed at the end of the simulation are depicted. (The ACASA-WRF model was run on the area inside the *grey rectangle*.)

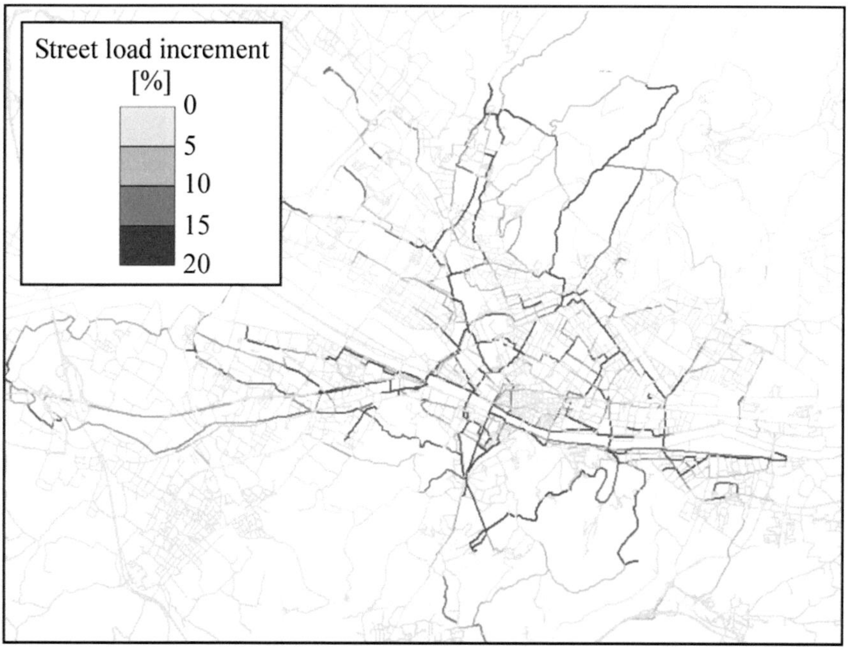

Fig. 7 Variation map of the street network loads which corresponds to the future land use projection depicted in Fig. 5

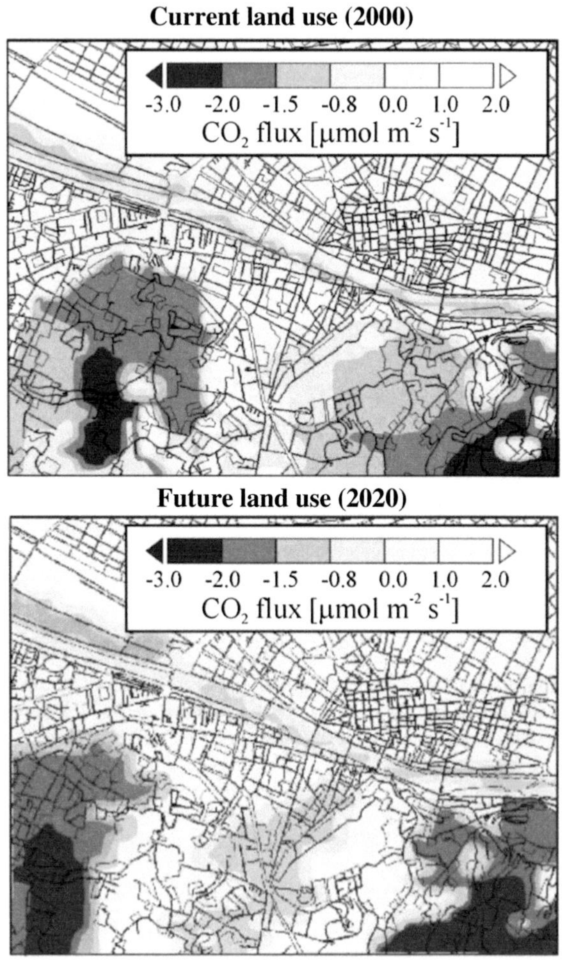

Fig. 8 WRF-ACASA simulated fluxes for current (*above*) and future 2020 (*below*) land-use scenario

can be seen in Fig. 8, many parts of the area under consideration changed from being a net sink to becoming a net source of CO_2, and points with values passing from −2 to 2.

This is partially due to the fact that many previously green areas have become urbanised under the land-use scenario (therefore emitting rather than contributing to the absorption of the local CO_2 fluxes), but the effect on the net variation of emissions is nonetheless wider than that strictly due to the newly urbanised zones from 2000 to 2020. This only shows that, according to our model, the relation between the extension of urbanisation and CO_2 emissions is not linear, and that increases in urbanisation, depending on its pattern and effect on transportation, can have more or less amplified impacts on CO_2 emissions.

4 Conclusions and Future Work

The complexities of urban systems and the interaction of carbon emissions with the environment and local meteorology make it a non-trivial task to support urban planning and management for a more sustainable metabolism of future cities.

In this context, we have presented an integration between different models with the purpose to link urban planning decisions to the estimates of their impact on the CO_2 emissions in the atmosphere. We have also briefly described an example application on the City of Florence to better explain the application protocol and workflow, to concretely show the necessary steps and possible issues (data acquisition, models calibration) involved in the application of our modelling framework, and to illustrate what kind of results may be expected.

As it often is the case, the calibration represents a crucial point for the application of this type of models. In particular, the CCA and the transportation models respectively require historical land-use data and traffic data. Also ACASA requires a preliminary calibration using the information about the real CO_2 concentration values and other input parameters.

While many aspects of the methodology and of the application protocol require further testing, we believe it has proven to be a promising perspective for developing decision support tools for sustainable urban planning.

Acknowledgments This research was funded by the Euro-Mediterranean Centre for Climate Change (CMCC) (Agreement CMCC-UNISS no 070115) within the FP7 European Project BRIDGE (Grant agreement no 211345).

We wish to thank CNR-IBIMET of Florence for providing the measurements of urban fluxes.

ACASA-WRF Model

The Advanced Canopy-Atmosphere-Soil Algorithm (ACASA) simulates the exchange of heat, water vapour and CO_2 within and above the canopy of above-ground portion of the plant community of the urbanised area (Pyles et al. 2000, 2003). To work properly in urban environments, the model has recently been adapted to account for the anthropogenic contribution to heat exchange and carbon production (Staudt et al. 2010a, b).

To account for local meteorology, the ACASA has been coupled with the Weather Research and Forecasting model (WRF) (Falk et al. 2010). WRF Skamarock et al. 2008) is a mesoscale numerical weather prediction system which, thanks to its flexibility and computational efficiency, is widely used for both operational forecasting and atmospheric research. WRF is particularly suitable for realistic representations of the surface-layer physics and physiology (Falk et al. 2010). The WRF model, in our case driven by North American Regional Reanalysis data (NCAR-NCEP), is run down to its planetary boundary layer, which is the lowest part of the atmosphere directly influenced by its contact with a surface, characterised by a rapid fluctuations (turbulence) of physical quantities such as flow velocity, temperature, moisture etc., where ACASA is called (see Fig. 3).

In ACASA, the canopy is represented as a horizontally homogeneous medium with all leaves and branches arranged with spherical symmetry. This includes both plant and urban (building) canopies. The surface-layer domain simulated by ACASA represents average conditions and fluxes for a 30 min time intervals at each spatial point in WRF. Twenty equally spaced atmospheric layers represented by a steady-state third-order turbulence extend to twice the canopy height (Meyers and Paw 1986, 1987), with the canopy occupying the lowermost 10 layers. All sources of heat and mass turbulent exchanges occur within the canopy and at the soil—(or snowpack)—atmosphere interface, while constant-flux assumptions prevail in the 10 layers above the canopy. Within the soil horizon, the model includes thermal and hydrologic diffusion among an adjustable number and depth of soil layers; varying by soil type. Surface temperatures, terrestrial-infrared radiative transfer, physiological conditions, and associated fluxes needed for the turbulence calculations are produced after iterative numerical convergence, and provided to WRF.

The required meteorological data from the lowest WRF atmospheric layer, applied to ACASA on a half-hourly basis, are: precipitation rate and form (kg m^{-2} $timestep^{-1}$), specific humidity (kg kg^{-1}), wind speed (m s^{-1}), downwelling short-wave radiation (W m^{-2}), downwelling long-wave radiation (W m^{-2}), air temperature (K), air pressure (hPa), and carbon dioxide concentration (ppm). Furthermore, initial soil temperature (K) and moisture (volumetric) profiles for the first model timestep are provided by the WRF initialisation routines. Surface morphological parameters that drive the physiological responses also have to be specified, which vary by WRF land use type (including CCA and/or satellite-derived data wherever possible). The set of key morphological parameters includes: total (green) leaf area index (m^2 m^{-2}), maximum canopy height (m), leaf-scale ideal photosynthetic potential (μmol m^{-2} s^{-1}), human population density (# people km^{-2}, currently keyed in by WRF urban land use type), and eventually vehicle flux density (# vehicles m^{-2}). Morphological parameters not represented in the WRF land-use parameter suite, quantities such as mean leaf diameter and basal respiration rates for plant tissues, are specified with constant near-cardinal values for all land points. The key set was chosen for a focus in this study, both due to model sensitivity, cogency with CCA architecture, and data availability. Staudt et al. (2010) provide additional background information on morphological parameters and model sensitivities.

The ACASA model output which feed information back to the WRF-simulated atmosphere include half-hourly vertical fluxes of heat (W m^{-2}), water vapour (kg m^{-2} s^{-1}), CO_2 (μmol m^{-2} s^{-1}), momentum (as friction velocity, m s^{-1}), and turbulence kinetic energy (m^2 s^{-2}). In addition, snowpack and/or soil and canopy thermal and hydrological states, needed for adequate simulation of the surface-layer, are updated at all land points and are stored between timesteps.

The initial conditions for a WRF run are pre-processed through a separate package called the WRF Preprocessing System (WPS). Among others, the input to

WRF-ACASA from WPS contains the vegetation/land-use type coded according to the U.S. Geological Survey (USGS) standard and to the Urban Canopy Model (UCM) (Kusaka et al. 2001). Other relevant input are represented by the CO_2 anthropogenic emissions which can be directly related to the estimated traffic load of the road network, to the prevalent land use of each cell and to the local population density.

References

Batty M, Xie Y (1994) From cells to cities. Environ Plan B 31–48

Blecic I, Cecchini A, Trunfio GA (2009) A general-purpose geosimulation infrastructure for spatial decision support. Trans Comput Sci 6:200–218

Blecic I, Cecchini A, Trunfio GA (2010) A comparison of evolutionary algorithms for automatic calibration of constrained cellular automata. LNCS 6016:166–181

Clarke KC, Gaydos LJ (1998) Loose-coupling a cellular automaton model and GIS: long-term urban growth predictions for San Francisco and Baltimore. Int J Geog Info Sci 699–714

Engelen G, White R (2007) Validating and calibrating integrated cellular automata based models of land use change. In: The dynamics of complex urban systems: an interdisciplinary approach, Springer, Berlin, pp 185–211

Erlander S, Stewart NF (1990) The gravity model in transportation analysis: theory and extensions. VSP, Utrecht

Falk M, Pyles RD, Marras S, Spano D, Snyder RL, Paw U KT (2010) A regional study of urban fluxes from a coupled WRF-ACASA model. American Geosciences Union Fall Meeting. San Francisco, California

Kusaka H, Kondo H, Kikegawa Y, Kimura F (2001) A simple single-layer urban canopy model for atmospheric models: Comparison with multi-layer and slab models. Bound Lay Meteorol 101:329–358

Marras S, Spano D, Pyles RD, Falk M, Snyder RL, Paw U KT (2010) Application of the ACASA model in urban environments: two case studies. AMS conference on ninth symposium on the urban environment, Keystone, Colorado, USA, Aug 2–6 2010

Marras S, Pyles RD, Sirca C, Paw U KT, Snyder RL, Duce P, Spano D (2011) Evaluation of the advanced canopy-atmosphere-soil algorithm (ACASA) model performance over Mediterranean maquis ecosystem. Agric Forest Meteorol 151:730–745

Meyers TP, Paw U KT (1986) Testing of a higher-order closure model for modelling airflow within and above plant canopies. Bound Lay Meteorol 31:297–311

Meyers TP, Paw U KT (1987) Modelling the plant canopy micrometeorology with higher-order closure principles. Agric Meteorol 41:143–163

Newman P (1999) Sustainability and cities: extending the metabolism model. Landscape Urban Plan 44

Pyles RD, Weare BC, Paw U KT (2000) The UCD advanced canopy-atmosphere-soil algorithm: comparison with observations from different climate and vegetation regimes. QJR Meteorol Soc 126: 2951–2980

Pyles RD, Weare BC, Paw U KT, Gustafson W (2003) Coupling between the university of California, Davis, advanced canopy-atmosphere-soil algorithm (ACASA) and MM5: preliminary results for July 1998 for Western–North America. J Appl Meteorol 42:557–569

Skamarock WC, Klemp JB, Dudhia J, Gill DO, Barker DM, Duda MG, Huang XY, Wang W, Powers JG (2008) A description of the advanced research WRF version 3. NCAR/TN–475 + STR–NCAR Technical Note

Staudt K, Falge E, Pyles RD, Paw U KT, Foken T (2010a) Sensitivity and predictive uncertainty of the ACASA model at a spruce forest site. Biogeosciences 7:3685–3705
Staudt K, Falge E, Pyles RD, Paw U KT, Foken T (2010) Sensitivity and predictive uncertainty of the ACASA model at a spruce forest site. Biogeosciences 7:3685–3705. www.biogeosciences.net/7/3685/2010/
Staudt K, Serafimovich A, Siebicke L, Pyles RD, Falge E (2011) Vertical structure of evapotranspiration at a forest site (a case study). Agric Forest Meteorol 151:709–729
Straatman B, White R, Engelen G (2004) Towards an automatic calibration procedure for constrained cellular automata. Comput Environ Urban Syst 28:149–170
Tsekeris T, Stathopoulos A (2003) Real-time dynamic origin-destination matrix adjustment with simulated and actual link flows in urban networks. Transp Res Rec—J Transp Res Board 1857:117–127
Tsekeris T, Stathopoulos A (2006) Gravity models for dynamic transport planning: development and implementation in urban networks. J Transp Geogr 14(2):152–160
White R, Engelen G (2000) High-resolution integrated modelling of the spatial dynamics of urban and regional systems. Comput Environ Urban Syst 2824:383–400
White R, Engelen G, Uljee I (1997) The use of constrained cellular automata for high resolution modelling of urban land use dynamics. Environ Plan B 24:323–343
Wolman A (1965) The metabolism of cities. Sci Am 213

A GIS-Based Performance Metrics for Designing a Low Energy Urban Agriculture System

Steven Jige Quan, John David Minter and Perry Pei-Ju Yang

Abstract This chapter explores a GIS-based performance metric for designing a low energy agriculture system in the City of Atlanta. A framework of a planning support system (PSS) is proposed, which contains dimensions of representation, performance and scenario planning. The specific PSS demonstrated how a performance-oriented urban design decision model reorganizes fragmented information and turns them into useful representational layers, analyses of performance measures, and scenario planning for making design decisions.

1 Introduction

The research is based on an urban design studio conducted in Spring 2012 in the School of City and Regional Planning and School of Architecture at The Georgia Institute of Technology on agrarian urbanism, in collaboration with Georgia Tech's School of Biology and Arkfab, an advocacy group on promoting urban farming in Atlanta. The studio started from a proposition that ecological footprints of a city have gone far beyond its geographic boundary. The resources that cities rely upon today, including energy, materials, water and food are more connected to the 'global hinterlands' rather than their adjacent local surroundings. The "food miles" concept, which considers how far food travels, is not sufficient for understanding its ecological effects. The embodied energy and carbon emission that the food takes to produce, process, transport and distribute is far greater than we can imagine.

This chapter introduces a low energy urban agriculture system in Atlanta that was conceptualized by a performance-based urban design model, a model that

S. J. Quan · J. D. Minter · P. P.-J. Yang (✉)
School of City and Regional Planning, College of Architecture, Georgia Institute of Technology, 245 Fourth Street, NW, Suite 204, Atlanta, GA 30332-0155, USA
e-mail: perry.yang@coa.gatech.edu

S. Geertman et al. (eds.), *Planning Support Systems for Sustainable Urban Development*, Lecture Notes in Geoinformation and Cartography,
DOI: 10.1007/978-3-642-37533-0_13,

suggests a metabolic process of energy, material and water flows in an urban setting. It explores how dimensions of *representation, performance and scenario planning* can be integrated into a planning support system, a system approach that connects urban form, ecological performance and future scenarios for design decision making.

Based on a test case in Atlanta, the proposed Planning Support System (PSS) model demonstrates how a GIS platform is designed for accommodating various sources of information, to include energy, material and water flows in the city, which is essential information for a local food production system that aims to grow local food using food waste and thus reduce the food desert. In this PSS model, we established a set of tools to manage the representational information, to use them for conducting performance energy flows and waste stream flows in four spatial scales, and to evaluate the overall performance of the system under different development scenarios. It extends a performance-based urban design dimension to contemporary PSS case studies in the book.

2 A Recent Development in Planning Support System

To manage and process the vast amount of urban information that may be potentially needed in urban design and planning, Geographic Information System (GIS) has been widely tested as a platform for turning information to planning support (Brail and Klosterman 2001). The developments of computerized urban applications in the earlier days emphasized the analytical dimension, as well as the predictive aspects of "what if" scenario planning. The development of Planning Support System initiated by Britton Harris provided significant evidences of this trend (Harris 1989; Klosterman 2001). However, with the rapidly increasing computational capacity following "Moore's Law", as well as the blooming new technologies for dissemination of information in digital format through internet, the question in front of PSS development changed from "what kind of models can we build" to "what kinds of models should we build" (Batty 2007; Klosterman 2008). With the planning theory developing from rationality, bounded rationality to collective negotiation and participation, PSS evolved from a tightly integrated structured system to reach "optimization" to a set of "Toolbox" containing a series of tools to be adapted to different situations, and these development stem from varies of ideals which shape planning practice and education: planning as design, planning as applied science and planning as reasoning together (Klosterman 2008). However, as Vonk and Geertman have argued, a large mismatch exists between the supply of PSS and the demand of PSS in planning practice (Vonk and Geertman 2008). To shorten the mismatch situation of PSS model and planning practice, it is important to develop a toolbox that is adaptable to different contexts in practice, especially in urban design decision making situations in which the goals are multiple and processes are complex.

GIS has experienced a tremendous amount of innovation in the recent decade, and so have the computing technologies which have enabled powerful supporting tools for urban design (van Leeuwen and Timmermans 2006; Ernstson et al. 2010). Recent development of GIS technology shows potentials for the design-oriented and interventional approaches. By articulating spatial database, spatial queries, spatial statistics, visualization, and simulation, GIS operations have been seen as a potential supporting tool for design decision making process. With planners' traditional tools such as generic and non-spatial tools like spreadsheets, innovative simulation tools, as well as decision support aids, GIS could play an even more powerful role in PSS (Klosterman 2001; Batty 2007).

However, the most fundamental problem of PSS doesn't concern a particular tool such as GIS, or the extent of the toolkits, but "a conception of planning" (Klosterman 2001). As Batty (2007) has pointed out, GIS is "theory-less" and "descriptive" rather than "predictive". Its development towards agent-based structure is driven by computation of fine-scale data rather than "theoretical advances in our understanding of the city" (Batty 2007). In this technique-dominant era with pervasive information and great computational capacity, without the purpose, we could achieve nothing meaningful other than collecting and storing them. At the same time, PSS models follow Box's Law which indicated that any model is wrong because it is a simplification of reality (Box and Draper 1987). The criteria for a good model is not "correct in the absolute sense", but in whether it is "useful for a particular purpose" (Box and Draper 1987; Klosterman 2008). Only driven by purpose, can the, tools be carefully developed and appropriately assembled into a PSS (Bishop 1998).

But in the process of identifying purposes of planning, due to the limitation of time and information available, we could hardly establish every possible future and get the most optimal one out of them as the "correct" one (Harris 1999; Klosterman 2008; Rickwood 2011). Instead, we can prepare scenarios to describe several possible futures (Avin 2007; Harwood 2007), and PSS models should be scenario-driven and policy-oriented which suggests what might happen if particular conditions exist and specific public policies are adopted (Klosterman 2008). For better understanding and easier interpretation of the models, they should clearly state the underlying assumptions about the scenarios and alternative policy choices before their functioning.

3 Modeling Urban Design Scenarios Based on Performance Metrics

The interweaving of technical and social systems in cities is "far more complex than most architects, engineers and planners are willing to admit", and this interaction has integrated the "twentieth-century discourses of scientific urbanism and technological modernism" into a highly complex model of urban spaces as "an efficient machine" (Gandy 2004).

There exist some models for making complex urban design decisions. They can be categorized as descriptive, predictive, explorative, or planning according to different purposes. Steinitz proposed a model that covers aspects of representation, process, evaluation, change, impact and decision making. It is a social communication that is experienced and understood, and is also an anthropocentric process of intentional change, which has the primary social objective of changing people's lives by changing their environment and processes, including ecological processes (Steinitz and Rogers 1970; Steinitz 1990, 2002).

Hopkins further simplified it into the representational model of data and informational layers, process and evaluation model of performance metrics and change model of scenario planning, among which the dimension of flows will be emphasized to evaluate the performance (Hopkins 1999). In this research, the three models were redefined to be dimensions of *representation, performance and scenario planning*:

The *representational dimension* is the premeditated way to organize and represent the information of urban reality in a spatial database. It is an abstraction of the real world that employs a set of data objects that support map display, query, editing and analysis (Zeiler 1999). Since all GIS-based analyses are supported by the spatial database, the representational model is regarded as the foundation for urban design decision support. The *performance dimension* deals with the process model and evaluation model which provide the information about the functional relationships between the components, and evaluate the effectiveness of the current models (Steinitz et al. 2002). The performance-based urban design relies on the process model and evaluation model to bridge the connection between the physical urban form and its function over processes that lead to its performance measure and consequences. The *dimension of scenario planning* is to indicate how the performance might be altered, under what conditions, and when (Steinitz 1990). Scenarios are normally set according to different propositions based on "what if" questions, and could be integrated in GIS platforms.

The model of a low energy urban agriculture system, the test case of the research, is designed according to the three dimensions of representation, performance and scenario planning. It is a model for "urban metabolism" at the neighborhood level, which deals with "the sum total of the technical and socioeconomic processes that occur in cities, resulting in growth, production of energy, and elimination of waste" (Wolman 1965; Hermanowicz and Asano 1999; Kennedy et al. 2011). The metabolic model of examining energy, material and water flows in cities and the system approach utilized has broadened the field of urban design by connecting urban form, ecological performance and future scenarios.

The application of a planning supporting system to performance-based urban design of this kind is still rare in current literature, while existing PSS applications are focusing on land use and transportation modeling, urban growth model, agent-based modeling…etc. (Brail and Klosterman 2001; Geertman and Stillwell 2004; van Leeuwen and Timmermans 2006). The accessibility to fine-scale data appears to be challenging for this application. The effectiveness of applying PSS approach to performance-based urban design is to be further explored and tested.

4 Problems Definition: A Low Energy Urban Agriculture and its Need of a Planning Support System

The idea of urban agriculture is emerging. The recent initiatives of urban farming, community gardens and local food systems have been observed in the City of Atlanta and elsewhere in American cities such as Chicago (Iron Street and Altgeld Gardens Urban Farm), Philadelphia (Greensgrow Farm), Detroit, and Seattle, to name a few. However, the current methods for planning a low energy urban agriculture system are somewhat fragmented. Most on-going operating urban farms were designed for business and educational purposes (Kaufman and Bailkey 2000; Richardson 2011; Power 2012; Project 2012), in which their planning processes were often driven by a business objective or learning process in communities. They normally lack a comprehensive approach for how surrounding renewable resources can be used to support the urban farm.

The studio selected The Wheat Street urban farm in Atlanta as the test case. The farm is located in a low-income neighborhood, historically significant, known as Sweet Auburn. The area is partially identified as a "food desert", in which there is low access to supermarkets or large grocery stores (US Department of Agriculture 2013) (Fig. 1). The research explored how a planning support system would facilitate a process of managing data of spatial representation of urban form, measuring their performance and then producing scenarios for an urban agriculture system that is driven by a sustainable agenda. A specific PSS for urban design was constructed based on the three dimensions of representation, performance and scenario planning for the objective of designing a low energy agriculture system.

We investigate a planning model for a low energy urban agriculture system, an emerging approach to reconstructing communities and cities ecologically, and its

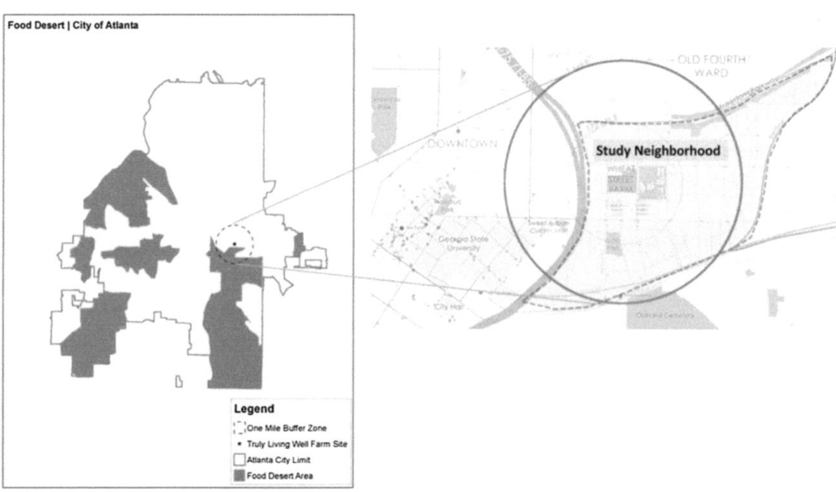

Fig. 1 The food desert and study area in Atlanta

design and policy framework for reorganizing neighborhoods and urban spaces by embedded sustainable food and agricultural system grounded in locality. The research contains an implicit design question of how flow generates form; a counter proposition to the modernist idea that form follows function. It suggests that contemporary urban forms and landscape patterns should emerge from sustainable approaches to designing flows of energy, materials, water and food.

5 Research Design

A research procedure was designed from data construction, boundary definition, and performance measures to facilitate scenario planning for design decision making.

5.1 Urban Informational Layers, System Boundary and Performance Measures

The first task for planners is to produce informational layers of the territories. A set of multi-scale urban informational layers were constructed to comprehend the geographic conditions for identifying potential resources such as solar, water and food waste for supporting the agriculture system in the urban neighborhood.

Four scales of territories were defined: the XL-scale, a meso-scale territory to cover the entire inner city neighborhood and surrounding areas, the L-scale to mean a district that is defined by the concept of Food Hub, the M-scale to include adjacent street blocks around the Wheat Street farm and the S-scale to mean the urban farm site (Fig. 2).

The delineation of territories is always problematic. It depends on how we define the system boundary, which is to be determined by issues for analysis. The solar availability, water reuse system and material recycle loop for the farm operation are examples that require appropriate scales for system designs respectively. Performance-based analysis was conducted according to the four-scale territories and boundary definition, including energy, material and water flows.

5.2 Urban Design Scenario Making

What constitutes design principles for a low energy urban farm system? How do we define a framework for performance analyses and mapping of energy, material and water flows to support planning the urban farm? How do we then connect the scenario making process to design decisions?

A performance-based and urban design oriented PSS model is suggested. It is grounded by the analysis and mapping of energy, material and water flows in

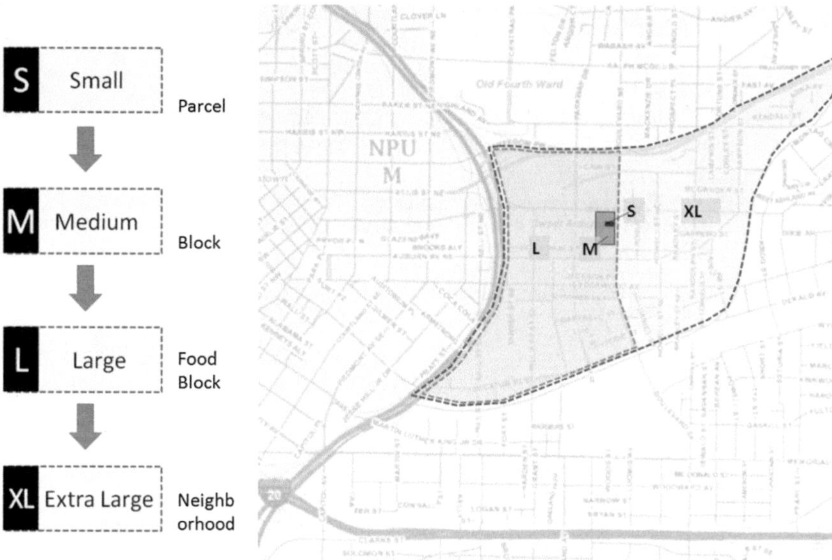

Fig. 2 Different scales of system boundaries

cities, which describes how physical urban form functions and operates ecologically which leads to speculation of its future transformation.

Three concepts were taken for generating scenarios of a low energy urban agricultural city: First, *the productive landscape scenario* advocates bringing productivity back to cities again by restoring top soil surface. It emphasizes the reconstruction of a symbiotic relationship in neighborhood networks of spaces and the identification of "green matches" of material exchange and food waste diversion and networking. Second, *the solar city scenario* tests how existing urban blocks surrounding the urban farm optimizes solar energy gain for supporting the operation of the farm by turning rooftops and building façade surfaces to renewable energy infrastructure. Third, *the hydropolis scenario* speculates the design of a sustainable irrigation system by organizing on-site water treatments and decentralized rainwater harvesting spaces.

Scenario setting is a creative exercise. They are both strategic and conceptual, and influential to what would emerge from the framework of representation-performance-scenarios, a set of organizational principles behind the PSS for a low energy urban agriculture system. The following model suggests a performance-based planning support system for urban agriculture system design, and to certain degree, it represents a form of simplification from Steinitz's model on landscape planning (Steinitz 2002) (Fig. 3).

The following steps illustrate how the model in processed in GIS platform: First, the representational model is constructed by identifying the informational layers. It may include vacant lands as available land resources, solar radiations as renewable energy resources, rainwater as potential renewable water resources,

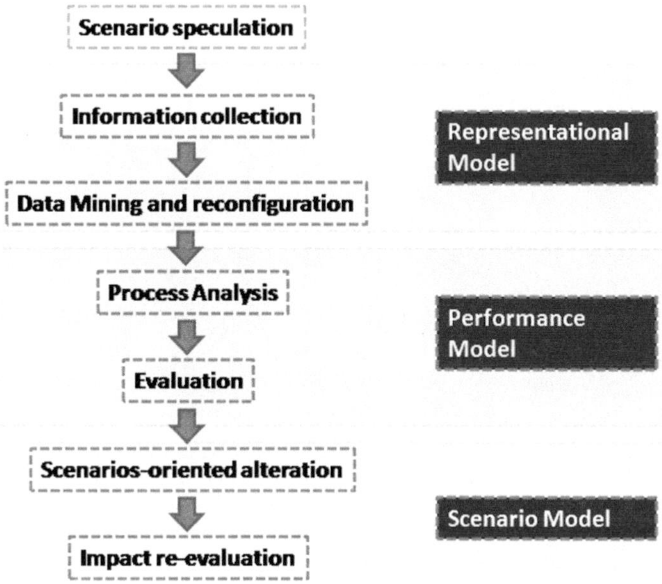

Fig. 3 Performance-based planning support system

food consumption and food waste as the output and input of the food system. Second, by organizing a set of analytical tools, such as ArcGIS, IES VE-Ware, Google Sketchup and E!Sankey (ESRI 2012; Ifu Hamburg GmbH 2012; SketchUp Team and Technology 2012; Integrated Environmental Solutions Limited 2012), the performance measures of selected areas are conducted in a systemic way, including flows of material (food waste), solar energy and water. Third, the performance measures are utilized to evaluate different scenarios, which define possible trajectories for future development. The scenario modeling will then provide implications for policy making.

6 Test Case: Performance Dimension of Food and Energy Flows for Food System Design

In the case Wheat Street Farm at Atlanta, a process was conducted from data processing and mapping, performance measurement to evaluation of scenarios to support planning decision making of the low energy urban agriculture system. Due to the data availability, our study focused on two categories of flows, food and energy, driven by the intent to reduce carbon emissions.

Table 1 Scenario criteria

Scenario	Criteria			
	GAR utilization (%)	Contribution of aquaponics system	Energy demand reduction (%)	Solar panel area ratio (%)
Scenario A	5	Low	30	20
Scenario B	10	Medium	50	50
Scenario C	20	High	80	100

6.1 Setting Scenarios

Three scenarios were set based on the goals of different levels of reduction in energy demands and renewable energy production from solar availability measurement from building rooftops.

According to the goal of U.S. Department of Energy's (DOE) Building America program, whole-house energy use in existing homes will be reduced by 30 % in all climates by 2014 and by 50 % before 2017 (US Department of Energy 2012), and some inspired individuals even aimed to save 80 % of pre-retrofit energy use (GPICHUB 2012). The goals for energy reduction were then set according to scenarios of 30, 50 and 80 %. For the solar panels coverage, three scenarios, 20, 50 and 100 % of solar panel installation on rooftops were set. From the reduction of energy demand and the solar energy supply, the amount of the carbon emissions could be estimated, which lead to the ratios of carbon mitigation (Table 1).

6.2 Data Mining on Vacant Lands

To support the low energy urban agricultural system, it is essential to identify resources such as vacant lands as potential sites for accommodating urban farms. The mapping of land resources provides representational component of the PSS model.

Using 2010 tax parcel GIS data obtained from the City of Atlanta, an indicator of "land value and improvement values ratio" was defined for land parcels within a 1 mile radius from the first farm site, in which parcels are considered "vacant" if the value ratio is greater than 3, "redevelopable" if between 1.5 and 3 and "stable" if less than 1.5. The categorization indicates a parcel's economic potential. By overlaying the permeable surface with land parcels under categories of "redevelopable" or "vacant", potential land availability for urban farm system can be estimated (Fig. 4).

Some indicators were developed for further describing characteristics of available land resources, including: the Impervious Surface Floor-Area-Ratio (IS-FAR) used for demonstrating the amount of impervious surface (structure and driveways) in relation to the total parcel area; the indicator Green Area Ratio (GAR) for defining the amount of yard space in relation to the total parcel area;

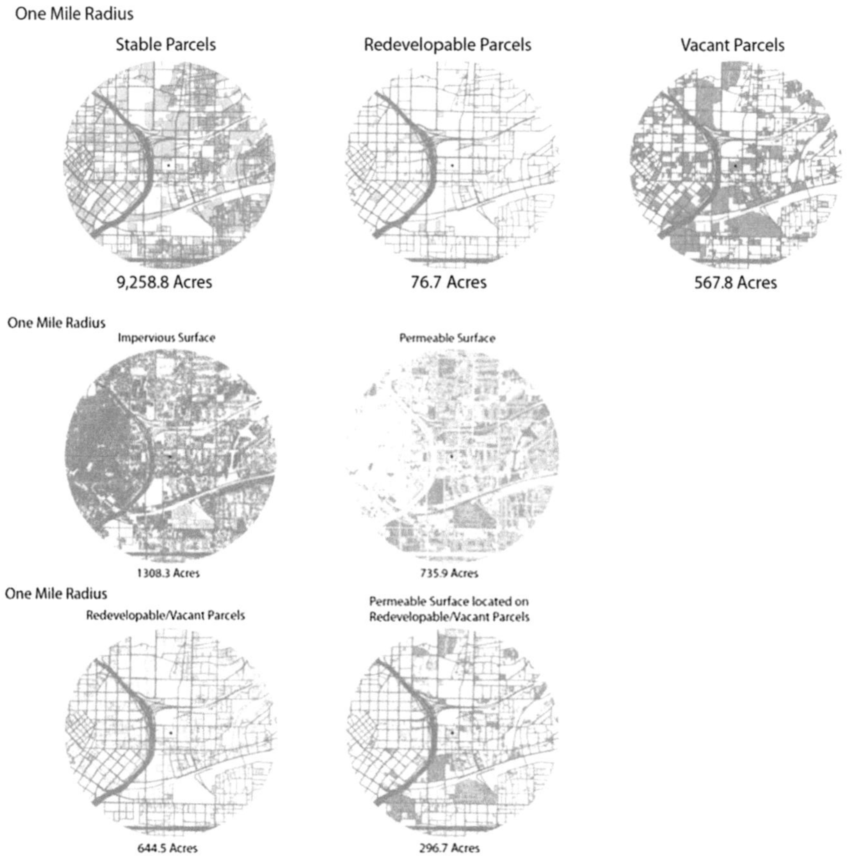

Fig. 4 Within one mile radius: **a** stable, redevelopable, and vacant parcels (*top*); **b** total area and permeable surface area (*middle*); **c** the redevelopable/vacant parcels and the overlay with permeable surface (*bottom*)

and the measure of building rooftops showing potential space for solar panel installations.

6.3 Food Flows

6.3.1 Performance Analysis of Food Flows

To track the flows of food and food waste in urban areas, the data of the average food consumption and food waste production per household were applied. According to the Food Availability (Per Capita) Data System (U.S. Department of Agriculture 2011) and the 2011 Population Survey from U.S. Census Bureau (2011), the daily food requirement for an average person is around 5 pounds. The amount of food waste produced per person annually is around 192 pounds (Agency EPA 2008).

There are differences in food types between the daily food requirement from households and food production from the urban farms. Urban farm productivity varies. The scope of research is limited to a singular food web based on the application of acquaponics systems and its food and waste flows in neighborhood network. A system of the aquaponics systems was designed in the Arkfab project at the Georgia Institute of Technology (ArkFab 2011) to estimate that a single aquaponics system with a size of twenty-four feet by seventeen feet would produce half a ton at each harvest, and 4 tons annually (Fig. 5).

ArcGIS was taken as the major platform in the performance model to integrate the food flow data with the spatial data at the four-scale territories defined as small, medium, large and extra-large within the study area. E!Sankey is applied for visualizing the temporal process for the food flows including the input, output and the close loops. The results of performance analysis are shown in Fig. 6.

6.3.2 Scenario Analysis of Food Flows

Three scenarios were set in each scale to represent the utilization of different levels of land availability and different types of urban agriculture (Table 2). As the scale increases in size, the productivity of urban farm gets more intensified and move to higher level. Small-scale examples are residential gardens within individual parcels. To move up to higher productivity of a medium-scale system would include a community garden together with a greenhouse of aquaponics system. A large-scale system could be organized around a foodhub, a local food market place and a community center to nurture farm activity and social wealth, as well as an education center for future farmer residents within the reach of the hub. Finally, an extra-large territory represents the surrounding environment that provides solar energy, food waste and water to support the urban farm operation. The following Fig. 7 shows the food flows across the four-scale territories under different scenarios.

6.3.3 Discussions on Food Flow

The results show that small-scale operations tend to be unproductive for all strategies, from the levels of the least intense to the most intense ones. The medium-scale system shows a dramatic increase in productivity by matching more food waste linkages to reduce the amount of nutrient input from outside. The utilization of vacant parcels within the area would provide social and economic benefits.

For the large-scale territory, the result shows the highest reduction up to 65 % of nutrient input from outside. The involvement of greatest number of vacant parcels for urban farms provides a complex framework for achieving high productivity. However, the creation of a self-reliant food hub network at this level requires more coordination among the participants in the neighborhood. The

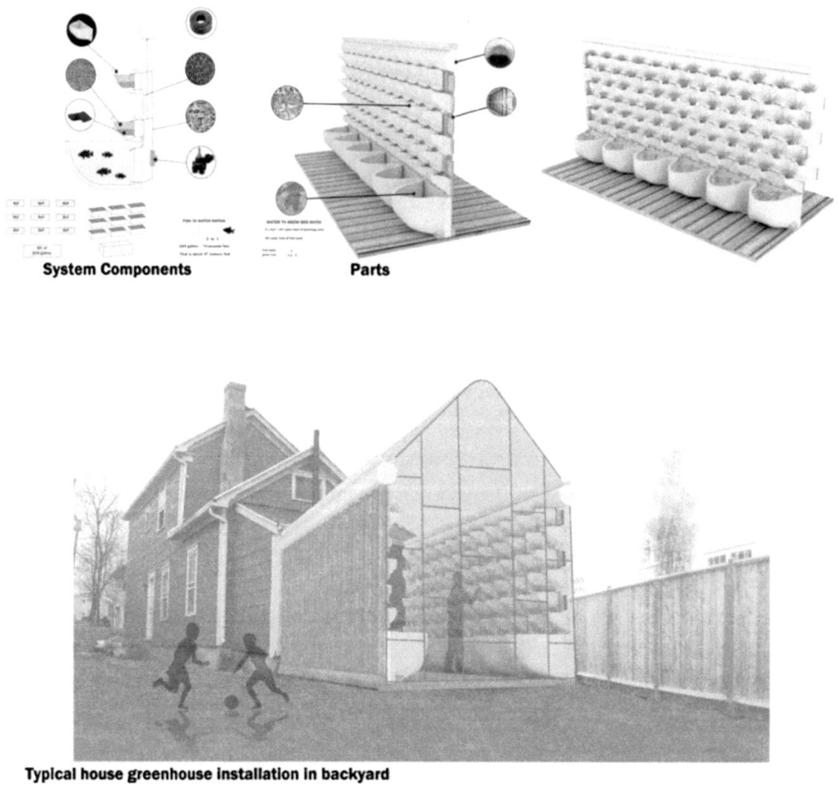

Fig. 5 Aquaponics system and greenhouse, images by Hrach Burtoyan (Yang et al. 2012)

scenario generated at the large-scale territory shows a clear picture of building the urban food hub framework and promising social and economic benefits that would allow community to take further action. The extra-large area demonstrated the larger the system, the higher complexity of the food web framework.

6.4 Energy Flows

Energy flows study in this chapter focuses on the measures of energy consumption of the residential buildings and the potential solar gain captured by solar panels on the rooftops. The purpose of this study is to find the potential to reduce carbon emissions by producing renewable energy from solar and reducing energy demanded through green building technologies.

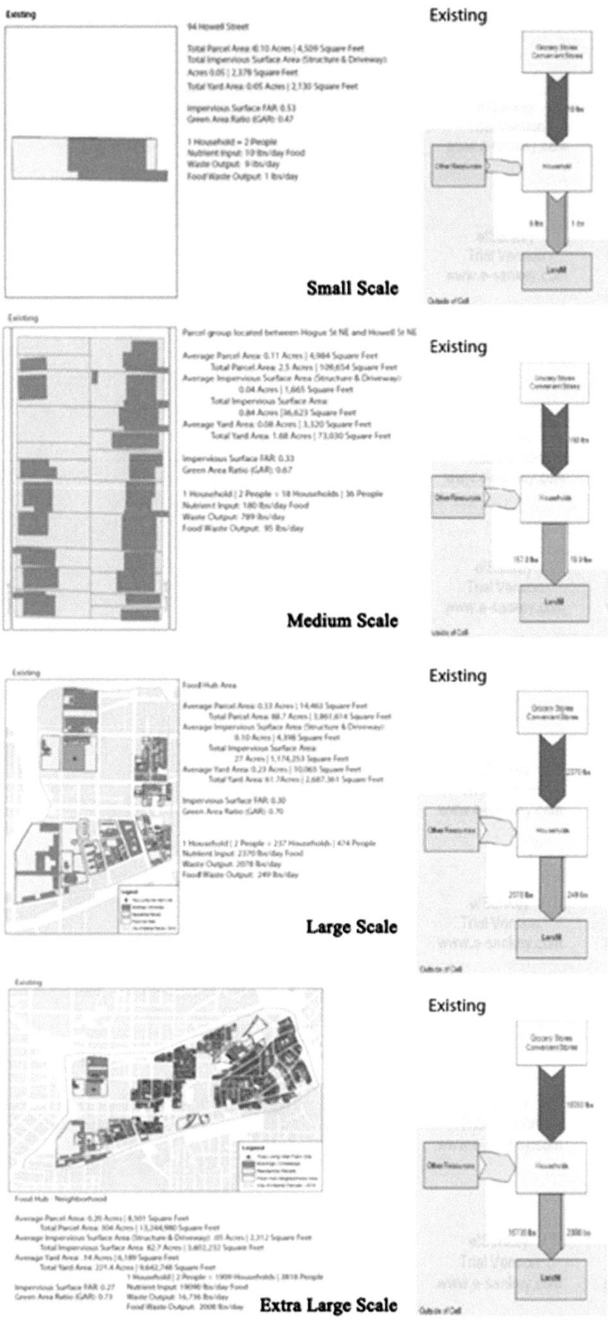

Fig. 6 Performance of food flows across four-scale territories: small, medium, large and extra-large scales at current conditions

Table 2 Scenario criteria of food flows

Scenario	Percentage of suitable lands utilization (or GAR) (%)	Scales that includes each type of development		
		Residential gardens	Community gardens/ aquaponics systems	Foodhub system
A	5	S, M, L, XL		
B	10	S, M, L, XL	M, L, XL	L, XL
C	20	S, M, L, XL	S, M, L, XL	L, XL

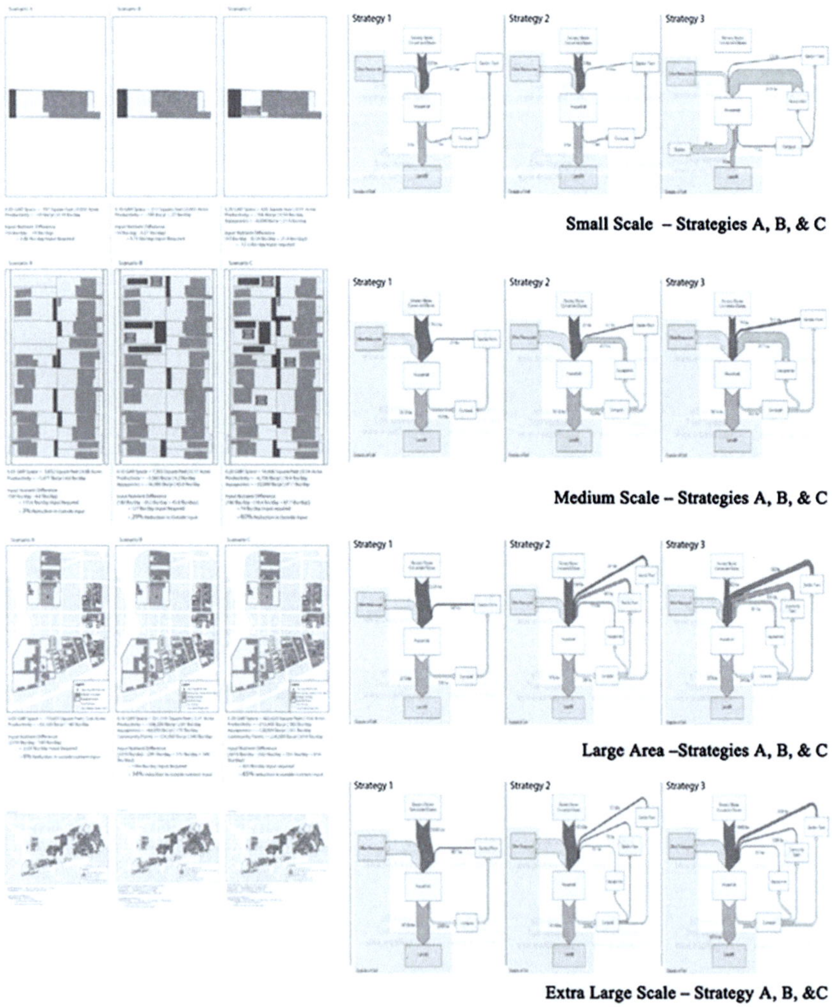

Fig. 7 The food flows across the four-scale territories under different scenarios

6.4.1 Performance Analysis of Energy Flows

A series of tools were integrated, including ArcGIS, Sketchup and IES-VE for modeling the urban environment in the study area and for measuring the energy performance including the potential solar energy supply and the energy demand.

The direct and diffuse solar energy are simulated using "Area Solar Calculation" Tool in ArcGIS 10.0. The annual solar energy that falls on the rooftops of all the buildings was calculated and converted to potential solar energy supply based on the assumption of the 17 % conversion rate in solar panels on the building rooftops. The energy demands of buildings were simulated on IES-VE platform based on the classification of building types, in which different usages of buildings generates various energy demand (Fig. 8). The result of energy demand simulation was defined based on energy intensity ratios, and then compare with the national average factors (U.S. Energy Information Administration 2003, 2005) (Table 3). Figure 9 shows how those individual building energy intensity ratio demand data

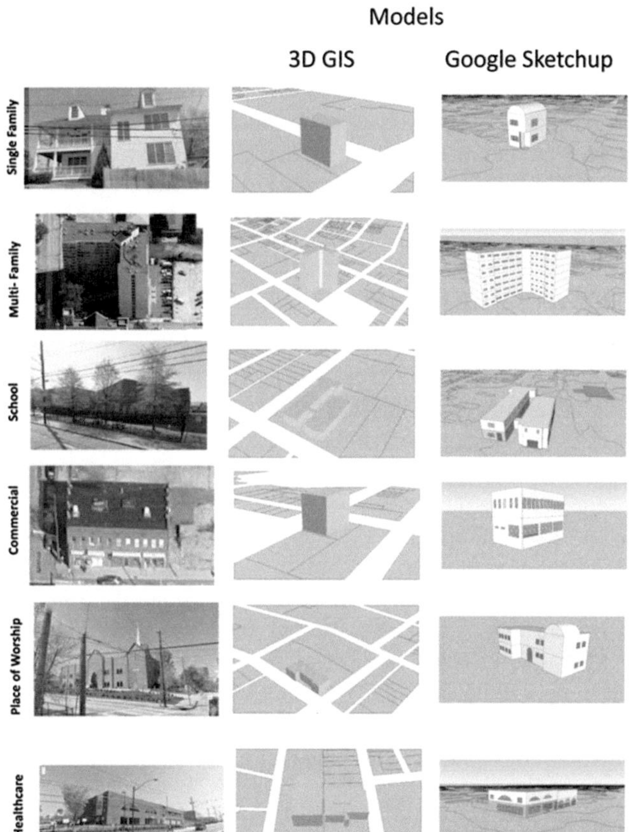

Fig. 8 Sketchup models with classification of building types (Yang 2012)

Table 3 Comparison of energy demands of different local building types using Sketchup-IES-VE approaches and national statistics

Building type		Estimated energy demands (Mbtu/1,000 sq.ft)	National factors for energy demands
Residential	Single-family	70	43.7
	Multi-family	85	
Schools		67	83.1
Commercial		102	89.9
Church		82	43.5
Healthecare		200	187.7

were applied to GIS platform and turned to mapping according to different levels of territories.

6.4.2 Scenario Analysis of Energy Flows

Three scenarios of energy flows were proposed based on different percentage of the rooftops covered with solar panels and various level of energy demand reduction (Table 4). The comparison of simulated result in energy demand and solar energy production are turned to GIS mapping based o scenarios generated at scales from small, medium, large and extra-large territories (Figs. 10, 11).

6.5 Total Carbon Mitigation

Based on the exemplar sites in the four scales, Parcel Area and GAR (Green Area Ratio) served as to characterize the four scales, and then food productivity, energy productivity as well as energy demand reduction were estimated by the combined toolbox of ArcGIS, Sketchup and IES-VE to represent the various performance of urban environment in current situations and three future scenarios, which consequently lead to carbon mitigation.

With the initiative of food flow reconfiguration by urban agriculture system at the four-scale territories, the food miles can be reduced from an average of 1,500–56 miles per year using the average food mile data from the study of Webber and Matthews (2008). Further carbon mitigation could be achieved by the improvement in agriculture technology, which could bring a considerable amount of carbon mitigation given that the food production accounts for about 12.5 % of the overall carbon emissions in US (Pimentel et al. 2005; Webber and Matthews 2008; Bomford 2009).

At the same time, converting food waste into food through community composting could reduce a great amount of carbon emissions comparing to landfills in common waste management. As indicated by the study of Selincourt (2008), for

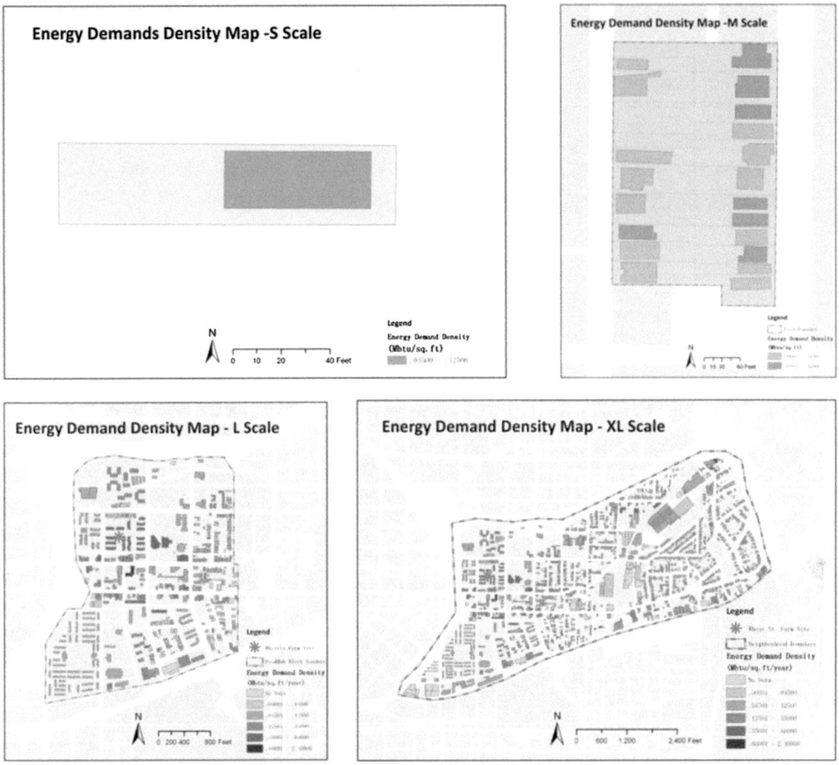

Fig. 9 Energy demand mapping in each scale

Table 4 Scenario settings in energy flows

Scenario	Percentage of rooftops covered by solar panels (%)	Percentage of energy consumption reduction (%)
A	20	30
B	50	50
C	100	80

every ton of food waste, there was 645 kg carbon emissions saved in UK. Although landfill techniques vary among different countries (Lundie and Peters 2005; Selincourt 2008), Selincourt's study in UK was taken to estimate the carbon mitigation of the food waste management in the test case of Atlanta.

The energy flows were reconfigured through two ways. One is to reduce the energy demand by building technologies and daily behavioral shifts. Another is to utilize solar energy to replace some of the carbon-related energy supply. The combination effect of the two ways results in the reduction of carbon emission in building sector.

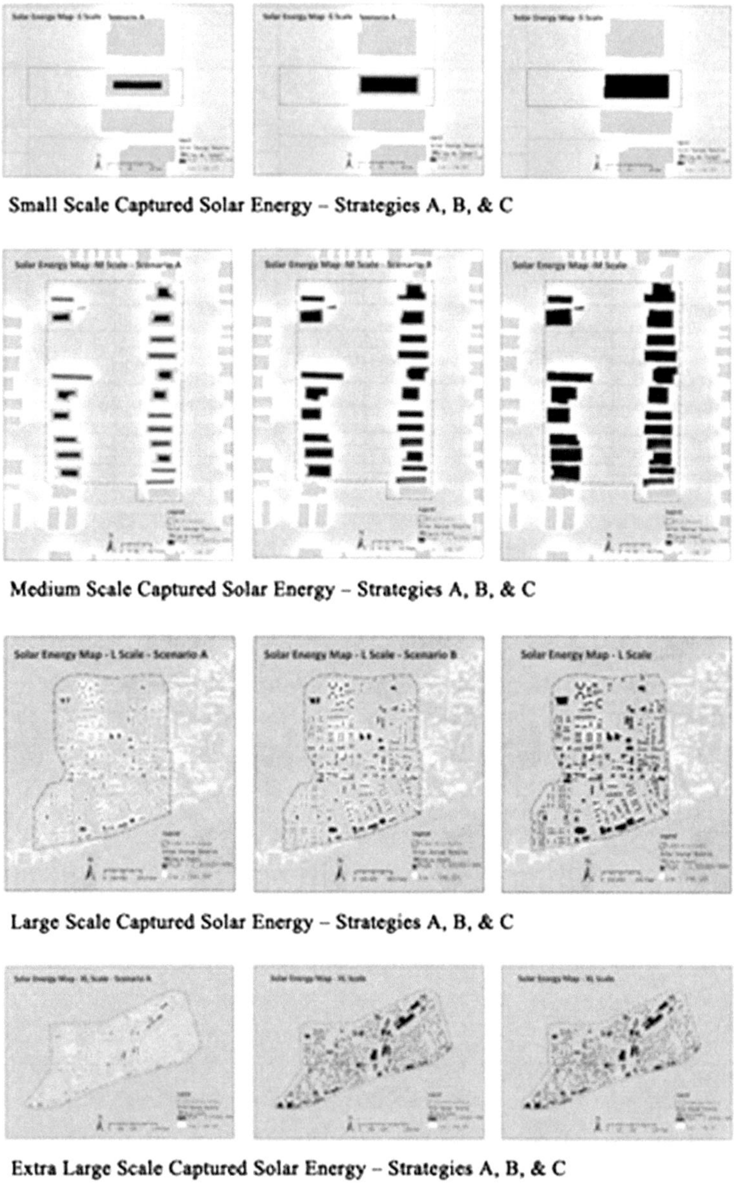

Fig. 10 **a**. GIS mapping of solar energy gain based on scenarios generated at four-scales territories: **a** small; **b** medium; **c** large; **d** extra-large

By adding up the carbon reduction through different ways, the findings and results were presented at the extra-large scale (Fig. 12). With different levels of planning interventions, the performance varies and carbon reduction can reach a

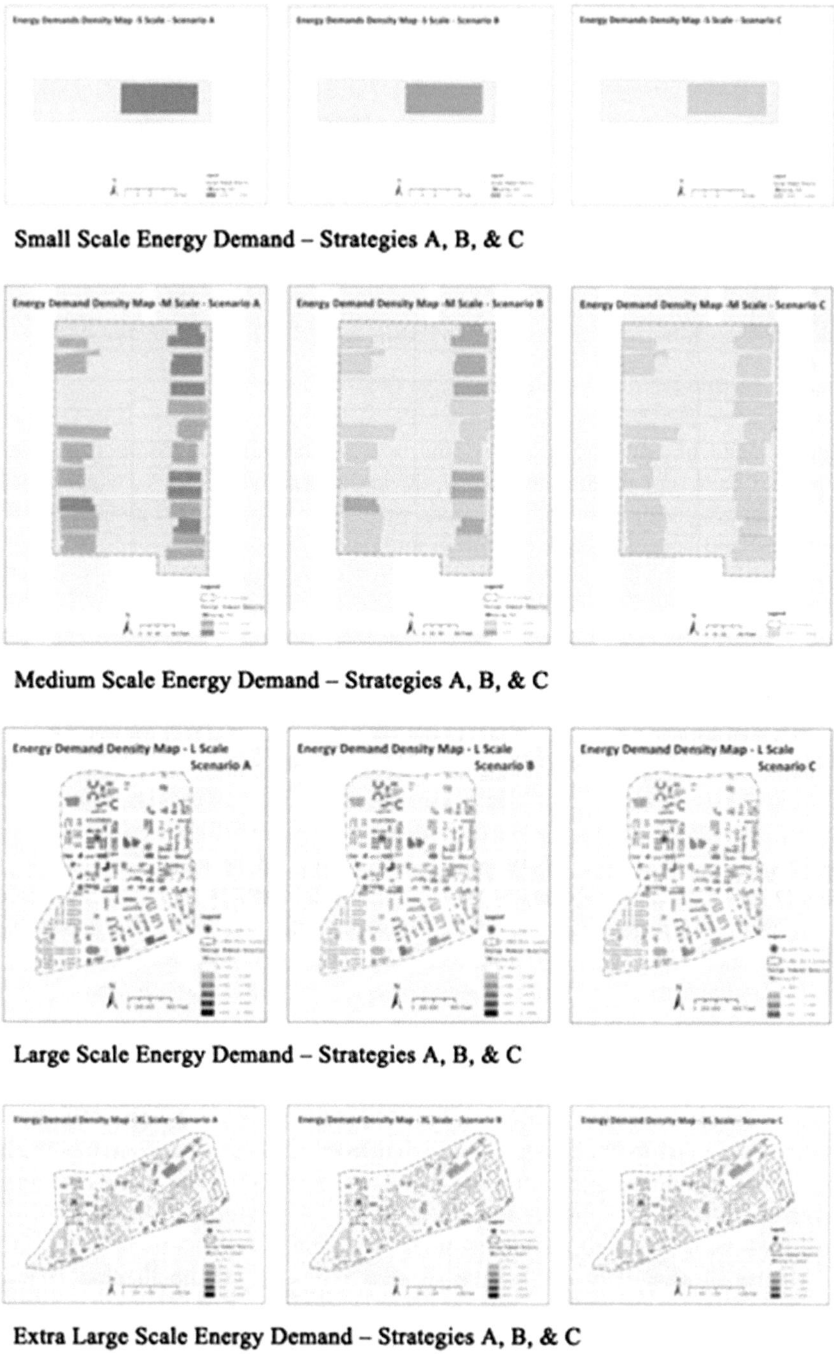

Fig. 11 GIS mapping of energy demand based on scenarios generated at four-scales territories: **a** small; **b** medium; **c** large; **d** extra-large

Scnario	Criteria				Performance & Productivity			
	GAR Utilization	Aquaponics Participatation	Energy Demand Reduction	Solar Panel Area Ratio	Food Productivity tons/year	Energy Productivity Mbtu/year	Carbon Reduction tons/year	Carbon Reduction Rate
Scenario A	5%	low	30%	20%	100.5	67920.11	42319.08	37%
Scenario B	10%	medium	50%	50%	387.7	169800.27	81301.77	72%
Scenario C	20%	high	80%	100%	722.7	339600.54	142266.01	126%

Fig. 12 Scenario generated performance and productivity measures

significant level. Scenario C shows how extreme the affect of planning interventions can have on carbon reduction, achieving a rate above 100 %, which means the maximum performance of such urban agriculture system could go beyond self-sustained and become a sink of carbon mitigation.

7 Conclusions

A PSS is proposed for designing a low energy urban agriculture system. The framework contains three dimensions: representation, performance and scenario making for informing a performance-based urban design process. The Wheat Street Farm was chosen as a test case to demonstrate the performance-based model of PSS. Using GIS as a platform to connect tools of performance analysis of energy and food flows, the research organizes four scales urban data as informational layers, performance analyses for measuring energy and food flows and such performance evaluation was conducted under different scenarios.

The findings are encouraging, which go beyond traditional solution-based urban design and provide a more comprehensive understanding of how an urban agriculture system functions and performs. Such results could not be achieved without a PSS framework. As Batty commented, urban designers need to broaden their fields of works to include flows and performance (Batty 2012). The PSS model shows an attempt of this kind to demonstrate how a GIS-based performance metric reorganizes fragmented information and turned them into useful representational layers, analysis of performance measures, and scenario planning for making design decisions for a low energy urban agriculture system.

The performance-based PSS model generates some policy implications to local governments, which could see the benefits of the low energy urban agriculture system in the form of increased productivity, and reduction of carbon emissions. Additionally, it adds economic vitality, new businesses and employment opportunities to inner city neighborhoods like Atlanta. By applying such a PSS model in its planning process, the city government could introduce performance evaluation and urban design scenarios to promote more productive sites in urban neighborhoods.

Non-governmental organizations can also use this PSS model to apply their agendas to broader urban context. For example, Truly Living Well, a non-profit urban farm organization in Atlanta could identify potential sites for their future projects and explore renewable resources from analyses of solar energy and food waste flows from a PSS model. The agenda for developing a large-scale foodhub will benefit from the PSS by identifying where the resources are and the paths toward the sustainable objectives based on the scenarios. It relies on more civic engagement to turn the objective from "for the public" to "by the public" (Klosterman 2008).

The development of such a PSS model for a low energy urban agriculture system should be dynamic and interactive. To make the performance metrics more comprehensive, the performance dimension could be further extended to other renewable resources such as water, other renewable energy and material flows, as well as flows of human and financial capital in future research.

References

ArkFab (2011) The ArkFab open source living machine series. http://arkfab.org/?tag=wheat-street-farm. Last date accessed 11.2012

Avin U (2007) Using scenarios to make urban plans. In: Hopkins L, Zapata M (eds) Engaging the future: forecasts, scenarios, plans, and projects. Lincoln Institute of Land Policy, Cambridge, pp 103–134

Batty M (2007) Planning support systems: progress, predictions, and speculations on the shape of things to come. In: Brail R (ed) (2008): planning support systems for cities and regions. Lincoln Institute of Land Policy, Cambridge, pp 3–30

Batty M (2012) Cities as flow systems: urban metabolism and complexity. In: Reports presented at the WasteMet Asia 2012 conference. Sands Expo and Convention Center Marina Bay Sands, Singapore, 1–4 July 2012

Bishop I (1998) Planning support: hardware and software in search of a system. Comput Environ Urban Syst 22(3):189–202

Bomford M (2009) How much greenhouse gas comes from food? http://energyfarms.wordpress.com/2009/04/16/how-much-energy-goes-into-our-food-system/. Last date accessed 08.2012

Box G, Draper N (1987) Empirical model-building and response surfaces. Wiley, New York

Brail R, Klosterman R (2001) Introduction. In: Brail R, Klosterman R (eds) Planning support systems: integrating geographic information systems, models and visualization tools. ESRI, Redlands, pp ix–xxi

Ernstson H, van der Leeuw S, Redman C, Meffert D, Davis G, Alfsen C et al (2010) Urban transitions: on urban resilience and human-dominated ecosystems. Ambio 39(8):531–545

ESRI (2012) ArcGIS 10.1. http://www.esri.com/news/arcnews/spring12articles/introducing-arcgis-101.html. Last date accessed 03.2013

Gandy M (2004) Rethinking urban metabolism: water, space and the modern city. City 8(3):371–387

Geertman S, Stillwell J (2004) Planning support systems: an inventory of current practice. Comput Environ Urban Syst 28(4):291–310
Greater Philadelphia Innovation Cluster for Energy Efficient Buildings (2012) GPIC goals. http://gpichub.org/about/gpic-goals. Last date accessed 03.2013
Greensgrow (2012) Greensgrow Philadelphia project. http://www.greensgrow.org/about-us/greensgrow-philadelphia-project/. Last date accessed 03.2013
Growing Power (2012) About US. http://www.growingpower.org/about_us.htm. Last date accessed 03.2013
Harris B (1989) Beyond geographic information-systems—computers and the planning professional. J Am Plan Assoc 55(1):85–90
Harris B (1999) Computing in planning: professional and institutional requirements. Environ Plan B: Plan Des 26(3):321–331
Harwood S (2007) Using scenarios to building planning capacity. In: Hopkins L, Zapata M (eds) Engaging the future: forecasts, scenarios, plans, and projects. Lincohn Institute of Land Policy, Cambridge, pp 135–153
Hermanowicz S, Asano T (1999) Abel Wolman's "The Metabolism of Cities" revisited: a case for water recycling and reuse. Water Sci Technol 40(4–5):29–36
Hopkins L (1999) Structure of a planning support system for urban development. Environ Plan B: Plan Des 26(3):333–343
Ifu Hamburg GmbH (2012) E!Sankey. 3.1. http://www.e-sankey.com/en/. Last date accessed 03.2013
Integrated Environmental Solutions Limited (2012) VE-Ware. http://www.iesve.com/software/ve-ware. Last date accessed 03.2013
Kaufman J, Bailkey M (2000) Farming inside cities: entrepreneurial urban agriculture in the United States, working paper. Lincoln Institute of Land Policy, Cambridge
Kennedy C, Pincetl S, Bunje P (2011) The study of urban metabolism and its applications to urban planning and design. Environ Pollut 159(8–9):1965–1973
Klosterman R (2001) Planning support systems: a new perspective on computer-aided planning. In: Brail R, Klosterman R (eds) Planning support systems: integrating geographic information systems, models and visualization tools. ESRI Press, Redlands, pp 1–23
Klosterman R (2008) A new tool for a new planning: the What if? TM planning support system. In: Brail R (ed) Planning support systems for cities and regions. Puritan Press Incorporated, Hampshire, pp 85–100
Lundie S, Peters G (2005) Life cycle assessment of food waste management options. J Cleaner Prod 13(3):275–286
Pimentel D, Hepperly P, Hanson J, Seidel R, Douds D (2005) Organic and conventional farming systems: environmental and economic issues. Department of Entomology & Section of Ecology and Systematics, New York State College of Agriculture and Life Sciences, Cornell University, Ithaca. http://ecommons.library.cornell.edu/bitstream/1813/2101/1/pimentel_report_05-1.pdf. Last date accessed 03.2013
Richardson M (2011) Identifying opportunities for urban agriculture in Atlanta. Technical Report, Georgia Tech Program in Science, Technology and Innovation Policy, Enterprise Innovation Institute, Georgia Institute of Technology, Atlanta
Rickwood P (2011) Modelling in urban and regional planning: past, present, and yet to come. In: Chan F, Marinova D, Anderssen R (eds) Sustaining our future: understanding and living with uncertainty, proceedings of MODSIM2011, the 19th international congress on modelling and simulation. Modelling and Simulation Society of Australia and New Zealand, Western Australia, pp 1652–1658, 12–16 Dec 2011. www.mssanz.org.au/modsim2011/D10/wongsosaputro.pdf. Last date accessed 03.2013
Selincourt K (2008) A comparison of greenhouse gas flux from composting compared to landfill and fertilizer production. Growing Heap 43:11–13
SketchUp Team and Technology (2012) SketchUp Pro 8. http://www.sketchup.com/intl/en/product/gsupro.html. Last date accessed 03.2013

Steinitz C (1990) Toward a sustainable landscape with high visual preference and high ecological integrity: the loop road in Acadia National Park, USA. Landscape Urban Plan 19(3):213–250
Steinitz C (2002) On teaching ecological principles to designers. In: Johnson B, Hill K (eds) Ecology and design frameworks for learning. Island Press, Washington, pp 231–244
Steinitz C, Rogers P (1970) A systems analysis model of urbanization and change: an experiment in interdisciplinary education. The MIT Press, Cambridge
Steinitz C, Arias H, Bassett S, Flaxman M, Goode T, Maddock T et al (2002) Alternative futures for changing landscapes: the Upper San Pedro River Basin in Arizona and Sonora. In: Ralph C, Rich T (eds) (2005): bird conservation implementation and integration in the Americas, proceedings of the third international partners in flight conference, Asilomar, 20–24 Mar 2002, U.S. Department of Agriculture, Forest Service, Pacific Southwest Research Station, Albany, pp 93–100
United States Environmental Protection Agency (USEP) (2008) Municipal solid waste generation, recycling and disposal in the United States: facts and figures for 2008. http://www.epa.gov/osw/nonhaz/municipal/pubs/msw2008rpt.pdf. Last date accessed 03.2013
US Department of Agriculture (2013) Food desert locator. http://www.ers.usda.gov/data-products/food-desert-locator.aspx#.UUsyzR1-lQA. Last date accessed 03.2013
US Department of Energy (2012) Building America: program goals. http://www1.eere.energy.gov/buildings/building_america/program_goals.html. Last date accessed 07.2012
U.S. Census Bureau (2011) Current Population Survey, 2011 Annual Social and Economic Supplement
U.S. Department of Agriculture (2011) Food Availability (Per Capita) DataSystem
U.S. Energy Information Administration (2005) Residential energy consumption survey (RECS). http://www.eia.gov/consumption/residential/data/2005/#undefined. Last date accessed 03.2013
U.S. Energy Information Administration (2008) Commercial buildings energy consumption survey (CBECS): 2003 detailed tables. http://www.eia.gov/emeu/cbecs/cbecs2003/detailed_tables_2003/detailed_tables_2003.html#consumexpen03. Last date accessed 03.2013
van Leeuwen J, Timmermans M (2006) Introduction. In: van Leeuwen J, Timmermans M (eds) Innovations in design and decision support systems in architecture and urban planning. Springer, Dordrecht, pp 13–20
Vonk G, Geertman S (2008) Improving the adoption and use of planning support systems in practice. Appl Spat Anal 1(3):153–173
Webber C, Matthews H (2008) Food-miles and the relative climate impacts of food choices in the United States. Environ Sci Technol 42(10):3508–3513
Wolman A (1965) The metabolism of cities. Sci Am 213(3):178–193
Yang P, Quan S, Minter D, Nguyen D, Hampton T, Brooks T, Brame C et al (2012) Agrarian urbanism studio report. Georgia Institute of Technology, Atlanta
Zeiler M (1999) Modeling our world: the Esri guide to Geodatabase design. ESRI Press, Redlands

A Comprehensive Review of Existing Urban Energy Models in the Built Environment

Saleh Mohammadi, Bauke de Vries and Wim Schaefer

Abstract This review considers previous researches that dealt with energy modelling in the built environment at different urban spatial scales. Up to now, modelling approaches focus on energy flow simulation in urban area from generation to distribution and finally consumption. Dependent on the energy flow direction and focus of these models in urban energy system, they are categorized in three distinct groups: supply, demand and integrated models. A critical evaluation of each category of these models based on the same criteria is provided. As a result, some of the main aspects that are missing in the existing models are highlighted, including: integrated multilayer models, spatial implications of renewable energy technologies and integration of simulation and optimization methods.

1 Introduction

Cities as a main consumer of energy resources need to transform from fossil fuel based energy systems to renewable energy systems to fulfill the renewable energy ambitions. The built environment with almost 40 % of the total energy demand has

S. Mohammadi (✉)
Department of Built Environment, Eindhoven University of Technology, Den Dolech 2, VRT 9.09, 5600 MB Eindhoven, The Netherlands
e-mail: s.mohammadi@tue.nl

B. de Vries
Department of Built Environment, Eindhoven University of Technology, Den Dolech 2, VRT 9.10, 5600 MB Eindhoven, The Netherlands
e-mail: b.d.Vries@tue.nl

W. Schaefer
Department of Built Environment, Eindhoven University of Technology, Den Dolech 2, VRT 8.08, 5600 MB Eindhoven, The Netherlands
e-mail: w.f.schaefer@tue.nl

S. Geertman et al. (eds.), *Planning Support Systems for Sustainable Urban Development*, Lecture Notes in Geoinformation and Cartography,
DOI: 10.1007/978-3-642-37533-0_14,

a substantial contribution in this transformation process. The energy characteristics of building sector are diffuse and complex. Therefore, comprehensive models are needed to capture the different aspects of them to support urban planners in developing integral city transformation plans. Before city councils commit to significant new energy policy initiatives, they require insight into the possible consequences of them and improve their understanding of the energy flow in urban areas. Different models aim to describe these processes and attempt to predict how it might be changed over time (Keirstead et al. 2009).

Advances to date have been based on predicting the energy demand of groups of buildings, mostly based on their functions—residential, commercial or industrial—as distinct categories. These models estimate the energy demand of stocks of buildings based on extrapolation of results from representative building typologies. A similar approach was developed for modeling demand in domestic building stock based on combinations of building type and household composition (Shimoda et al. 2004). On the supply side, modeling availability of renewable energies, particularly, solar and wind in urban area has a reasonably long history, dating back to simple computer simulation models to assist site layout planning (Robinson et al. 2007a, 2008). More recently, supply models have focused on evaluation of renewable energy potentials with geographical information system (GIS) technologies and remote sensing data. Lately, researchers tend to develop integrated models. Integrated models such as; SEP, LT-Urban, SUNtool, CitySim and SINCITY focus on the whole urban energy system. The typical goal of these models is developing design tools for planners to support evaluation of alternative scenarios.

This review provides an up to date overview of the modeling approaches that are developed to simulate the energy flow in urban built environment. A vast variety of modeling techniques are used to simulate energy production, distribution and consumption. These models have different characteristics, strengths, weaknesses, capability and applicability. They are usually constructed based on the different assumptions, input data and level of details (Swan and Ugursal 2009). Accordingly classification of them can be done in different ways. Based on the spatial and temporal dimensions, cities are composed of several spatial levels, such as building, neighborhood, district or city that in each level specific factors influence the energy flow. Therefore, the categorization of energy models can be done based on these spatial scales (Owens 1986). The most common way of categorization of urban built environment energy models is based on the application focus of these models in the urban energy system. Supply and demand models are most frequently used. Recently, researchers are interested in developing integrated models that include supply and demand models. Each of these models relies on different levels of input data, calculation techniques and different applicability. A critical review of each of these models is provided based on the specific criteria including: spatial features (scale and dimension), temporal resolution, simulation approach and analytical method. In this review, in Sects. 2, 3, 4 we discuss the supply, demand and integrated models subsequently. Then we discuss aspects that are missing in current urban energy models. We finish with some conclusions.

2 Supply Models

The first category of energy simulation models in the urban built environment is related to the supply side of urban energy system. The typical goal of these models is estimating the potential availability of different kinds of renewable energy resources in urban fabrics at different spatial scales. This type of models has a quite long history and come back to the simple computational and mathematical techniques to aid planners in large scale urban layout planning (Robinson et al. 2008). In recent years the main focus of urban designers and planners in sustainable design of urban environment have been based on the concepts such as bioclimatic and smart design, building orientation and daylight access that are more dealing with the efficient use of solar energy either in terms of solar active or passive design strategies. In this regard modeling availability of solar radiation in urban fabrics for use by converting technologies or natural lighting has gained much attention. Accordingly in this part, the main focus of supply models will be on solar energy simulation models.

Mardaljevic and Rylatt (2003) proposes an approach for evaluating the total annual irradiation incident on building facades in urban settings based on a physical-based rendering approach and data-visualization techniques to generate maps of annual irradiation. The sun and sky irradiation maps are evaluated and the sky contribution is calculated using realistic, non-isotropic models for the sky radiance distribution. Compagnon (2004) used a similar approach, but based on pre-processing the sky radiance distribution rather than post-processing results for individual skies/suns. Moreover, they proposed a methodology to quantify the potential of facades and roofs located in urban areas for active and passive solar heating and photovoltaic electricity production in London. Solar irradiation and illuminance values obtained through numerical simulations form the core part of the method. In an integrated approach with GIS, Gadsden et al. (2003) developed a general methodology of solar energy planning for energy advisers that determines the potential for generating solar energy. The methodology identifies suitable dwellings for installing solar panels and quantifies the potential energy savings and reductions.

In a European project, Sustainable Urban Neighborhood Tool (SUNTool) was developed using a slightly more computationally accurate way of simulating irradiation with a cumulative sky. The researchers proposed a technique, which produces annual irradiation images using the popular ray-tracing tool RADIANCE. This involves the use of a pre-processor, which generates a cumulative sky radiance distribution that the ray-tracing program can reference to at run time (Robinson and Stone 2004). The cumulative sky may be described in terms of either a global radiance distribution for a discretized sky vault or a diffuse discretized radiance distribution with either hourly or a statistical subset of suns.

Recently new data extraction and acquisition technologies from aerial images such as Remote Sensing and Light Detection and Ranging have been used in urban planning domain. Agugiaro et al. (2011) use these technologies for estimating solar radiation on building roofs in a complex landscape area. The approach was

Table 1 Characteristics of reviewed supply models

Features Models	Years	Spatial scale	Spatial analysis	Temporal dimension	Renewable energy sources	Analysis method
Mardaljevic	2003	City	3D	Annual	Solar radiation and natural light	Simulation
Gadsden	2003	City	2D	Annual	Solar thermal energy and electricity	Simulation
Compagonen	2004	Neighborhood	3D	Monthly	Solar radiation and day light	Simulation
Robinson	2004	Neighborhood	3D	Hourly	Solar radiation, Water, Waste	Optimization
Agugiaro	2011	Regional	3D	Daily	Multi scale solar radiation	Estimation
Carlson	2011	City	2D	Monthly	Solar reflectivity	Simulation
Vettorato	2011	Regional	2D	Annual	Solar radiation, biomass, geothermal	Simulation, Optimization
Strzalka	2012	District	2D	Monthly	Solar electricity energy	Optimization

based on a multi-scale solar radiation estimation methodology that combines 3D data, ranging from regional scale to the architectural scale. Moreover, they present a very high-resolution geometric model of the building roofs by advanced automated image matching methods. Models are combined with raster and vector data sources to estimate the incoming solar radiation hitting the roofs (Agugiaro et al. 2012). Based on the POLYCITY project, Strzalka et al. (2012) develop a four-step methodology that uses Laser Scanning data along with Geo-Media for finding the potentials of rooftops for the PV-system. These steps include; collect the laser scanner data, introduction of a GIS software tool as an interface to visualize data, estimate the suitable roof areas and calculate the solar roof potential for producing the electrical energy from the PV-systems. Estimating the potential of urban scale implementation of reflective roof materials in built environments for estimating the potential energy saving is another research field, which is considered in recent years. Jo et al. (2010) develop a methodology for assessing the potential capacity of installing reflective roofs in an urbanized area. In this methodology, remote sensing image data combined with a building energy models to quantify the rooftop reflectivity and predict the potential electricity savings. Based on matching energy supply and demand in urban areas, Vettorato et al. (2011) developed a three-step framework to compare the potential supply of renewable energy technologies and the energy demand of buildings at local scales.

Table 1 shows the main characteristics of the reviewed supply models.

3 Demand Models

Around 40 % of the total energy demand in urban area is consumed in the built environment (Sunikka 2006). Hence fulfilling any energy ambitions requires a considerable transformation of the current pattern of energy use in this sector. In general energy demand models focusses on the evaluation and estimation of energy consumption in built environment. Building energy characteristics are very broad and complex issue and in simulation and estimation need to consider a wide range of aspects such as physical, constructional and behavioral characteristics in different spatial and temporal levels. This review based on the econometric and engineering approaches attempts to highlight more physical and constructional aspects of urban built environment. Behavioral considerations and policy implications associated with it are beyond the scope of this study. To date, researchers use two fundamental different classes of modeling techniques for estimating energy use in the built environment; the top-down and bottom-up techniques (Bohringer et al. 2007). Recently spatial evaluation of the energy demand has been added incorporating new data collection techniques. In the next subsections, each of these approaches is elaborated in more detail.

3.1 Top-Down Approach

The top-down modeling approach works at the aggregated and macro level, and treat the built environment as energy sink (Swan and Ugursal 2009). These models tend to be used to investigate the interrelationships between the energy sector in urban area and other aspects such as economy and socio-demographic characteristics of the large scale (Kavgic et al. 2010). A variety of variables can be used in these models, such as socioeconomic variables, climatic conditions, stock building construction rates etc.

These models are categorized into two main subgroups, econometric and technological models. Econometric models as evident from their names are based on interconnection between energy use with socioeconomic variables to examine the connection between the energy sector and economic output. Technological models examine the energy consumption to more detailed technological and physical characteristics of the built environment that influence energy use, such as: saturation effects, technological progress, and structural change (Johnston 2003). The use of these approaches come back to the energy crises of the late 1970s, when researchers try to investigate the impacts of the energy price and changing in consumption patterns on the end user behaviors in high spatial levels (Swan and Ugursal 2009). The first types of these models were more econometric models based on statistical data that energy consumption expressed as a function of simple econometric variables. Hirst (1978) proposed a U.S. residential annual energy consumption model that is based on the economic and housing ownership changes, and the model has developed into an economic model including housing ownership and technical parameters. His work was developed over the following years and resulted in an econometric model, which had both housing and technology components (ONeal and Hirst 1980). Saha and Stephenson (1980) developed an engineering-economic model to simulate residential energy use in New Zealand that is sensitive to major technological, economic and demographic determinants. Based on Genetic Algorithm (GA) the Energy Input Estimation Model was developed by Ozturk et al. (2004) that used to estimate Turkey's future residential-commercial energy input demand based on GDP, population, import, export, house production, cement production and basic house appliances consumption figures. This model estimates the future projections on energy consumptions and examines the effect of the design parameters on the energy input of the sector. To support the decision-making process by benchmarking of the environmental performance of buildings between real-estate managers and municipal administrations, Tornberg and Thuvander (2005) present a GIS implemented energy model for the building stock in the city of Goteborg. Energy data comprises energy source and amount of energy used for heating, hot water, and electrical appliances.

3.2 Bottom-Up Approach

The bottom-up models are built up based on the data on an individual level to investigate the contribution of end user on energy usage in urban and regional level. These models can be used for simulation of energy use for individual occupants, building or groups of buildings and then extrapolate results to represent the city or region, based on the representative weight of the modeled samples (Yamaguchi et al. 2007). As the bottom-up models work on a micro level, thus for evaluating the energy use, they need extensive databases of empirical data to support the description of each user (Shorrock and Dunster 1997). Advances to date in the use of this modeling approach have been based on evaluating the energy consumptions of groups of buildings (residential, commercial or industrial buildings) as distinct categories. Michalk et al. (1997) developed a structural model for energy use in the residential sector whose customers were segmented into four main categories. The model provides energy-use patterns for each category of customers for selected customer samples. Lariviere and Lafrance (1999) present a statistical model that establishes a relationship between the annual electricity consumption per capita and some variables which characterize a sample of Canadian cities, include, urban density, demography and meteorological data. In the UK Jones et al. (1998, 2007) develop a model to estimate the energy demands and associated emissions for a municipality's stock of buildings based on the extrapolation of results from 100 representative building typologies for the case study site of Neath and Port Talbot. Using a same approach, Shimoda et al. (2004) proposed an energy demand model for Osaka city based on the occupants' schedule of living activities, weather data and energy efficiencies of appliances and dwellings. By summing up the simulation results for various household categories, total energy consumption in the residential sector estimated. A significant difference in this model is incorporating occupational characteristics such as occupants' presence and use of appliances in the modeling using detailed survey statistics. Yamaguchi et al. (2003) develop a bottom-up approach based on the constructional features and occupancy characteristics. It provides insights into the changes required in all the components of urban energy systems from the equipment level to the entire building and systems level. Moreover, Yamaguchi et al. (2007) increased their scale of analysis to the city scale and proposed a general methodology based on the prototypical approach. In this methodology, in general building stock is divided into several categories based on the physical or behavioral characteristics and then energy use for each category is quantified using prototypical buildings that representing a whole building stock category. The model revealed how the classification of building stock according to a number of building properties improves the accuracy of the simulation model. To optimize the generation and distribution of energy in district heating/cooling systems and to predict accurate energy demand profiles, Kim et al. (2009) developed a detailed building simulation technique based on the location, orientation and configuration variables that was employed for the prediction of individual building energy load patterns. The model shows that the effect of model resolution is more evident in the winter, the effect of building orientation is clearer in summer, and the effect of shade from adjacent buildings is greater in winter.

3.3 Spatial Evaluation of Energy Demand

The energy demand of building sector can be evaluated quickly and robustly without costly on-site measurements (Rylatt et al. 2003). This part reviews integration of urban energy demand models with new technologies in data extraction and management in urban planning domain i.e. Geographical Information Systems (GIS) and Remote Sensing. It demonstrates how these technologies can be employed to facilitate the acquisition and extraction of building data and spatial parameters from the plan form of buildings. These tools fulfill the large data requirements of urban models without the need of visual inspection and long survey of the properties (Jones et al. 2007). Rylatt et al. (2003) introduced a prototype general domestic energy modeling approach to the estimation of so-called baseline energy consumption. It shows how useful data relating to the plan form of dwellings such as the plan form and its area, the size of the perimeter, the exposed perimeter size, the built form type and the orientation can be derived from digital maps to use in energy analysis. Heiple and Sailor (2008) combine annual building energy simulations for city-specific prototypical buildings and available Geospatial data in a GIS framework for estimating hourly energy consumption profiles in the building sector at spatial scales down to the individual parcel. Based on a systematic approach, Mavrogianni et al. (2009) combines GIS databases and a modified version of the Standard Assessment Procedure algorithm to estimate the space heating demand of urban domestic energy users for exploring the impact of urban built form on the levels of domestic energy consumption in London. Comparison of the model outputs with top down energy statistics for annual household energy consumption indicates that the model ranks areas based on their domestic energy demand with relative success. Recently most researchers tend to use sophisticated spatial models in their simulations i.e. 3D modeling. Beginning in the early 1990s, first 3D city models were built to have a 3D representation of a city (Lippold 2010). Until today, the establishment of a comprehensive 3D model is still the focus of many researchers around the world. Nowadays by employing high quality remote sensing data, technology capability has reached a level where 3D analysis can be achieved much easier. Beside that cloud computing infrastructures and software applications technology provides a solid and improvable base for 3D and sophisticated analysis (Robinson et al. 2007a, 2008). As a result, 3D city models can be applied as a promising tool for energy evaluation on large-scale areas. Strzalka et al. (2010) develop an urban energy management tool to evaluate the heating demand of urban district and analysis appropriate measures to improve the building energy performance. For this purpose, a method was developed that enables an automatic extraction of the building's heating volume from a Geo-information system. LAIDR technology was used to extract the buildings volume, building footprints and the measured building heights and finally allows generating a 3D city model of the analyzed area. Based on this 3D model required data for analysis in terms of input file is generated. Table 2 shows the main characteristics of the reviewed demand models.

4 Integrated Modeling Approach

Modeling of urban energy system into separate parts will not cover all aspects of the issue; therefore, modelers tend to develop a more holistic model of this system. In urban scale, due to the complex structure of urban systems and plurality of constituent elements, simulation of system and their subsystems are involved with a variety of complicated and interrelated features. As a result, for comprehensive modeling in urban level even on a small scale, a large number of aspects such as spatial, functional, socioeconomic and technological variables should be taken into account. Hence the number of permutations is overwhelmingly large and the probability of identifying an optimal combination of these variables, increasingly reduced (Robinson et al. 2007a).

To date, the use of integral scenarios and models in evaluating urban energy performance are in the initial stages of development. Recently, some preliminary energy-economy optimization models-encoded with a set of structured, self-consistent assumptions and decision rules-have emerged as a key tool for analysis of energy at the national and international scales. For investigating the interrelationships between buildings, transport and industry in urban area based on the

Table 2 Characteristics of reviewed demand models

Features Models	Years	Spatial scale	Spatial analysis	Temporal dimension	Variables	Approach	Analysis method
Hirst	1978	National	–	Annual	Socioeconomic, Technical	Econometric	Estimation
O'Neal	1980	National	–	Annual	Socioeconomic, Technical	Econometric	Estimation
Saha	1980	National	–	Annual	Socioeconomic	Engineering, Econometric	Estimation
Ozturk	2004	National	–	Annual	Socioeconomic, Design	Genetic algorithm	Simulation, Optimization
Tornberg	2005	City	2D	Monthly	Socioeconomic, Physical	Engineering	Estimation
Michalk	1997	Regional	–	Monthly	Physical, Occupational	Engineering	Estimation
Lariviere	1999	City	–	Monthly	Physical	Engineering	Simulation
Shimoda	2004	City	–	Daily	Physical, Occupational	Prototypical approach	Simulation
Jones	2007	Municipality	2D	Monthly	Physical	Prototypical approach	Simulation
Yamaguchi	2007	District	–	Daily	Physical, Occupational	Prototypical approach	Simulation
Kim	2009	Building	–	Hourly	Physical	Building simulation	Optimization
Rylatta	2003	Neighborhood	3D	Weekly	Physical	Image processing	Optimization
Heiple	2008	Building	3D	Hourly	Physical	Prototypical approach	Optimization
Mavrogianni	2009	District	3D	Monthly	Physical	Prototypical approach	Simulation
Strzalka	2010	District	3D	Hourly	Physical	Image processing	Simulation

energy supply and demand, Jones et al. (1998, 2007) develop Energy and Environmental Prediction (EEP) model. EEP is a kind of environmental auditing and decision-making tool, which is based on GIS techniques and unified urban submodels—housing, non-domestic, industry and traffic—to establish energy use by these sectors. Solar Energy Planning (SEP) was proposed by Gadsden et al. (2003) to establish the baseline energy consumption of domestic properties. It determines the possible potential for reducing demand using the three key solar technologies; passive solar design, solar heating panels and photovoltaic panel systems. As in these models, the analysis performed based on the 2D maps, the spatial variables in analysis were partially superficial. As part of European Union funded project, Ratti et al. (2000) develop LT-Urban model as a tool to support urban-scale building energy modeling by interfacing a simplified building energy model with image processing techniques to improve the scale of evaluations. By employing techniques from image processing and the Geo-sciences, an algorithm has been developed to derive the necessary urban form variables such as built up area, facade orientation and obstruction angle of the sky from digital elevation models. Although in this model 3D spatial model were used to assess the impacts of urban form on the acquisition of solar energy, but diversity of variables for identifying an optimal configuration of urban layout was very limited. Brownsword et al. (2005) based on linear programming develop an urban energy optimization model using energy supply data for solar photovoltaic and postal code information that simulates spatial and daily variations in energy demand. A linear programming optimization module was used to identify the most cost-effective measures to achieve specified energy reduction targets.

In a collaborative effort and in terms of a European project, SUNtool was developed as a decision support system for designers to optimize the environmental sustainability of master planning proposals (Robinson et al. 2007b). It was based on the integrated resource flow modeling of buildings of disparate uses. For considering of all energy aspects in neighborhood level, four sub models; radiation exchange modeling, reduced thermal modeling, stochastic modeling of occupant presence and urban plant modeling was embodied in it. Later, CitySim was developed to provide more sophisticated simulation of resource flows at neighborhood levels (Kampf and Robinson 2008). Its aim was support designers to optimize the performance of new and existing urban developments based on more spatial parameters. In this model, as the spatial variables of an urban development is infinitely large, a microsimulation model of energy flow are coupled with an evolutionary algorithm to identify a subset of urban design variables that have been parameterized. Such method was robust, but spatial scale of the model was still limited. In a more technical model, Girardin et al. (2010) developed EnerGIS as a GIS based urban energy model that simulate with sufficient detail the energy service requirements of a given geographical area to allow the evaluation of the integration of advanced integrated energy conversion systems.

Various studies have shown that energy flow in urban area is associated with city structure and urban form. However, this connection makes a paradox between demand and supply sides. On one hand, densification of cities reduces transportation and buildings consumptions whereas on the other hand, densification has a negative impact on urban microclimate and renewable energy potential. Based on this assumption, Bonhomme et al. (2011) develop MORPHOLOGIC model that is a GIS platform, which calculates solar potential using a simplified model of shadows and evaluates the energy consumption of city blocks and allow urban planners to evaluate the best urban form to reduce GHG emissions. Energy is essential to the delivery of urban services. Integrating the urban layout models with the socioeconomic structure of the city to evaluate energy performance is a new challenge in integrated models that has a great potential for future researches. As an initial effort, SynCity was developed for integrating these two urban modeling approaches. In this model different aspects of urban energy system including; the layout of a city, the socioeconomic structure of it, activities, energy carriers and technologies are integrated. This synthetic toolkit, facilitate the integrated modeling of urban energy systems across all of the issues that related to the energy supply and demand in the city (Keirstead et al. 2009). Table 3 shows the list of these models along with the main criteria that used for evaluation of them.

5 Discussion of the Reviewed Models

From this review, it is evident that already there is a wide range of urban energy models. They are quite diverse in terms of spatial and temporal scales, energy side focus, analysis variables and methodological approaches that used for simulation. This review, without going too much into detail provides an overview of these models. By disaggregating, the energy models based on the energy flow direction and in terms of their main distinctive features, a consistent framework was provided for examining the different energy models. In this section their applications and development status is discussed.

5.1 Supply and Demand Models

The initial supply models were simple models for evaluation of solar passive strategies in urban built environments. But gradually, and with the emergence of other new renewable energy technologies, these models became sophisticated. Recently, a new computational technique is used, which produces daily/monthly/annual irradiation images using ray-tracing tools such as RADIANCE. In addition, by employing technologies such as Remote Sensing (RS) and Light Detection and Ranging (LIDAR), urban modelers could utilize a huge data extraction and management in evaluation of the availability of renewable energy resources in complex urban forms and morphologies. However, the sophisticated supply models, particularly based on the three-dimensional spatial models, are still in its early stages.

Table 3 Integrated models

Features Models	Years	Supply	Distribution	Demand	Spatial scale	Spatial analysis	Temporal dimension	Analysis method
LT urban	2000	+	–	+	Neighborhood	3D	Daily	Optimization
SEP	2003	+	–	+	City	2D	Monthly	Simulation
Brownsword	2005	+	–	+	City	–	Daily	Optimization
EEP	2007	+	–	+	City	2D	Monthly	Simulation
SUNtool	2007	+	+	+	Neighborhood	3D	Daily	Optimization
CitySIM	2008	+	+	+	Neighborhood	3D	Daily	Optimization
SYNCity	2009	+	+	+	City	2D	Daily	Optimization
EnerGIS	2010	+	+	+	District	2D	Daily	Simulation
MORPHOLOGIC	2011	+	–	+	Neighborhood	3D	Monthly	Optimization

Each of the top-down and bottom-up energy demand models has their own strengths and weaknesses. Top-down approaches are comparatively easy to develop; as their developments are more based on the limited historical information about energy use pattern and macro-socioeconomic variables. However, these models are not able to provide an explicit representation of the final end-users in details and give a coarse analysis of energy use in the urban area. Bottom-up techniques can provide a connection between end-users on individual level with macro-socioeconomic indicators on urban level. As these models are based on the details on individual level, it is possible to consider micro variables related to the physical and behavioral characteristics of buildings and households, along with macroeconomic variables in different scenarios. However, these analyses require an extensive database of empirical data on energy use which the quantity and quality of these data could threaten the validity of these approaches. Acquisition and extraction of the physical and constructional data of buildings from the digital maps and aerial images has established a great chance to fulfill the large data requirements of urban bottom-up models without the need of visual inspection and long survey of the properties. This approach provides a quick and robust method, without costly on-site measurements, but still needed sophisticated 3D models and advanced techniques to extract appropriate spatial and geometrical variables.

5.2 Integrated Models

This review shows that each of the existing integrated models only addresses some specific parts of the whole urban energy system. Despite the diversity of approaches that are highlighted by the review, a number of common challenges such as sophisticated spatial variables, problem specific and data intensive, availability and credibility of data, policy relevancy and integrated multi-layer models can be identified.

5.2.1 Towards Integrated Multilayer Models

One of the major challenges involved in the development of integrated models is capturing the interaction and dynamics between different spatial scales and developing appropriate models that can capture these dynamics in macro and micro levels. It was observed that current integrated models are strictly constrained to one spatial scale such as building, neighborhood, district or city. In reality, energy consumption depends on different variables at different spatial scales. By constraining the simulation to one of these scales, all other variables that have effects at other scales, are faded out. Nowadays, for modeling at both macro and micro levels, reference models are available. On macro level, Top-down approaches such as Econometric or Statistical models have provided aggregate results. On micro level, Building Energy Performance Simulation Tools such as ESP-r,

ECOTECT and DOE-2 more deal with single buildings and do not address the upper spatial levels. Therefore, an integrated approach that allows examining the interaction of these levels is still missing.

There are two promising approaches for modeling this complexity; on the macro level, aggregate dynamics models and on the micro level, microsimulation and agent-based models. Aggregate dynamics models such as System Dynamics help us to understand the behavior of urban systems over time with use of internal feedback loops, time delays and stocks and flows, demonstrate how even seemingly simple systems display complexity and nonlinearity. Agent Based Models are computational methods that enable researchers to create, analyses and experiment with models composed of agents that interact within an environment. Agents perceive their environment and other agents; make decisions following some rules, and act, possibly changing the environment in the process. Most of the complex phenomena that happen in cities due to the interactions between agents and also environment can be addressed by this approach. In urban area, the interactions of agents at the micro scales and emergent patterns from them on the macro scale (bottom-up), and interactions from the upper scales into the agents (top-down) can be simulated with these approaches. While these approaches integrate with the energy models, it is possible to simulate energy flow in urban areas at different spatial levels.

5.2.2 Spatial Implications of Renewable Energy Technologies

The close interrelation between energy and space becomes obvious by looking at the historical development of cities, where almost 85 % of the worldwide energy is consumed. Today, these interactions have accelerated with the growing trend in the adoption of renewable energy technologies in supply side. Unlike the fossil fuels that are more based on the underground sources and have no significant spatial effects, renewable energy technologies have spatial and functional implications. Urban energy transition requires implementation of huge volume of these technologies in the urban area and consequently affects the spatial and functional structure of cities. Current models address the energetic effect of these technologies. However, this is just part of the analysis and it is of equal interest to consider the spatial aspects of these technologies as well. Recently, much progress has been made in urban spatial modeling. Spatial transition models concentrate on predicting how a city evolves over time under different policies and are commonly found in the area of land use/land cover and urban growth modeling such as OPUS/UrbanSIM, CommunityViz and TRANUS. Through these approaches, the urban system is considered as a complex system which is focused on space and time. Such models can establish robust platforms to incorporate energy models to demonstrate spatial implications.

5.2.3 Simulation and Optimization Methods

Simulation on urban scale involved with various aspects such as spatial, social, economic, technological and environmental issues over time. Therefore, simulating any issues, even at a small scale requires considering large number of variables. Analyzing of these variables and the possibility of identifying an optimal configuration to them based on the conventional computational methods such as statistical, mathematical or manual trial and error are not too realistic. As shown in Tables 1, 2 and 3, different simulation and optimization analysis methods already have been applied. Simulation methods such as System Dynamics, Statistical methods on macro level and Agent Based Models (ABM) or Microsimulation methods on micro level are the main candidates. Optimization methods such as Econometric approaches and Multi Objective Optimization Models including linear or nonlinear programming and genetic algorithms are also applied.

6 Conclusions

In this review, three main approaches that are used to simulate energy flow in urban area have been examined. The goal is to provide an up to date overview of them. The evaluation has been done based on the energy flow direction and the application focus of these models in the urban energy system. Accordingly, three main distinct categories of models were identified, including: supply, demand and integrated models. Each of these categories uses different assumptions and simulation techniques and relies on different levels of data. A critical review of these models is provided. Finally, some of the main issues that are missing in these models are highlighted, which open up new lines of research for researchers. These issues should be considered and integrated in developing new urban energy models in built environment to help designers and planners in simulating and optimizing their urban energy proposals in a more comprehensive way. These new challenges are:

- Integrated multilayer models.
- Spatial implications of renewable energy technologies.
- Integrated simulation and optimization methods.

References

Agugiaro G, Remondino F, Stevanato G, De Filippi R, Furlanello C (2011) Estimation of solar radiation on building roofs in mountainous areas. IAPRS&SIS 38:155–160

Agugiaro G, Nex F, De Remondino F, Filippi R, Droghetti S, Furlanello C (2012) Solar radiation estimation on building roofs and web-based solar cadaster. ISPRS Ann Photogrammetry, Remote Sens Spat Inf Sci 1–2:177–182

Bohringer C, Rutherford F, Arbor A (2007) Combining bottom-up and top-down. Energy Econ 30(2):574–596

Bonhomme M. Haddou A, Adolphe L (2011) Energy and urban morphology: a decision support tool for urban energy paradox. In: Proceedings of the eighteenth international seminar on urban form, Concordia University, Montréal, Canada, 26–29 August 2011

Brownsword A, Fleming D, Powell C, Pearsall N (2005) Sustainable cities—modeling urban energy supply and demand. Appl Energy 82(2):167–180

Compagnon R (2004) Solar and daylight availability in the urban fabric. Energy Buildings 36(4):231–328

Gadsden S, Rylatt M, Lomas K, Robinson D (2003) Predicting the urban solar fraction: a methodology for energy advisers and planners based on GIS. Energy Buildings 35(1):37–48

Girardin L, Marechal F, Dubuis M, Darbellay C, Favrat D (2010) EnerGis: a geographical information based system for the evaluation of integrated energy conversion systems in urban areas. Energy 35(2):830–840

Heiple S, Sailor J (2008) Using building energy simulation and geospatial modeling techniques to determine high resolution building sector energy consumption profiles. Energy Buildings 40(8):1426–1436

Hirst E (1978) A model of residential energy use. Simulation 30(3):69–74

Jo H, Carlson J, Golden S, Bryan H (2010) Sustainable urban energy: development of a mesoscale assessment model for solar reflective roof technologies. Energy Policy 38(12):7951–7959

Johnston D (2003) A physically based energy and carbon dioxide emission model of the UK housing stock. Ph.D. thesis, Leeds Metropolitan University

Jones P Williams J, Lannon S (1998) An energy and environmental prediction tool for planning sustainability in cities. In: Proceedings of the 2nd European Conference REBUILD, Florence, April 1998

Jones P, Patterson J, Lannon S (2007) Modelling the built environment at an urban scale energy and health impacts in relation to housing. Landscape Urban Plan 83:39–49

Kampf J, Robinson D (2008) A hybrid CMA-ES and DE optimisation algorithm with application to solar energy potential. Appl Soft Comput 9(2):738–745

Kavgic M, Mavrogianni A, Mumovic D, Summerfield A, Stevanovic Z (2010) A review of bottom-up building stock models for energy consumption in the residential sector. Build Environ 45:1683–1697

Keirstead J, Samsatli N, Shah N (2009) SynCity: an integrated tool kit for urban energy systems modeling. In: Proceedings of the 5th urban research symposium—cities and climate change, Marseille, France, 28–30 June 2009

Kim M, Yook I, Kim D, Kim L, Kim J (2009) Study on simulation modeling strategy for predicting thermal energy demand profiles of residential complex. In: Proceedings of the building simulation '09, University of Strathclyde, Glasgow

Lariviere I, Lafrance G (1999) Modelling the electricity consumption of cities: effect of urban density. Energy Econ 21(1):53–66

Lippold A (2010) 3D city-scale modeling applied to sustainable design: history, example, and future. In: Proceedings of the policity conference, Stuttgart's Wagenhallen, 16–17 September 2010

Mardaljevic J, Rylatt M (2003) Irradiation mapping of complex urban environments: an image-based approach. Energy Buildings 35:27–35

Mavrogianni M, Davies Z, Chalabi P, Wilkinson M, Kolokotroni J (2009) Space heating demand and heat wave vulnerability: London domestic stock. Build Res Inf 37(5–6):583–597

Michalik G, Khan E, Bonwick J, Mielczarski W (1997) Structural modeling of energy demand in the residential sector. Energy 22(10):937–947

ONeal L, Hirst E (1980) An energy use model of the residential sector. IEEE Trans Syst Man Cybern 10(11):749–755

Owens S (1986) Energy, planning and urban form. Pion Limited, London

Ozturk H, Canyurt E, Hepbasli A, Utlu Z (2004) Residential-commercial energy input estimation based on genetic algorithm (GA) approached. Energy Buildings 36(2):175–183
Ratti C, Robinson D, Baker N, Steemers K (2000) LT Urban: the energy modelling of urban form. In: Proceedings of the PLEA 2000, Cambridge, July 2000
Robinson D, Stone A (2004) Irradiation modelling made simple: the cumulative sky approach and its applications. In: Proceedings of the PLEA 2004, Eindhoven 2004
Robinson D, Filchakova N, Kämpf J, Rasheed A, Scartezzini J (2007a) Towards an evolutionary model of city sustainability. In: Proceedings of the Holcim Forum, Beijing, China, 17–24 April 2007
Robinson D, Campbell N, Gaiser W, Kabel K, Le-Mouele A, Morel N, Page J, Stankovic S, Stone A (2007b) SUNtool: a new modelling paradigm for simulating and optimising urban sustainability. Sol Energy 81(9):1196–1211
Robinson D, Giller C, Haldi F, He F, Kämpf J, Kostro A (2008) Towards comprehensive simulation and optimisation for more sustainable urban design. In: Proceedings of the PLEA 2008, Dublin, 22–24 October 2008
Rylatt M, Gadsden S, Lomas K (2003) Methods of predicting of urban domestic energy demand with reduced datasets, a review and a new GIS-based approach. Build Eng Serv Res Technol 24:92–102
Saha G, Stephenson J (1980) A model of residential energy use in New Zealand. Energy 5(2):167–175
Shimoda Y, Fujii T, Morikawa T, Mizuno M (2004) Residential end-use energy simulation at city scale. Build Environ 39(8):959–967
Shorrock D, Dunster E (1997) The physically-based model BREHOMES and its use in deriving scenarios for the energy use and carbon dioxide emissions of the UK housing stock. Energy Policy 25(12):1027–1037
Strzalka A, Bogdahn J, Eicker U (2010) 3D City modelling for urban scale heating energy demand forecasting. In: Proceedings of the policity final conference, Stuttgart, Germany, 15-17 September 2010
Strzalka A, Alam N, Duminil E, Coors V, Eicker U (2012) Large scale integration of photovoltaics in cities. Appl Energy 93:413–421
Sunikka M (2006) Policies for improving energy efficiency in the European housing stock. IOS Press BV, Amsterdam
Swan G, Ugursal V (2009) Modeling of end-use energy consumption in the residential sector: a review of modeling techniques. Renew Sustain Energy Rev 13(8):1819–1835
Tornberg J, Thuvander L (2005) A GIS energy model for the building stock of Goteborg. In: Proceedings of the 25th annual ESRI user conference
Vettorato D, Geneletti D, Zambelli P (2011) Spatial comparison of renewable energy supply and energy demand for low-carbon settlements. Cities 28(6):557–566
Yamaguchi Y, Shimoda Y, Mizuno M (2003) Development of district energy simulation model based on detailed energy demand model. In: Proceedings of the building simulation 2003, Eindhoven 2003
Yamaguchi Y, Simoda Y, Mizuno M (2007) Transition to a sustainable urban energy system from a long-term perspective: case study in a Japanese business district. Energy Buildings 39(1):1–12

Part III
Traffic and Network Modelling

A Procedure Using GIS to Analyze the Access by Non-Motorized Transport to Transit Stations

Fernanda Borges Monteiro and Vânia Barcellos Gouvêa Campos

Abstract This paper presents a procedure to analyse the quality of the urban space on the vicinity of public transport stations. This procedure evaluates walk and bicycle paths near the stations based on indicators mainly related to the physical characteristics of these paths. To define these indicators, a bibliographical review and a survey research in subway stations in Rio de Janeiro, Brazil were developed. From this survey, the distances travelled by users to the stations were evaluated and the attractive factors of walk and bicycles trips around them were identified; finally, based on these indicators, a spatial analysis of paths around a transit station is presented. For the spatial analysis, a Geographic Information System (GIS) was used. The resulting maps indicate the quality levels of the walk and bicycle paths.

1 Introduction

For achieving sustainable mobility, it is important to encourage the use of non-motorized and public transport. The quality of urban space around public transport stations can be an important factor to encourage the access of non-motorized transportation, especially by foot or by bicycle. In many situations, the best way to increase the use of public transport, instead of automobile use, is to improve pedestrian facilities and cycling conditions to access the stations.

Thus, the integration of public transport systems with non-motorized transports could be the basis for increasing sustainable mobility. However, in order to achieve a successful integration, it is necessary to analyze aspects associated with

F. B. Monteiro (✉) · V. B. G. Campos
Instituto Militar de Engenharia, Rio de Janeiro, Brazil
e-mail: arqnandy@gmail.com

V. B. G. Campos
e-mail: vania@ime.eb.br

S. Geertman et al. (eds.), *Planning Support Systems for Sustainable Urban Development*, Lecture Notes in Geoinformation and Cartography,
DOI: 10.1007/978-3-642-37533-0_15, © Springer-Verlag Berlin Heidelberg 2013

non-motorized transport modes like the quality of walking and cycling paths. It is also important to identify factors that encourage people walk or bicycle to station in their daily journey for work, and the characteristics of urban environment that generate and incentive the walking habit.

In this context, a procedure was developed for spatial analysis of the urban space on the neighborhood of transport stations. This procedure evaluates different pedestrian and cyclist paths to access the stations, based on criteria related to the physical characteristics of these paths. For defining such criteria, a literature review about walkways and bicycle facilities was developed and a survey research based on these criteria was conducted in subway stations in Rio de Janeiro, Brazil.

2 Literature Review

The spaces for the traffic of pedestrians and cyclists should be planned and designed in order to maximize their safety and comfort during their diary trips. The quality of non-motorize transport path includes its continuity, attractiveness and convenience of routes, including elements that involve several factors as the distance to be traveled, the gradient of the track, the conditions of sidewalks, the straightness of the route and any other factor to facilitate walking and cycling.

The main difficulty related to the quality of urban spaces is the definition of instruments that can be applied to evaluate the adequacy of the conditions of walk and cycle paths. In the literature, there are important methods and criteria for assessing these spaces. For assessing pedestrian space, Fruin (1971a, b) has proposed a method, based on the methodology for roadway level service. This author proposes to qualify the pedestrian facility by studying parameters such as: human anatomy, field of view, comfortable distance between bodies, walking down stairs and psychological perception of urban space. Years later, in 1985, the reference manual on practical design of roads, the Highway Capacity Manual—HCM (TRB 1985) was supplemented with a guide for the pedestrian to elaborate the design of sidewalks, based on the methodology of Fruin (1971a).

In 90 years, Sarkar (1995) tried to make sidewalks safe for urban intersections and user groups considered vulnerable. Dixon (1996) developed a method to evaluate the spaces allocated for pedestrians on roadways. This work include qualitative methods to measure some subjective variables, such as security, safety, comfort, convenience, continuity, system coherence and attractiveness.

The methodology of Kristy (1995) Ferreira and Sanches (2001) and Leslie et al (2006) have in common the evaluation of the qualitative elements of pedestrian spaces, according to the perspective of users. In these methods, performance indicators were used to measure the quality of the sidewalk, based on subjective variables such as visual attractiveness, comfort, system continuity, safety and security.

In the case of cyclists, most of the methods are based on the work of Landis (1994), Davis (1987), Sorton and Walsh (1994), Epperson (1994) and Landis et al. (1997). Until 1980 the factors commonly used to quantify the quality level of service offered to cyclists were: speed, freedom to maneuver, traffic interruptions, comfort, convenience and safety (Epperson 1994). After 1980, some works have been developed based on the conditions of the roads (Epperson 1994; Sorton and Walsh 1994; Dixon 1996; Landis et al. 1997) using the following evaluation criteria: traffic volume, bandwidth, speed limit, pavement condition and location of the pathway (Turner et al. 1997). Only with the methods of Hunter et al. (1995) and Wang and Nihan (2004) the accident risk has been considered on the analysis.

Gatersleben and Appleton (2007), reviewing the literatures about cycling to work, verified that the common factors that prevent people for cycling are travel distance, traffic safety, heavy traffic, physical condition, pollution, bad weather and the working clothes that many times are not appropriate for cycling.

A well-developed cycle ways infrastructure combined with bicycle parking at stations may improve the accessibility of public transport for cyclists, and hence, provide a attractive alternative to the use of car (Martens 2004, 2007).

In the handbook "Cycling-inclusive Policy Development" (Kuijper and Braakman 2009, Chap. 5), the authors define five main requirements for bicycle infrastructure:

1. Perception and being able to ride side by side create requirements in terms of *attractiveness and comfort.*
2. Minimizing resistance requires *comfort and directness.*
3. Optimizing mental capacity and allowing enough free space requirements in terms of *comfort and safety.*
4. Cyclists' vulnerability creates requirements in terms of *safety.*
5. The need for complete, understandable cycling infrastructure creates requirements in terms of coherence (or consistency).

The authors of this handbook consider that, if the minimum level of one or more of the five requirements cannot be met, the infrastructure must be modified.

3 The Survey Analysis and the Criteria for Spatial Analysis

Based on the literature review, a survey was prepared to interview subway users. The main reason for this survey was identify the transport used to access stations, the distances and time spent and the user's perceptions walking or cycling to the station. Studies developed by subway showed that 65 % of subway users walk to the stations, and only 0.2 % cycling.

The survey was performed in three subway stations in Rio de Janeiro: Ipanema, Pavuna and Colégio. Ipanema and Pavuna stations are located at the end of two lines and Colegio is one of line 2 stations between Pavuna station and the downtown. The map on Fig. 1 shows the subway lines and stations. At these subway stations, between 7:30 a.m. and 11:30 a.m., 390 users were interviewed at the board platforms. About 125 of the interviewed people arrived at the station by foot.

Fig. 1 Subway lines in Rio de Janeiro

Part of the survey was designed to identify difficulties or impossibilities of travel by foot or by bike. Table 1 presents the main difficulties in using each of these modes. The common problem pointed by users was the distance to the station or the conditions of the path, as expected, based on the literature review. It was also noted that most of those who walk to the station had its origin no more than 1.5 km far from the station and those who use bike mostly traveled a distance of 5.5 km.

These distance values of walking and cycling are important to define the area around the station that should be improved in order to incentive the subway users to access the station by foot or by bicycle.

Table 1 Difficulties for accessing the station by foot or by bicycle

For pedestrians	Poor sidewalk pavement (holes, puddles, etc.)
	Lack of safe crossings (crosswalk and traffic lights)
	Public insecurity (assaults)
	Insufficient lighting
	Physical disability (age, disease, etc.)
	Distance to the station
For cyclist	Bad conditions of the cycle path (holes, puddles, etc.)
	Public insecurity (assaults)
	Cannot take the bike in the wagon
	Lack of cycleways
	Distance to the station

4 Proposed Indicators

From the literature review summarized on Sect. 2, and based on survey answers, four indicators were defined as those who have major influence in the decision to walk and cycle to the station (Monteiro and Campos 2012). These proposed indicators are related to the infrastructure and environment of the sidewalk and bicycle paths around stations. Tables 2 and 3 present these indicators and the criteria proposed to be used in their evaluation.

5 The Spatial Analysis Procedure Using GIS

This paper proposes a procedure, based on the use of GIS, to evaluate the paths used by pedestrians and cyclists to access the stations. This procedure aims to qualify the segments of the possible paths to stations checking those with low

Table 2 Pedestrian space: proposal indicators and criteria

Indicators	Criteria
Facilities for pedestrians	Presence of sidewalks on both sides of the street
	Effective width of the sidewalk
	Quality of sidewalk pavement
	Presence of trees
	Street lighting
Accessibility and mobility	Access ramps on the pedestrian crossroads
	Sidewalk with tactile floor
	Visual and audible signaling at pedestrian crossings
	Gradient of the sidewalk
Safety	Safe pedestrian crossings
Security	Presence of police officers on the streets

Table 3 Cyclist space: proposal indicators and criteria

Indicators	Criteria
Facilities for cyclists	Presence of bicycle lanes
	Effective lane width of the bicycle lanes
	Quality of lane pavement
	Presence of trees
	Street lighting
Accessibility and mobility	Access ramps on the curb at intersections and near stairways
	Gradient of the bicycle lanes
Safety	Safe crossings
	Distance from de vehicle flow
	Flow vehicle density
Security	Presence of police officers on the streets

quality and will need some improvement to attract walking or biking access. The analysis must be done separately for pedestrians and cyclists.

The procedure for spatial analysis of walking and cycling paths comprehends four steps:

Step 1 Obtain a map of the neighbourhood region where stations are located

Step 2 Define a radius around the stations and evaluate each criteria for each segment inside the radius. For walking path, each variable should be measured in each segments of street in the 1.5 km radius from the station. For cycling path, the segments should be assessed in the 6.0 km radius from the station. These radius values were defined based on the results of the survey.

Step 3 Normalize each criteria value on a scale from 0 to 3, and definition of the indicators values and quality index for each segment.

For the normalization, the process proposed by Eastman and Jiang (1996) using interval range [0.3] should be applied. These authors present a simple form for criteria standardization, which is a linear variation, defined by equation:

$$X_i = \frac{(R_i - R_{\min})}{(R_{\max} - R_{\min})} \times range\,value$$

where:

X_i	normalized value
R_i	criteria value to be normalized;
R_{min}	minimum value of the criterion
R_{max}	maximum value of the criterion;
Range value	3,0 (to have values between 0 and 3)

To define a quality index for each street segment, first the value of each indicator should be calculated. This is obtained by the average of all criteria values, based on the normalized values of the 11 variables for

each segment. Finally, the quality index per segment is defined by the average of the values of the four indicators (see Tables 2, 3).
Quality index values from 0 to 1 means that the segment has bad conditions, values between 1 and 2, good condition, and more than 2 indicated that it has optimum conditions for the pedestrian or cyclist.

Step 4 Development of a map representing the index of each street segment. It is necessary a color scale for each value to be represented.

For index value less than or equal to 1 (one), the street segment is highlighted in red, for indices between 1 and 2, the color is blue and for indices larger than 2 the color is green.
The Tables 4 and 5 present each criterion and its respective measures. Some criteria are binary variables (value 1 or 0), indicating if the characteristic related to the variable does exist or not. For others, such as the

Table 4 Pedestrians variables and respective measures

	Pedestrians	Measures	
Indicators	Criteria	max	min
Facilities for Pedestrians	Sidewalks on both sides of the street	2	0
	Effective width of the sidewalk	2.0	0
	Quality of sidewalk pavement	100	0
	Presence of trees	100	0
	Street lighting	100	0
Accessibility and mobility	Access ramps on the curb at pedestrian crossings	1	0
	Sidewalk with tactile floor	1	0
	Visual and audible signaling at pedestrian crossings	1	0
	Sidewalk with low gradient	1	0
Safety	Safe pedestrian crossings	1	0
Security	Police officers on the streets	1	0

Table 5 Cyclists' variables and respective measures

	Cyclists	Measures	
Indicators	Criteria	max	min
Facilities for cyclists	Presence of bicycle lanes	2	0
	Effective lane width the bicycle lanes	2.5	0
	Quality of lane pavement	100	0
	Presence of trees	100	0
	Street lighting	100	0
Accessibility and mobility	Access ramps on the curb at intersections and near stairways	1	0
	Street segment with low grade	1	0
Safety	Safe crossings	1	0
	Cycle way segregated from the traffic flow	1	0
	Low traffic of vehicles	1	0
Security	Police on the streets near the bicycle lanes	1	0

Table 6 Result of the analysis on Jangadeiros street

Pedestrians—Indicators and Measurements												
Urban Design		Ipanema Jangadeiros Street										
		Parameters		Values							Result	
Indicators	Criteria	Max	Min	QD 01	Nor.	R/Q	QD 02	Nor.	R/Q	average	Sum	Result
Facility for pedestrians	Sidewalk on both sides of the street	2	0	2	3,00	2,22	1	1,50	2,31	2,25	2,27	Excellent
	Sidewalk width	2	0	2	3,00		1,5	2,25		2,63		
	Regularity of sidewalk pavement	100	0	100	3,00		100	3,00		3,00		
	Presence of trees	100	0	0	0,00		80	2,40		1,20		
	Effective public illumination	100	0	70	2,10		80	2,40		2,25		
Accessibility and mobility	Access ramps on the crossing	1	0	1	3,00	1,88	1	3,00	1,88	3,00	1,88	Good
	Sidewalks with tactile floor	1	0	0	0,00		0	0,00		0,00		
	Visual and audible signaling on the crossing	2	0	1	1,50		1	1,50		1,50		
Safety	Sidewalk with low gradient	1	0	1	3,00	3,00	1	3,00	3,00	3,00	3,00	Excellent
	Safe pedestrian crossings	1	0	1	3,00		1	3,00		3,00		
Security	Police officers present on the streets	1	0	0	0,00	0,00	0	0,00	0,00	0,00	0,00	Bad
Final Result for segment (QD)				Result		1,77	Good		1,80	Good		

regularity of the pavement, the values are defined as the percentage of the segment of walk and cycle ways on good conditions. A detailed description of each criterion is presented by Monteiro and Campos (2012).

6 An Example of Application

The proposed procedure was applied in the region around Ipanema/General Osório station. First, the segments around the station were analyzed *in loco* and all variables (for walking and biking) were measured. An example of this analysis is showed on Table 6. This table presents the analysis of the segments on Jangadeiros street in neighborhood of Ipanema station.

Figure 2 presents the map around Ipanema (Gal. Osório) station and 1.5 km surrounding streets. Figure 3 shows the map resulted by the procedure, e.g., the quality index of each analyzed street segment for walking, represented by colors, from ArcGis Desktop 9.3 software.

On Fig. 3, we can note that almost all streets around Ipanema station present good condition for walking. Only three street segments present bad quality. In fact, this map allows a visualization and easy identification of the situation for walking access around the station.

Figure 4 show the result of the analysis for cyclist. In Ipanema neighborhood, there is a cycle way which has a good infrastructure (green segment). However, the existent cycle way is not connected to the subway station and the others street segment have bad condition to cyclist.

Fig. 2 Street map around Ipanema station

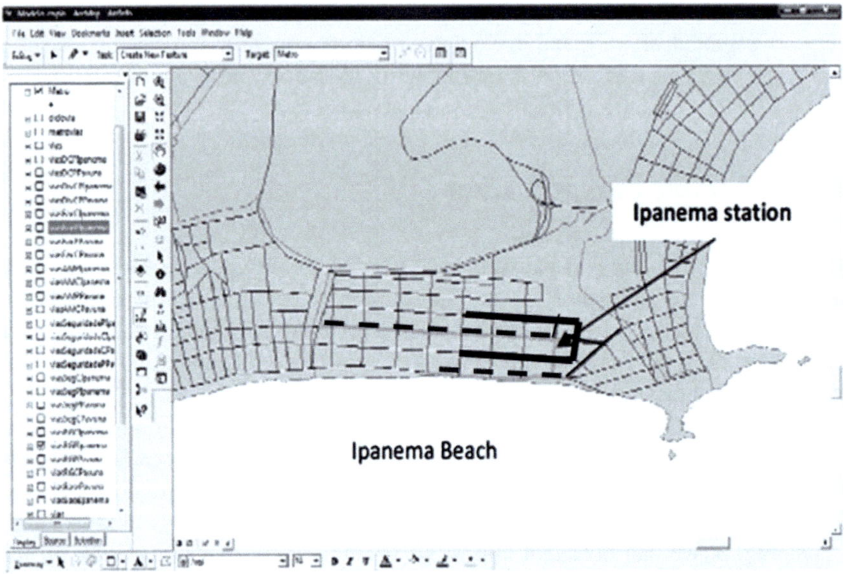

Fig. 3 Spatial analysis of the street segment in access by foot to Ipanema Station

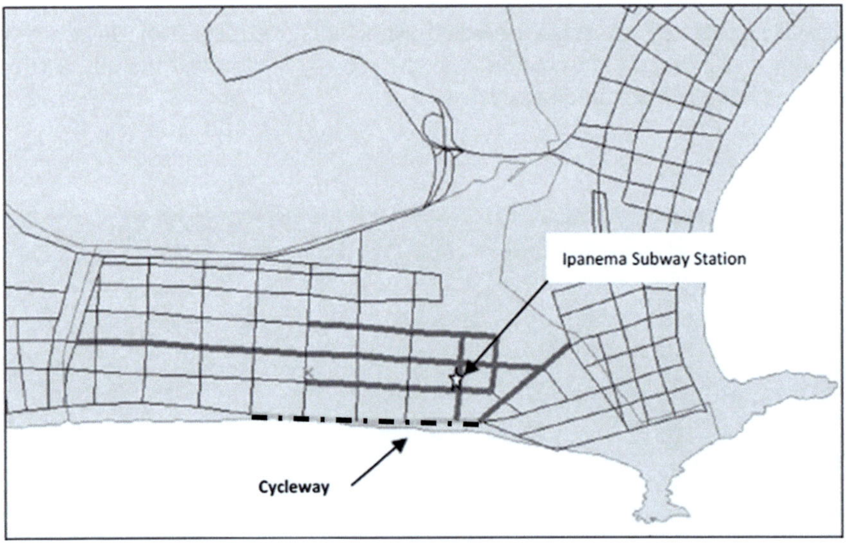

(red ———) (green – · – · –)·

Fig. 4 The result of the analysis of Urban Spaces for Cyclists—Ipanema

We emphasize that this example application was only made, around Ipanema/General Osorio station with the objective to evaluate the applicability of the procedure.

7 Conclusions

This paper presents a procedure for analyzing the pedestrian and cyclist access to public transport stations. This method intends to use indicators for evaluating the quality of walkways and bicycle facilities used to access stations. The proposed procedure permits an objective analysis of the segment streets, measuring the quality of these segments for walking or for cycling.

The importance of urban spaces study arises from the need to create a pleasant space to encourage non-motorized transport, creating sustainable alternatives for cities. Today, urban mobility is mainly performed by car due the inefficiency of public transportation, the lack of infrastructure for cyclists and inadequate spaces for pedestrian.

The proposed method support future urban spaces and transport decisions identifying regions that have problems and proposing solutions to improve the urban environment, encouraging people to access station by foot or by bicycle.

Through the use of GIS software (ArcGIS), it is possible to represent the values of each indicator on maps. Such maps easily identify which street segments shall get more attention and clearly define which criteria/indicators should be improved to offer a safe and enjoyable walking or biking trip to the station.

It is important to note that the measures for each criteria can be changed according each study as well as its maximum and minimum values.

Acknowledgments The authors acknowledge CNPq and CAPES, organizations of supporting research in Brazil, for support in developing this work.

References

Davis J (1987) Bicycle safety evaluation. Auburn University, Auburn

Dixon LB (1996) Bicycle and pedestrian level-of-service performance measures and standards for congestion management systems. Transp Res Rec 1538:1–9

Eastman JR, Jiang H (1996) Fuzzy measures in multicriteria evaluation. In: Proceedings of 2nd international symposium on spatial accuracy assessments in natural resources environmental studies, Fort Collins, Colorado, pp 527–534

Epperson B (1994) Evaluating suitability of roadways for bicycle use: toward a cycling level of service standard. Transp Res Rec 1438:9–16

Ferreira MA, Sanches SP (2001) Índice de Qualidade das Calçadas—IQC. Revista dos Transportes Públicos 91:47–60

Fruin J (1971a) Designing for pedestrians: a level-of-service concept. Highway Res Rec 355:1–15

Fruin JJ (1971b) Pedestrian planning and design. Metropolitan association of urban designers and environmental planners

Gatersleben B, Appleton KM (2007) Contemplating cycling to work: attitudes and perceptions in different stages of change. Transp Res Rec Part A 41:302–312

Hunter W, Wayne E, Stutts J (1995) Bicycle-motor vehicle crash types: the early 1990s. Transp Res Rec 1502:68–74

Khisty CJ (1995) Evaluation of pedestrian facilities: beyond the Level-of-Service concept. Transp Res Rec 1438:45–50
Kuijper D, Braaskman B (2009) Cycling-inclusive policy development: a handbook, Chapter Five: Five main requirements for cycling-inclusive infrastructure
Landis B (1994) Bicycle intersection hazard score: a theoretical model. Transp Res Rec 1438:3–8
Landis BW, Vattikuti VR, Brannick MT (1997) Real-time human perceptions: toward a bicycle level of service. Transp Res Rec 1578:119–126
Leslie E et al (2006) Measuring the walkability of local communities using Geographic Information Systems data. Paper presented at Walk21-VII, "The Next Steps", The 7th international conference on walking and liveable communities, Melbourne, Australia in: www.walk21.com
Martens K (2004) The bicycle as a feeding mode: experiences from three European countries. Transp Res Part D 9:281–294
Martens K (2007) Promoting bike-and-ride: the Dutch experience. Transp Res Part A 41:326–338
Monteiro FB, Campos BGV (2012) A proposal of indicators for evaluation of the urban space for pedestrians and cyclists in access to mass transit station. Procedia—Soc Behav Sci 54:637–645
Sarkar S (1995) Evaluation of safety for pedestrians at macro and microlevels in urban areas. Transp Res Rec 1502:105–118
Sorton A, Walsh T (1994) Bicycle stress level as a tool to evaluate urban and suburban bicycle compatibility. Transp Res Rec 1438:17–24
TRB (1985) Highway capacity manual. Transportation Research Board, National Research Council, Washington, D.C.
Turner S, Shafer C, Stewart W (1997) Bicycle suitability criteria: literature review and state-of the practice survey. Texas Transportation Institute, Dallas
Wang Y, Nihan L (2004) Estimating the risk of collisions between bicycles and motor vehicles at signalized intersections. Accid Anal Prev 36:313–321

Locations with Frequent Pedestrian-Vehicle Collisions: Their Transportation and Neighborhood Environment Characteristics in Seattle and King County, Washington

Junfeng Jiao, Anne V. Moudon and Yuan Li

Abstract Improved pedestrian safety is integral to walkable, sustainable neighborhoods. This study aggregated 2,944 pedestrian-vehicle collisions in the Seattle-King County area using a 500 m^2 grid overlay that captured related road and neighborhood characteristics. Collision cells were concentrated in 17.4 % of the extent, and 17.0 % of the collisions took place in high-frequency collision locations clustered in a 3.5 km^2 area. A negative binomial model estimated that the frequency of collisions in grid cells correlated to higher volumes of vehicles and bus riders (boarding and alighting counts), higher intersection and traffic signal densities, higher densities of residential units and jobs, and several known pedestrian activity generators. These proxy measures of pedestrian activity and exposure to vehicles confirmed a lack of pedestrian safety in the metropolitan areas. The relatively small number of locations with high and very high pedestrian-vehicle collisions should facilitate targeted, effective, and inexpensive pedestrian safety improvement programs.

J. Jiao (✉)
Department of Urban Planning, Ball State University, Muncie, IN 47304, USA
e-mail: jjiao@bsu.edu

A. V. Moudon
Department of Urban Design and Planning, University of Washington,
Seattle, WA 98105, USA
e-mail: moudon@u.washington.edu

Y. Li
School of Architecture and Civil Engineering, Xiamen University, Xiamen 361005, China
e-mail: yako79@gmail.com

S. Geertman et al. (eds.), *Planning Support Systems for Sustainable Urban Development*, Lecture Notes in Geoinformation and Cartography,
DOI: 10.1007/978-3-642-37533-0_16, © Springer-Verlag Berlin Heidelberg 2013

1 Introduction

The design of transportation systems has overwhelmingly focused on motor vehicles, mostly ignoring the pedestrians who also use these systems (Retting et al. 2003). Yet pedestrians make up a significant part of the travelling public especially in metropolitan areas. Their safety matters. In 2010, 4,280 pedestrians were killed and an estimated 70,000 were injured in the United States (NHTSA 2012). In 2004 (the time of this study's data), 1,769 pedestrians were involved in vehicle collisions and 60 died within 3 weeks of the events in the State of Washington alone, representing 10.6 % of the total traffic-related fatalities in the state (NHTSA 2006). As cities and regions seek to accommodate modes of travel other than driving, it is imperative to create roads and streets that can be safely shared between all users (Retting et al. 2003). Traffic safety enhancement programs should seek to reduce both the frequency of collisions and the severity of injuries sustained in these collisions, taking into account that in addition to human injury, collisions produce inefficiencies in transportation systems, induce loss of productivity, and create significant property damage.

Knowing what factors relate to the occurrence of collisions and to the severity of injuries resulting from collisions is essential to guide the designs and policies shaping future transportation systems. Previous research already determined that the likelihood of a pedestrian dying as the result of a collision increases by 35 % when the vehicle speed increases from 20 to 30 mph. When the vehicle reaches 40 mph, the likelihood of a fatality reaches 85 % (Limpert 1984; Leaf and Preusser 1999). It is also well-known that older and younger pedestrians are overly represented in collisions, especially those that end in fatalities (Kong et al. 1996; Campbell et al. 2004; Lee and Abdel-Aty 2005). Road designs (e.g., the number of lanes, the presence of crosswalk markings, etc.) are also related to pedestrian collisions and injuries (Zegeer et al. 2002). In contrast, traffic calming policies focusing on narrower roads encourage slower speeds (Zajac and Ivan 2003), and on-street parking has been associated with lower collision rates (Marshall et al. 2008).

1.1 Objective

This study sought to identify locations where vehicle–pedestrian collisions were frequent, and to isolate the road and neighborhood environment characteristics of locations that were associated with frequent collision occurrences. Successfully identifying these characteristics would help researchers develop a better planning support system to locate, predict, or remove collision hot spots within a city and provide safer environments for its residents. Previous research using the same data in Seattle and King County, WA, had modeled the odds of a collision occurring, based on the characteristics of collision locations (Moudon et al. 2008). This case–control study, which used binomial logistic models, considered state routes but not city streets. A second study modeled pedestrian injury severity as a result of a collision (Moudon et al. 2011). Binary and ordinal logistic models, including collisions on both state routes and city street, predicted injury severity based on the socio-demographic characteristics of the pedestrian and the characteristics of the locations where the collision occurred. The present study adds to this past work by modeling the factors related to the frequency of collisions in specific locations.

2 Methods

2.1 Research Design

Pedestrian-vehicle collisions are known as "rare" events, which need to be studied over multiple years in order to include a sufficient number of observations. Collisions also cluster in space, because most pedestrian travel today takes place in the more densely developed parts of metropolitan areas. An area-based approach was taken to frame the collision locations spatially and to capture the characteristics of their surrounding transportation facilities and built environment. To do so, a small-area grid was laid over the King County area contained within the Urban Growth Boundary (KCUGB). Each grid cell was used to aggregate pedestrian collisions, and the concomitant road and neighborhood environment characteristics. The grid cell was also the spatial unit of analysis. Three different grid structures were tested that determined the spatial allocation of collisions: 305 by 305, 500 by 500, and 1000 by 1000 m. The 500 m cell was selected because its statistical advantages. Compared to other cell sizes, the number of collisions within each 500 m cell was more evenly distributed. Further, less dispersion was observed in the pedestrian collision distribution represented by the 500 m cell.

Collision locations were characterized by the road design and the surrounding neighborhood built environment. Five conceptual domains governed the selection of variables: transportation infrastructure (e.g., street-block size, traffic signals, sidewalks, bus stops, etc.); traffic conditions (e.g., vehicular volumes, bus boarding and alighting counts, etc.); regional location (distance to Central Business District); neighborhood environment (e.g., densities of development); and land use and activity generators (e.g., various destinations known to be associated with pedestrian travel). Road design being a standard focus in transportation safety, its characterization with regard to pedestrian collisions will provide important information for future transportation safety programs. On the other hand, many variables of the neighborhood built environment act as proxy measures for exposure. For example, residential and employment densities are correlated with the share of pedestrian travel in an area. Together with vehicular volumes and bus boarding and alighting counts, they capture the magnitude of potential conflicts between vehicles and people. The question posed by the present research is the extent to which these conflictual conditions explain the frequency of pedestrian-vehicle collisions.

2.2 Data Sources

Collision data came from the Washington State Department of Transportation (WSDOT). They included all police records of collisions involving pedestrians from 2001 to 2004, and occurring on city streets (CS) and state routes (SR) of the KCUGB. This area had a total of 3,182 collisions over the four years, or more than 90 % of pedestrian collisions in the county; 2,952 collisions were successfully geo-coded, and 2,944 were within the KCUGB. Collision records comprised the socio-demographic and behavioral characteristics of pedestrians and drivers; road class, conditions, and design where the collision occurred; time of day, year, and weather conditions when the collision occurred.

Data on traffic conditions came from the Puget Sound Regional Council (PSRC), and included average annual daily traffic (AADT) figures and estimated speed on major routes (AADT and estimated speed were EMME2 modeled data). King County Metro provided bus boarding and alighting counts data. WSDOT also provided data on traffic signals, intersections, crosswalks, sidewalks, and number of traffic lanes on state routes. Objective data on the neighborhood environment came from the King County Assessor's office, which included land use, property assessment values, and residential density at the parcel or tax lot level. Employment data were generated at the Urban Form Lab at the University of Washington.

2.3 Measurements

The dependent variable was the number of collisions within each cell. The independent variables were the road and neighborhood environment variables, which described the transportation infrastructure and traffic conditions for both vehicles and pedestrians; and activity generators in the built environment. Variables, which were used in previous studies of pedestrian travel, included sidewalks, street intersections, traffic signals, bus stops, bus boarding and alighting counts, and speed limits; residential and employment density, and the number of grocery and convenience stores, general retail facilities, fast food restaurants, eating/drinking establishments, and schools. Road and neighborhood environment variables were measured at the cell level (Table 1).

2.4 Statistic Analysis

Counts of the number of pedestrian collisions within each cell were analyzed using a Negative Binomial model to take into account the over-dispersion of the count data. Modeling was done in three steps. First, the variables found to be significant in logistic regressions previously carried out in models estimating the risk of a

Table 1 Independent variables and measurements

	Domain	Variables	Measurements	City streets model (Moudon et al. 2011)	State route 99 models (Moudon et al. 2011)	State routes (all other than SR 99) (Moudon et al. 2008)	Proxy variable for exposure
Road environment	Transportation infrastructure and road design	Intersection	# of intersections (Cate.)	–	N.I.	N.I.	
		Road lane length	Sum of each lane length (<=45 mph) (LN1.)	N.I.	N.I.	N.I.	Yes
		Traffic signal	Having traffic signal or not (BI.)	+	–	+	
		Sidewalk	Having sidewalk or not (BI)	N.I.	X	X	Yes
		Speed limit (posted)	Average posted speed of cell/buffer (normalized by length of road) (LN1.)	–	X	X	Yes
		Bus stops	Having bus stops or not (BI.)	+	X	X	
	Traffic conditions	Estimated AADT	Sum of the estimated 24 h AADT data on city street (no day/week provided) (LN1.)	X	+	X	Yes
		Bus boarding and alighting	Having boarding and alighting within the cell or not (BI.)	–	X	+	Yes
		Estimated Speed	Average estimated speed from EMME2 model (LN1.)	X	X	X	Yes
Built environment	Regional location	Downtown	Distance to downtown (LN.)	N.I.	N.I.	N.I.	
		Seattle	Within city of Seattle or Not (BI.)	–	X	–	
	Neighborhood environment	Residential unit value	Median residential unit value [land +improvement] (LN1.)	–	X	X	
		Residential density	Sum of residential units (LN1.)	+	X	+	Yes
		Employment density	Having jobs or not (BI.)	+	–	X	Yes
	Land use and activity generators	NC2	Having NC2 or not (BI.)	+	X	X	
		Retail	Having the parcel or not (BI.)	X	+	+	
			Distance to the closest retail store (LN1.)	–	X	X	

Table 1 (continued)

Domain	Variables	Measurements	City streets model (Moudon et al. 2011)	State route 99 models (Moudon et al. 2011)	State routes (all other than SR 99) (Moudon et al. 2008)	Proxy variable for exposure
	Office	Having the parcel or not (BI.)	X	+	X	
	Grocery store	Having the parcel or not (BI.)	X	X	X	
	Fast food restaurant	Having the parcel or not (BI.)	X	X	X	
	Convenience store	Having the parcel or not (BI.)	X	X	X	
	Drinking and eating	Having the parcel or not (BI.)	+	X	X	
	Primary school	Having the parcel or not (BI.)	X	X	X	
	Middle school	Having the parcel or not (BI.)	X	X	X	
	High school	Having the parcel or not (BI.)	+	X	X	
	College	Having the parcel or not (BI.)	X	X	X	

NC2: Neighborhood commercial center, defined as a cluster of at least one grocery store, one restaurant, and one retail outlet within 50 m of each other
N.I: Not Included
BI: Dichotomized into binary variable
X: Included but not significant
Cate: Transformed into categorical variable
+: Positively related
LN: Natural log (original value)
−: Negatively related
LN1: Natural Log (original value + 1)

collision occurring at a location (Moudon et al. 2008) and the odds of the pedestrian incurring severe injuries or dying were included in a base model (Moudon et al. 2011). These variables were traffic signals, bus boarding and alighting counts, being within Seattle or not, residential density, and employment density. Second, the remaining variables shown in Table 1 were added to the base model one by one. A final model was run which included the base model and the variables found to be significant in the one by one test.

3 Results

3.1 Descriptive Analysis

The KCUGB area contained 5,506 cells and had 2,944 pedestrian collisions (Table 2). The average number of collisions per cell was 0.5, with a maximum of 65; the latter cell was located in Downtown Seattle (Fig. 1). Cells with at least one collision numbered 957 (17.41 %). The remaining 82.59 % of the cells did not have any collision. Almost two-thirds of the collisions were concentrated in 5.3 % of the region's extent. The spatial over-dispersion rate was 10.38.

For the independent variables, 36,684 intersections were included in the analysis; the average number of intersections per cell was 6.7. There were 1,964 traffic signals at these intersections, for an average of 0.36. Total road lane length per cell averaged 2.4 miles. The average length of sidewalks within each cell was 1,658 ft; 51.5 % of the cells had a value of zero as the sidewalk data only included those on major roads. Routes with posted speed limits over 45 mph were excluded from the analyses as they were assumed to restrict pedestrian access. The average posted speed limit in each cell was 25.3 mph, and the average estimated speed in each cell was 15.7 mph. The average sum of estimated AADT in each cell was 42,650. Because AADT data only included major streets, 43.8 % of the cells had a value of zero. The average number of bus stops per cell was 1.7, and 3,242 (58.9 %) of the cells did not have any bus stops. The average number of bus boarding and alighting counts per cell was 104 per day and 64.1 % of the cells that did not have such information.

Table 2 Distribution of pedestrian collisions in 500 by 500 m cells within the KCUGB

Cell class	# of collisions per cell	# of cells	% of cells	Area (km^2)	Area (mi^2)	# of collisions	% of collisions
1.	40–65	3	0.06	0.75	0.29	181	6.15
2.	21–39	11	0.22	2.75	1.06	318	10.80
3.	3–20	277	5.04	69.25	26.73	1632	55.43
4.	1–2	666	12.09	166.5	64.27	813	27.62
5.	0	4549	82.59	1137.25	438.98	0	0
Total	2944	5506	100.00	1376.5	531.33	2944	100

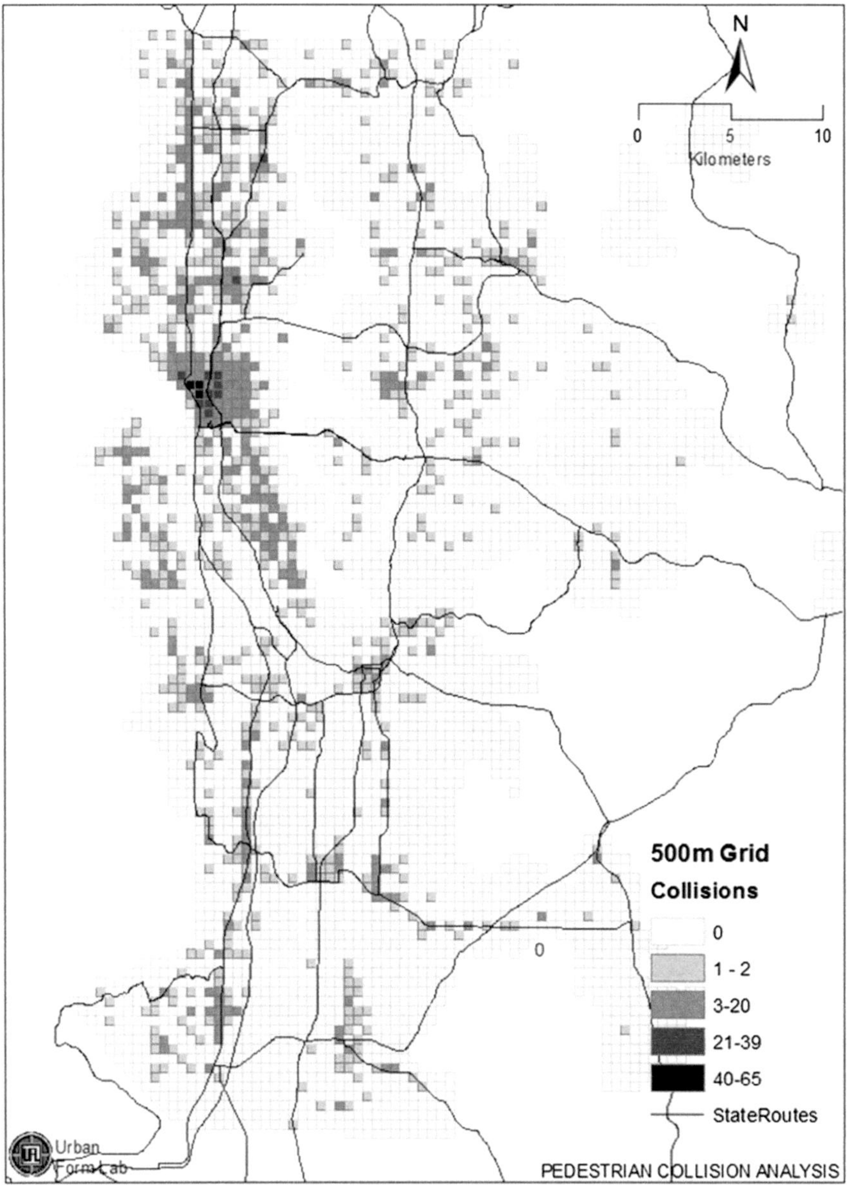

Fig. 1 Spatial distribution of pedestrian collisions in a 500 m grid within the KCUGB

As for regional location, 4,641 (84.3 %) cells were not within Seattle's boundaries. The average distance from each cell's centroid to Downtown Seattle was 13 miles. The average number of residential units per cell was 128; the maximum value was 10,532 and 17.3 % of the cells that did not have residential units. The average assessed property value per residential unit was $158,460. The average employment density was 223 jobs per cell, with 41.1 % of the cells having no employees. At least a part of a neighborhood center (NC2, defined as a cluster of at least one grocery store, one restaurant, and one retail outlet within 50 m of each other) was found in 23.3 % of the cells.

There were 4,458 retail parcels within KCUGB. They were highly clustered with 82.9 % of the cells have a value of zero. The average number of retail parcels per cell was 0.81. The average distance from the cell's centroid to the closest retail parcel was 3,176 ft, with a maximum of 18,523 ft. Office parcels was also highly clustered. The 4,534 office parcels were located in 22.5 % of the cells, with an average of 0.82. Convenience stores, fast food restaurants, groceries, and drinking and eating places were similarly distributed in space. There were 474, 342, 332, and 1,159 parcels corresponding to each one of these activity generators, with an average of less than 0.1, except for drinking and eating places, which averaged 0.21 per cell. More than 90 % of thc cells did not include any of these four activities. Education uses were few; they included 508 elementary schools, 22 middle schools, 47 high schools, and 19 colleges. The average number of education parcels per cell was less than 0.1.

3.1.1 Model Results

Of the five built environment variables included in the base model, four remained significant in the final model and positively associated with the frequency of collisions: the number of traffic signals, bus boarding and alighting counts, residential density, and employment density. Being inside and outside of the City of Seattle's boundaries lost significance, but the distance to the city's downtown was significant. An additional 11 variables were significant for a total of 16 (Table 3). In the domain of transportation infrastructure, intersections, road lane length, and traffic signals were significantly correlated with the frequency of collisions in a cell. In the domain of traffic conditions, average posted speed limit, estimated AADT, and having a bus boarding or alighting within each cell were all positively correlated with the frequency of collisions. For all transportation related variables, the number of intersections and traffic signals within each cell, the average posted speed limit, and having a bus boarding and alighting within each cell were the four strongest predictors for the frequency of collisions. Similar results were reported by other researchers (Dumbaugh and Li 2010; Moudon et al. 2011). In the neighborhood built environment domain, having a commercial center (NC2), or having residential units, jobs, office parcels, fast food restaurant, drinking and eating establishments, and high schools within the cell were all positively correlated with the number of collision happened in that cell. Differently, distance to Seattle Downtown, to the closest retail store, and the median value of a residential unit within the cell, were all negatively correlated with the frequency of collisions.

Table 3 Negative Binomial model results

	Domain	Variables	Measures	Results			
				Sig.	B	95 %CI. of B	
Road environment within each cell	Tranportation infrastructure and Road design	**# of intersections**	10 = 0		R		
			11 = 1–3	0.209	0.495	−0.278	1.267
			12 = 4–6	0.136	0.604	−0.189	1.396
			13 = 7–12	**0.033**	0.875	0.070	1.679
			14 = 13+	0.060	0.809	−0.035	1.654
		Sum of road lane length (<=45 mph)	Log (value + 1)	**0.003**	0.347	0.117	0.577
		Having traffic signals or not	1 = 1+	**0.000**	0.548	0.389	0.708
			0 = 0		R		
		Having sidewalks or not (major streets)	1 = 1+	0.183	0.161	−0.076	0.399
			0 = 0		R		
		Having bus stops or not	1 = 1+	0.092	0.338	−0.056	0.731
			0 = 0		R		
	Traffic condition	**Having bus boarding and alighting count or not**	1 = 1+	**0.017**	0.438	0.079	0.798
			0 = 0		R		
		Average posted speed limit (normalized by road length)	Log (value + 1)	**0.012**	0.569	0.127	1.012
		Sum of estimated AADT	Log (value + 1)	**0.000**	0.074	0.046	0.102

(continued)

Table 3 (continued)

	Domain	Variables	Measures	Results			
				Sig.	B	95 %CI. of B	
Neighborhood (NB.) environment within each cell	Regional location	**Dist. to Seattle Downtown**	Log (value)	**0.000**	−0.348	−0.457	−0.240
		In Seattle or Not	1 = yes	0.645	−0.048	−0.252	0.156
			0 = no		R		
	NB. Wealth	**Median value of residential unit**	Log(value + 1)	**0.001**	−0.054	−0.085	−0.023
	Devel−opment	**Having a residential parcel or not**	Log(value + 1)	**0.000**	0.189	0.126	0.251
		Having an employment parcel or not	1 = 1+	**0.001**	0.436	0.183	0.689
			0 = 0		R		
	Land use	**Having a NC2 or not**	1 = 1+	**0.003**	0.286	0.099	0.474
			0 = 0		R		
		Having a retail parcel or not	1 = 1+	0.301	0.125	−0.112	0.362
			0 = 0		R		
		Dist. to the closest retail store	Log (value + 1)	**0.000**	−0.239	−0.355	−0.124
		Having an office parcel or not	1 = 1+	**0.002**	0.280	0.099	0.460
			0 = 0		R		
		Having a grocery store or not	1 = 1+	0.888	0.014	−0.187	0.216
			0 = 0		R		
		Having a fast food restaurant or not	1 = 1+	**0.032**	0.225	0.020	0.430
			0 = 0		R		
		Having a convenience store or not	1 = 1+	0.069	0.163	−0.013	0.339
			0 = 0		R		
		Having a drinking or eating place or not	1 = 1+	**0.017**	0.224	0.040	0.407
			0 = 0		R		
		Having a high school or not	1 = 1+	**0.031**	0.568	0.052	1.084
			0 = 0		R		
		Constant		**0.011**	−4.135	−7.330	−0.939

NC2: Neighborhood commercial center, defined as a cluster of at least one grocery store, one restaurant, and one retail outlet within 50 m of each other
R: Reference category
Bold: variable significant at <0.05 level

4 Discussion

Collisions between pedestrians and vehicles concentrated in a small portion of the Seattle-King County region. More than 80 % of the region's extent experienced no collisions at all over the multiyear assessment period. Slightly more than one third of the collisions were in cells with 1–2 collisions, which covered 166.5 km^2 (64.3 mi^2), an area smaller than the City of Seattle. Almost 50 % of the collisions were in cells with 3–20 collisions, covering 69.3 km^2 (26.7 mi^2). And as many as about 17 % of the collisions were in cells with large numbers of collisions (21–65), concentrated in a tiny area of 3.5 km^2 (1.35 mi^2). In detail, 181 (6 %) collisions and 318 (11 %) collisions happened within 0.75 and 2.75 km^2, respectively. The implication of these findings are clear: first, significant road and built environment variables need to be included in the future planning support system to better identify and predict collision hot spots within a region. Secondly, not only does pedestrian safety need to be addressed at the small area level, but also the high level of spatial clustering suggests that interventions can be easily and likely inexpensively directed to small even very small portions of a metropolitan region.

The analyses demonstrated a consistent effect of the road and neighborhood environment on the frequency of pedestrian collisions: the more intensely developed and traveled environments were also the locations where pedestrian collisions were more frequent. A direct association was noted between such proxy measures of pedestrian volumes as the density of residential units and jobs, and bus boarding and alighting counts with the number of collisions. This same direct association was noted with higher vehicular volumes. There was no "safety in numbers" (Bhatia and Wier 2011): areas with more pedestrians also had more vehicular traffic, resulting in more collisions. The same lack of safety in numbers was found in an earlier study where the risk of severe injury increased with area density and traffic volumes (Moudon et al. 2011). This is in contrast to a study done in the U.K. where a quadratic relationship was found between urban density (population and employment captured at the scale of a ward) and pedestrian casualties: pedestrian casualties increased with density, but decreased in the most dense wards (Graham and Glaister 2003). Such a quadratic relationship should be expected where pedestrian safety is a priority in transportation policies and programs, and where such policies and programs focus not on entire regions, but on where there actually is pedestrian traffic.

Specific variables used to capture both the transportation and neighborhood environment explained the characteristics of locations with higher frequency of collisions. The smaller street-blocks with more intersections, lower vehicular speeds, higher number of traffic signals, which are all indicators of vehicular traffic calming through engineering solutions, were nonetheless associated with more frequent collisions. On the built environment side, neighborhood commercial centers and pedestrian attractors such as retail establishments, restaurants, and schools were also related to more frequent collisions. These findings suggested that engineered safety solutions were insufficient to guarantee pedestrian safety. It is possible that the behaviors ostensibly projected by the engineered road environment were not followed or enforced, and that, for example, actual vehicular speeds might be higher than those posted, and signalization might not be respected by either drivers or pedestrians.

A more detailed review of specific predictors of collision frequency follows.

4.1 Transportation Infrastructure and Traffic Conditions

A higher number of intersections and traffic signals and a longer road length in each cell corresponded to a high number of pedestrian collisions, suggesting that the dense street network advocated as supportive of pedestrian travel did not provide a safer environment for pedestrians. Similarly, traffic signals that aimed to regulate vehicular traffic at intersections did not appear to protect pedestrians. On the other hand, lower posted speed limits (but not modeled speeds) were as expected related to lower numbers of pedestrian collisions. Previous research based in Australia estimated that up to a 48 % reduction in collisions could be realized with speeds being lowered by 10 km/h (Anderson et al. 1997). Speed management offers the most potential for not only for injury prevention but also for collision prevention, as lower speeds give pedestrians and motorists a chance to successfully react to a conflictual situation (Retting et al. 2003).

The fact that the length of sidewalks was not significant in the final models might be due to limitations in the sidewalk data which only covered major streets. The number of bus boarding alighting counts in a cell was a significant predictor of collision frequency, but not the number of bus stops. The latter are not related to actual bus service: about 90 % of the ridership is found within the City of Seattle, which has 43 % (4,082 out of 9,570) of the region's bus stops. Finally, AADT as a measure of possible conflicts between pedestrians and vehicles had a significant positive relationship to pedestrian collision frequency.

4.2 Regional Location

The distance from each cell's centroid to Downtown Seattle showed negative significance, but collisions being within the City of Seattle did not. Figure 1 displays the highest concentrations of collision locations as being in Downtown Seattle and the nearby neighborhoods. For the remaining portions of the region, higher collision cells followed main circulation spines.

4.3 Neighborhood Environment and Activity Generators

Residential and employment density, two variables that act as proxy measures for pedestrian volumes, were positively related to the frequency of the pedestrian collisions. The median value of the residential units as a measure of neighborhood wealth was negatively related to collision frequency, confirming the protective effect of wealth on pedestrian safety (Moudon et al. 2008, 2011). Possible explanations might be better pedestrian facilitation, more concerned drivers, or even a lower pedestrian volume in the wealthier neighborhoods.

The presence of a neighborhood commercial center (NC2), measured as the proximate combination of one grocery store, one restaurant, and one retail outlet, had a positive relationship with the frequency of pedestrian collisions. Living near such neighborhood commercial centers was associated with the probability of walking in the neighborhood in other studies (Moudon et al. 2007). Also significant were the distance to the closest retail parcel, the presence of fast food and other restaurants (drinking/eating places), and the presence of office and high school parcels, all proxy measures of pedestrian activity. Streets and roads along and near these facilities would thus require special safety treatment (Hess et al. 2004). Previous research found that the presence of recreational facilities was positively associated with crash occurrences surrounding public schools in the city of Baltimore (Clifton and Kreamer-Fults 2007). On the other hand, however, the presence of parcels with retail or education uses such as colleges, convenience and grocery stores was not significant, suggesting that the need to carefully use individual activity generators to predict high frequency collision locations.

4.4 Methodology

Since no data exist on the actual number of pedestrians at the locations identified, this study, like all studies of pedestrians in relatively large areas and particularly in metropolitan regions, cannot estimate the individual-level risk of a pedestrian being involved in a collision with a vehicle. Indeed, if such pedestrian count data were available, an individual's risk of a collision could well be the same in high collision frequency locations than in their low frequency counterparts. From a transportation policy perspective, however, individual-level risk for pedestrians is an inacceptable criterion for initiating safety programs given the need to promote alternatives to driving in order to reduce not only congestion, but also the environmental effects of motorized travel. Fortunately, finding that collision frequency is associated with relatively few locations that have high numbers of pedestrians suggests that mitigation strategies and countermeasures can be effectively targeted to these locations, and by implication, should facilitate the swift improvement of pedestrian safety

Analyzing aggregated spatial data is always subject to the threat of the Modifiable Areal Unit Problem (MAUP) (Fotheringham and Wong 1991), where spatial units of analysis of different sizes or shapes can influence the model results (Wong and Amrhein 1996). In terms of the zonal effect, Zhang and Kukadia (2005) found that the grid approach to aggregating data produced more tractable and stable results than an approach based on irregular areal units (such as the ones found in Transportation Analysis Zones and Census tracts or block groups). Given that the MAUP is inherent in data aggregation, the key question is how to minimize its impact (Siddiqui et al. 2012). This research used a uniform grid structure made of small cells which reduced potential MAUP effects due to aggregation. Future research should test the results using even smaller grid cells.

5 Conclusion

The study benefited from a unique data set of pedestrian-vehicle collisions on all streets and roads in almost 1,400 km^2 (531 mi^2) of King County, WA between 2001 and 2004. Cells with more than 21 collisions contained 17 % of the collisions recorded over the 4-year assessment period. Yet these "dangerous" cells were few, concentrating in a very small 3.5 km^2 (1.35 mi^2) area or 0.31 % of the region. In contrast, more than 80 % of the region had no collisions at all. Such extreme spatial clustering means that improving pedestrian safety should be readily achievable by focusing on the few high-collision frequency locations in the region.

High-frequency collision locations were consistently characterized by similar road and neighborhood environments. Of the 16 significant variables in the model, six road environment variables showed that the frequency of collisions was related to higher vehicular and pedestrian volumes; but lower posted speed were protective of collision frequency. Road environment variables were the number of intersections, traffic signals, and bus boarding and alighting counts; the sum of road lane lengths and daily estimated AADT, and the average posted speed limit within each cell. The remaining ten significant variables were in the neighborhood environment domain. They suggested that locations with denser development and more pedestrian activity generators had a higher frequency of collisions. Neighborhood environment variables were the distance to downtown Seattle, residential and employment density, the median value of the residential units, the presence of a neighborhood commercial center, the distance to the closest retail store, the number of office parcels, fast food and other restaurants, and high schools within each cell.

Densely developed locations should support safe pedestrian travel. This study showed that this is not the case in the Seattle-King County region. Yet the fact that locations with a high frequency of pedestrian-vehicle collisions were few and spatially concentrated indicated that safety programs could effectively be targeted to these locations.

Acknowledgments This research was supported by the U.S. and Washington State Departments of Transportation and the Department of Urban Design and Planning at the University of Washington.

References

Anderson RWG, McLean AJ, Farmer MJB, Lee BH, Brooks CG (1997) Vehicle travel speeds and the incidence of fatal pedestrian crashes. Accid Anal Prev 29(5):667–674

Bhatia R, Wier M (2011) "Safety in Numbers" re-examined: can we make valid or practical inferences from available evidence? Accid Anal Prev 43(1):235–240

Campbell BJ, Zegeer C, Huang H, Cynecki M (2004) A review of pedestrian safety research in the United States and abroad
Clifton KJ, Kreamer-Fults K (2007) An examination of the environmental attributes associated with pedestrian-vehicular crashes near public schools. Accid Anal Prev 39(4):708–715
Dumbaugh E, Li W (2010) Designing for the safety of pedestrians, cyclists, and motorists in urban environments. J Am Plan Assoc 77(1):69–88
Fotheringham AS, Wong DWS (1991) The modifiable areal unit problem in multivariate statistical analysis. Environ Plan A 23:1025–1044
Graham DJ, Glaister S (2003) Spatial variation in road pedestrian casualties: the role of urban scale, density and land-use mix. Urban Stud 40(8):1591–1607
Hess PM, Moudon AV, Matlick JM (2004) Pedestrian safety and transit corridors. J Publ Transp 7(2):73–93
Kong LB, Lekawa M, Navarro RA, McGrath J, Cohen M, Margulies DR, Hiatt JR (1996) Pedestrian-motor vehicle trauma: an analysis of injury profiles by age. J Am Coll Surg 182(1):17
Leaf WA, Preusser DF (1999) Literature review on vehicle travel speeds and pedestrian injuries. US Department of Transportation, national highway traffic safety administration
Lee C, Abdel-Aty M (2005) Comprehensive analysis of vehicle–pedestrian crashes at intersections in Florida. Accid Anal Prev 37(4):775–786
Limpert R (1984) Motor vehicle accident reconstruction and cause analysis, 2nd edn. Michie Co, Charlottesville
Marshall WE, Garrick NW, Hansen G (2008) Reassessing on-street parking. Transp Res Rec: J Transp Res Board 2046(1):45–52
Moudon A, Lee C, Cheadle AD, Garvin C, Johnson DB, Schmid TL, Weathers RD (2007) Attributes of environments supporting walking. Am J Health Promot 21(5):448–459
Moudon A, Lin L, Hurvitz P, Reeves P (2008) "Risk of pedestrian collision occurrence: case control study of collision locations on state routes in King County and Seattle, Washington. Transp Res Rec: J Transp Res Board 2073(1):25–38
Moudon AV, Lin L, Jiao J, Hurvitz P, Reeves P (2011) The risk of pedestrian injury and fatality in collisions with motor vehicles, a social ecological study of state routes and city streets in King County, Washington. Accid Anal Prev 43(1):11–24
NHTSA (2006) Traffic safety facts 2005. Overview Washington, DC, national highway traffic safety administration, national center for statistics and analysis, 12p
NHTSA (2012) Traffic safety facts 2010 data. Washington, DC, national highway traffic safety administration, national center for statistics and analysis, 7p
Retting RA, Ferguson SA, McCartt AT (2003) A review of evidence-based traffic engineering measures designed to reduce pedestrian–motor vehicle crashes. Am J Public Health 93(9):1456–1463
Siddiqui C, Abdel-Aty M, Choi K (2012) A comparison of geographical unit-based macro-level safety modeling. In Proceedings of transportation research board 91st annual meeting, no. 12-0243
Wong D, Amrhein C (1996) Research on the MAUP: old wine in a new bottle or real breakthrough. Geogr Syst 3:73–76
Zajac SS, Ivan JN (2003) Factors influencing injury severity of motor vehicle–crossing pedestrian crashes in rural Connecticut. Accid Anal Prev 35(3):369–379
Zegeer CV, Stewart JR, Huang HH, Lagerwey PA (2002) Safety effects of marked vs. unmarked crosswalks at uncontrolled locations: executive summary and recommended guidelines
Zhang M, Kukadia N (1902, 2005) Metrics of urban form and the modifiable areal unit problem. Transp Res Rec: J Transp Res Board 1:71–79

A Computer-Aided Approach for Planning Sustainable Trips to Large Trip Generators: The Case of Cycling Routes Serving University Campuses

Thais de Cássia Martinelli Guerreiro and Antônio Nélson Rodrigues da Silva

Abstract This manuscript describes a computer-aided approach for the design and evaluation of potential impacts of new cycling routes connecting a university campus with the city's street network. The routes were based on the analyses of actual trips reported in a survey carried out with over 900 undergraduate students, and field information obtained from Google Earth (GE). Whereas GE helped in the identification of suitable paths, GIS tools were used to assess the potential impacts of the proposed routes on bicycle, car and walking trips. The results show that the proposed methodology is adequate for exploratory studies and transferable to a Planning Support System (PSS). Such a system could be designed for a variety of goals, such as design of paths, management of bicycles flows, etc. Although this study focuses on the particular case of a university campus, the approach can be easily applied to other large trip generators.

1 Introduction

Congestion and mobility problems are no longer limited to large cities. They are now affecting medium-sized cities and, in some cases, even smaller towns. Under several circumstances, however, only planning more efficient travel patterns could substantially mitigate the problems. This is the case, for instance, of large trip generators located in urban areas without the adequate transport infrastructure.

T. de Cássia Martinelli Guerreiro · A. N. Rodrigues da Silva (✉)
São Carlos School of Engineering, Department of Transportation Engineering,
University of São Paulo, São Carlos, SP 13566-590, Brazil
e-mail: anelson@sc.usp.br

T. de Cássia Martinelli Guerreiro
e-mail: guerreiro_ec@yahoo.com.br

S. Geertman et al. (eds.), *Planning Support Systems for Sustainable Urban Development*, Lecture Notes in Geoinformation and Cartography,
DOI: 10.1007/978-3-642-37533-0_17,

University campuses are often large trip generators served by several transport modes. Given the profile of the largest group of regular visitors of the campuses (i.e., the students), an intense use of non-motorized modes can be expected everywhere. However, whereas this is actually observed in some developed countries, it is not a worldwide pattern. In many cities of developing countries, this is frequently not the case. To a certain extent, this intense use of motorized modes aggravates the mobility problems found in the urbanized areas. Part of the problem is attributed to the planning process, at least in developing countries. The lack of reliable information about the travel patterns and appropriate and affordable planning tools to deal with the problem are often pointed as causes of the problem, when it refers to planning.

However, these explanations make no sense if one considers the resources available in most university campuses. Information about the travel patterns of the students can be obtained by the students themselves, for example, as an academic task demanded by distinct subjects. Furthermore, Geographic Information Systems (GIS) are found in the educational toolboxes currently used by nearly all universities of the world. Free online programs and data sets can provide the additional geographic information required for planning activities at a strategic level. This is the case, for example, of Google Earth, Google Maps, and many other free packages available online. The challenge is to combine them in an integrated strategy for tackling the complex problem of planning for sustainable trips.

This study aims to set up the basis of a planning support system for the design and evaluation of potential impacts of new cycling routes connecting a major trip generator (in this case, a university campus) with the city's street network. The proposed paths are based on the analyses of actual trips reported in surveys conducted with regular visitors of the campus. The approach relies on field information obtained from Google Earth (GE) in order to identify suitable cycling paths. GIS tools are applied to assess the potential impacts of the proposed routes not only on the bicycle trips, but also on car and walking trips.

The paper is organized in five sections, as follows. After this introduction, the next section contains a brief literature review. The studies reviewed essentially discuss factors affecting the use of bicycles as a regular transport mode. Mobility problems associated with university campuses are also covered by some selected studies. As the approach proposed is tested in a Brazilian city, contributions of local authors are also included in the literature review. The other sections subsequently contain the proposed method, the results and conclusions, followed by the references.

2 Literature Review

Litman et al. (2002), Rodríguez and Joo (2004), CROW (2007), Zahran et al. (2008), and Heinen et al. (2011) are some of the authors that have discussed the requirements of infrastructures for sustainable trips. Litman et al. (2002) described the characteristics of the demand and the different types of infrastructures to serve it when planning a cycling network. CROW (2007) reports the key factors of such a network: cohesion, directness, safety, comfort and attractiveness.

The planning and provision of facilities for cycling have been central topics of studies conducted in different parts of the world. An overview of Europe, for instance, was provided by Pucher and Buehler (2008). Dixon (1996), Turner et al. (1997), Federal Highway Administration (1998), North Central Florida Regional Planning Council (2001), Dill and Carr (2003), Alta Planning and Design (2005), NCHRP (2006), Pucher et al. (2011) have conducted studies in the USA; CERTU (2000), in France; NAASRA (1993), in Australia; Providelo (2006), in Brazil. Howard and Burns (2001), in the USA, Ortúzar et al. (2000), in Chile, and Pezzuto (2002), in Brazil, have focused on the preferences and the behavior of cyclists.

Studies about the mobility patterns towards and within university campuses can also be found in different parts of the world (Balsas 2002; Shannon et al. 2006; Miralles and Domene 2010; and Delmelle and Delmelle 2012). In particular, the modal split is a valuable piece of information in the study of the travel patterns to and from university campuses. This point was highlighted by Herz et al. (2007) in a study conducted in Argentina. The authors suggested that university campuses have to be treated as major trip generators. As a consequence, both trip generation patterns and modal split have to be identified on a case by case basis, given that the international literature available may not be applicable to the particular conditions found in different locations.

Among the studies developed in Brazil are Parra (2006), Kuwahara et al. (2008), Souza (2007), Paoli et al. (2008), and Schiavon and Barbosa (2011). It is important to emphasize that these studies differ from the case analyzed here at least for one important reason. The city where the studied campus is located has two large university campuses. If counted together, the number of students enrolled in these two institutions represents nearly 10 % of the city population. Although there exists other smaller private institutions for higher education in the city, this figure refers only to two public universities (i.e., USP—University of São Paulo and UFSCar—Federal University of São Carlos).

Given the importance of the USP campus in the urban context and the point raised by Herz et al. (2007), which was highlighted in the previous paragraph, this is a suitable environment to develop and apply a strategy for planning sustainable trips to large trip generators. This was done for the case of new cycling routes connecting the campus with the city's street network, as discussed in the next section.

3 Method

The method developed for this study was set up in two steps. The first step focused on the identification of network links potentially interesting for the implementation of new cycle paths. This identification was carried out in the specific urban context formed by the streets and avenues in the area surrounding the university campus. In this case, the campus is a major trip generator in a Brazilian medium-sized city. In this city with over 220,000 inhabitants, around 10,000 people visit this campus on a regular basis. The second step of the method was developed after the design of the network of cycling routes had been completed. It consisted in evaluating the impacts of the proposed routes not only on the bicycle trips, but also on car and walking trips.

The procedures involved in the first phase (i.e., design of a cycling network able to meet the potential demand of the university campus) were:

1. Selection of a representative sample of regular users of the different transport modes for data collection, with details of origins, destinations and travel schedules.
2. Analysis of data consistency to avoid incomplete or incorrect information, particularly regarding the travel origins.
3. Storage of all trips recorded in a Geographic Information System.
4. Identification of the actual cycling trips to the campus, for a further analysis of their origins in the city and destinations within the campus.
5. Definition of four alternative paths from each origin to the respective destination (identified in item 4) by means of the shortest path routines available in the GIS package (TransCAD was used in this case, but other GIS packages, including free online mapping programs, can be employed).
6. Identification of links in the city street network that have concentrated the largest number of shortest path overlaps.
7. Design of a network of cycling routes based on the selection of the links described in item 6, their traffic volumes and directions, and the gradients of the streets.

The definition of the network of cycling routes permitted the evaluation of the impacts on trips by different transport modes to the campus. In order to do so, the following additional steps were required:

8. Assessment of the existing bicycle trips that would benefit from the proposed network of cycling routes, as well as an estimate of walking and car trips that could be potentially switched to the cycling alternative because of the accessibility of the new network. This assessment was based on the number of trips originated within bands of 100, 200, 300, and 400 m created around the proposed infrastructure.
9. Revision of the proposed network design, if necessary.

The method was applied to the campus of the University of São Paulo, located in the city of São Carlos. The results and conclusions are discussed in the next two sections.

4 Results

The presentation of the results carefully follows the same order of the steps described in the previous section, in order to facilitate the comprehension of the method. Regarding the sample of campus users, approximately 20 % of all undergraduate students enrolled in 2011 at the São Carlos campus were interviewed and their daily trips registered. The following information was requested in the questionnaires used for data collection: trip origins (identified as a street crossing close to the actual address), trip destinations (represented by the campus accesses), trip modes (walking, bicycle, auto, and public transport—in this case, only bus), and trips hours (recorded as separate trips to and from the campus in a table of fixed intervals).

Trips by public transportation were not considered in the analyses. As a result, the modal split observed in the 1,217 trips registered in the busiest day was 865 walking trips (71.1 %), 291 car trips (23.9 %), and 61 bicycle trips (5.0 %). The points of access of these trips to the oldest campus area in the city, named Area 1, are five gates. However, the São Carlos campus also has an expansion area, named Area 2, which is not contiguous to Area 1, as shown in Fig. 1. The access points of Area 1 have been named after nearby buildings, such as the Institute of Architecture and Urbanism (i.e., IAU). The other access point names are: the institute of Physics (i.e., IFSC), the Institute of Mathematics and Computer Science (i.e., ICMC), the Department of Production Engineering of the São Carlos School of Engineering (i.e., Produção), and the Astronomical Observatory (i.e., Observatório).

The busiest access point to the campus was the gate named IAU, which received 56 % of the car trips and 39 % of the bicycle trips. In the case of walking trips, the largest proportion of trips accessed the campus through the gate named

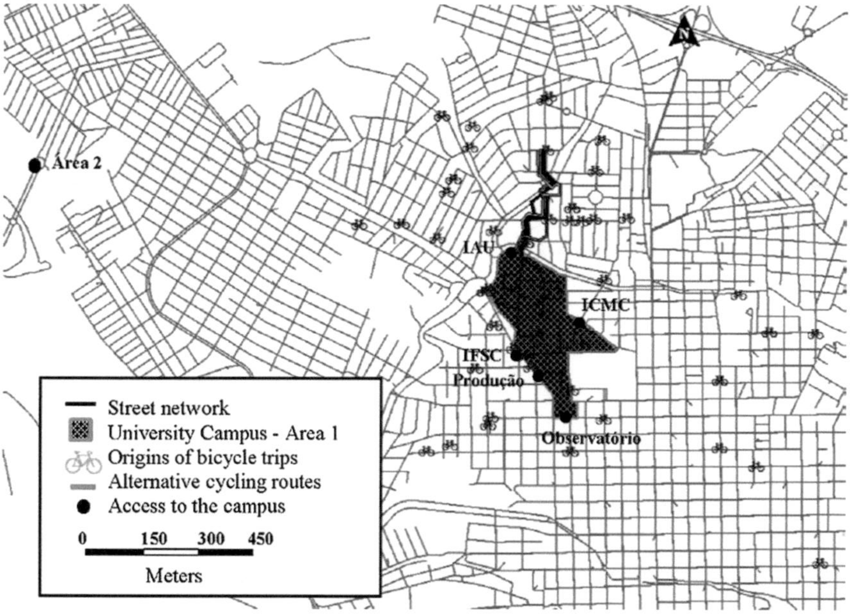

Fig. 1 Origins of bicycle trips reported, points of access to the campus, and example of bicycle routes connecting a specific origin to the campus

ICMC, with 41 % of the total walking trips. This is probably a direct consequence of the large concentration of college students living in the surroundings of this access. Figure 1 also shows the origins of the bicycle trips reported.

As the information of the origins was based on street crossings and given that most respondents have not been city residents for a long time, a careful analysis of those data was carried out to avoid inconsistencies (item 2 of Method). During this process, all origins were georreferenced and linked to the respective campus access points reported. All data were stored in a GIS database (item 3 of Method) along with additional information required by the study, such as the streets network. Together, these elements permitted the identification of the actual cycling trips to the campus (item 4 of Method) and definition of alternative paths from each origin to the respective destination (item 5 of Method).

Regarding the sixth step of the method, four possible routes were considered for connecting each origin–destination pair, as shown in the example of Fig. 1. The objective of this procedure was not exactly to find the shortest paths between origins and destinations, but alternative paths that could eventually be used to connect these points. Some of these alternatives shall match the routes actually used by the respondents. After all alternative routes connecting the origins and the destinations had been identified, the street segments with the highest number of overlapping routes were selected. Although they totalized 33 km, only 19 km were considered in the design of the network of bicycle routes (item 7 of Method), as

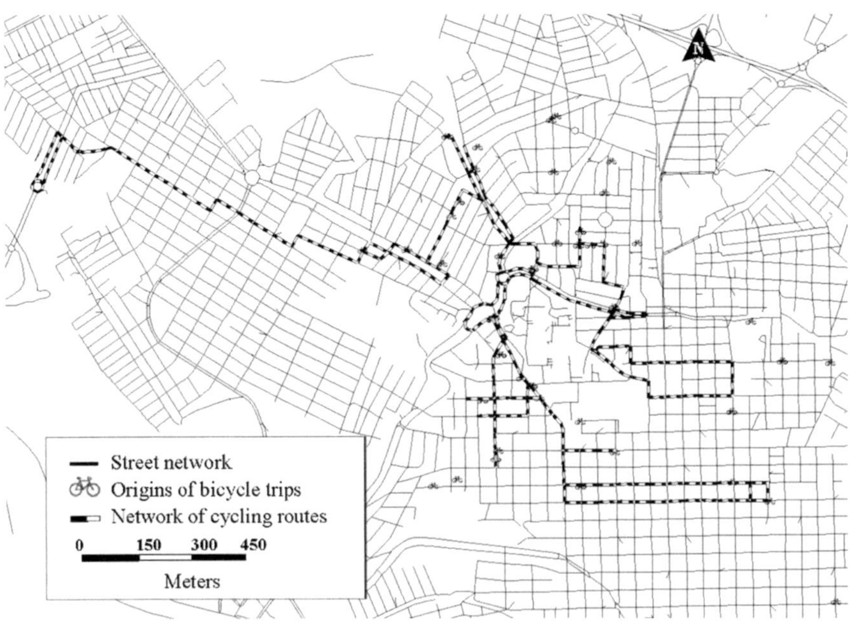

Fig. 2 Proposed network of cycling routes

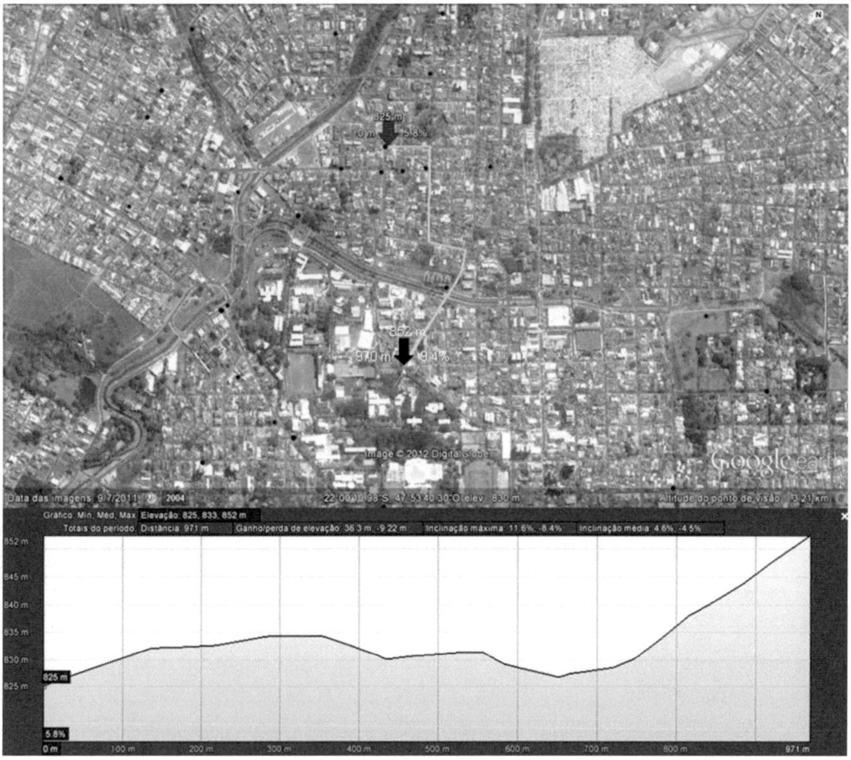

Fig. 3 Longitudinal profile, obtained with Google Earth, of a selected stretch of the proposed network of cycling routes

shown in Fig. 2. The exclusion of several links can be explained by either unfavorable topography or traffic conditions.

Regarding topography, the analysis of the terrain for the definition of the cycling routes was performed with the online program Google Earth. This freeware package permits not only the identification of the total length of the routes, but also the grades along them. The gradient information can be retrieved in different ways: at any point, maximum and minimum values, or in selected stretches (as shown in the example of Fig. 3). This is not a reliable source of information for construction purposes, but it provides a reasonable view of the critical parts of the network for cycling activities. In the particular case of this study, as the actual field information regarding the elevations of the street crossings was available, the actual data were compared with the estimates obtained from GE. In a sample of 211 points, the correlation coefficient 'r' was 0,996, which seems to indicate that the information of the profiles provided by GE can be used for planning analyses at a strategic level, as suggested.

In the analyses of the routes selected with the GIS tools, the street links with the highest number of overlapping routes were naturally close to the campus.

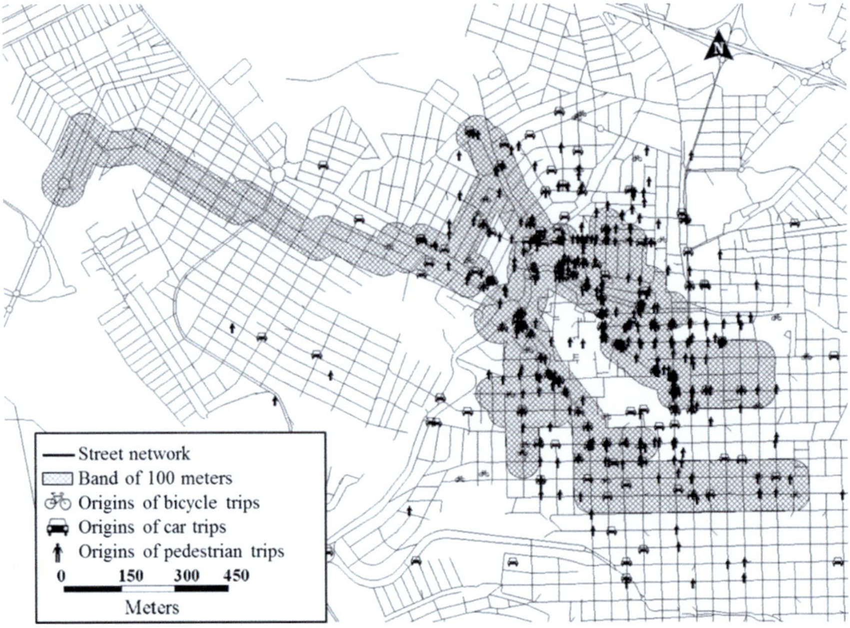

Fig. 4 100 m band around the network of cycling routes and origins of trips undertaken by different transport modes

However, some of them are not convenient in terms of slope and some are coincident with the urban arterial streets, which can be partially explained by the central location of the university campus (particularly Area 1) in the urbanized area. In these parts of the network, the intense flow of motorized vehicles may generate unsafe and uncomfortable conditions for cyclists. Both constraints were avoided whenever possible, but they were not totally eliminated in the case studied. However, it was possible to create a quite reasonable network of cycling routes with the proposed approach, which was the starting point of the following analytical steps.

A straightforward GIS procedure was used to evaluate the potential impacts of the proposed network of cycling routes on other transport modes serving the campus (item 8 of Method). Bands with variable widths (i.e., 100, 200, 300, and 400 m) were created around the proposed network. Walking trips, cycling trips and car trips originated within the different bands were separately computed. Figure 4 provides an example of the procedure, in which a band of 100 m was created around the network.

The proportions of trips within the different bands are summarized in Fig. 5. They give an indication of the trips that could be potentially attracted to the cycling mode, given the existence of a new facility (i.e., the network of cycling routes) not far from the actual origins and destinations reported. In the case of car trips, for example, approximately 60 % of the origins of the existing trips are

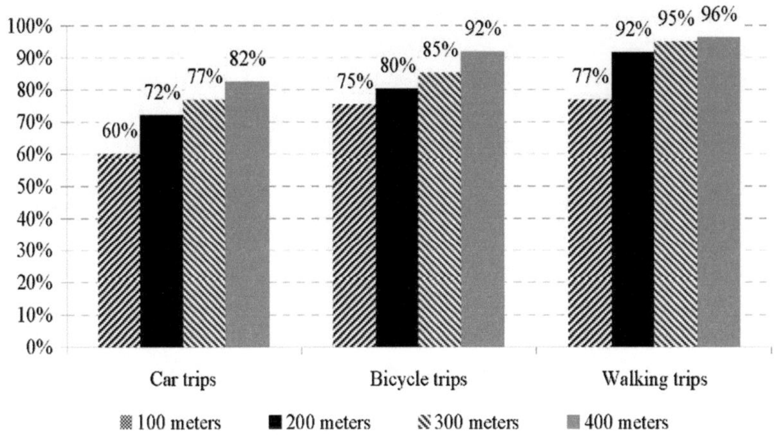

Fig. 5 Proportions of trips whose origins are 100, 200, 300 and 400 m from any point along the proposed network of cycling routes

contained in a 100 m band from some point along the cycling network. Furthermore, this value reaches 82 % if a 400 m band is considered. In the case of a mode shift, even the lowest value could represent a significant improvement in the mobility conditions around and within the campus. Unused parking spaces could be then transformed into additional infrastructure for cycling, attracting even more cyclists, which can be seen as a positive feedback loop.

The values are higher in the case of bicycle trips, given that their origins were the starting point of the network design. They reach, for example, 92 % in the case of band widths of 400 m, which means that almost all origins of bicycle trips are not far from the proposed facility. Another value, however, is even more meaningful here, because it indicates a potential for mode shift. It is the proportion of walking trips computed within the bands, which reached 96 % in the 400 m band. A significant share of these walking trips would certainly benefit from the proposed network of cycling routes.

5 Conclusions

This study focused on the development a computer-aided approach for the design and evaluation of a network of cycling routes to serve a large trip generator. The central idea was to define a simple and straightforward method based on the identification of the actual paths already used by the cyclists and to explore GIS and Google Earth resources in the process. Two groups of conclusions were drawn from the case study conducted in a university campus. The first one refers to the method, whereas the second focuses specifically on observations of the case study.

Although simple, the proposed method could identify the routes probably used by the existing cyclists (as suggested by the percentages of bicycle trips in Fig. 5). The combination of these routes enabled the design of a basic network of cycling routes, which also included a connection to the expansion area of the campus (Area 2). Elevation profiles of the cycling routes were also identified, although with a low level of accuracy. This is not necessarily a problem at this stage of the planning process. On the other hand, the information of path gradients may be valuable when associated with the total lengths between origins and destinations. Together, they provide a measure of the total impedances along the routes.

Regarding the identification of potential cyclists, the data collection procedures can be improved if other questions are added to the questionnaire used. As proposed by Akar and Clifton (2009), it could be interesting to ask about the reasons that motivate the use of a bicycle as a regular transport mode (or otherwise) to and from the trip generator. In the case of the university campus studied here, this refinement in the data collection process is currently underway. Over 2,000 respondents answered the survey conducted in 2012, but the results are not available yet.

Regarding the outcomes of the case studied, the largest potential impact of the proposed network of cycling routes seems to be on walking trips, for all bands analyzed. Almost all origins of the walking trips recorded (96 %) are within the 400 m band around the proposed network, which can be explained by the fact that most undergraduate students live around the campus. Such a percentage suggests that at least part of this demand can be attracted to the cycling mode.

On the other hand, car trips could also benefit from the implementation of a network of cycling routes around the campus, although probably not in the same proportion. In the case of the car trips originated within the bands, 60 % of the origins reported are contained in the 100 m band. Considering a total estimate of 1,455 car trips (obtained from the expansion of the sample to represent the population), 875 daily trips could be transferred to the cycling mode. Assuming a minimum parking space of 12.5 m^2 per car, the total area could represent approximately 10,940 m^2 of unused parking space released for other uses, including cycling infrastructure. This is a conservative estimate, if one considers that this calculation was performed for the 100 m band. Estimates based on the percentages found within the other bands would be obviously higher.

The effects would be positive even for the existing bicycle trips. As shown in Fig. 5, 80, 85 and 92 % of the origins of these trips are contained in band widths of 200, 300 and 400 m around the proposed network, respectively. The new infrastructure would certainly improve the conditions of safety and comfort for the users.

The results show that the proposed methodology is adequate for exploratory studies and transferable to a Planning Support System (PSS). Such a system could be designed for a variety of goals, such as the design of paths, the management of bicycles flows, etc. Also, although this study focuses on the particular case of a university campus, the approach can be easily applied to other large trip generators. A final remark must be done regarding the importance of studies like this in a university setting. These initiatives help to consolidate the role of the university as a source and disseminator of knowledge, by showing that it can also contribute for solving the problems created to the city.

Acknowledgments The authors would like to acknowledge the Brazilian agencies CNPq (Brazilian National Council for Scientific and Technological Development), FAPESP (Foundation for the Promotion of Science of the State of São Paulo) and CAPES (Post-Graduate Federal Agency) for the financial support provided to this research. They are also indebted to the Civil Engineering undergraduate students enrolled in the "*Transport systems planning and analysis*" course in 2011. They were essential to the success of the data collection process.

References

Akar G, Clifton KJ (2009) Influence of individual perceptions and bicycle infrastructure on decision to bike. Transp Res Rec 2140:165–172

Alta Planning and Design (2005) Pedestrian and bicycle facilities in California—a technical reference and technology transfer synthesis for Caltrans planners and engineers. Califórnia, http://www.dot.ca.gov/hq/traffops/survey/pedestrian/TR_MAY0405.pdf. Accessed 01 2012

Balsas CJ (2002) Sustainable transportation planning on college campuses. Transp Policy 10(1):35–49

CERTU (2000) Recommandations pour des Aménagements Cyclables. Centre d'Etudes sur les Reseaux, les Transports, l'Urbanisme et les constructions publiques, Lyon, France

CROW (2007) Design manual for bicycle traffic. Record no. 25. Ede, The Netherlands

Delmelle EM, Delmelle EC (2012) Exploring spatio-temporal commuting patterns in a university environment. Transp Policy 21(1):1–9

Dill J, Carr T (2003) Bicycle commuting and facilities in major U.S. cities: if you build them, commuters will use them. Transp Res Rec 1828:116–123

Dixon LB (1996) Bicycle and pedestrian level of service performance measures and standards for congestion management systems. Transp Res Rec 1538:1–9

Federal Highway Administration (1998) The guidebook on methods to estimate non-motorized travel. Publication FHWA-RD-98-166. FHWA, U.S. Department of Transportation, Washington, D.C

Heinen E, Maat K, Van Wee B (2011) The role of attitudes toward characteristics of bicycle commuting on the choice to cycle to work over various distances. Transp Res Part D 16(2):102–109

Herz M, Galarraga J, Pastor G (2007) Centros Universitários como Polos Geradores de Viagens. In: Proceedings of the 21st Conference on Transport Research and Education, Rio de Janeiro, Brazil

Howard C, Burns EK (2001) Cycling to work in phoenix: route choice, travel behavior, and commuter characteristics. Transp Res Rec 1773:39–46

Kuwahara N, Balassiano R, de Santos MPS (2008) Alternativas de Gerenciamento da Mobilidade no Campus da UFAM. In: Proceedings of the 22nd Conference on Transport Research and Education, Fortaleza, Brazil, pp 800–811

Litman T, Blair R, Demopoulos B, Eddy N, Fritzel A, Laidlaw D, Maddox H, Forster K (2002) Pedestrian and bicycle planning: a guide to best practices. Victoria Transport Policy Institute, http://www.vtpi.org/nmtguide.doc. Accessed 10 2012

Miralles CG, Domene E (2010) Sustainable transport challenges in a Suburban University: The case of the Autonomous University of Barcelona. Transp Policy 17(6):454–463

NAASRA (1993) Guide to traffic engineering practice, part 14—bicycles. AUSTROADS Publication no AP-11.14/93, National Association of Australian State Road Authorities, Sydney, Australia

NCHRP (2006) Guidelines for analysis of investments in bicycle facilities. NCHRP Report 552, Transportation Research Board, Washington, D.C., http://onlinepubs.trb.org/onlinepubs/nchrp/nchrp_rpt_552.pdf. Accessed 03 2012

North Central Florida Regional Planning Council (2001) Alachua countywide bicycle master plan. Tampa Washington-Baltimore: technology and innovation planners and engineers

Ortúzar JD, Iacobelli A, Valeze C (2000) Estimating demand for a cycle-way network. Transp Res Part A 34(5):353–373

Paoli M, Rosa M, Kronka RC M (2008) A bicicleta no Campus da USP/SP: Proposta de ciclorede e mobiliário urbano. In: Proceedings of the 7° Seminário Internacional "Espaço Sustentável: Inovações em Edifícios e Cidades", São Paulo, Brazil. http://www.usp.br/nutau/CD/145.pdf. Accessed 10 2012

Parra MC (2006) Mobility management in university campuses: problems, difficulties and possible solutions in the case Ilha do Fundão (in Portuguese). Master's thesis, Federal University of Rio de Janeiro

Pezzuto CC (2002) Factors influencing the use of bicycle (in Portuguese). Master's thesis, Federal University of São Carlos

Providelo JK (2006) A method for the definition of cycling routes in urban areas (in Portuguese). Master's thesis, Federal University of São Carlos

Pucher J, Buehler R (2008) Cycling for everyone: lessons from Europe. Transp Res Rec 2074:58–65

Pucher J, Buehler R, Seinen M (2011) Bicycling renaissance in North America? An update and re-appraisal of cycling trends and policies. Transp Res Part A 45(6):451–475

Rodríguez DA, Joo J (2004) The relationship between non-motorized mode choice and the local physical environment. Transp Res Part D 9(2):151–173

Schiavon AF, Barbosa HM. (2011) Rotas cicláveis no Campus da UFMG—Um estudo com simulação de tráfego. In: Proceedings of the 25th Conference on Transport Research and Education, Belo Horizonte, Brazil

Shannon T, Giles-Corti B, Pikora T, Bulsara M, Shilton T, Bull F (2006) Active commuting in a university setting: assessing commuting habits and potential for modal change. Transp Policy 13(3):240–253

Souza SCF (2007) Trip generation models for higher education institutions (in Portuguese). Master's thesis, University of Brasilia

Turner S, Hottenstein A, Shunk G (1997) Bicycle and pedestrian travel demand forecasting literature review. Publication 1723-1. Texas Transportation Institute, Texas A&M University, and Texas Department of Transportation

Zahran S, Brody SD, Maghelal P, Prelog A, Lacy M (2008) Cycling and walking: explaining the spatial distribution of healthy modes of transportation in the United States. Transp Res Part D 13(7):462–470

Walk Route: A New Methodology to Find the Optimal Walking Route in the City of Atlanta

Subhrajit Guhathakurta, Ge Zhang, Manoj K. Panguluru and Ramachandra Sivakumar

Abstract This paper demonstrates a new methodology for finding optimal walking routes according to user specified conditions selected from a set of attribute choices. The pedestrian network planning methodology discussed in this paper reflects the influence of environmental factors facilitating or impeding pedestrians' propensity to walk. The principle tasks involved in applying this method include identifying attributes of walkability, weighting the importance of each attribute, evaluating the composite walking cost of each street segment, and identifying the optimal route by aggregating segments that minimize the total cost. A case study of the city of Atlanta is presented to demonstrate the application of this method and discuss its limitations.

1 Introduction

There has been a lot of interest in evaluating walkability, given that walkable environments have been associated with urban social life, economic regeneration, public health, and overall quality of life. Urban planners, urban designers, and architects have long advocated for walkable neighborhoods as a means of creating "a sense of community" (Calthorpe 1993; Duany and Plater-Zyberk 1994). Walking is an important aspect of environmental health, given its ability to improve cardiorespiratory and metabolic fitness and reduce the risk of coronary

---

S. Guhathakurta (✉) · R. Sivakumar
School of City and Regional Planning, Georgia Institute of Technology, Atlanta, USA
e-mail: subhro.guha@coa.gatech.edu

G. Zhang
College of Life Sciences, Zhejiang University, Hangzhou, China

M. K. Panguluru
School of Computer Science, Georgia Institute of Technology, Atlanta, USA

S. Geertman et al. (eds.), *Planning Support Systems for Sustainable Urban Development*, Lecture Notes in Geoinformation and Cartography,
DOI: 10.1007/978-3-642-37533-0_18,

heart disease. Public health scholars and practitioners have found measurable health benefits for individuals residing in walkable areas (Frank and Engelke 2001; Saelens et al. 2003a; Lathey et al. 2009). Walking is also critical for environment quality given that it is a substitute for vehicular travel and thus reduces negative impacts from vehicles, such as noise and automobile exhaust. Transportation planners have noted that, under some circumstances, walkable areas tend to encourage several alternative modes of mobility that reduce the dominance of single-occupancy automobiles (Boarnet and Crane 2001). The support for walking also comes from environmentalists who contend that these neighborhoods are energy efficient with a lower carbon footprint (Norman et al. 2006). Given the multiple benefits of walking, urban planners and policy-makers are keen to find ways to make walking more popular among urban inhabitants. Developing tools that measure and provide critical information about walkability of places is increasingly becoming an important aspect of planning support systems (PSS).

Walking facilities, such as sidewalks and underground passageways, have been introduced into the urban pedestrian network system for satisfying the demands of walking. However, simply building more walking facilities in order to increase the occurrence of walking is not necessarily the most effective solution. To induce walking, the entire pedestrian system should be well designed by providing end-to-end solutions between the potential origins and destinations. Additionally, a paradigm shift is occurring in transportation planning that moves away from a mobility-oriented evaluation (which evaluates transport system performance based on quantity and quality of physical travel) toward an accessibility-based evaluation (which considers the ability to obtain goods, services, and activities, and places people at the center of the transportation system, rather than automobiles) (Todd 2012). An efficient pedestrian system is usually a fundamental component of transportation systems that score high on accessibility metrics.

Various walkability scores for neighborhoods and streets already exist (e.g., www.walkscore.com, walkshed.org). However, various travel surveys document that walkability factors and their impacts vary from person to person, and a growing number of studies discuss the influence of built environment on walking behavior through such factors as access to amenities, residential density, street connectivity, land use diversity, land use accessibility, traffic safety and crime safety (Leslie et al. 2007; Cerin et al. 2006). Established walkability applications do not provide a comprehensive range of user selectable routing algorithms that take these myriad factors influencing walkability into account.

Walk Score has been demonstrated as a valid and reliable tool for estimating the accessibility of nearby facilities (Carr et al. 2010; Duncan et al. 2011). However, the impact factors in its algorithm have lacked some variables (i.e. population density, traffic safety) to describe built environment characteristics. The potential pedestrians are unable to select the impact factors and cannot determine the most walkable route between two points in the city. The function of Walkshed is very similar to Walk Score and they are designed based on the same premise, which is the accessibility of nearby amenities and their impacts on walkability. The Walkshed application provides additional functionality that enables the pedestrian to select impact factors for walkability, but it also lacks some key built environment factors (i.e. crime safety, aesthetics) in its walkability calculation, and cannot generate an optimal route between two points in the network.

The objective of this chapter is to describe a methodology for routing walking paths based on user selected attributes and to demonstrate its potential through an application developed for the city of Atlanta. This application is now being served over the web and is now in the process of being implemented in both iOS and Android based handheld devices. The rest of the chapter is organized as follows: The next section, Sect. 2, explains the methodology for generating walkability scores and outlines the algorithms used for determining the optimal routes based on user preferences. Section 3 demonstrates how the methodology was applied to the city of Atlanta and provides sources of data used in this application. The following section, Sect. 4, discusses the various ways in which this application can also be used urban planning and policy-making and points to some of its limitations. The chapter then concludes with Sect. 5 highlighting some of the key objectives achieved in designing this application.

2 Methodology

2.1 Selecting Attributes of Walkability Cost

Several characteristics of the built environment have been found to influence an individual's propensity to walk. High-density neighborhoods support greater retail and service variety, resulting in shorter, walkable distances between facilities (Leslie et al. 2007). As a result, residents of high walkable neighborhoods are reported to be in areas with high residential density (Cerin et al. 2007; Saelens et al. 2003b); and residential density has been positively related to walking for transport (Cerin et al. 2006). Walkable neighborhoods provide more options for destinations where goods and services may be purchased and where more local employment opportunities can be reached by walking (Leslie et al. 2007). Further, employment density is also reported to significantly affect walking patterns (Canepa 2007).

In addition to density, diversity of land uses and built forms have been also known to increase the probability of walking (Leslie et al. 2007). Residents of high-walkable neighborhoods rated relevant attributes of land-use mix (access and diversity) consistently higher than did residents of the low-walkable neighborhoods (Leslie et al. 2005). Both density and diversity of the built environment influence distances between activities and land uses. Expectedly, shorter distances to different land uses also increases walking (Shigematsu et al. 2009; Merom et al. 2009; Leslie et al. 2005; Cerin et al. 2007). Similarly, a study by Moudon et al. (2007) shows that distance to neighborhood destinations is strongly associated with pedestrian activity.

Higher intersection densities provide people with a greater variety of potential routes, easier access to major roads where public transport is available, and shorter times to get to destinations (Leslie et al. 2007). Residents of highly walkable neighborhoods have rated relevant attributes of street connectivity higher than have residents of less walkable neighborhoods (Leslie et al. 2005).

Lower crime rates, lower physical barriers, higher quality pedestrian infrastructure, less traffic volume and higher aesthetic quality increase propensity for walking (Leslie et al. 2005; Kerr et al. 2006; Gauvin et al. 2005). Conversely, McDonald (2008) found significant negative association between violent crime and minutes walked per day.

Residents' perceptions of neighborhood walkability also have been related to the themes of safety, comfort or convenience, and aesthetics (Lu et al. 2011). Barriers can reduce walkability and accessibility to school (Bejleri et al. 2011). Poorly maintained or missing sidewalks, absence of crosswalks, intersecting bike paths or lanes, and traffic safety are issues that discourage pedestrian activity (Strath et al. 2007).

Given extensive literature on the attributes of walkability and their associated measures, the selected attributes for this application fall under ten categories. These are: (1) residential density; (2) business density; (3) land use diversity; (4) land use accessibility; (5) street connectivity; (6) crime safety; (7) traffic safety; (8) physical barriers; (9) aesthetics; and (10) pedestrian infrastructure. The application offers users the ability to select and rank these relative attributes of walkability according to their requirements.

2.2 Developing an Analytical Hierarchy Process for Deriving Network Weights

Application of the Analytical Hierarchy Process (AHP) provides a convenient means of deriving weights by comparing each attribute with all others one at a time (Arrington et al. 1984; Cheng et al. 2005; Piltan et al. 2012). This process was initially developed by Saaty (1980) to organize and analyze complex decisions (Satty and Vargas 1980). The process involves a stratified system of ranking each attribute with respect to all others. The matrix of relative ranks is then used to calculate eigenvectors, which are, in turn, normalized to derive relative weights for the various attributes. The Analytical Hierarchy Process (AHP) has been applied in an evaluation model to assign weights to various metrics (Shen 2010; Crossman et al. 2011; Javadian et al. 2011).

For our application, two kinds of weights were required; one for the categories themselves and the other for each of the attributes within the categories. We first perform a pairwise importance comparison between variables. Second, we normalize each matrix element by the sum of elements in each column, calculate the sum for each row and then normalize the sum of the rows for the weights for each variable. Finally, we calculate the consistency ratio of the selection to validate its consistency, which according to Saaty should be less or equal to 10 % (according to his famous rule of thumb, the consistency or uncertainty of the decision should be an order of magnitude less than decision) (Satty and Vargas 1980).

2.3 Applying the Weights to Generate Walkability Score

The walkability score is measured according to the built environment attributes and the walking distance. The walkability scores are assigned to each street segment for quantifying the likelihood that walkers would prefer to walk along this segment. Various factors such as crime safety, street connectivity, traffic safety and aesthetics contributing to the walkability score of each street segment, are combined in a linear function to derive the final walkability score. For each street segment, the overall score can be calculated as (see Eq. 1):

$$WS_j = D_j \times \sum_{i=1}^{n} (V_i W_i) \tag{1}$$

where WS_j is the walkability score of the street segment j, D_j is the length of the street segment j, n is the number of the attributes of walkability and V_i and W_i are the value and the weight for the attribute i, respectively.

2.4 Develop and Implement a Routing Algorithm

Network analysis has been widely used to analyze the possible routes from one location to another (Muraleetharan and Hagiwara 2007; Ericsson et al. 2006). This tool can get the shortest route between two points based on network distance. In addition, other criteria, such as travel speed, can be built into the analysis to obtain the optimal route for a specific objective (Hudecek 2008; Devlin et al. 2008).

Walkability is an important characteristic for the pedestrian travel route. There are numerous possible routes between two locations in the city. The optimal route is determined by maximizing the aggregate average walkability score for the route. The routing algorithm for optimal route can be described as (see Eq. 2):

$$O_r = \max \sum_{j=1}^{m} \frac{1}{D_j} \times \sum_{j=1}^{m} (WS_j) \tag{2}$$

where O_r is the optimal route between two points, WS_j is the walkability cost of the street segment j and m is the number of the street segments of the route.

3 An Application in the City of Atlanta

To demonstrate the methodology described above, we developed a web-based application to generate walkable routes for the city of Atlanta in the U.S. A walkability cost network was developed using a routing algorithm to identify optimal pedestrian routes. These routes were then used to guide walk route selection between two locations within the city.

3.1 Variable Selection and Spatial Transfer

Street segments were weighted according to the resistance they offered to walkers based on various combinations of street attributes. Ten categories of attributes (as described earlier) were applied to determine the network cost of walkability. Each of these categories are weighted combinations of various attributes (See Table 2) with some category having only one attribute variable while others having ten or more attributes.

The data for these attributes were stored in GIS format and analyzed using ESRI's ArcGIS suite, version 10.0. We first cleaned the 2007 street network data to eliminate the expressways, which did not include legal walking routes. The data were then categorized into two spatial units—blocks and street segments. Based on the spatial location, all street lines were interrupted and distributed to each block. Each segment was given the attribute of the block. If one segment intersected two blocks, the segment was given the average value of the attributes of both blocks. All data for 32 variables were assigned to the street network shapefiles. Such variables as population density, housing density, household density, and multi-family units' density were generated for each block, while the distance from street to the nearest credit intermediation related activities, education services, food beverage stores and similar destinations were created for each street segment. The relationships between variables and walking were either positive (e.g., population density and employee density) or negative (e.g., crime rate and traffic volume). The characteristics of each variable are listed in Table 1.

We used the street network data from 2007 Georgia Department of Transportation and the Atlanta Regional Commission street file. Block boundary data were obtained from 2010 U.S. Census Bureau TIGER/Line shapefiles. A measure of intersection density was calculated using the ArcGIS network analysis tool based on the street network data. The variables in the residential density group were calculated using the ArcGIS operator tool based on the 2010 U.S. census block data. Similarly, the variables in the business density, land use mix diversity and land use mix accessibility categories were calculated using the ArcGIS operator tool, zonal analysis tool and Euclidean distance tool, based on the 2012 reference USA business database. Crime data were obtained from 2009–2011 police department report, provided by the Atlanta Police department. Slope was calculated using the ArcGIS

Table 1 The characteristics of each variable used in developing walkability cost

Variable	Spatial unit	Relationship to walking
Population density	Block	Positive
Housing density	Block	Positive
Household density	Block	Positive
Multi-family units' density	Block	Positive
Employee density	Block	Positive
Credit intermediation related activities density	Block	Positive
Educational services density	Block	Positive
Information services density	Block	Positive
Professional scientific technical services density	Block	Positive
Religious grant making civic professional density	Block	Positive
Food service drinking places density	Block	Positive
Amusement gambling recreation services density	Block	Positive
Food beverage store density	Block	Positive
Park services density	Block	Positive
Land use variety	Block	Positive
Distance from street to the nearest credit intermediation related activities	Street segment	Negative
Distance from street to the nearest food service drinking places	Street segment	Negative
Distance from street to the nearest educational services	Street segment	Negative
Distance from street to the nearest information services	Street segment	Negative
Distance from street to the nearest park services	Street segment	Negative
Distance from street to the nearest professional scientific technical services	Street segment	Negative
Distance from street to the nearest religious grant making civic professional	Street segment	Negative
Distance from street to the nearest food beverage store	Street segment	Negative
Distance from street to the nearest amusement gambling recreation services	Street segment	Negative
Distance from street to the nearest transit	Street segment	Negative
Street intersection density	Block	Positive
Crime rate	Block	Negative
Slope	Block	Negative
NDVI	Block	Positive
Traffic volume	Street segment	Negative
Sidewalk perimeter divide by street length	Street segment	Positive
Sidewalk area divide by street length	Street segment	Positive

3D analysis tool, based on the 1999 digital elevation model (DEM) data (30 meters pixel resolution) provided by the U.S. Geological Survey (USGS). We used ERDAS 9.3 Indices tool to calculate NDVI (A simple graphical indicator that can be used to measure vegetation coverage. The higher the value of indicator, the higher the vegetation coverage) based on the 2011 Landsat image data provided by the U.S. Geological Survey (USGS). Traffic volume data were obtained from the

2009 average annual daily traffic (AADT) data provided by the Georgia Department of Transportation. Public transit data were collected from the 2010 Georgia regional transportation authority file. Finally, sidewalk data were obtained from the Georgia Department of Transportation.

3.2 Calculation of Variable Weights and Generation of Walkability Cost

The weights of the variables were calculated through the AHP process discussed earlier. Prior to the analysis, the influence of each variable in the walkability model needed to be ascertained. Weights for each variable was assigned based on prior studies. The more often they appear in current walkability research papers, the higher the value received. In our case, it is reasonable to assume that the crime safety is the most important factor related to walking. Physical barriers were also assigned a high importance value to walkers. Traffic safety was considered less important than slope. The details of the weight of each variable are listed in Table 2.

Using the selected walkability variables and their respective weights as defined in Table 2, the walkability cost map was developed using the GIS operator tool. The street segments were reclassified based on the accessibility of street segment to the point of origin of individual walkers (Fig. 1).

3.3 Optimal Route Identification

Using the walkability cost map developed in the above section the shortest route and the optimal route can be identified among all possible routes using the ArcGIS network analysis tool (Fig. 2).

Figure 2 showcases a feature of ArcGIS network analyst. However, the network analyst methodology will only work for a static network. In order to achieve our goal, which is giving users the choice of the selecting their own parameters, the network characteristics need to be changed on the fly. The walk score of a street segment varies based on user provided criteria. To reconstruct the network each time a selection is made or changed we used an open source framework called the OpenTripPlanner. The OpenTripPlanner has an inbuilt routing mechanism, which gives the shortest path but this was not of much value to us. We needed to get a path that maximizes the walk score by aggregating the scores on all possible street segments. So we developed another routing algorithm, which uses the walk score as the routing parameter rather than the length of the street segment. In this routing algorithm we identify the route which has the maximum aggregate average walk score.

Table 2 The weights of each variable used in developing walkability cost

Variable group	Weight of the group	Variable	Weight within the group
Residential density	0.0091689	Population density	0.54
		Housing density	0.27
		Household density	0.13
		Multi-family units' density	0.06
Business density	0.0091689	Employee density	0.57
		Credit intermediation related activities density	0.058
		Educational services density	0.058
		Information services density	0.058
		Professional scientific technical services density	0.058
		Religious grant making civic professional density	0.058
		Food service drinking places density	0.035
		Amusement gambling recreation services density	0.035
		Food beverage store density	0.035
		Park services density	0.035
Land use mix diversity	0.0349971	Land use variety	1
Land use mix accessibility	0.0506621	Distance from street to the nearest credit activities	0.134
		Distance from street to the nearest food service	0.0825
		Distance from street to the nearest educational services	0.134
		Distance from street to the nearest information services	0.134
		Distance from street to the nearest park services	0.0825
		Distance from street to the nearest scientific services	0.134
		Distance from street to the nearest religious	0.134
		Distance from street to the nearest food beverage store	0.0825
		Distance from street to the nearest recreation services	0.0825

(continued)

Table 2 (continued)

Variable group	Weight of the group	Variable	Weight within the group
		Distance from street to the nearest transit	0.134
Street connectivity	0.0738547	Street intersection density	1
Crime safety	0.31211	Crime rate	1
Physical barrier	0.222342	Slope	1
Aesthetics	0.0247238	NDVI	1
Traffic safety	0.107506	Traffic volume	1
Pedestrian infrastructure	0.155465	Sidewalk perimeter divide by street length	0.5
		Sidewalk area divide by street length	0.5

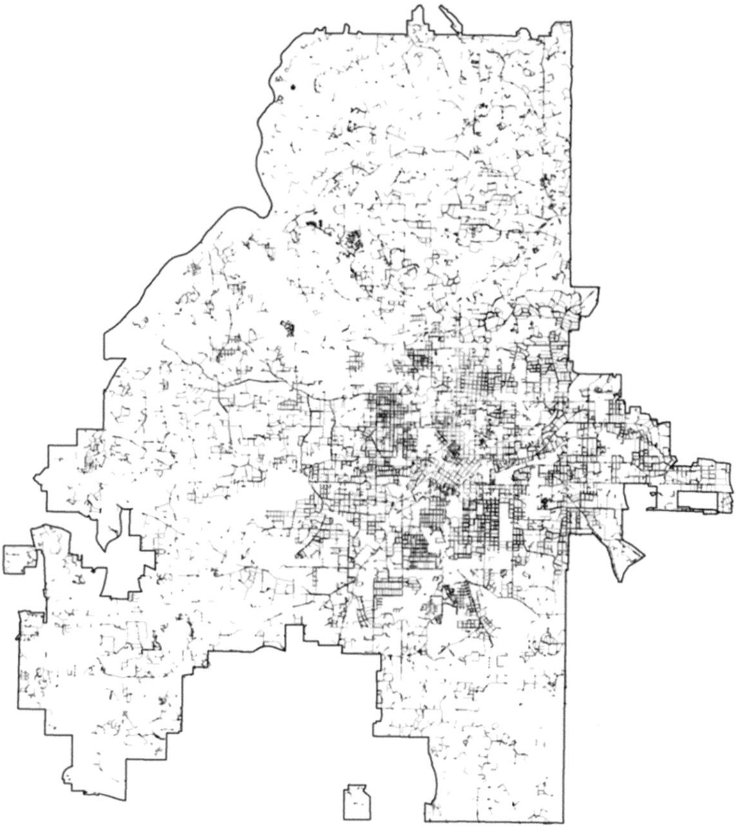

Fig. 1 The map of walkability cost in the city of Atlanta

The routing algorithm explores all possible paths to the destination to maximize the walk score. However, exploring all possible paths might not be efficient as this will include examining routes that are not practically feasible for a person to walk. For example we might get a route which is an order of magnitude longer than the shortest path. So we restrict the search by defining an upper boundary. As of now we have defined the upper boundary in which all the places which are more than twice the shortest path distance will not be considered while trying to maximize the walk score. This eliminates paths which are practically not feasible to walk and also makes the computation efficient and quick (Fig. 3).

The next change we made in the OpenTripPlanner framework was to ensure that the network characteristics vary based on the user preference. This framework maintains the network in a graph format. When the application gets a request from a user it calculates the walk score based on the weights and applies the new walk score to all the street segments present in the graph. Hence, the network characteristics are altered based on the user preference. The following two figures showcase the difference between the routes returned by the two routing algorithms.

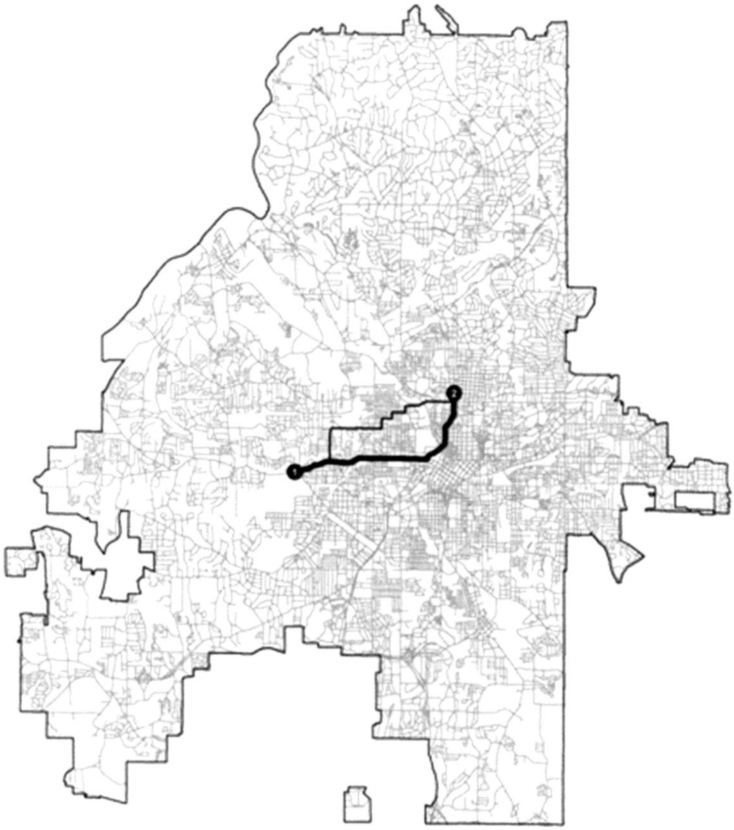

Fig. 2 The map of comparing the shortest route with the optimal route

The area shown in the map is from the midtown of Atlanta. Notice that the route shown in Fig. 4 has a longer route. The route takes the longer Peachtree street segment because the walk score in that area is higher than the shortest path as it has more shops and restaurants, which the default user rates highly.

4 Discussion

The shortest route may not always be the most walkable route if the user is not time constrained. People walk for various reasons and many of these reasons relate to enjoying the experience of walking through places that please the senses. Also, different persons respond differently to similar places, hence the walk route that is pleasing to one may not be the same to another. Indeed, the same individual might choose to walk along different routes depending upon the specific activities she is pursuing or her mood at that time. The application described in this paper caters to

Fig. 3 Shortest path using the inbuilt routing algorithm of OpenTripPlanner

the diversity of reasons that make people walk and provides the best path according to the specified choices individuals make at a particular time.

The model implemented in the walk route application can be easily reproduced for other regions as long as similar data are available. The model is also flexible in adapting as many or as few user selectable options as necessary. For example, parents could provide kids with a walk route application that sets the safety feature to high and non-selectable so that the kids are choosing only among the safe options when planning a walk route. Similarly, different versions of this application can be designed to cater to different groups such as tourists, joggers, and shoppers using different weights for the attributes (or by adding new attributes).

Walk route could also be used to evaluate different streets, neighborhoods, or places comprehensively or along particular attributes. If a typical weighting system is used for scoring walk routes, the overall scores for all street segments in a neighborhood or place can provide a measure of walkability of that place. These scores can then be used to compare among places or neighborhoods according to how walkable they are. Also, the walkability scores can be used to plan appropriate infrastructure and built-form interventions to make places more walkable. In this respect, this application can be used as a decision support tool for planning.

Although this study is geared towards assessing the standard pedestrian's walking experience, this methodology could also be applicable to optimizing

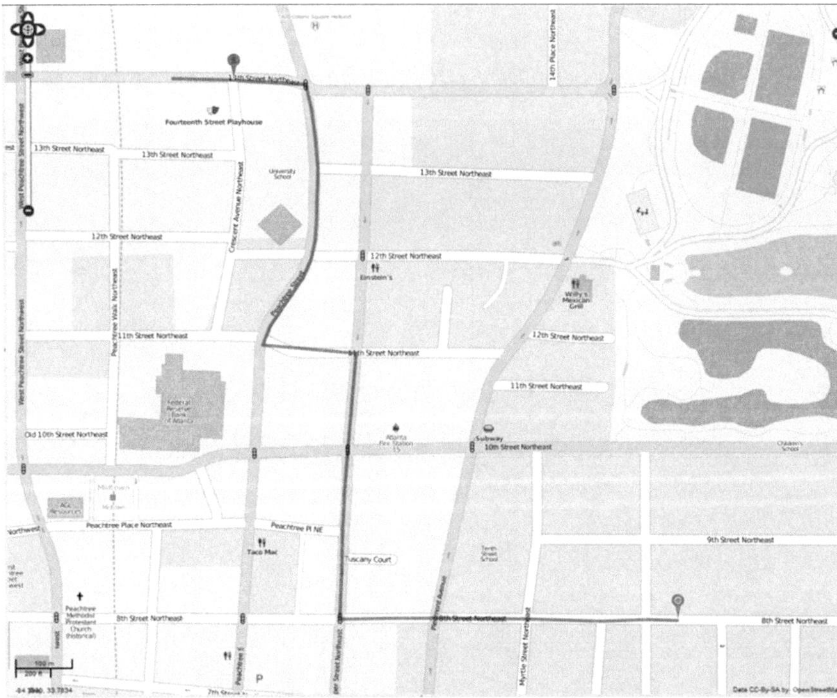

Fig. 4 Route returned by the walk score based routing

routes for individuals with disabilities. Physical elements in the travel path such as low overhanging signs, insufficient width of the sidewalk, obstacles and protruding elements in the path of travel, and uneven curbs with obstacles and holes, could be included as variables for an adapted Walk Route model. Many of the original variables would still be relevant, and could reference a more detailed street network that includes Americans with Disabilities Act (ADA) facilities.

Like most models, this application too has its limitations. One limitation is the quality and the timeliness of the data. Given that most of the data is obtained from publicly available or secondary sources, there are several data quality issues. For example, the quality of the sidewalks in terms of cracks and buckling, which may be hazardous to pedestrians, is not available in this data. Also, the data is somewhat dated given that much of it was collected 2+ years ago and will be updated at relatively long intervals. Hence some of the built environment changes will not be captured in this data.

A second limitation of the walk route methodology is that it does not consider the effects of subjective measures of walkability. In this study, only objective measures of walkability were used in the model, but much of the walkability research has shown that subjective measures of walkability significantly impact walking behavior. Given the same walking environment, people will have different propensity to walk. Future steps to remedy this shortcoming will likely include administering travel surveys to compile subjective data from local residents that can then be added to the model to improve its performance.

Another difficulty in constructing the walk route model is the determination of the weights of each attribute of walkability. In this study, the variable weights were assigned by the frequency of their occurrence in the walkability literature. There was no research about the rank of importance of different variables, which were important to walking and the direct relationship between these variables and propensity to walk. If the actual amount of walking at various street segments is known, statistical methods such as regression analysis could be used to calculate the weight of each variable of walkability. The observed data are as yet unavailable to validate the model. Therefore the application of the walk route in an actual city environment still needs to be tested at this stage.

5 Conclusion

This paper presents a new methodology for ascertaining the optimum pedestrian travel route based on individual preferences. Several characteristics of the built environment have been found to influence walking behavior in the literature and are thus employed as variables in the model. The Analytical Hierarchy Process (AHP) is used to weigh each attribute, which were then combined to calculate a walk score. The optimum route was then determined by applying a routing algorithm based on the walkability cost of the network.

The methodology was applied to the city of Atlanta. The application included 32 variables in 10 attributes. While the weights for each of the variables combined to form attributes were based on the review of literature, the importance of the attributes themselves were selected by the users and determined through the AHP process. The most walkable route between two locations in the city was found using the network analysis tool in the OpenTripPlanner framework, based on the walkability cost map in Atlanta.

Walk Route is designed to address multiple objectives: (1) It offers a means for pedestrians to identify routes that meet their requirements; (2) it provides framework to measure the quality of the pedestrian infrastructure in the city; and (3) it serves as a decision support tool for urban planning to make neighborhoods and places more walkable. Although this study focused on the pedestrian's experience, the methodology could also be modified and applied as a route optimization model for people with disabilities.

References

Arrington CE, Hillison W, Jensen RE (1984) An application of analytical hierarchy process to model expert judgments on analytical review procedures. J Acc Res 22(1):298–312

Boarnet M, Crane R (2001) The influence of land use on travel behavior: specification and estimation strategies. Transp Res a-Pol 35(9):823–845

Bejleri I, Steiner RL, Fischman A, Schmucker JM (2011) Using GIS to analyze the role of barriers and facilitators to walking in children's travel to school. Urban Des Int 16(1):51–62
Canepa B (2007) Bursting the bubble—determining the transit-oriented development's walkable limits. Transp Res Rec 1992:28–34
Carr LJ, Dunsiger SI, Marcus BH (2010) Walk score (TM) as a global estimate of neighborhood walkability. Am J Prev Med 39(5):460–463
Calthorpe P (1993) The next american metropolis: ecology, community, and the american dream. Princeton Architectural Press, New York
Cerin E, Saelens BE, Sallis JF, Frank LD (2006) Neighborhood environment walkability scale: validity and development of a short form. Med Sci Sport Exer 38(9):1682–1691
Cerin E, Macfarlane DJ, Ko HH, Chan KCA (2007) Measuring perceived neighbourhood walkability in Hong Kong. Cities 24(3):209–217
Crossman ND, Bryan BA, King D (2011) Contribution of site assessment toward prioritising investment in natural capital. Environ Model Softw 26(1):30–37
Cheng SC, Chou TC, Yang CL, Chang HY (2005) A semantic learning for content-based image retrieval using analytical hierarchy process. Expert Syst Appl 28(3):495–505
Devlin GJ, McDonnell K, Ward S (2008) Timber haulage routing in Ireland: an analysis using GIS and GPS. J Transp Geogr 16(1):63–72. doi:10.1016/j.jtrangeo.2007.01.008
Duncan DT, Aldstadt J, Whalen J, Melly SJ, Gortmaker SL (2011) Validation of walk score (R) for estimating neighborhood walkability: an analysis of four us metropolitan areas. Int J Env Res Pub He 8(11):4160–4179
Duany A, Plater-Zyberk E (1994) The neighborhood and the district. In: Katz Peter (ed) The new urbanism. McGraw Hill, Inc, New York
Ericsson E, Larsson H, Brundell-Freij K (2006) Optimizing route choice for lowest fuel consumption—potential effects of a new driver support tool. Transp Res C-Emer 14(6):369–383
Frank LD, Engelke PO (2001) The built environment and human activity patterns: exploring the impacts of urban form on public health. J Plan Lit 16(2):202–218
Gauvin L, Richard L, Craig CL et al (2005) From walkability to active living potential—an "ecometric" validation study. Am J Prev Med 28(2):126–133
Hudecek T (2008) Model of time accessibility by individual car transportation. Geografie-Prague 113(2):140–153
Javadian M, Shamskooshki H, Momeni M (2011) Application of sustainable urban development in environmental suitability analysis of educational land use by using AHP and GIS in Tehran. Int Conf Green Buildings Sustainable Cities 21:72–80
Kerr J, Rosenberg D, Sallis JF, Saelens BE, Frank LD, Conway TL (2006) Active commuting to school: Associations with environment and parental concerns. Med Sci Sport Exer 38(4):787–794
Lathey V, Guhathakurta S, Aggarwal RM (2009) The Impact of Subregional Variations in Urban Sprawl on the Prevalence of Obesity and Related Morbidity. J Plan Educ Res 29(2):127–141
Leslie E, Cerin E, duToit L, Owen N, Bauman A (2007) Objectively assessing 'walkability' of local communities: using GIS to identify the relevant environmental attributes. Lec Not Geo Carto:90–104
Leslie E, Saelens B, Frank L, Owen N, Bauman A, Coffee N, Hugo G (2005) Residents' perceptions of walkability attributes in objectively different neighbourhoods: a pilot study. Health Place 11:227–236
Lu ZP, Rodiek SD, Shepley MM, Duffy M (2011) Influences of physical environment on corridor walking among assisted living residents: findings from focus group discussions. J Appl Gerontol 30(4):463–484
McDonald NC (2008) The effect of objectively measured crime on walking in minority adults. Am J Health Promot 22(6):433–436
Merom D, Bauman A, Phongsavan P, Cerin E, Kassis M, Brown W, Smith BJ, Rissel C (2009) Can a motivational intervention overcome an unsupportive environment for walking-findings from the step-by-step study. Ann Behav Med 38:137–146

Muraleetharan T, Hagiwara T (2007) Overall level of service of urban walking environment and its influence an pedestrian route choice behavior—analysis of pedestrian travel in Sapporo, Japan. Transp Res Record 2002:7–17

Moudon AV, Lee C, Cheadle AD, Garvin C, Johnson DB, Schmid TL, Weathers RD (2007) Attributes of environments supporting walking. Am J Health Promot 21:448–459

Norman J, MacLean HL, Kennedy CA (2006) Comparing high and low residential density: life-cycle analysis of energy use and greenhouse gas emissions. J Urban Plan D-Asce 132(1):10–21

Piltan M, Mehmanchi E, Ghaderi SF (2012) Proposing a decision-making model using analytical hierarchy process and fuzzy expert system for prioritizing industries in installation of combined heat and power systems. Expert Syst Appl 39(1):1124–1133

Saaty TL, Vargas LG (1980) Hierarchical analysis of behavior in competition—prediction in chess. Behav Sci 25(3):180–191

Saelens BE, Sallis JF, Frank LD (2003a) Environmental correlates of walking and cycling: Findings from the transportation, urban design, and planning literatures. Ann Behav Med 25(2):80–91

Saelens BE, Sallis JF, Black JB, Chen D (2003b) Neighborhood-based differences in physical activity: an environment scale evaluation. Am J Public Health 93(9):1552–1558

Strath S, Isaacs R, Greenwald MJ (2007) Operationalizing environmental indicators for physical activity in older adults. J Aging Phys Activ 15(4):412–424

Shigematsu R, Sallis JF, Conway TL, Saelens BE, Frank LD, Cain KL, Chapman JE, King AC (2009) Age differences in the relation of perceived neighborhood environment to walking. Med Sci Sport Exer 41:314–321

Shen SZ (2010) The application of analytical hierarchy process in evaluating healthcare systems. Model Simul 6:93–96

Todd AL (2012) Evaluating accessibility for transportation planning. Victoria Transport Policy Institute

Part IV
Web-Based Support Systems

Access to UK Census Data for Spatial Analysis: Towards an Integrated Census Support Service

John Stillwell, Justin Hayes, Rob Dymond-Green, James Reid, Oliver Duke-Williams, Adam Dennett and Jo Wathan

Abstract In the absence of a comprehensive population registration system in the United Kingdom (UK), the decadal Census of Population provides a crucially important source of demographic and socio-economic data both for academic research as well as planning and policy making for urban sustainability. Data from recent UK censuses have been used to produce a variety of products in digital form at different spatial scales, ranging from the counts of individuals or households with particular characteristics captured by the census questionnaire and often referred to as aggregate statistics, to samples of the population at individual or household level, known collectively as micro data. Access to these often large and complex data sets for social science research has been facilitated through the development of a set of services funded by the Economic and Social Research Council (ESRC) under the Census Programme. In this chapter, we review the current Census Support Service at a point of transition to a more integrated system

---

J. Stillwell (✉)
School of Geography, University of Leeds, Leeds LS2 9JT, UK
e-mail: j.c.h.stillwell@leeds.ac.uk

J. Hayes · R. Dymond-Green
The University of Manchester, Mimas, Roscoe Building (5th Floor), Oxford Road, Manchester M13 9PL, UK

J. Reid
EDINA, Causewayside House, 160 Causewayside, Edinburgh, Scotland EH9 1PR, UK

O. Duke-Williams
Department of Information Studies, University College London, Gower Street, London WC1E 6BT, UK

A. Dennett
Centre for Advanced Spatial Analysis, Faculty of the Built Environment, University College London, Gower Street, London WC1E 6BT, UK

J. Wathan
Cathie Marsh Centre for Censuses and Surveys, School of Social Sciences, Humanities Bridgeford Street Building University of Manchester, Manchester M13 9PL, UK

S. Geertman et al. (eds.), *Planning Support Systems for Sustainable Urban Development*, Lecture Notes in Geoinformation and Cartography,
DOI: 10.1007/978-3-642-37533-0_19,

allowing users 'one stop shop' access to a range of different data sources, including those associated with the census, through the facility known as the UK Data Service.

1 Introduction

A 'Census of Population' has been taken in the UK in every decade over the last 200 years (with the exception of 1941 and the inclusion of 1966). The results of the latest round of censuses in 2011, coordinated in England and Wales, Scotland and Northern Ireland by the respective census agencies—the Office for National Statistics (ONS), the National Records for Scotland (NRS) and the Northern Ireland Statistics and Research Agency (NISRA)—are now being processed and quality checked and census estimates are being released into the public domain. These estimates are a hugely important source of information for those interested in the sustainability of urban areas because they provide rich socio-demographic information for relatively small geographic areas across the whole of the UK.[1] Consequently they are used to create the geodemographic profile of local populations, to identify health, educational and occupational characteristics, to measure levels of deprivation and affluence, as well as to establish the characteristics of households and the housing which they inhabit. The census provides a benchmark set of data against which progress towards sustainability can be measured. Norman (2010), for example, uses data from the 1991 and 2001 Censuses at ward level in Great Britain to indicate how socio-economic deprivation measured by the Townsend index eased during the 1990s due to better access to a car, increasing home ownership, declining household overcrowding and reduced levels of unemployment.

Census data are also a unique resource of vital importance to planning and administrative practitioners in national, regional and local agencies. Age and sex specific census population counts underpin the annual mid-year sub-national population estimates produced by the ONS which in turn form one of the key drivers for the allocation of public finance to local authorities. Boundary data enable immediate visualisation of census data at different spatial scales. Origin–destination migration and commuting flows provide benchmark data for the decade that follows and, through multivariate cross-tabulations, microdata provide detailed insights into individual and household characteristics that are impossible to glean from aggregate counts. The data are also of immense value for social science researchers in the UK, particularly for those interested in geographical variations at the local level, as noted in the House of Commons (UK Parliament) Select Committee report on *The Census and Social Science*, which concluded that the absence of a 2021 Census would have serious ramifications for social science

---

[1] The 2001 Census provides data for 223,060 output areas across the UK.

unless surveys and administrative sources could be identified that would "scale to a nationwide coverage"(Science and Technology Committee 2012). The feasibility of these sources is currently the focus of attention of the 'Beyond 2011' programme (ONS 2011).

The changing information environment over the next 5 years will present new challenges for those who collect, supply and use data. Data collection agencies want to improve the efficiency of their operations and gain better value from their data production and dissemination mechanisms. Census outputs will include data generated in new multi-dimensional forms and supplied using new transfer methods to devices such as mobile phones. Census users will have increased expectations of the ease of access to census and related data; they will expect more immediate access through delivery mechanisms that are more intuitive and user friendly and they will seek increasingly to maximise the impact of their research. Data linkage and integration will be a key challenge, particularly in the context of 'Beyond 2011', with attention increasingly being paid to administrative, survey and commercial data. More specifically, the 2011 Census data may assume much greater significance in future years if the traditional form of the census is discontinued.

The last decade has witnessed the development of a set of services designed to provide members of the academic community in the UK with quick and easy access to census data sets, primarily for research purposes but also for teaching. These services were funded by the Economic and Social Research Council (ESRC) under the Census Programme up until August 2012. This chapter is designed to provide a review of the services which, over the next 5 years, will become part of a more integrated Census Support (CS) service which itself will become an integral part of a new ESRC funded UK Data Service, aiming to allow researchers access to a wider range of data from census, administrative and survey sources through a single point of entry. Sections 2–6 of this chapter introduce the data and outline the current services on offer. Section 7 provides an example of how different types of census data are used in the analysis of net migration in London. The last section discusses some of the issues relating to the transition towards service integration and the implications of the UK Government Licensing Framework (UKGLF) that now standardises the licensing principles for government information and recommends that the Open Government License (OGL) is the default license for public sector information (The National Archives 2011).

2 Census Data Products and Census Support

The 2011 Census took place on 27 March 2011, with all households in the UK being issued with a questionnaire that asked a series of questions about the demographic characteristics of household inhabitants and visitors, including age, sex, marital status and ethnicity, but also a series of questions that allow the census agencies to provide statistics on topics such as health, occupation, migration, length of stay, national identity, religion qualifications, employment, housing tenure and commuting. Before the data are released, they must undergo processes of cleaning, validation and quality assurance whilst imputation and disclosure control adjustment provide robust census estimates that preserve confidentiality. The ONS is confident that the 2011 Census achieved a national household response rate of 95% in England and Wales overall, and a person response rate of over 80% from every local authority (ONS 2012). As with previous censuses, the 2011 Census will provide a cross-sectional snapshot of the nation's population, revealing important changes in society that have occurred since 2001.

The first release of statistics by ONS from the 2011 Census took place in July 2012 and included rounded counts of usually resident census population estimates at single year of age and sex at country level, usually resident population estimates by 5 year age bands and sex at regional and local authority level, and estimates of occupied households at country, regional and local authority level. The second release of *Key and Quick statistics,* timetabled from November 2012 to February 2013 included data for small areas, including output areas and lower and middle layer super output areas (SOAs), wards and other geographies including parishes (England) and communities (Wales), health areas, postcode sectors, national parks, political constituencies and postcodes. The variables in this release correspond with 2001 Census Key Statistics (KS) tables produced for univariate information (counts and percentages) and a look-up table has been developed to facilitate comparison with 2001 data. The third release, timetabled for March–June 2013, contains multivariate tables that generally correspond to the 2001 Census Area Statistics (CAS) tables, e.g. age by sex by resident type, or sex and age and ethnic group, for the output area hierarchy and wards, plus unitary and local authority areas, regions and health areas at more aggregate scales. The fourth release is timetabled for July–October 2013 and content includes more detailed characteristics, theme tables and Armed Forces tables with data available for higher level geographies including middle layer SOAs, wards, unitary and local authorities, regions, health areas and national parks but not for output areas or lower layer SOAs for confidentiality reasons. The statistics correspond to the 2001 Census Standard Tables (ST), standard and CAS theme tables and armed forces tables.

This schedule of release refers specifically to the aggregate statistics products and will be accompanied by the release of a hierarchical set of digital boundaries corresponding with the attribute data. Separate prospectuses are available for data release in Scotland and Northern Ireland. The release of UK interaction or flow data sets and national micro data sets is likely to follow the last release of the aggregate statistics. Once the data are released, they will be captured from the ONS Application Programming Interface (API) by UK Data Service Census Support (UKDS-CS), the 'value added services' that have been developed under the ESRC's Census Programme over the last decade and which provide user-friendly query, extraction and analysis facilities for researchers in the UK academic community. UKDS-CS involves four of the 'data support units' operating under the Census Programme that provided expert support and online access to different datasets: (1) the Census Dissemination Unit (CDU) based within Mimas at the University of Manchester provides support for the *aggregate outputs*; (2) the UKBORDERS service within

EDINA at the University of Edinburgh provides access to census-related digital *boundary data*; (3) the Centre for Interaction Data Estimation and Research (CIDER) at University College London (formerly at the University of Leeds) provides access to census and related *interaction data* sets; and (4) the Cathie Marsh Centre for Census and Survey Research (CCSR) at the University of Manchester provides guidance and access to census *microdata* sets known collectively as the Samples of Anonymised Records (SARs).

In addition to the collection, preservation and dissemination of census data, the units collectively hold and provide access to a wide range of valuable related data including: geographical look-up tables; yearly updates of Experian datasets; deprivation scores; gridsquare data; geographical boundary data for a wide range on non-census geographies; inter-district migration flow estimates supplied by ONS; and estimated migration matrices from ESRC funded research projects. These four units, together with the three units supporting longitudinal data in England and Wales (Celsius), Scotland (LSCS) and Northern Ireland (NILS-RSU) respectively, have constituted the Census Programme together with the Census Registration Service (CRS) under the Census Portal which also acts as focal point for information about the census and training events. From August 2012, these four units and the CRS will gradually transition into the new Census Support Service (CSS) which is part of the UK Data Service (Fig. 1) and branded as UK Data Service—Census Support (UKDS-CS), with the longitudinal studies being supported by a network of dedicated services and a new UK Census Longitudinal Study Development Hub. While remaining separate, the UK census longitudinal studies services and their hub will collaborate closely with the UKDS-CS and the UK Data Service as a whole.

The next four sections outline services that will constitute the new UKDS-CS and introduce some of the software systems that are currently in operation that provide users with routes of access to the respective sets of data.

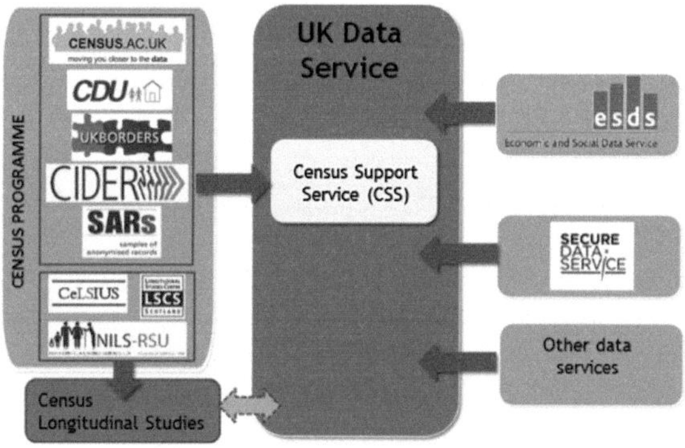

Fig. 1 From Census Programme to Census Support Service

3 Access to Aggregate Statistics Through InFuse

InFuse is an innovative and intuitive web interface that makes it easier for users to find, understand, extract and make use of data from aggregate outputs from the census. In this section, we explain and exemplify the development of InFuse from its predecessor, Casweb.

Census aggregate outputs have historically been sets of tables containing various combinations of variables (e.g. age by sex by religion) that are specific to different information requirements and user communities. Unfortunately, this table-focussed approach has allowed tables to be designed independently without central control on structure being enforced, resulting in inconsistencies in the categorisation of variables and in the labelling of the same variable from one table to another. Thus, for example, an age grouping of '24 years and under' may be labelled inconsistently as '0-24'. Having variations in terms for identical concepts causes problems for an application. These inconsistencies make it difficult to compare and search for data of interest across all of the outputs, effectively splitting them into many separate datasets. Census users have to look through many tables in Casweb to see if they contained the variables they were interested in.

Moreover in Casweb, metadata—the descriptions of items (e.g. economic activity)—were divorced from the data. In practice, users had to look through the census definitions volume or at the original census questionnaire to find this metadata. The underlying data were not associated with the labels. For example, if a Casweb user selected data from the first two cells of Table CT003: Theme Table on Ethnic Group (Fig. 2), only the label (e.g. ct0030001) associated with the count in each cell would appear in the output file (Fig. 3). The first four characters of the label refer to the table ID, and the last four digits correspond to the cell ID. To find out what the code referred to, a user would have to look at the interface and manually type in their own more meaningful description.

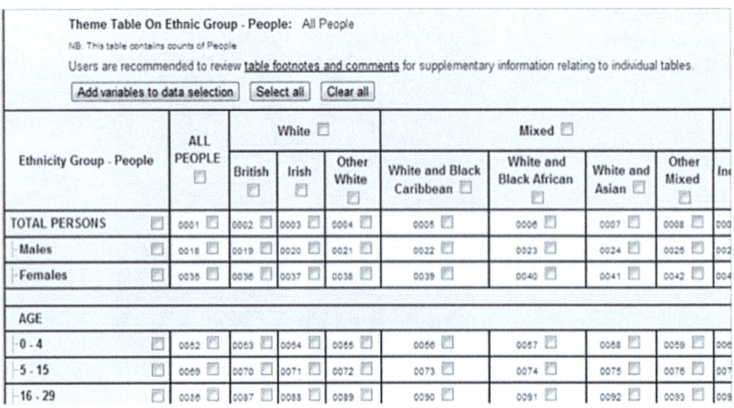
Theme Table On Ethnic Group - People: All People

NB: This table contains counts of People

Users are recommended to review table footnotes and comments for supplementary information relating to individual tables.

Add variables to data selection | Select all | Clear all

Ethnicity Group - People	ALL PEOPLE	White			Mixed				In
		British	Irish	Other White	White and Black Caribbean	White and Black African	White and Asian	Other Mixed	
TOTAL PERSONS	0001	0002	0003	0004	0005	0006	0007	0008	000
Males	0018	0019	0020	0021	0022	0023	0024	0025	002
Females	0035	0036	0037	0038	0039	0040	0041	0042	004
AGE									
0 - 4	0052	0053	0054	0055	0056	0057	0058	0059	006
5 - 15	0069	0070	0071	0072	0073	0074	0075	0076	007
16 - 29	0086	0087	0088	0089	0090	0091	0092	0093	009

Fig. 2 Part of table CT003 in Casweb

Fig. 3 Output from Casweb after selecting the first two cells in table CT003

	A	B	C	D
1	Zone Code	Zone Name	ct0030001	ct0030002 c
2	00AAFA	Aldersgate	1603	1194
3	00AAFE	Bishopsgate	106	67
4	00AAFQ	Cripplegate	3008	2200
5	00AAFS	Farringdon Within	180	130
6	00AAFT	Farringdon Withou	691	480
7	00AAFX	Portsoken	1072	508
8	00AAFY	Queenhithe	282	175

So as to improve variable label consistency and metadata provision, it was necessary to change the structure of the data to make it more usable. The format of the census tables was an obstacle to the creation of a more usable application. In order to enable creation of a more flexible interface (InFuse), the database was reformatted into a multi-dimensional structure based on SDMX,[2] a data model which has been adopted by non-governmental agencies around the world as the standard for data and metadata exchange for aggregate time series data. Thus, users of census aggregate data in InFuse are able to more easily and intuitively locate data without hunting through a multitude of tables. The example in Fig. 4 shows the first screen of 46 topic combinations associated with the selected variable (Ethnic group).

Infuse interface users are able to access data from multiple geographical levels at the same time, as shown by the inclusion of London GOR as well as CAS wards in the InFuse output shown in Fig. 5. Output from Infuse for the same information shown in Fig. 3 is shown in Fig. 5, indicating the metadata attached to each variable in the second row as well as information about the type of geographical unit in the second column.

Initial data logs from users of InFuse during the period September 2010 to October 2012 indicate that there were 676 unique users of InFuse, of which 115 did not use the Casweb interface. Though a relatively small sample, the data shows that the new approach has helped reduce barriers to use among the UK academic community and broadened the reach of census data. There is evidence of greater use with university staff and postgraduates since 69% of InFuse users were postgraduate students or HE/FE staff as opposed to 40% of Casweb users. There is also evidence of greater range of disciplines accessing census data. While Geographers still form the largest proportion of users (52% for Casweb, 36% for InFuse), other disciplines have increased evidence of usage, including Sociology (Casweb 2.3%, InFuse 6.2%) and Medicine-related studies/professions (Casweb 2.6%, InFuse 5.8%). Some examples of non-traditional social science use are: Drama; Dance and Performing Arts, English Language and Literature, Physics and Sports-related Subjects. While these proportions remain small, the increases amongst a small number of 'early adopters' suggest that once InFuse reaches

[2] See Statistical Data and Metadata eXchange (SDMX) web site at sdmx.org.

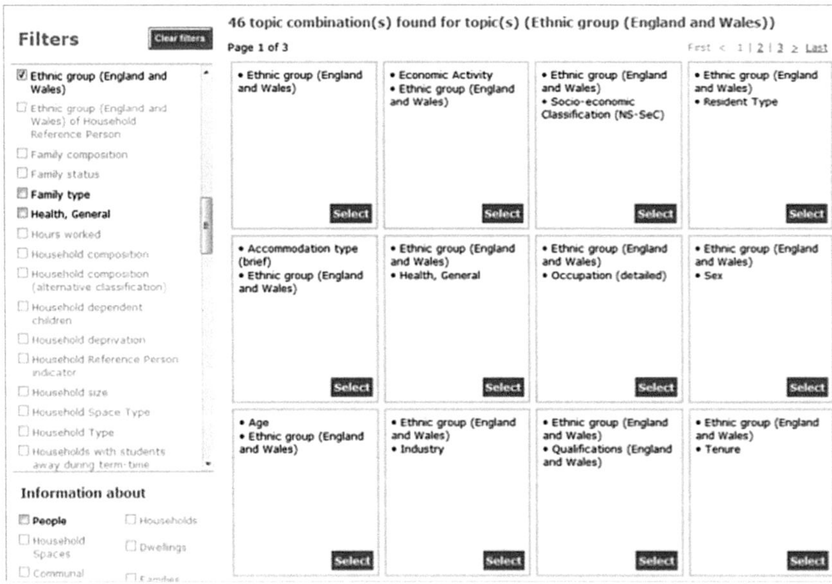

Fig. 4 The filtering facility in InFuse

GEOGRAPHY_LABEL	GEOGRAPHY_TYPE	F12714	F12715
		Ethnic group (England and Wales) : All people - Sex : All People - Unit : People	Ethnic group (England and Wales) : White \ British - Sex : All People - Unit : People
H : London	Government Office Regions	7172091	4287861
00AAFA : Aldersgate	Census Area Statistics (CAS) wards	1603	1194
00AAFE : Bishopsgate	Census Area Statistics (CAS) wards	106	67

Fig. 5 Output from InFuse after selecting populations by ethnic group

greater penetration, it has the potential to promote the use of census aggregate data to a wider academic audience and therefore will become the main delivery vehicle for the 2011 Census aggregate data.

4 Digital Boundary Data from UKBORDERS

One of the more notable features of the UK census is its granularity in terms not just of the wealth of personal and household information attributes that it gathers but also the *spatial* units at which the aggregate results are released. Academics in the UK have access to an unrivalled time series of digital boundary data (DBD) at a very high level of spatial resolution which are typically accessed as GIS/CAD files for use in desktop software. Whilst the spatial building blocks for each decennial census have tended to vary both in respect of the types and nature of how those spatial units are derived, it is anticipated that the 2011 Censuses in England and Wales, Scotland and Northern Ireland will provide a more coherent and cross-comparative set of geographical outputs than its predecessors. Since 2001, there has been increasing recognition of the need for consistency in the spatial units used for recording counts and enabling backwards compatibility for comparative purposes. Both inter- and intra-zonal comparisons over time continue to be a preoccupation of academic research since it is the changes in society and their spatial manifestations which are frequently more informative than cross-sectional maps.

The UKBORDERS suite of services provides mechanisms for online delivery of the geography outputs of the last five decennial censuses as well as a range of historical census geographies, associated look-up tables, postcode directories and non-census geographical boundary data. From 1993, the service has delivered mechanisms whereby users of the census geography outputs can search, filter and extract (download data) by spatial and temporal extent. Figure 6 is a screenshot of the 'Boundary Data Selector' interface of UKBORDERS which enables a user to build a query under the 'What and Where?' tab by initially selecting the country of interest (i.e. England), the type of data required (i.e. census boundaries) and the reference period (i.e. post 1999). Once these parameters have been chosen, the user can make a specific selection of spatial units; in this example, the user is preparing a query to extract the 2001 Census Area Wards for Greater London in ESRI shapefile format. Once the data have been extracted they are held in a zip file for downloading and inputting to a GIS on the user's personal computer.

As already noted, the UK has a distinctively rich geography as indicated in ONS' *Beginners Guide to Geography* (ONS undated), and the digital boundaries have changed over time out of necessity as administrative and pragmatic requirements dictate. UKBORDERS endeavours to provide a comprehensive 'digital library' archival snapshot of these changing digital boundary files along with metadata which can assist in their interpretation and use. Furthermore, as the digital boundary data are widely used across a range of disciplines beyond the conventional social sciences in areas as diverse as archaeology, land-use planning and zoology, a wide range of data output formats are supported in order to meet these broad user requirements. Service users may choose to drill-down to specific individual spatial units or take larger areal extents incorporating many spatial units. In providing a consistent user interface to a diverse and rich collection of data, UKBORDERS attempts to strike a balance between usability and functionality whilst also needing to consider performance and scalability when doing real time data extraction and delivery. The latter is pertinent as some of the spatial datasets can be rather large (even by modern standards), running to several hundred megabytes. As mitigation against this, an 'Easy Download' service is also available which provides access to original resolution and resampled lower resolution generalised (simplified) sets of digital boundaries as 'pre-canned', ready

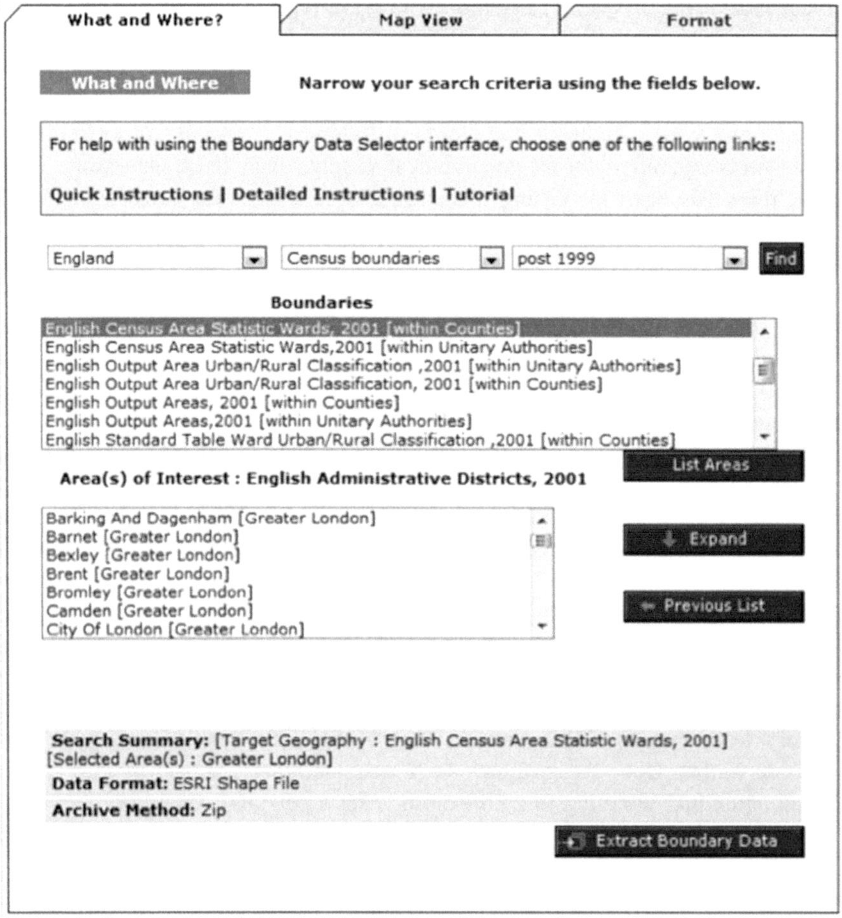

Fig. 6 The boundary data selector interface in UKBORDERS

to use datasets in a variety of the most popular data formats (ESRI shape, DXF, MapInfo) for quick download without the need to perform live data extractions. Further services include equivalent tooling onto the extensive holdings of current and historic postcode directory datasets permitting custom bespoke look-up and drill-down facilities and an online thematic mapping tool for visualisations. The relative balance between the ‘grab and go’ requests and bespoke, tailored data extraction requests is roughly equal and runs to several thousand unique data requests per month.

It is this ‘value added’ dimension of these services that make them so valuable to their academic users, as well as the availability of digital boundary data for five censuses. By centralizing access to and standardizing mechanisms that enable authorized access to census resources, UKBORDERS (and other services) has reduced the barriers to use by removing the need for each potential user of the resources to manage, curate and maintain local copies of the data. This implicit need is reflected in the fact that despite over 20 years of continuous operation, the services remain in heavy demand and users return repeatedly to access and use the data. At inception, it had been suggested that users would simply take resources once and cache local copies which would diminish the need for the services over time. This proved unfounded and the data management and curation role continues to play a significant role in academic census provision by providing continuity of service and user expectation of on-demand web based services.

Whilst user interfaces are tuned to meet simple and complex data selection and extraction needs, the UKBORDERS services have been designed to permit not just human-centric interaction but also provide standards-based machine interfaces such as Web Mapping Services (ISO19128) and Web Feature Services (ISO19142). These machine interfaces are expected to become increasingly important in emergent computing paradigms where distributed, ‘mashable’ data sources are becoming increasingly pervasive.

5 Interaction Data from WICID

Interaction data have been produced from the modern censuses from 1981 onwards, termed the Special Migration Statistics (SMS), Special Workplace Statistics (SWS), and Special Travel Statistics (STS). These data tabulate information about flows within and between locations, and take the conceptual form of matrices of origins by destinations for migrants, or residential locations by workplace locations for commuters. These matrices are often sparsely populated; the highest level of spatial detail for the 2001 Census outputs was the output area (OA) and there were over 220,000 OAs defined across the UK, giving a total of close to 50 billion possible origin–destination pairs. In a population of 60 million persons, any such matrix is going to be very sparsely filled. For those combinations of origin, destination, age and sex for which a non-zero value was observed, that value was likely to very small. One consequence of this was that the data were strongly affected by disclosure control mechanisms (Duke-Williams and Stillwell 2007), but more general observations may be made: firstly, that it can be difficult for users to explore such a large data space, and find the parts in which they are interested; and secondly, that few data users actually wish to use data at that fine scale—the utility of highly spatially detailed data is that they can be re-aggregated to a wide variety of coarser geographies that are appropriate for the analysis being undertaken.

The online access system developed by the Centre for Interaction Data Estimation and Research (CIDER) provides a number of routes into the data that aim to address the practical problems for users of sparse data. The main system is an interface called WICID (Web Interface to Census Interaction Data), which allows users to flexibly build and execute a query. There are two main parts to a query: the data elements and the geography terms. In both cases, there is a variety of ways of defining the terms of interest, reflecting differing user requirements. There are various approaches to selecting data items of interest. Users can: select a fixed total (migrants or commuters) from a dataset; select by navigating a tree of data set names, table names and finally table layouts; or search by chosen variable(s) (e.g. 'ethnic group') to be given a list of tables that contain a version of that variable, together with a menu of cross-classifying variables. Figure 7 illustrates the data selection page of the WICID query interface, indicating how the user can select the data items (cells in the table), in this case total persons in total and by ethnic group.

In the case of 'Geography', users might be interested in a single area only, or a potentially large number of areas, and different search tools are appropriate in different circumstances. Users can: select one or more areas from a fixed list; select all areas in one geography simultaneously; select by a free text search on place name or containing postcode; or select using a map. Some approaches permit a separation of selection and output geographies and thus a user has the capability to automatically select, for example, all wards in a chosen region. A distinctive characteristic of interaction data is that there are two sets of spatial constraints and the interface allows users the quick option of mirroring origins as destinations, or permits them to select an independent set, possibly at a different geographic scale.

In addition to the main holdings of interaction data, the WICID interface also provides access to a set of population counts derived from the Census Area Statistics or other sources (as methodologically appropriate) that can be included in a query by users in order to generate flow rates. These counts are also available

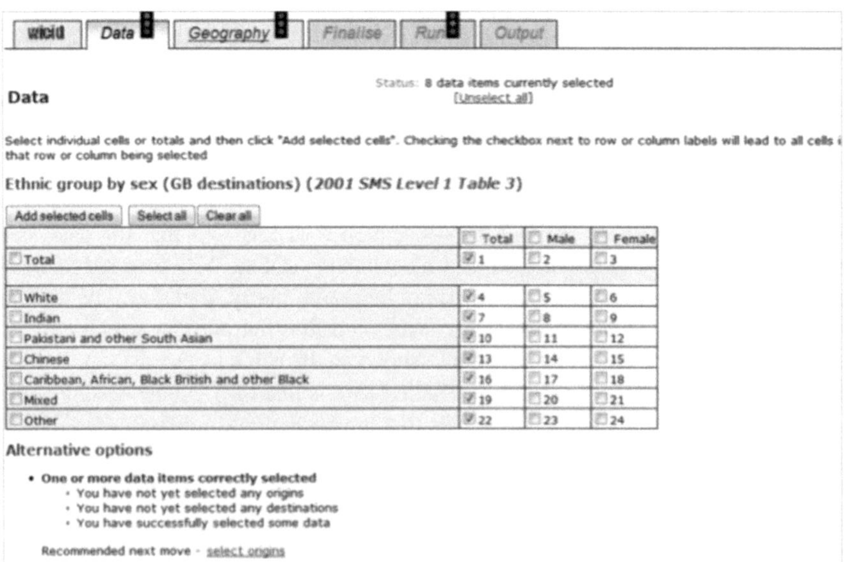

Fig. 7 Data selection page in WICID

via the data search tools, and can be browsed or can be added via an automated tool than identifies the relevant population count for each interaction data term in a query.

6 Access to the Samples of Anonymised Records

Thus far, the sources described have provided data broken down into geographical areas, either as boundaries for those areas, or as tables of summary statistics relating to those areas or flows within and between them. These are, of course, ideal resources if one is interested in differences across areas, but less helpful if one is interested in variation between individuals. There are risks involved in attempting to understand individual characteristics on the basis of areal evidence (Tranmer and Steel 1998) rather than using individual level data that allows analysts to explore differences at this level directly either on their own or alongside area analysis.

By using microdata, one is able to work with individual characteristics much more flexibly than is the case with pre-tabulated data. It is possible to generate tables that do not already exist because the data typically contain a full range of census topics which can be tabulated against each other any way that the user sees fit. Similarly, it is possible to generate sub-samples based on some constraint to limit the population to which the analysis applies. Furthermore, a large number of variables may be included in a single analysis permitting individual level multivariate models to be produced.

However, there are some major shortcomings of using microdata rather than aggregated data which can be traced to the need to maintain the confidentiality of respondents. Because the microdata contain a large number of person characteristics in each record, the potential for disclosure is greatly increased if the data are not treated. Accordingly, data are released only for non-overlapping samples and the amount of detail available for any record, particularly geographical detail, is restricted.

The structure of the data is relatively simple which means that while the data have been made available only to users who satisfy authentication, the data can be distributed using standard tools rather than necessitating the production of bespoke ones. Typically, the data are stored as simple matrices which contain rows for individual records and where each column relates to an individual characteristic. In the microdata context, it is these characteristics that vary from individual to individual that are referred to as variables. These data are stored either as delimited text files for importing into an analysis package, or as preformatted files for reading into the commonly used statistical packages, Stata and SPSS. This comes about as it is inevitably the case that a computer package is required to analyse the data. This is an internationally common practice amongst microdata archives and access services and the approach used by the Economic and Social Data Service (ESDS) and its predecessors (Cole et al. 2008).

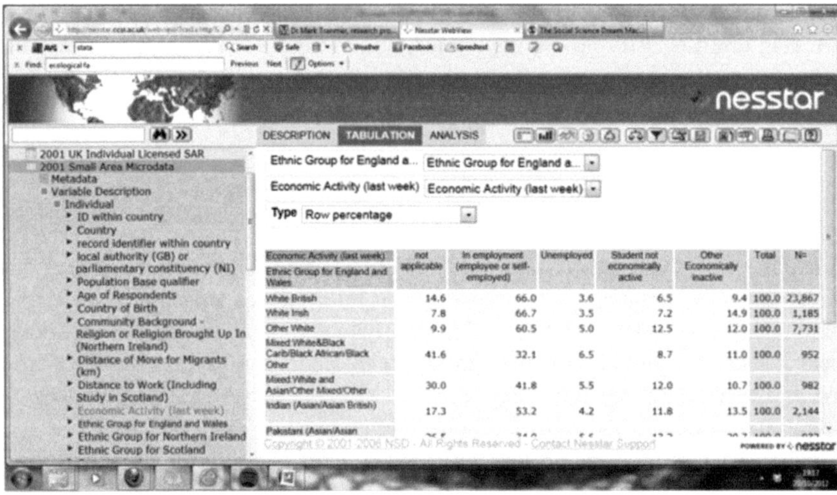

Fig. 8 Nesstar tabulation of economic activity by ethnic group for London migrants, small area microdata 2001 (5% sample of individuals)

The microdata supported by CCSR are also made available through an online discovery tool called nesstar (Fig. 8) which was created by a consortium of European data archives (Ryssevik and Musgrave 2001). This tool has the advantage of bringing data and metadata together in a browsable format, so that even unregistered users can view population-wide univariate distributions. Additionally, registered users can undertake simple analyses online and download subsets to their own requirements.

7 Spatial Analysis of Ethnic Migration in Greater London

The development of communities that are sustainable from an ethnic perspective is an issue of major importance in many cities in the UK. Ethnic migration in particular has been a focus of attention by social scientists in recent years, in the context of increasing volumes of immigrants and the perceived socio-economic consequences of their arrival in localised communities and their subsequent redistribution. The search for reliable and detailed geographical information on ethnic migration begins and ends with the census; there is simply no alternative data source that provides the possibility to identify geographical patterns of ethnic population concentration and movement of whites versus minority non-white ethnic groups.

The internal migration propensities and spatial patterns of migration in London have been investigated by Stillwell (2010a, b) using data on ethnic populations drawn from the aggregate statistics. Figure 9a illustrates the number of usual residents in each of six ethnic minority categories represented as pie charts and this distribution is superimposed upon a ranged map of the percentage of the population of each London borough that is white.

Two boroughs, Brent and Newham, had majority non-white populations in 2001 and the higher proportions of whites are found in the boroughs of Outer London, for the most part, with exceptions such as the City of London, Islington and Westminster and Chelsea. There are distinct concentrations of certain non-white groups with blacks being the majority non-white groups in Haringey, Hackney, Lambeth, Southwark and Lewisham, and large numbers of Asians being present in western boroughs and in Newham and Redbridge. The ethnic population counts have been downloaded from InFuse (using the query shown in Fig. 4) and linked in MapInfo Professional with the polygons representing the 32 London boroughs and the City of London whose boundaries have been downloaded from UKBORDERS (using query shown in Fig. 6).

The map shown in Fig. 9b shows the patterns of net internal migration within London of whites. More specifically, it shows for each ward, the difference between the total number of white in-migrants from other wards of London and the total number of white out-migrants. The pattern of gains and losses is very distinctive with most Inner London wards losing net migrants and most Outer London wards gaining net migrants. Whilst Fig. 9b clearly shows the decentralisation or dispersal of white migrants in London, Stillwell (2010b) exposes the reverse net pattern for wards when the variable plotted is the net migration for each ward with the rest of England and Wales outside London; in this case, the central wards are gaining through net in-migration and the peripheral wards and losing in net terms as whites in the outer suburbs move further out into the rest of the South East, to East Anglia and to the South West. Figure 9 is based on migration data

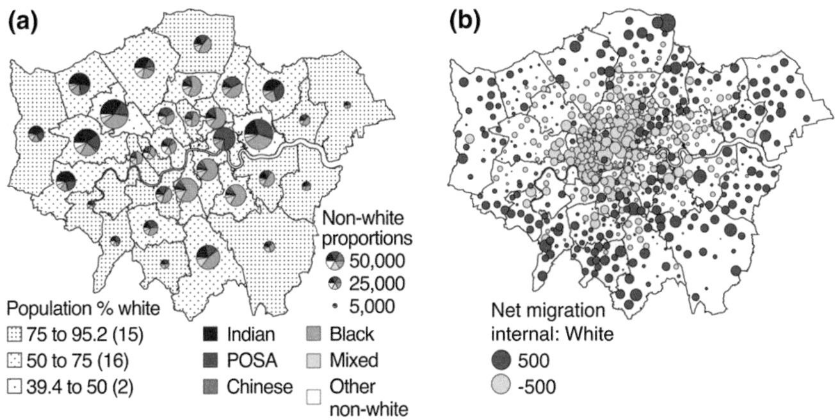

Fig. 9 Maps of London's **a** ethnic population by borough and **b** white ethnic net migration by ward, based on data drawn from InFuse, WICID and UKBORDERS. *Sources* 2001 Census: census area statistics theme tables (England and Wales); commissioned table C0723; district and ward boundaries. Crown copyright 2003

extracted from WICID (using the query shown in Fig. 7) and linked with ward boundary data downloaded from UKBORDERS. Although data from the Samples of Anonymised Records (SARs) have not been not used in this example, parallel work on ethnic migration using microdata has been reported in Finney and Simpson (2008) and Simpson and Finney (2009).

8 Conclusions: Towards an Integrated Census Support Service

The resources created and the services developed under the Census Programme that have been outlined in previous sections have provided the academic community in the UK with the facilities to access and extract data from large and complex data sets using state-of-the-art tools via user-friendly interfaces. This level of support for researchers in social and health sciences to access data from the last five censuses is unique across the world; the results of user surveys suggest that the services (available 24/7) are well used and much appreciated. However, as the data deluge grows at an ever increasing rate, as technology rapidly changes and as user requirements reflect the new research challenges of an increasingly dynamic society, so it becomes necessary for the services that have been developed under the Census Programme to respond. The ESRC have recognised that part of their Data Strategy is the creation, over the next five years, of a more integrated service of Census Support (UKDS-CS) within a new UK Data Service that also contains a range of other non-census services that include the existing Economic and Social Data Service (ESDS), the Secure Data Service and the UK Data Archive.

There are a number of key requirements associated with this integration challenge, including:

- development of a brand for both the UK Data Service and its component services (including UKDS-CS) that will provide the framework for a new web site and the integration of the existing services into an easy to use, innovative, trusted ‘one-stop shop’ for suppliers and users of social science data;
- ingest of the disseminatable Samples of Anonymised Records into the mainstream UK Data Service collection
- transfer of the census databases and software systems held at MIMAS (at the University of Manchester) and EDINA (at the University of Edinburgh) onto a new UK Data Service platform (at the University of Essex) and their operationalization;
- integration of the UKDS-CS services with the other data services already in operation at Essex;
- maintenance of the current level of service provision to users in the academic community but opening up the services to non-academic users as a function of new open government licensing arrangements; and

- development of a new query tool that will enable users to perform true topic-based searches across the range of census data supported by the service and thus be able to extract data on a particular theme or topic from multiple sources using one query (rather than several queries, one for each census service as at present).

Each of these challenges presents considerable problems to solve and difficulties to overcome. In relation to branding for the UK Data Service as a whole, this task was outsourced to a company, Conran Design Group, and the recommendations of these consultants have provided the designs for the front end of the integrated web site and the service logos, thereby creating the brand framework for subsequent service development.

It is envisaged that transfer of existing UKDS-CS infrastructure to a new Windows platform will present a range of technical difficulties and begs a number of questions about whether the development work using new technologies and new approaches should be undertaken in the existing host environments (at Manchester and Edinburgh) or in the new host environment (at Essex). If the former approach is chosen, then the updated services might be transferred to an intermediary server at Essex for testing and configuration before final release. Alternatively, it might be appropriate to create a virtual machine that would be used for development and configuration before being transferred to the new environment for operationalisation by technicians at Essex who would undertake to prepare the new system for the existing setup.

Distinct from the issues of a technical nature are the cultural differences that exist between the organisations involved and the procedures and protocols associated with the way in which projects are planned and carried out. The transfer of each census service may well be an independent 'project' that will adopt the existing ESDS/Data Archive project management procedures and fit within the hierarchical governance structures of the new UK Data Service. The basic project requirements are founded on best practice and principles set out in PRINCE2 in order to assure successful initiation (involving preparation of a Project Initiation Document or PID), monitoring of progress, quality assurance and product delivery. Each UKDS-CS project will require formal approval by the UK Data Service senior management team so as to ensure that technical service resources at Essex are available to incorporate UKDS-CS services within the common interface that is being proposed. In the first instance, the UKDS-CS services will be 'reskinned' with minimal change to the underlying code in each case. In addition to the technical challenges associated with moving the UKDS-CS software and data to a new environment, there will also be a requirement for UKDS-CS services to meet the data ingest and preservation requirements of the UK Data Service using a standard set of procedures.

One thorny issue that arises when projects involve collaboration and partnership is that of the ownership of intellectual property rights. In the case of the UKDS-CS, this is complicated because of the background intellectual property associated with the software development work undertaken over the previous decade that is owned by the individual institutions involved. Thus, a Collaboration Agreement has been drawn up between the parties involved stating that each party has to grant the other parties royalty-free, non-exclusive licence for the duration of the current 'project' to use its background intellectual property for the sole purpose of carrying out the 'project'. Moreover, the role of the funding agency is important here; intellectual property rights (IPR) arising from the 'project' are governed by the conditions set out by the ESRC who will expect to have the right to 'use' the IPR in any products generated through ESRC grant funding for the UK Data Service, which includes the value added services including the UKDS-CS. The ESRC has to have these rights in order to ensure continuity of the services beyond the end of the current 'project' and to grant the successor licenses to the relevant IPR were the successor not to be the current partner institutions.

One important change in the service provision involves the potential opportunity to open up of the services to allow any user in the world to have access to the data sets available under OGL. The provision of open data is something that the UKDS-CS is eager to embrace but transition to an open service is complicated because of the two issues in particular. The first of these is the licensing arrangements associated with several of the data sets associated with previous censuses; a Concordat between the ONS and the UK Data Service has been drafted in which the ONS grants rights to the UK Data Service to distribute ONS data under the various licensing arrangements and it is the intention for this Concordat to be replicated for NRS and NISRA. The second is the reduction in the information recorded by the current registration and monitoring procedures that provide a fairly detailed picture of who the users are, which institutions they belong to and which services and data sets they are using. This level of detail is valuable in reporting usage or performance statistics to the ESRC and making judgements about the impacts that the services are making on the academic community.

A first step towards open data is for the ESRC to agree that the services developed under the Census Programme that constitute the CS can be utilised to provide open access to OGL data holdings. This decision has been made and will pave the way for the technical exercise required to actually provide open access to OGL data whilst retaining necessary registration procedures for enabling access to data available under more stringent license conditions. The Concordat refers to three levels of dissemination: public access through OGL; safeguarded access where researchers can use the data under either an end user licence (EUL) or a special user licence (SUL); and controlled access where access is only possible via a secure data service (SDS) or a virtual microdata lab (VML). Whilst it has been widely reported that OGL applies to 2001 Census, the disclosure control methods used to adjust the counts unfortunately mean the likelihood that certain data sets, e.g. multivariate interaction data, will only be available via safeguarded or controlled access, whereas the equivalent data in 2001 was in the public domain.

Finally, the development of a UKDS-CS integrated access tool is a specific technical challenge which will provide benefits to users by making it easier for them to find, understand and make use of census information of relevance to their needs. It will more easily facilitate cross-product working, including multilevel models. In fact, a unified model for the 2001 aggregate outputs has already been developed as part of the CAIRD projects at MIMAS, and work is already underway to develop a model for the 2011 aggregate outputs. An initial objective will be to extend the 2011 model to incorporate the 2011 interaction outputs using existing structures shared with the aggregate outputs with additions specific to the interaction outputs, including twin origin and destination versions of the geography structures. It will be possible to incorporate 2011 boundary outputs into the model as attributes of the geography structures. Further work will then be necessary to develop models for the 1991, 1981 and 1971 Census data sets, and then to effectively 'merge' all of the conceptual models using shared concepts and structures where they exist, as well as measures of similarity where matches are not exact. The result of this work will be a single, unified conceptual model which comprehensively and consistently describes all of the outputs from the 1971–2011 Censuses, together with all published metadata and additional metadata to quantify the degrees of similarity between different values.

In conclusion, there is much work to be done over the next 5 years and there is considerable uncertainty surrounding many of the tasks that need to be accomplished. However, there is much less doubt that the creation of a new UK Data Service embracing the census services that have become such an important part of the UK 'access to data' landscape over the past decade provides a huge opportunity to construct a data discovery and extraction service that will provide many benefits to future generations of social science researchers whose interests will include urban sustainability.

Acknowledgments Census output is Crown copyright and is reproduced with the permission of the Controller of HMSO and the Queen's Printer for Scotland. The Census Programme has been funded by the Economic and Social Research Council (ESRC) and the UK Data Service—Census Support is currently being funded by the ESRC.

References

Cole K, Wathan J, Corti L (2008) The provision of access to quantitative data for secondary analysis. In: Fielding N, Lee RM, Blank G (eds) The sage handbook of online research methods. Sage, London

Duke-Williams O, Stillwell J (2007) Investigating the potential effects of small cell adjustment on interaction data from the 2001 Census. Environ Plann A 39(5):1079–1100

Finney N, Simpson L (2008) Internal migration and ethnic groups: evidence for Britain from the 2001 Census. Popul, Space Place 14(2):63–83

Norman P (2010) Identifying change over time in small area socio-economic deprivation. Appl Spat Anal Policy 3:107–138

ONS (2011) Beyond 2011, Available at http://www.ons.gov.uk/ons/about-ons/what-we-do/programmes—projects/beyond-2011/index.html

ONS (2012) Response rates in the 2011 census, 2011 census methods and quality report. ONS, December

ONS (undated) A beginners guide to geography, Available at http://www.ons.gov.uk/ons/guide-method/geography/beginner-s-guide/index.html

Ryssevik J, Musgrave S (2001) The social science dream machine: resource discovery, analysis and delivery on the web. Soc Sci Comput Rev 19(2):163–174

Science and Technology Committee (2012) Third report the census and social science, Available at http://www.publications.parliament.uk/pa/cm201213/cmselect/cmsctech/322/32202.htm

Simpson L, Finney N (2009) Spatial patterns of internal migration: evidence for ethnic groups in Britain. Popul, Space and Place 15(1):37–56

Stillwell J (2010a) Internal migration propensities and patterns of London's ethnic groups. In: Stillwell J, Duke-Williams O, Dennett A (eds) Technologies for migration and commuting analysis: spatial interaction data applications. IGI Global, Hershey, pp 175–195

Stillwell J (2010b) Ethnic population concentration and net migration in London. Environ Plann A 42:1439–1456

The National Archives (2011) UK government licensing framework for public sector information, Available at http://www.nationalarchives.gov.uk/information-management/government-licensing/the-framework.htm

Tranmer M, Steel DG (1998) Using census data to investigate the causes of the ecological fallacy. Environ Plann A 30(5):817–831

The Online What if? Planning Support System

Christopher J. Pettit, Richard E. Klosterman, Marcos Nino-Ruiz, Ivo Widjaja, Patrizia Russo, Martin Tomko, Richard Sinnott and Robert Stimson

Abstract The chapter introduces the Online What if? (OWI) GIS-based planning support system, which is being made available through the Australian Urban Research Infrastructure Network (AURIN). AURIN has been established to provide an advanced information infrastructure to support discipline-specific and multi-disciplinary research and promote sustainable urban development in Australia. OWI is an open source online version of the widely used desktop What if? planning support system developed by Klosterman (1999). OWI enables a range of end users to create and explore what if? land use change scenarios. This chapter discusses OWI in the context of a demonstrator case study in Hervey Bay,

C. J. Pettit (✉) · P. Russo · M. Tomko · R. Stimson
Faculty of Architecture Building and Planning, University of Melbourne,
Victoria 3010, Australia
e-mail: cpettit@unimelb.edu.au

P. Russo
e-mail: russop@student.unimelb.edu.au

M. Tomko
e-mail: tomkom@unimelb.edu.au

R. Stimson
e-mail: rstimson@unimelb.edu.au

R. E. Klosterman
What if? Incorporated, Hudson, OH, USA
e-mail: dick.klosterman@gmail.com

M. Nino-Ruiz · I. Widjaja · R. Sinnott
Department of Computing and Information Systems, University of Melbourne,
Victoria 3010, Australia
e-mail: marcosnr@unimelb.edu.au

I. Widjaja
e-mail: ivow@unimelb.edu.au

R. Sinnott
e-mail: rsinnott@unimelb.edu.au

S. Geertman et al. (eds.), *Planning Support Systems for Sustainable Urban Development*, Lecture Notes in Geoinformation and Cartography,
DOI: 10.1007/978-3-642-37533-0_20,

Queensland, and introduces future applications of this collaborative planning tool to support the sustainable planning of cities in Australia.

1 Introduction

Dramatic advances in the integration of eResearch infrastructure (also known as cyberinfrastructure) and geographic information systems (GIS) promise to fundamentally improve public sector planning and policy making in an increasingly complex and urbanised world. They will do this by providing understandable and easy to use on-line tools that will allow specialist users (planners, public officials, institutional stakeholders, and researchers), as well as private citizens to quickly and easily evaluate the potential effect of alternative public policies on the sustainable future growth of cities and regions. In particular, the migration of specialist tools into the Web environment reduces the obstacles faced by potential users and thus reduces the threshold of access in terms of cost (hardware, software and data licenses), and enables collaborative analysis.

The paper describes the development and applications of the online What if? planning support system (OWI) that is being developed as a component of the Australian Urban Research Infrastructure Network (AURIN—www.aurin.org.au) project. The core component of AURIN is a rich online eResearch Infrastructure platform that is being developed to provide urban researchers, planners and policy makers' access to diverse sources of data, data integration capabilities, and advanced capabilities for analysing and visualizing those data through a user-driven Web-based environment offering a one stop shop lab-in-a-browser. The project will facilitate research activity in order to enhance our understanding of key issues relating to the sustainable development of Australia's national settlement system and improve the evidence base for informed public policy making and business operations. This is particularly important as the nation debates the implications of population growth, how the population will be distributed across space, the urban environments in which people will live, and how the nation can promote the sustainable development of its cities and towns (AURIN EIF Final Project Plan)AURIN 2011.

OWI is an innovative, open source GIS-based planning support system (PSS) that can be used to prepare conditional what if? scenarios for an area's future land use, population and employment patterns. It is being developed by porting the widely used desktop What if? system (Klosterman 1999, 2008) into the AURIN environment. The OWI tool builds on a state-of-the-art eResearch Infrastructure realised through a service-oriented architecture (SOA) that offers seamless access to diverse data sets, with advanced user-interface capabilities and transparent access to high-performance computational resources. OWI is one of a broad range of advanced computational and visualization capabilities for supporting urban research and planning that are being integrated within the AURIN environment.

This chapter begins by describing how OWI aims to overcome the limitations of current GIS for planning. It then introduces the technical architecture of the system and its integration within the wider AURIN e-infrastructure. It then briefly describes the application of OWI in the context of a previous research application in Hervey Bay, Queensland. The paper concludes by discussing OWI and its potential for supporting sustainable urban planning in Australia.

2 Background

2.1 From GIS to Planning Support Systems

The ability of GIS to combine easy-to-understand maps, standard database operations, and sophisticated spatial analysis tools has proven to be extremely useful for a wide range of scientific, public-, and private-sector applications. However, the general applicability of current GIS has not met the needs of particular users such as public sector planners and urban researchers. Rather, GIS has traditionally been used almost exclusively to analyze current and/or past conditions and has not addressed planners' unique concerns with helping public and private decision-makers deal with an uncertain future. The limitations of current GIS to meet the particular needs of planners has led to the development of a large number of planning support systems (PSS) which combine the spatial analysis and display capabilities of GIS with specialized forecasting and visualization tools that serve the needs of planners (see, for example, Brail and Klosterman 2001; Geertman and Stillwell 2003, 2009; Brail 2008.

PSS have been developed and applied literally around the world. Most PSS are academic prototypes or "one off" professional applications but a small number of systems such as Community Viz (http://placeways.com/communityviz), Sleuth (Clarke 2008), UrbanSim (Waddell 2002, 2011), What if? (Klosterman 1999, 2008), and, in Australia, the LSUM-SEQ (Large Scale Model for the Brisbane-South East Queensland Region) (Stimson et al. 2012) have gained a level of acceptance and uptake. However, a number of fundamental problems have limited the use of PSS in academic research and professional practice (Vonk et al. 2005). A particularly important obstacle to widespread PSS use is the difficulty of implementing and accessing current PSS which have generally only been available as proprietary desktop applications. OWI directly addresses this limitation by providing any time any place online access to a state of the art PSS that users can freely use to access a rich variety of data and create a range of what if? scenarios to meet their own particular needs.

2.2 Online What if? Description

As its name suggests, OWI does not attempt to predict future conditions exactly. Instead, it is a policy-oriented planning support tool that can be used to determine *what* could happen *if* clearly defined policy choices are made and assumptions concerning the future prove to be correct. Policy choices that can be considered in the model include the staged expansion of public infrastructure and the implementation of alternative land use plans or zoning ordinances. Assumptions for the future that can be considered in the model include future population and employment trends, assumed household characteristics, and anticipated development densities.

OWI is a bottom-up growth allocation model which starts with vector overlay-generated homogeneous land units, applies alternative policy choices to these units, allocates projected land use demands to them, and then derives small area land use, population and employment projections by aggregating the values for these land units. The model projects future land use patterns by balancing the supply of, and demand for, land suitable for different uses at different locations. Alternative visions for an area's future can be explored by defining alternative suitability, growth, and allocation scenarios.

OWI does not include measures of spatial interaction, represent the interlinked markets for land, labor and infrastructure or model the behavior of actors such as households, businesses, and developers. Instead, it provides an easy-to-use and understand GIS-based policy-oriented model that can accommodate the spatially-referenced data and policy choices for a particular study area, providing a readily available foundation for community dialog and decision making.

OWI augments current efforts such as public participation GIS (PPGIS) and volunteered geographic information (VGI) which attempt to use the power of GIS to empower private citizens and marginalized groups (see, for example, http://www.ppgis.net/). It does this by, for the first time, allowing urban professionals, public officials and members of the public to use the Web to explore what if? scenarios which describe alternative futures under different assumptions about the future and alternative public policy choices.

2.3 eResearch Infrastructure: The Way of the Future

Development of an online PSS has much to benefit from advances in eResearch infrastructure. Seamless and secure access to digital data, tools and computational resources has been at the fulcrum of major efforts in the UK through the UK e-Science Core Program, in the US through the cyberinfrastructure initiative and in a range of European-wide efforts such as the Enabling Grids for e-Science (EGEE). The Australian Government has also made considerable investments in this area as part of its Aus$1bn Education Investment Fund Super Science

Initiative. Major programs include the Aus$47 m National eResearch Collaboration Tools and Resources (NeCTAR—www.nectar.org.au) project, the Aus$50 m Research Data Storage Initiative (RDSI—https://rdsi.uq.edu.au) project, and the Aus$20 m AURIN—http://aurin.org.au/) project (amongst many others).

The AURIN eResearch infrastructure has been established to support the urban research community across Australia by providing better access to data, interrogation and visualization tools. AURIN is end-user driven and includes ten 'aspirational' strategic implementation streams or 'lenses', which relate to issues of national urban policy priority. The lenses are: (1) population and demographic futures and benchmarked social indicators; (2) economic activity and urban labour markets; (3) urban health, wellbeing and quality of life; (4) urban housing; (5) urban transport; (6) urban water and energy supply and consumption; (7) city logistics; (8) urban vulnerability and risks; (9) urban governance, policy and management; and (10) innovative urban design.

AURIN is providing online access to 2001, 2006, and 2011 census data provided by the Australian Bureau of Statistics (ABS) and a number of census-derived data products including: location quotients, demographic and social-economic percentages, and shift-share statistics. AURIN is also providing access to the national cadastral boundaries and geocoding facilities made available via the Public Sector Mapping Agency (PSMA), along with crowd-sourced Twitter data. These are seamlessly delivered through a one-stop shop web-based portal offering access to distributed (federated) data sets with accompanying analytical tools such as OWI, a Walkability tool, and a library of spatial statistical routines. Once users have completed their analysis they are able to visualize their results as choropleth maps, graphs and charts. Security plays a key role in this and the AURIN platform has been fully integrated into the Australian Access Federation (www.aaf.edu.au), which provides decentralized (federated) access control, whereby end users are able to access distributed resources available through AURIN through authentication at their home institution.

3 Technical Architecture

As shown in Fig. 1, OWI has been developed as a core component of the AURIN infrastructure. The AURIN eResearch infrastructure is accessible through the single entry point—the AURIN portal. This portal provides access to a wide variety of heterogeneous authoritative data from a variety of data sources, and provides the ability to combine, analyze and visualize these data within the portal, or download them if users have permission to do so.

The AURIN internal technical architecture is based on a loosely-coupled design that includes a collection of independent web services communicating through documented Application Programming Interfaces (API). Internal AURIN services are mostly communicating through REST-based APIs, but the architecture supports also Open Geospatial Consortium (OGC)-compliant and SOAP services.

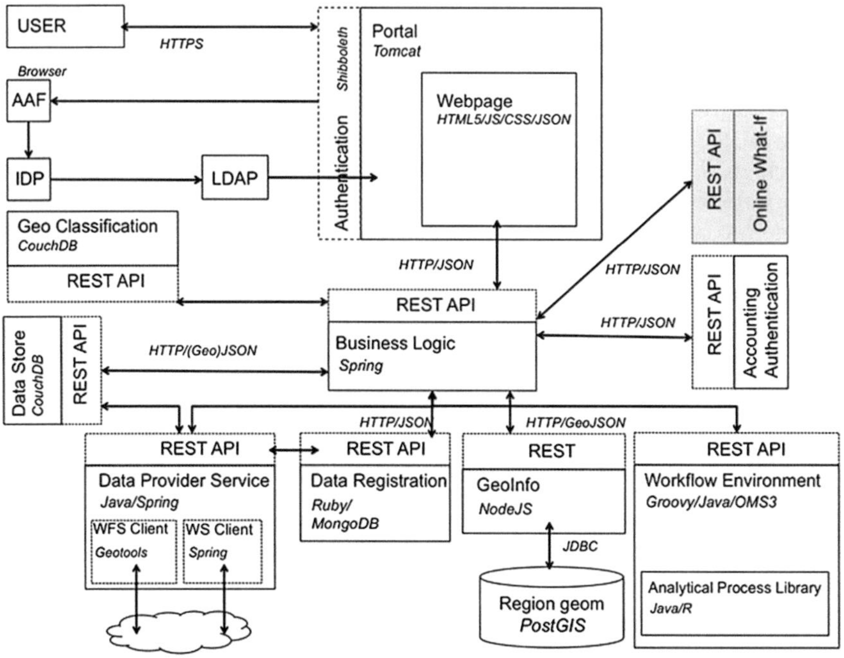

Fig. 1 The high-level schema of the AURIN technical architecture, with the OWI functionality integration highlighted in *grey* (adapted from Sinnott et al. 2012)

This architecture enables AURIN to flexibly respond to the evolving requirements of the project and integrate data and systems from diverse sources. The ability to adapt to technological changes during the lifetime of the project and the long-term maintainability of the platform were among the key requirements for implementing the AURIN framework.

AURIN's infrastructure supports the discovery and acquisition of data from federated data sources and allows user-based data upload and download (subject to security based on data custodian-defined rules). The data are internally stored in the AURIN Data Store, which offers persistent storage. Interim data results, such as those generated during the testing of multiple OWI scenarios can also be stored in the Data Store. The data stored in AURIN can also be protected and have associated restrictions on their use. The user interface enables interactive visualization of data, enhanced by advanced cartographic brushing capabilities Pettit et al. (2012). A workflow environment facilitates complex data analysis and processing such as the GIS analysis required to implement the OWI model. More complex analytical tools, including advanced user interface interaction modalities and data processing functions, such as OWI, are exposed as self-contained applications.

The AURIN infrastructure needs to ensure system scalability, enforce component isolation, provide communication through publicly documented interfaces, and support a heterogeneous software stack that adapts to the technological requirements of the individual components.

AURIN leverages the support of high-performance computing infrastructure, such as the Cloud-enabled computing environment of the sister NeCTAR project. This enables computationally intensive applications, such as the analysis of OWI scenarios for large regions, to run efficiently, providing better response times than is possible with a desktop tool. It also reduces the barriers to access for researchers, policy makers and ultimately members of the public who do not have computational infrastructure support.

4 The Online What if? Technical Architecture and Software Component

Re-engineering the existing desktop What if? tool as a core module of the AURIN eResearch infrastructure required recoding the existing desktop implementation into a web-based, service-oriented environment. The implementation has reused the standard GIS visualization components implemented in the main AURIN infrastructure to provide a user-friendly experience. The OWI model has been coded entirely with open source software and will in due course be available under an open source license.

Figure 2 illustrates how the various OWI sub-modules interact with each other. In this architecture, when the user requests a What If? analysis within the AURIN portal. The AURIN environment delegates this request to the What If? Controller, which performs a REST request with the What If? information that is acknowledged by the What If? Analysis Services, where the algorithm processing takes place. The Analysis Services can be located physically on any server. After the appropriate information is obtained, this service processes the appropriate What If? Suitability, Demand, or Allocation Service. When the analysis is completed,

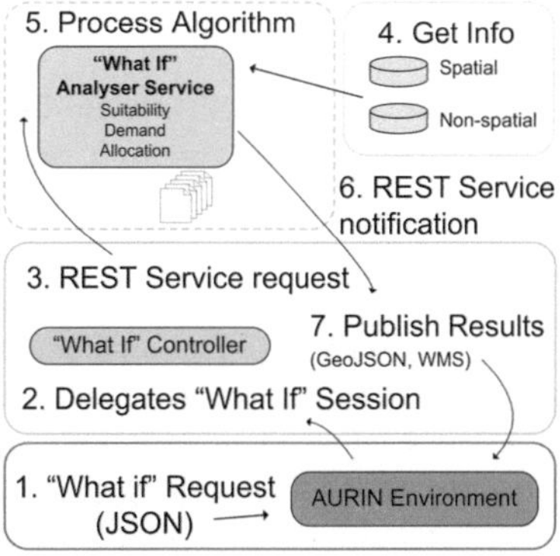

Fig. 2 AURIN online What If? module architecture

this module notifies the What If? Controller (Geertman and Stillwell 2009), which then informs the AURIN environment that the analysis results are ready to be shown in the view layer (Klosterman 1999).

4.1 Implementation Strategy

The OWI PSS uses core Java Enterprise Edition (version 5.0) APIs to access the required geospatial and non-geospatial information. The persistence technology used for the OWI model is integrated with the AURIN portal (noSQL CouchDB and postGIS database). All of the model entities are stored in the database as business objects, which comply with the database guidelines for ACID (atomic, consistent, isolated and durable) transactions. By using the AURIN data store standard connectors, the persistence layer provides direct access to any database, through a standard JDBC library, or the REST APIs to access noSQLs storage services. This flexibility allows most of the AURIN capabilities to be quickly and easily migrated to any major database implementation in the future.

4.2 GeoTools—The 2D Spatial Data Handling API

The geospatial API used for the OWI model is the Java-based GeoTools Open Source (LGPL) library (http://www.geotools.org/) which provides extensive modules to support spatial data access, manipulation and remote interrogation in compliance of the Open Geospatial Consortium (OGC) specifications (in particular the Simple Feature Specification compliant data models). This allows OWI to transfer information between database objects and create spatial features stored and analysed in Java. The implementation also leverages the Java Topology Suite library (JTS), abstracting the geometry functions in a numerically robust, scaleable and performance manner.

At the core of the analysis lies a multi-criteria weight based comparison between diverse attributes that share the same spatial geometry. This analysis is straightforward thanks to the FeatureType interface provided by GeoTools. The suitability analyzer component factors in all the relative importance set up by the user simultaneously for each spatial unit/feature, and then produces a suitability score that is stored with the proper spatial reference in a GeoTools DataStore. This very flexible interface allows querying for relevant features in the complex allocation stage of the analysis. Using another flexible module, the GeoTools Filter and Expression interface, the analyser uses the common query language (CQL) to perform the relevant sorting and filtering according to allocation controls at each projection year, thus being able to produce several what if? outcomes efficiently without having to reload the complex setup and geospatial information.

4.3 Persisting Spatial Data and Scenario Configurations

Two types of data are accessed by the OWI application in order to analyse the land use scenarios: spatial data themselves, and the configuration of the scenario, as set-up by the expert planners or during community consultation.

Raw spatial data that are used by OWI include information about land-use, population, employment, terrain characteristics, and others. These conflicting data layers are processed (usually offline, in a desktop GIS software) in an overlay process resulting in homogenous land units, also called Uniform Analysis Zones (UAZ) file. Each UAZ contains all the relevant information necessary to perform a what if? analysis, ensuring that all features will be properly analysed and the outcomes will be consistent. In summary, this is a spatial dataset that is homogeneous with respect to all of the factors considered in the model (Klosterman 2007).

The configuration information is required for the interpretation of the UAZ file's attributes and the classification of their values, and defining scenario and analysis options. This setup information is encoded in a Java Script Object Notation (JSON) object and persisted in the Business Logic layer of the AURIN environment as illustrated in Fig. 1. For example, a land suitability class is defined as a JSON object. These user-defined objects will be parsed, persisted and translated into a land use suitability entity that will be attributed a projection and geometry to constitute a GeoJSON object. This GeoJSON can then be understood by GeoTools for performing geoprocessing, such as spatially overlay driven land suitability analysis. This approach in formulating (Geo)JSON objects is also undertaken in setting up the OWI Demand and Allocation routines—see Fig. 2.

4.4 User Interface

Figures 3, 4, 5, 6 show the AURIN portal and OWI interfaces for analysing the suitability of different locations for accommodating different land uses and viewing the resulting suitability scenario maps for Hervey Bay. Figure 3 shows the AURIN portal interface and mapping window, which highlights the Hervey Bay study area and can be used to show different input data layers. Figure 4 demonstrates the procedure for selecting a previously defined suitability scenario or creating a new one. Figure 5 shows the interface for specifying the weights for different suitability factors (e.g., slopes and prime agricultural soils), the rating for each factor type (e.g., different slope values), and the land uses that can be converted to other uses to accommodate future land use demands. Figure 6 shows the suitability maps for one suitability scenario and the system's ability to display multiple suitability scenarios.

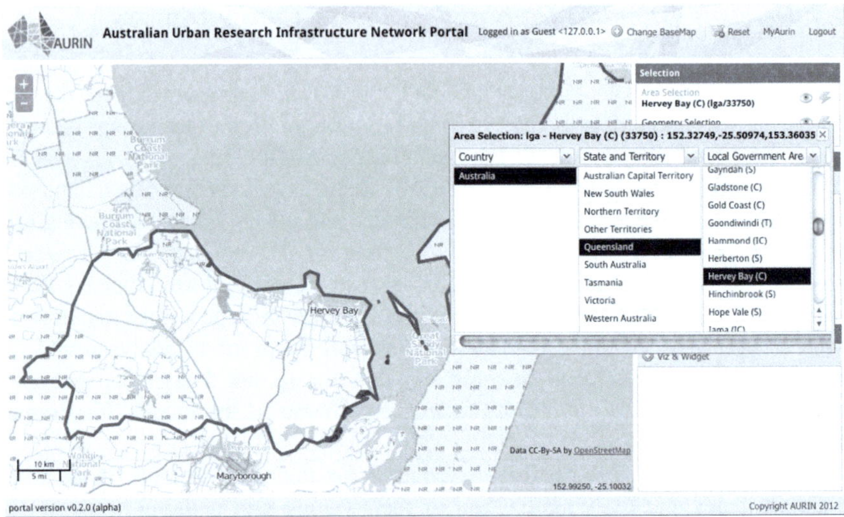

Fig. 3 AURIN portal interface showing selection of the Hervey Bay case study area

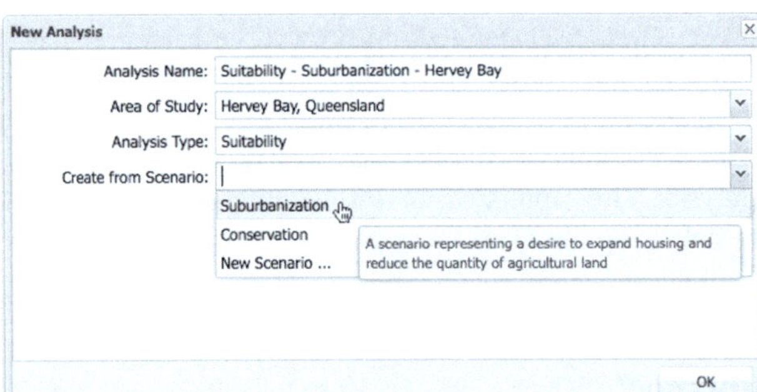

Fig. 4 Suitability scenario selection form

5 Case Study Demonstrator

The data set for Hervey Bay, Queensland, a previous application of the What if? model (Pettit 2005), is being used to develop the initial OWI application and provide an exemplar for the model implementation. The model is also being implemented in Perth, which is experiencing rapid urban growth due to the Australian mining boom, and in Townsville, which is one of first twenty-four cities in the world to be involved in the IBM Smarter Cities Challenge.

Hervey Bay, including Fraser Island (a World Heritage area), is a popular destination for tourists along the east coast of Australia because of its natural amenities and traditional rural characteristics. Furthermore, Hervey Bay is characterised by strong population growth and high levels of unemployment. As a

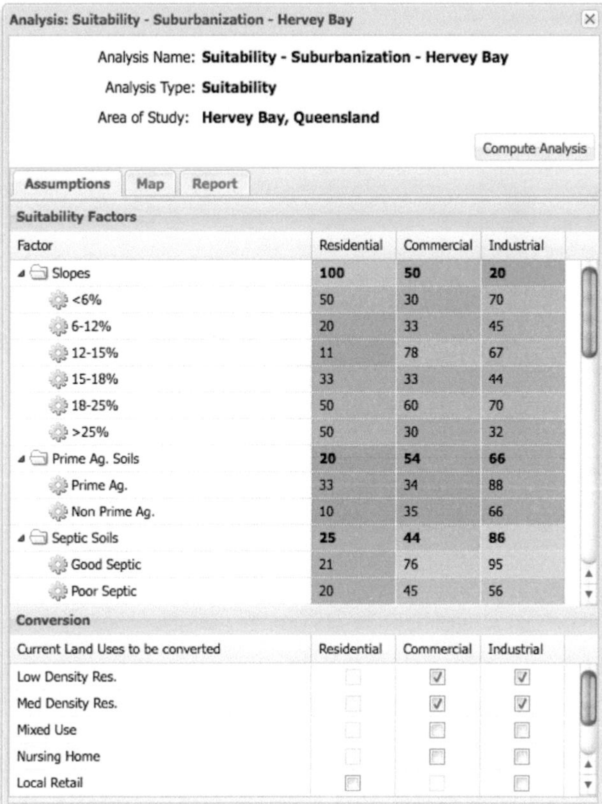

Fig. 5 Suitability assumptions specification form

result, the local government faces a need to promote future growth while preserving the area's environmental quality and tourism industry planning. A number of land use scenarios were initially developed to explore the issue of competing land uses (Pettit and Pullar 2004). The 'sustainable development' driven scenarios initially generated in the What if? PSS (Pettit 2005) have been re-implemented in OWI.

The OWI application allows the planners, policy-makers and citizens of Hervey Bay to examine scenarios and projections that urban planning experts have generated and made freely accessible via the AURIN portal. More importantly, it permits stakeholders and citizens in Hervey Bay to access the AURIN portal and use OWI to develop their own scenarios and determine the implications that their assumptions and forecasts may have for their community. The tool also increases their understanding of the land use development process and the complex public policy issues that land use experts confront.

The AURIN OWI application has initially been set-up with the Hervey Bay data so that the re-coding of the desktop What if? model can be validated with a

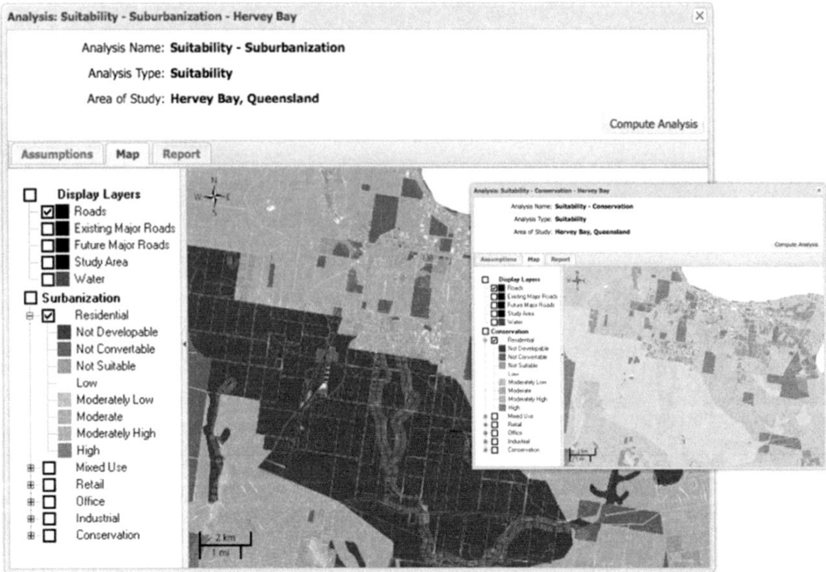

Fig. 6 Suitability map for suitability scenario with functionality to view and compare multiple scenarios

known case study example. This allows the OWI set-up, suitability, demand and allocation modules to be tested against previously published sustainable urban development scenarios (Pettit 2005). Future applications of OWI will examine the likely implications of greenfield urban development around the City of Perth in Western Australia and the future growth in and around the city of Townville in North East Queensland, which is considering novel ways to promote sustainable urban development. OWI can help improve planning in Australia by providing a framework for bringing together social, economic, environmental and physical datasets, and support more evidenced-based decision-making. Analyzing these data in a GIS-based planning support system will also allow what if? scenarios for alternative sustainable futures to be explored by policy-makers, planners, stakeholders and citizens throughout Australia.

6 Conclusions

The AURIN OWI application offers several benefits for planners, public officials and private citizens. For the first time, it provides direct online access to a state of the art PSS that allows for both professionals and non-professionals to create and view alternative scenarios for their communities. Making the software systems available online from a single source will both encourage uptake, and leverage the integration and extensions of the system together with state of the art development tools and processes available through AURIN. This includes for example feature and bug tracking, software updates and patches, and configuration management and deployment support (Sinnott 2013). The software will be released under an open source license which will allow it to be adopted elsewhere.

However, there are also challenges in adopting an online open source approach. OWI is being developed as an open source tool, which means that its underlying logic and algorithms can be freely extended over time by anyone. However, OWI is part of a complex AURIN technical architecture that utilizes a number of dependencies on different system components and libraries. This means that implementing OWI outside the AURIN eResearch infrastructure will require substantial software engineering experience, which is not traditionally the strength of planners and GIS experts.

Nevertheless, OWI illustrates the potential that advanced eResearch infrastructure can provide for helping planners, decision-makers and private citizens explore alternative futures for cities and regions around the world. Further research and development will see the OWI applied in exploring sustainable urban futures in Australia and these results and the lessons will be reported in due course.

Acknowledgments The authors would like to thank the AURIN technical team who are building the AURIN portal infrastructure, which underpins the OWI application. The authors also acknowledge the Australian Government, Department of Industry, Innovation, Science, Research and Tertiary Education who are funding the AURIN project through the Australian Education Investment Fund.

References

AURIN (2011) AURIN EIF final project plan, Apr. Available on www.aurin.org.au

Brail RK, Klosterman RE (eds) (2001) Planning support systems: integrating geographic information systems, models and visualization tools. ESRI Press, Redlands

Brail RK (ed) (2008) Planning support systems for cities and regions. Lincoln Institute of Land Policy, Cambridge

Clarke KC (2008) A decade of cellular urban modeling with SLEUTH: unresolved issues and problems. In: Brail RK (ed) Planning support systems for cities and regions. Lincoln Institute of Land Policy, Cambridge

Geertman S, Stillwell J (eds) (2003) Planning support systems in practice. Springer, Berlin

Geertman S, Stillwell J (eds) (2009) Planning support systems: best practices and new methods. Springer, New York

Klosterman RE (1999) The what if? collaborative planning support system. Environ Plan B: Plan Des 26:393–408

Klosterman, RE (2007) What if?™ 2.0 user's guide. What If? Inc, Ohio, USA

Klosterman RE (2008). A new tool for a new planning: The What If?™ planning support system. In: Brail RK (ed) Planning support systems for cities and regions. Lincoln Institute of Land Policy, Cambridge

Pettit CJ (2005) Use of a collaborative GIS-based planning support system to assist in formulating a sustainable-development scenario for Hervey Bay, Australia. Environ Plan B: Plan Des 32(4):523–546

Pettit C, Pullar D (2004) A way forward for land use planning to achieve policy goals using spatial modeling scenarios. Environ Plan, B: Plan Des 31:213–233

Pettit C, Widjaja I, Russo P, Sinnott R, Stimson R, Tomko M (2012) Visualisation support for exploring urban space and place, XXII ISPRS Congress, Technical Commission IV 25 Aug – 01 Sept 2012. Melbourne, Australia Editor(s): M Shortis, J Shi, E Guilbert, ISPRS Annals 1–2:153–158

Sinnott RO, Bayliss C, Galang G, Greenwood P, Koetsier G, Mannix D Morandini L, Nino-Ruiz M, Pettit C, Tomko M, Sarwar M, Stimson R, Voorsluys W, Widjaja I (2012) A data-driven urban research environment for Australia. IEEE e-Science Conference. Chicago, Illinois: IEEE

Sinnott RO, Bayliss C, Morandini L, Tomko M (2013) Tools and processes to support the development of a national platform for urban research: lessons (being) learnt from the AURIN project. 11th Australasian symposium on parallel and distributed computing, Jan 2013

Stimson R, Bell M, Corcoran J, Pullar D (2012) Using a large scale urban model to test planning scenarios in the Brisbane-South East Queensland region. Reg Sci Policy Pract 4:372–392

Vonk G, Geertman S, Schot P (2005) Bottlenecks blocking widespread usage of planning support systems. Environ Plan, A 37:909–924

Waddell P (2002) UrbanSim: modeling urban development for land use, transportation and environmental planning. J Am Plan Assoc 68:297–314

Waddell P (2011) Integrated land use and transportation planning and modeling: addressing challenges in research and practice. Transp Rev 31:209–229

A Web-Based Fuzzy CA Model for Urban Growth Simulation

Yan Liu

Abstract Cellular automata (CA) as an emerging technology has been adapted increasingly by geographers and planners to simulate the spatial and temporal processes of urban growth. While the literature reports many applications of cellular automata models for urban studies, in practice, the operation of the models and the calibration of the parameters in use were only known to the modellers. This is largely due to the constraint that most CA models were developed based on desktop computer programs, either by incorporating the model within a desktop GIS environment, or developing the model independent of a desktop GIS. Consequently, there is little input from the user to test or visualise the actual operation or evaluate the applicability of the model under different conditions. This chapter presents a methodology to implement a fuzzy constrained urban CA model within a web-based GIS environment using Gold Coast City in Southeast Queensland, Australia as a case study to simulate its spatio-temporal processes of urban growth. With the web-based CA model, users can visualise and test the operation of the model; they can modify or calibrate the model's parameters and evaluate its simulation accuracies, or even feed the model with 'what-if' conditions to generate alternative outcomes. This web-based modelling platform provides a useful channel to foster public participation in urban planning and management.

1 Introduction

Cellular automata as an emerging technology has been adapted increasingly by geographers and planners to simulate the spatial and temporal processes of urban growth (White and Engelen 1993; Couclelis 1997; Batty 1997; Batty et al. 1997,

Y. Liu (✉)
School of Geography, Planning and Environmental Management, The University of Queensland, St. Lucia Campus, Brisbane QLD 4072, Australia
e-mail: yan.liu@uq.edu.au

S. Geertman et al. (eds.), *Planning Support Systems for Sustainable Urban Development*, Lecture Notes in Geoinformation and Cartography,
DOI: 10.1007/978-3-642-37533-0_21, © Springer-Verlag Berlin Heidelberg 2013

1999; Torrens and O'Sullivan 2001; Liu 2008). Various cellular automata models have been developed and applied as an analytical tool to explore the driving forces of urban growth and, or as a projection tool to foresee future development of a region (White et al. 1997; Wu 1998, 2002; Wu and Webster 1998; Yeh and Li 2002; Liu and Phinn 2003; Ward et al. 2003; Stevens and Dragicevic 2007; Al-kheder et al. 2008; Almeida et al. 2008; He et al. 2008; Vancheri et al. 2008; Yang et al. 2008; Al-Ahmadi et al. 2009; Vliet et al. 2009; Pinto and Antunes 2010; Feng et al. 2011; Li et al. 2011; Mitsova et al. 2011; Tang 2011; García et al. 2012; Liu 2012; Feng and Liu 2012).

While the literature reports many studies on the development and applications of the CA models for urban growth simulation, with convincing results demonstrating the usefulness of the models, in practice, the operation of the models and the calibration of the parameters used in the models were only known to the model developers. This is largely due to the fact that most cellular automata models were developed based on desktop computer programs, either by incorporating the model within a desktop GIS environment, or developing the model independent from a desktop GIS. Consequently, there is little input from the user to test or visualise the actual operation of the model and evaluate the applicability of the model under different conditions.

The advance of web-based GIS technology has made it possible to develop simulation models over the Internet. With web-based GIS services, users can visualise and test the operation of the model; they can also modify or calibrate the model's parameters to evaluate its simulation accuracies, or even feed the model with 'what-if' conditions to generate alternative outcomes. Such web-based urban modelling platform can provide a useful channel for both government and stakeholders to evaluate different urban growth scenarios. It can also provide an interactive environment to foster public participation in urban planning and management.

This chapter builds on the research by Liu (2012). It presents a methodology to implement the fuzzy constrained CA model of urban growth within a web-based GIS environment using Gold Coast City in Southeast Queensland, Australia as a case. The following section presents the fuzzy constrained CA model and its application to simulate urban growth dynamics. Section 3 discusses the implementation of the model on the web using ESRI's ArcGIS Server technology. Section 4 concludes the current research and identifies future research directions.

2 The Fuzzy Constrained CA Model

A fuzzy constrained CA model was developed in previous studies to simulate the spatio-temporal process of urban growth (Liu 2008, 2012). The model considers urban development as a fuzzy process which is constrained by both internal and external forces. Internally, an area tends to continue its development once it has started to develop from rural to urban land use, especially if this growth tendency is also driven by development from within its neighbourhood. Externally, factors such as the physical settings of the area, its socio-economic conditions as well as the institutional controls also affect the process of its development. For instance, the physical constraints such as water bodies and steep terrain may restrict or slow down the process of urban development, as do the institutional controls that may accelerate or prohibit further urban growth. The socio-economic factors such as land availability and the demand on available land, the accessibility to nodes of employment and other services and facilities such as schools, shops, public transport and the contiguity to existing urban areas also play important roles in driving or restricting urban development. Such internal and external forces can be represented in the CA model through a set of primary and secondary transition rules (Liu 2008, 2012).

2.1 Primary Transition Rules

Based on the principles of cellular automata, the internal forces driving the process of urban development can be represented by a set of primary transition rules. That is, at a homogeneous urban space with no variation in either the physical conditions or socio-economic or institutional constraints, if a cell has the propensity to develop into an urban state and it can receive sufficient support for such development from its neighbouring cells, the cell will develop at a certain speed towards the urban state. Such a development process can be slowed down if its own propensity for development is weak, or if the cell does not receive sufficient support for development from its neighbourhood. On the other hand, the development can also be speed up if the cell has a strong propensity for development and/or it receives strong support for development from its neighbourhood.

Three sets of fuzzy linguistic terms were proposed to implement the primary transition rules, reflecting the non-deterministic nature of the driving factors for urban development. One set of such terms is used to quantify the speed of development, such as ‘slow’, ‘medium’ or ‘fast’; one set to quantify the propensity of cells for further development, such as ‘weak’, ‘moderate’ or ‘strong’; and the third to quantify the extent of neighbourhood support, such as ‘weak’, ‘sufficient’ or ‘strong’. These fuzzy linguistic terms can be modified by the user during the simulation process.

2.2 Secondary Transition Rules

While the primary transition rules deal with an ideal situation where the area under study is of homogeneous nature, which hardly exists in the real world, the secondary transition rules add in the effects of other factors such as topographic constraints, transportation networks, socio-economic status as well as planning and human decision-making behaviours to the simulation process. Some factors, such as the provision of urban infrastructure and transportation networks, will function as accelerators to speed up the process of urban development; others, such as a mountainous landscape, or an area with a lack of provision of urban infrastructure may function as constraints to slow down the development process. These factors are built into the model as secondary transition rules to modify the primary rules on the transition of cell states.

For instance, if there is an accelerating factor within the neighbourhood of a cell in question, the speed of urban development of the cell will be faster, that is, from slow to medium, or from medium to fast. On the other hand, if there is a constraining factor within the neighbourhood of the cell in question, the speed of urban development of the cell will be slowed down, that is, from medium to slow, or from slow to very slow development. With more than one accelerator or constraint, the speed will be upgraded or downgraded two steps up or down. On the other hand, if both an accelerator and a constraint exist at the same time, the effect of both factors will be cancelled; hence, the speed of development will remain unchanged.

2.3 The Case Study Area and its Development Scenarios

Gold Coast, the sixth most populous city and the most populous non-capital city in Australia, was selected as a case study site to implement the fuzzy constrained CA model. Gold Coast City has been growing rapidly in the last two decades with an increase in population from 375,000 in 1996 to 494,500 in 2011 (Australian Bureau of Statistics 1996, 2011). While the rapid population increase has led to a phenomenal economic development in the State of Queensland, it has also resulted in challenging issues for this ecologically diverse city to achieve social, environmental and economic sustainability (Ward et al. 2003; Liu 2012).

The Fuzzy CA model was configured with a 250 m cell size and a rectangular neighbourhood of 5 by 5 cells (including the cell itself in the centre) to simulate its urban development from 1996 to 2016 under a set of primary and secondary transition rules. By using the actual urban growth data from 1996 to 2006 to calibrate the model's transition rules and parameters, the model was run from 2006 to 2016 to generate future scenarios of urban growth of the region. Three urban growth options—compact smart growth, planned eco-growth and unconstrained natural growth—were proposed. The compact smart growth scenario aimed for promoting infill urban development, limiting the outward expansion and preserving open space. This growth option was implemented in the model through strong support for growth on partly-urban cells but strictly constrained new development of non-urban cells. The planned eco-growth scenario was proposed based on the current planning scheme of the region which aimed for ecological sustainability (Gold Coast City Council 2011). This growth option protects the city's coastal, estuarine, riverine and hinterland environments and ensures the maintenance of the city's biodiversity and natural landscape values; it also strengthens the economic development of the city and ensures the maintenance of the cultural, economic, physical and social wellbeing of people and communities (Gold Coast City Council 2011). This growth option was implemented in the model by strengthening the impact of urban and town centres connected through existing and proposed transportation networks and tightening the constraints of planning control factors. The unconstrained natural growth option represents a scenario that the future urban growth follows the previous trends with little or no institutional control over the development process. Detailed description of the fuzzy-CA model can be found in Liu (2012).

3 A Framework for Web-based CA Modelling

Web-based GIS technology has been developing rapidly, with many GIS software vendors providing powerful platforms for users to develop web applications. Using web GIS technology, the fuzzy constrained CA model of urban growth can be deployed over the Internet for users to interact with the model and test different growth scenarios.

3.1 Model Implementation

The fuzzy constrained CA model of urban growth can be implemented either within or outside a GIS platform. Using the built-in functions of ArcGIS program, the model can be realised as a geoprocessing tool to operate within the ArcGIS platform. Although this method is convenient and easy to implement, the model cannot operate without the ArcGIS program. In addition, the coding of complex algorithm of the CA model cannot be optimized as it is difficult to change the underlying operations of the ArcGIS source codes. For instance, the model generates a new raster dataset each time an operation is executed, resulting in numerous intermediate and temporary datasets which slow down the operation of the program. In fact, when implementing the model as a geoprocessing tool in ArcGIS, the computation time of the model was considerably long even for a small number of iterations.

Using the Microsoft Visual Studio as a development tool, an independent model named Fuzzy CA was developed to implement the fuzzy constrained CA model. By developing the CA model outside a GIS platform, the model is self-contained and independent of any specific GIS platform and component. This approach provides flexibility in model construction although it requires higher expertise of the developer in computer programming. The coding of the CA algorithm and transition rules were optimized which saves the computation time; and improves the operational efficiency of the model.

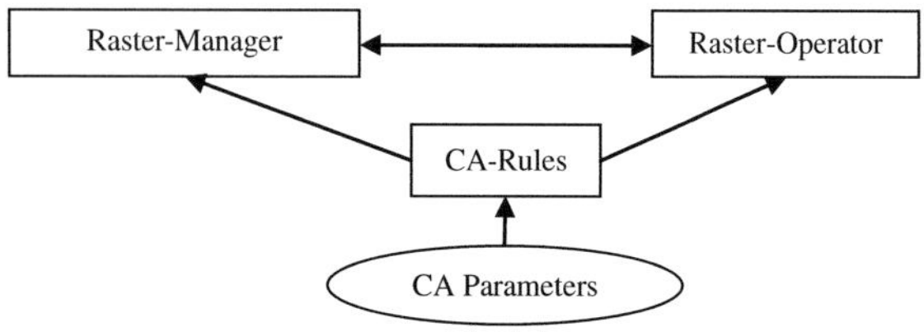

Fig. 1 Three core components of the Fuzzy CA model

The Fuzzy CA model has three components; a Raster-Manager module, a Raster-Operator module, and the CA-Rules module. The Raster-Manager module manages all raster datasets, including generating, copying, saving and releasing the raster data. The Raster-Operator module provides a number of raster operation functions, including arithmetic, conditional and logical operations. The core component of the model is the CA-Rules module, which encapsulates all transition rules of the fuzzy constrained CA model. The CA-Rules module receives parameter values from the user's input, and invokes the functions of the Raster-Operator module to process data in the Raster-Manager module and generate results (Fig. 1).

3.2 Model Deployment Over the Internet

ESRI's ArcGIS Server was used to deploy the Fuzzy CA model over the Internet. ArcGIS Server allows users to share GIS resources across an enterprise and over the web. It supports large potential users and the services can be consumed by many clients using different web service standards. Using the Software Development Kit (SDK) for .NET Developers supported by ArcGIS Server, ArcObjects was used to operate the raster datasets and execute the map operations such as adding raster data to the map document. In addition, the Web Application Developer Framework (ADF) was also used to build a web application of the model and deploy the model over the Internet. The Web ADF Task Framework provides a set of interfaces and abstract classes to create custom tasks. A custom task named CA-Task was created to encapsulate the CA functionalities as a distributable component. This custom task is a web control which offers an interface for users to interact with the model. The communication between the web user on the client side and the Fuzzy CA model on the server side is illustrated in Fig. 2.

Initially, users encounter a graphic user interface (GUI) through a web browser to select the transition rules to implement. They can also change or calibrate the parameters of the model. Once a user commits to execute the model, input from the user will be processed and transmitted to a Web ADF. These input will be sent to the CA-Task on the server side asynchronously. The CA-Task will then commit the values to the Fuzzy CA model to start the simulation process. The computation of the model with user defined transition rules and parameters is carried out by the Fuzzy CA model. It generates a new raster dataset representing the simulated state

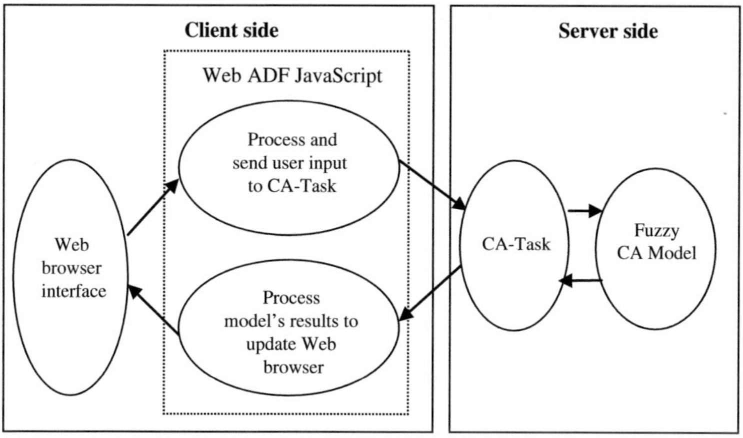

Fig. 2 Communications between the user on the client side and the Fuzzy CA model on the server side

of cells at the specified time. The model also generates a set of statistics illustrating the various simulation accuracies. Results are sent back to the CA-Task. Subsequently, the CA-Task module send the results to the web controls for processing and delivering to the user through the web browser. Upon receiving the updated information, the web browser refreshes the screen display on the user's machine. The map results generated by the model are added to the map document as a new data layer and displayed; the simulation accuracy statistics are also displayed as a tabular view on the user's web interface.

3.3 User Interface

A graphic user interface was designed and deployed over the Internet for users to interact with the model and calibrate the model's parameters (Fig. 3). For instance, users can choose to set the size of the cells to 100, 250, or 500 m, representing a high, medium or low cell resolution. They can also set the size of the neighbourhood to 3, 5 or 7 cells, representing a small, medium or large neighbourhood. The start and stop dates of the model can be reset by the user; users can also choose which transition rules to implement in the model, being primary or secondary transition rules. For instance, if no secondary rules are selected, the model will only operate based on the primary transition rules to simulate a scenario of natural urban growth within a homogeneous landscape. Users can also add in one or more secondary transition rules such as the DEM or slope factors, the transportation network or the urban planning factor to constrain the natural urban growth process. The initial configuration of the model's parameters can be launched through user input, which may be modified during the model calibration process.

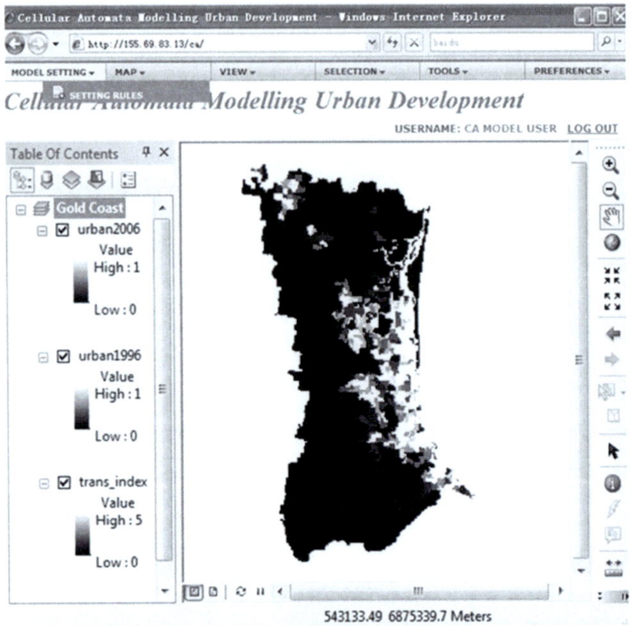

Fig. 3 The graphical user interface

Inputs from the user are encapsulated into the model as a custom task; it is provided as a floating panel within the web-based CA modelling environment. Users can launch the model configuration panel by selecting the 'MODEL SETTING' menu and then click 'SETTING RULES'. The simulation process can be activated once the model's rules and parameters are configured by the user (Fig. 4).

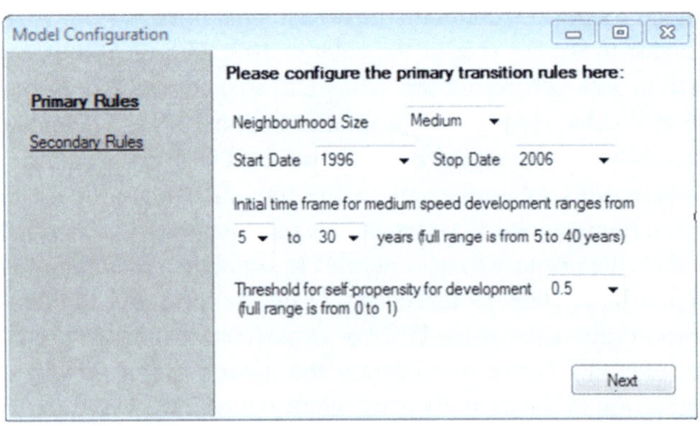

Fig. 4 Configuration of the model's transition rules

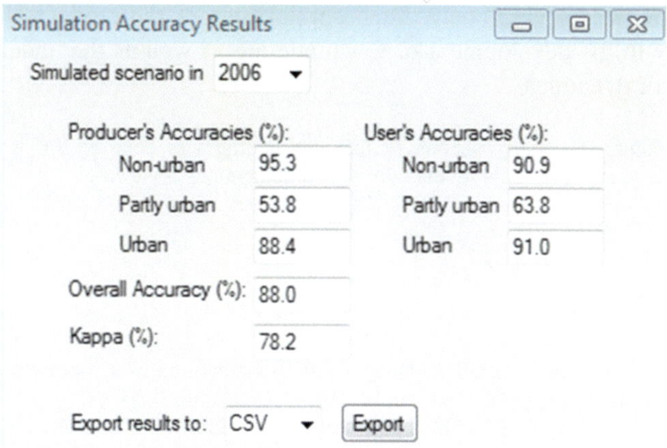

Fig. 5 Simulation accuracies generated by the model

After the operation of the model, a new set of data will be generated and added to the map document. This is the simulated scenario of urban growth under the conditions of the users' choice. The simulation accuracies of the model can also be computed by selecting the Assessment tool, and the results including the users' and producers' accuracies, the overall accuracy and Kappa coefficient of the model will be displayed in the Simulation Accuracy Results panel (Fig. 5).

4 Conclusion

So far the fuzzy constrained urban CA model has only been developed and released to an Intranet where the author works. Once the model is fully calibrated and tested, it will be deployed to the Internet and become openly accessible by anyone with an Internet connection. Vigorous testing and evaluation of the model on its usage over the Internet will need to be carried out. Specifically, the usability of the model, including the amount of time required to complete each iteration of the model with user's input, the stability of accessing the service, the flexibility of the model in responding to users' queries, and the number of users that can be supported simultaneously and how this may affect the processing and transmission time, needs to be evaluated. The results from the usability testing will be reported and discussed in a separate paper when data become available.

Furthermore, the Fuzzy CA model developed in this research needs to be advanced as a web-based modelling tool for urban researchers, planners and policy makers to engage in their own studies and practices. To this end, users will be able to not only interact with the datasets provided by the developer for the selected case study site, but also utilise the model with their own data to evaluate the urban growth scenarios of other regions. With this applications, the re-applicability of the model can be tested vigorously. Such applications of the model will contribute significantly to the development of CA modeling as well as the understanding of urban growth dynamics.

Acknowledgments The early version of the web coding was done by Jin He. The author appreciates the comments made by the editors and anonymous reviewers.

References

Al-Ahmadi K, See L, Heppenstall A, Hogg J (2009) Calibration of a fuzzy cellular automata model of urban dynamics in Saudi Arabia. Ecol Complex 6(2):80–101

Al-kheder S, Wang J, Shan J (2008) Fuzzy inference guided cellular automata urban-growth modelling using multi-temporal satellite images. Int J Geogr Inf Sci 22:1271–1293

Almeida CM, Gleriani JM, Castejon EF, Soares-Filho BS (2008) Using neural networks and cellular automata for modelling intra-urban land-use dynamics. Int J Geogr Inf Sci 22(9):943–963

Australian Bureau of Statistics (1996) 1996 Census data. http://www.abs.gov.au/websitedbs/censushome.nsf/home/historicaldata1996?opendocument&navpos=280. Last date accessed 01 Feb 2013

Australian Bureau of Statistics (2011) 2011 Census data. http://www.abs.gov.au/websitedbs/censushome.nsf/home/Census?opendocument#from-banner=GT. Last date accessed 01 Feb 2013

Batty M (1997) Cellular automata and urban form: a primer. J Am Plan Assoc 63:266–274

Batty M, Couclelis H, Eichen M (1997) Urban systems as cellular automata. Environ Plan B: Plan Des 24:159–164

Batty M, Xie Y, Sun Z (1999) Modelling urban dynamics through GIS-based cellular automata. Comput, Environ Urban Syst 23:205–233

Couclelis H (1997) From cellular automata to urban models: new principles for model development and implementation. Environ Plan B: Plan Des 24:165–174

Feng Y, Liu Y (2012) A heuristic cellular automata approach for modelling urban land-use change based on simulated annealing. Int J Geogr Inf Sci. doi:10.1080/13658816.2012.695377

Feng Y, Liu Y, Tong X, Liu M, Deng S (2011) Modeling dynamic urban growth using cellular automata and particle swarm optimization rules. Landscape Urban Plan 102(3):188–196

García AM, Santé I, Boullón M, Crecente R (2012) A comparative analysis of cellular automata models for simulation of small urban areas in Galicia, NW Spain. Comput Environ Urban Syst 36(4):291–301

Gold Coast City Council (2011) Gold Coast Planning Scheme 2003. http://www.goldcoast.qld.gov.au/t_standard2.aspx?pid=1395. Last date accessed 01 Feb 2013

He C, Okada N, Zhang Q, Shi P, Li J (2008) Modelling dynamic urban expansion processes incorporating a potential model with cellular automata. Landscape Urban Plan 86(1):79–91

Li X, Lao C, Liu X, Chen Y (2011) Coupling urban cellular automata with ant colony optimization for zoning protected natural areas under a changing landscape. Int J Geogr Inf Sci 25(4):575–593

Liu Y (2008) Modelling urban development with geographical information systems and cellular automata. CSC Press, New Yrok

Liu Y (2012) Modelling sustainable urban growth in a rapidly urbanising region using a fuzzy-constrained cellular automata approach. Int J Geogr Inf Sci 26(1):151–167

Liu Y, Phinn SR (2003) Modelling urban development with cellular automata incorporating fuzzy-set approaches. Comput Environ Urban Syst 27:637–658

Mitsova D, Shuster W, Wang X (2011) A cellular automata model of land cover change to integrate urban growth with open space conservation. Landscape Urban Plan 99(2):141–153

Pinto NN, Antunes AP (2010) A cellular automata model based on irregular cells: application to small urban areas. Environ Plan B: Plan Des 37:1095–1114

Stevens D, Dragicevic S (2007) A GIS-based irregular cellular automata model of land-use change. Environ Plan B: Plan Des 34:708–724

Tang J (2011) Modeling urban landscape dynamics using subpixel fractions and fuzzy cellular automata. Environ Plan B: Plan Des 38:903–920

Torrens PM, O'Sullivan D (2001) Editorial: cellular automata and urban simulation: where do we go from here. Environ Plan B: Plan Des 28:163–168

Vancheri A, Giordano P, Andrey D, Albeverio S (2008) Urban growth processes joining cellular automata and multiagent systems. part 2: computer simulations. Environ Plan B: Plan Des 35:863–880

Vliet J, White R, Dragicevic S (2009) Modeling urban growth using a variable grid cellular automaton. Comput Environ Urban Syst 33(1):35–43

Ward DP, Murray AT, Phinn SR (2003) Integrating spatial optimization and cellular automata for evaluating urban change. Annu Reg Sci 37(1):131–148

White RW, Engelen G (1993) Cellular automata and fractal urban form: a cellular modelling approach to the evolution of urban land use patterns. Environ Plan A 25:1175–1193

White R, Engelen G, Uljee I (1997) The use of constrained cellular automata for high-resolution modelling of urban land-use dynamics. Environ Plan B 24:323–343

Wu F (1998) Simulating urban encroachment on rural land with fuzzy-logic-controlled cellular automata in a geographical information system. J Environ Manage 53:293–308

Wu F (2002) Calibration of stochastic cellular automata: the application to rural-urban land conversions. Int J Geogr Inf Sci 16(8):795–818

Wu F, Webster CJ (1998) Simulation of land development through the integration of cellular automata and multi-criteria evaluation. Environ Plan B: Plan Des 25:103–126

Yang Q, Li X, Shi X (2008) Cellular automata for simulating land use changes based on support vector machines. Comput Geosci 34(6):592–602

Yeh AGO, Li X (2002) A cellular automata model to simulate development density for urban planning. Environ Plan B: Plan Des 29:431–450

Flexible Geospatial Platform for Distributed and Collaborative Urban Modelling

Yi Zhu, Mi Diao, Joseph Ferreira, Weifeng Li and Shan Jiang

Abstract This chapter presents the framework for and implementation of a flexible, loosely coupled geoportal platform for supporting distributed urban modelling with spatially detailed land use and transportation interactions. The proposed framework is built upon open source applications, such as Apache, PostgreSQL, MapServer, Drupal, Flex, and MediaWiki, with minimal custom code and some integration of proprietary tools such as ArcGIS Server. It provides collaborative tools to support parallel development and interconnection of complex model components in a manner that facilitates the visualization, discussion, and sharing of intermediate results. Large-scale land use, transportation and environment modelling contains many components which are themselves complex, in various stages of development, and distributed among different research groups. Many of the datasets needed for model calibration are incomplete, more aggregated than desired, or yet to be acquired. The geoportal platform is aimed at accelerating the realistic evolution of modelling data and cross-group learning as well as enhancing sense-making during early stages of model exploration and

Y. Zhu (✉) · J. Ferreira · S. Jiang
Department of Urban Studies and Planning, Massachusetts Institute of Technology, 77 Massachusetts Avenue, Cambridge, MA 02139, USA
e-mail: zhuyi@mit.edu

J. Ferreira
e-mail: jf@mit.edu

S. Jiang
e-mail: shanjang@mit.edu

M. Diao
Department of Real Estate, National University of Singapore, 4 Architecture Drive, Singapore 117566, Singapore
e-mail: rstdm@nus.edu.sg

W. Li
Department of Urban Planning and Design, University of Hong Kong, Hong Kong, China
e-mail: wfli@hku.hk

S. Geertman et al. (eds.), *Planning Support Systems for Sustainable Urban Development*, Lecture Notes in Geoinformation and Cartography,
DOI: 10.1007/978-3-642-37533-0_22,

integration. The development and sharing of origin/destination matrices and modelling results from four scenarios are used to illustrate the types of visualization and cross-group interactions that are facilitated.

1 Introduction

The efforts to develop large-scale urban models for simulating the interconnections among land uses, transportation, environment and pertinent activities within a metropolitan area can be traced back to 1950s, when regional economics theories were coupled with social physics ideas to give rise to the rudimentary urban models (Batty 2007). Since then, urban modelling practice has moved from macroscopic toward microscopic scales, from equilibrium toward dynamic, and from trip-based toward behavior- and activity-based (Bowman and Bradley 2008; Waddell et al. 2008). To accommodate the size and complexity of the evolution, the traditional large, monolithic modelling environments have to be decomposed into more loosely-coupled distributed modelling systems. The trend also necessitates more collaboration among research groups with different expertise as they face rapidly growing knowledge and data requirements (Geertman and Stillwell 2009). Recent efforts such as the Open Platform for Urban Simulation (OPUS) for UrbanSim (Waddell et al. 2005) have begun to adopt a modular, extensible and interactive open source framework for developing urban models. However, these frameworks are still limited in terms of the connections among models and distributed model builders. Specific programming skills are usually needed for model expansion and integration, and information and data exchange among spatially distributed model developers rely on relatively traditional tools such as email, forums and wikis.

In parallel, advances in GIS, ICT technology and the explosion of spatially detailed data have given rise to geospatial portals. Starting from the US Federal Geographic Data Committee's (FGDC) Clearinghouse web sites in 1994, a majority of geoportals have focused on geographic data sharing by providing online mapping with a growing list of additional functions such as direct access to raw data in multiple formats, complete metadata, and online visualization tools and commenting mechanisms. However, none of them are explicitly built for large-scale urban modelling, where extensive geographic information and associated spatial analysis and modelling tools are required.

Having noticed the advantages and limitations of existing distributed modelling frameworks and geoportals, we prototyped and demonstrated a GeoPortal to support limited visualization and exchange of data relevant to Land Use, Transportation and Environment (LUTE) modelling. The initial version of the GeoPortal is published in Transportation Research Record (Ferreira et al. 2010). This chapter explains the efforts to further improve the GeoPortal with expanded visualization and data exchange capabilities.

2 Conceptual Framework and Previous Efforts on GeoPortal

The conceptual framework of the GeoPortal proposed by Ferreira et al. (2010) consists of four layers with a service-oriented architecture (SOA) as diagrammed below (Fig. 1). The data layer stores original and model-derived data. The middleware layer contains map server (ArcGIS server and MapServer), web server (Apache) and other application services to support data cataloging, processing and visualization at the presentation layer. The presentation layer contains portals to provide user interfaces for accessing, viewing, querying and interacting with data and models, and provides room for idea and information exchange among researchers.

The analytic layer includes a series of modular scalable backend applications (like UrbanSim, R and TransCAD) for analyses and simulations. A single research group does not need to own or learn each of the complex and/or proprietary software packages since the standardized original data and intermediate model results can be explored and shared through the data server.

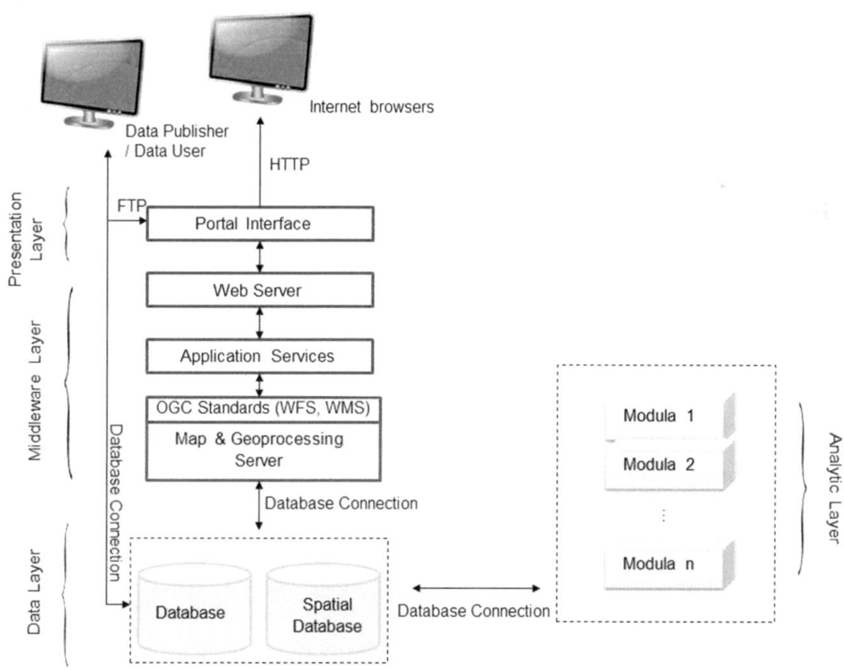

Fig. 1 Conceptual GeoPortal framework for collaborative modelling (modified from Ferreira et al. 2010)

2.1 Motivation for Designing an Improved Modelling Infrastructure

Version one of the GeoPortal was developed and implemented within the Transportation Systems focus area of the MIT-Portugal Program (MPP) (The MIT/Portugal Program 2010). The Strategic Options for Integrating Transportation Innovations and Urban Revitalization (SOTUR) project within the Transportation Systems focus area aims to promote sustainable urban development patterns and travel patterns by leveraging transportation innovations such as intelligent transportation systems (ITS) and their interactions with land use and urban growth patterns, to assist in the revitalization of the city centers such as those in Lisbon, Portugal. A range of research activities have been undertaken within the SOTUR project, namely: (1) scenario building with stakeholders, (2) developing modelling tools for analyzing the effectiveness of policies at achieving urban regeneration in Portuguese cities, and (3) designing a policy evaluation framework for policy comparison and ranking.

Besides the work on policy design and survey, different approaches toward modelling land use and transportation interactions have been developed including an OPUS UrbanSim application (Guevara 2009), an AnyLogic agent-based model (Martínez and Viegas 2009; Martínez et al. 2010) and a cellular automata model for the City of Coimbra (Pinto and Antunes 2007). The ultimate aim of the parallel modelling efforts is to accelerate the development of generalizable model via cross-validation and compare among different approaches.

However, different models use different data sets and parameters, implemented at different spatial aggregation levels such as Freguesia (a census geographical subdivision in Portugal), block group or traffic zone. Even within each model, travel demand component is integrated in different ways. In the UrbanSim application, TransCAD is initially used for applying the traditional four-step travel demand model to estimate accessibility between origins and destinations. The Anylogic agent-based model has the land use and transportation parts tightly integrated within its own framework, but shares data about land use, demographics, road network, and the like with parallel modeling efforts (Martínez and Viegas 2009; Martínez et al. 2010).

As each model is developed independently, it is not only critical to distribute data and resources among groups in an explicit and convenient way, but also important to have a common platform for the distributed researchers to investigate and discuss modelling results during the scenario and model development process. Although our previous GeoPortal provided useful supports for cataloging and sharing geospatial data, a number of practical obstacles to effective collaboration were exposed in the course of implementation:

- The system appeared insufficient in coordinating research processes and dataset versions. When a shared dataset was manipulated, it would be difficult for each group to be immediately aware of the changes.
- The functions of the visualization tools were limited to web-based mapping and thus could not support the level of data exploration that researchers require.

- Exploring and simulating urban land use and transportation system requires recurrent runs of the models with varying parameters and settings. It was inefficient and inconvenient for developers to use a system without a work flow management system pipelining the modelling and results reporting process.
- The variations in data types, spatial and temporal resolutions, study areas, and data sources set a barrier to the interoperability of the datasets for comparison among scenarios, and for recycling among different models.

In this chapter, the applications of a database structure and two web-based interactive visualization tools on asynchronous collaborative research are highlighted to address the complication of the infrastructure needs to facilitate sharing and visualizing the modelling results, and to help in the calibration and tuning of the model.

3 Modelling Infrastructure, Components and Applications

3.1 System Infrastructure Design

Version two of the GeoPortal platform is improved via enhanced federated database structures, new web-based visualization applications, and a content management system. Figure 2 shows the system architecture of the latest version of the GeoPortal, which is built upon open source applications, such as Apache, PostgreSQL, MapServer, Drupal, Flex and MediaWiki, with minimal custom code and some integration of proprietary tools such as ArcGIS Server. Currently, the data, middleware, and presentation layer components are located on a single (virtual) server running Red Hat Linux and the analytical layer components like UrbanSim, TransCAD and R are located on distributed computers.

3.1.1 Federated Database Server

The core of the system remains the open source relational database management server PostgreSQL with its geospatial extension PostGIS. It stores selected raw data and modelling results and exchanges data with analytical models, application servers, and privileged user via SQL queries across Open Database Connectivity (ODBC) and web services. Other raw data and modelling results having formats not compliant with the PostgreSQL database are stored in the FTP storage space.

3.1.2 Post-Processing and Visualization Tools

We developed two types of tools for interactive data visualization: (1) an AJAX-based web interface linked to MapServer; and (2) an Adobe Flex-based interface with a backend ArcGIS server. Mapserver provides flexibility and open source

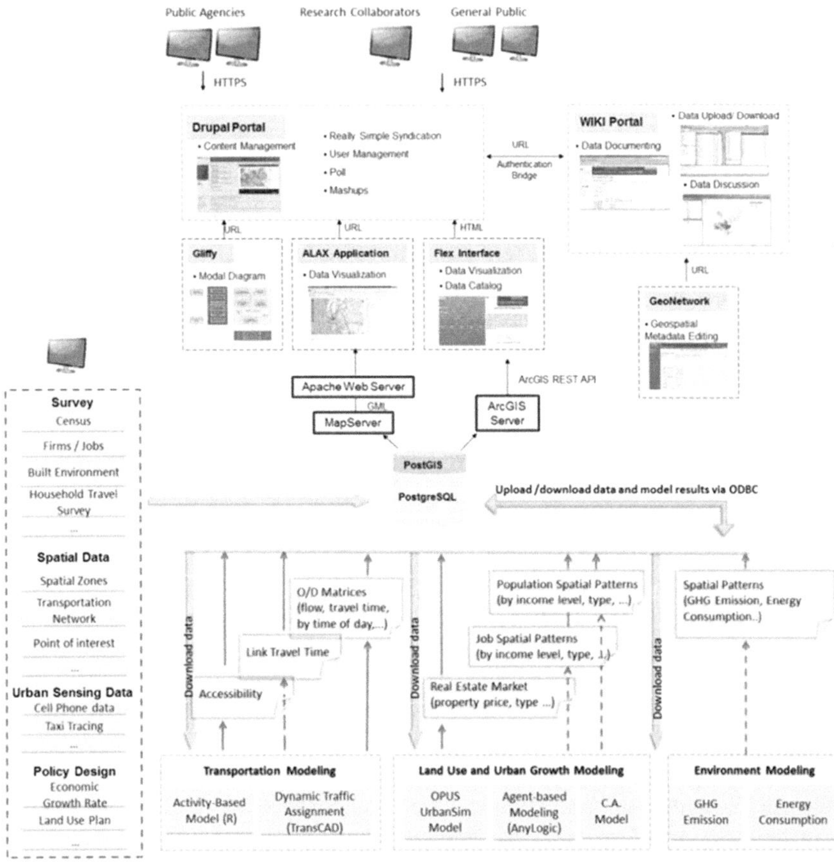

Fig. 2 Structure of the distributed LUTE modelling system for the MPP-SOTUR project

while ArcGIS server has powerful geoprocessing and model building functions along with some open standards support and interoperability with common desktop GIS packages. In addition, to repurpose the data in useful and meaningful ways for researchers, we employed an open-source statistical software package R for post-processing model results by automating certain filtering, sampling and aggregation of the data.

3.1.3 Content Management System

We employ Drupal as the content management system for orchestrating all separated components, and use MediaWiki as a data catalog for storing information about datasets. The Drupal interface provides a manageable framework for linking the Mediawiki for data cataloging to the Flex and Ajax applications for visualizing model results, while allowing easy navigation across components and enhanced scalability and flexibility of the whole system.

The Drupal portal interface contains four main categories of web pages: "Urban Modelling", "Visualization", "Data Catalog" and "Event". Each set of webpages includes an introduction, a list of applications, logs of research discussion, and provide links to corresponding portals (Fig. 3).

With extra functionalities contributed by the numerous third-parties and tools compatible with many Web 2.0 components, the content management portal can be tailored for additional collaborative needs. For example, web-based diagramming tools, such as Gliffy, can be integrated to the content management portal to present the complicated modelling structure in a clear and simple way. RSS feeds keep researchers aware of other's latest process. A "poll" module facilitates coordination and events scheduling through simple online surveying and balloting.

3.1.4 Users and Access Control

In order to enable the different levels of user participation and sharing data resources, different levels of security and access control management are needed. In the geoportal, we differentiated the access permissions of three groups of people: (1) researchers from different universities and institutions (in the U.S. and Portugal) across the Atlantic Ocean; (2) involved government agencies and industrial stakeholders; and (3) the general public. The access control system addresses data confidentiality and limited sharing of proprietary files and software.

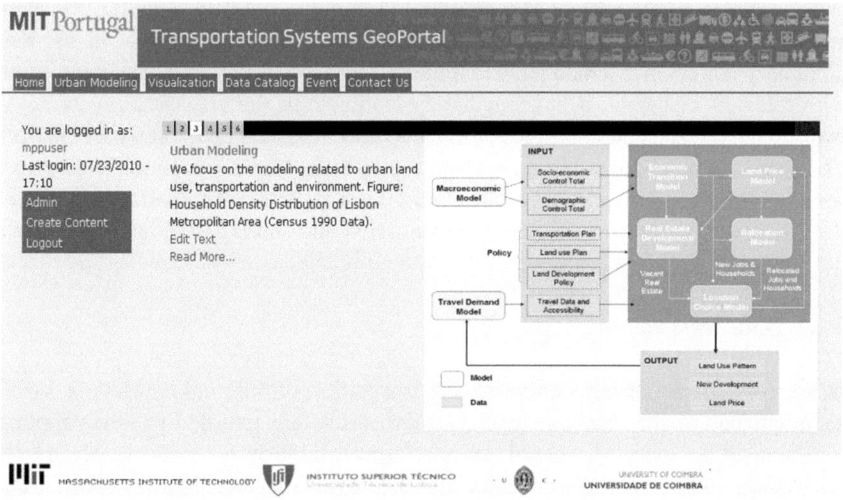

Fig. 3 Druple system interface screenshot. The header menu leads to the introductory pages of urban modelling, visualization, data catalog and event

3.2 PostgreSQL Database Structure/Schema

To achieve the desired level of mix and match flexibility among the distributed model components and datasets, it is critical to find appropriate database structure and data conversion schemes that not only can be straightforward for data visualization, but also can allow sharing common datasets for different models and comparing different modelling results under the same scenario settings. Version two of the GeoPortal allows individual researchers to publish their modelling results by uploading data tables to the PostgreSQL database management server and demonstrating the spatial patterns in the visualization application tools. The results of a particular simulation are saved as data tables in the PostgreSQL database. Data tables can be imported into the PostgreSQL database from widely used desktop database software such as Microsoft Access using ODBC. Each data table includes a geographic ID field used for mapping the various model output indicators at varying levels of spatial detail using the (Ajax- and Flex-based) visualization applications.

3.2.1 Metadata Structure

Metadata are important components of the new database structure that are also saved as tables in the PostgreSQL database. A table named "datatables" contains metadata about data tables, including table name, user name, study area, spatial unit, a brief description of the associated model run, name of the model, date of creation, and date of change. Each model output table constitutes one row in the "datatables" table. A "data dictionary" table provides a description of each indicator report in a 'datatable.' This description includes the data table ID, the name and description of each indicator, etc. Each indictor takes one row in the data dictionary table. Once a data table is imported into the PostgreSQL database, along with relevant metadata, it is immediately available to the visualization tools. An Ajax-based interface has also been created to make the data uploading process easy. The proposed database structure enables the visualization applications to automate some of the data delivery, speed up data updating, and make spatial aggregation flexible, thus greatly streamlining the discussion and sharing of simulation results.

3.2.2 Data Integration

In the version two of the GeoPortal, the integration of different datasets is mostly accomplished at the database level and the results are returned to other applications via SQL query. A spatial cross-reference table is provided to ensure the integration of the modeling results at different spatial aggregation level. More complicated spatial join or geo-processing tasks can be achieved by taking advantage of the spatial functions and spatial indexing enabled by PostGIS. For the temporary data, in this case, we created one database schema for each simulation year so that it is easy to locate the inputs and outputs of a particular simulation year. However, if the models become more temporally fine-grained and different levels of temporal data are generated, then SQL routines are needed to transform modeling inputs and results to comparable temporal scales.

In a nutshell, the backend database server handles the aggregation, joining, filtering and sampling of the data based on SQL requests and returns formatted results to the front-end applications to avoid heavy loading.

3.3 Ajax-Based Visualization Application

The AJAX application is implemented based on web services and open source software (Postgres/PostGIS, Apache, MapServer, PHP, and Linux) with access control for users and groups. To generate maps, it uses the OpenLayers javascript library to request WMS-compliant images and then overlay them on top of Google Maps. The client is built using AJAX (Asynchronous Javascript and XML) programming techniques to provide interactive capability with improved performance.

3.3.1 State Saving and Sharing

The AJAX application also provides state saving and downloading functions. During the interactive visualization, when a map with interesting patterns is identified, the parameter for the WMS call can be recorded as a row in a table in the PostgreSQL database. The map can then be published to other research groups for further discussion, and restored on demand by resending the WMS calls using the saved parameters. When a researcher with proper privilege finds a data table suitable for his/her own research, he/she can download the dataset using the web service.

3.3.2 Application Examples

Two examples are provided here to showcase the AJAX application in scenario planning. O/D data are ubiquitously useful for multiple types of modelling (e.g., four-step modelling, activity-based modelling, traffic simulation, land use and urban growth modelling, and environment and energy modelling). Modelers can use the AJAX application to visualize and compare O/D matrices from others' modelling results.

Figure 4 displays the commuting sheds into a downtown (Sao Nicolau) and a suburban (Almada) Freguesia in metro Lisbon based on an O/D matrix generated by a TransCAD four-step model using 1994 survey data and related census data. The different commuting patterns (with Freguesias viewed as workplaces or as

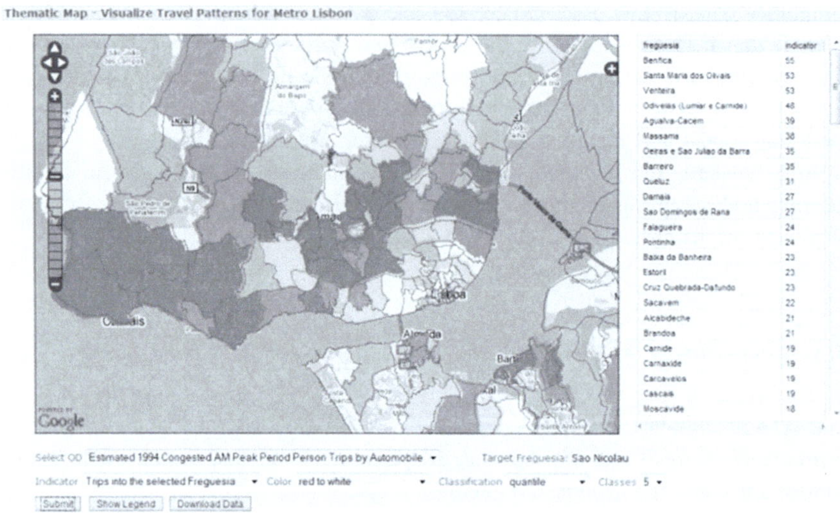

(a)

(b)

Fig. 4 Commuting sheds to a downtown Freguesia and a suburban Freguesia in Lisbon. **a** Commuting shed to a downtown Freguesia (Sao Nicolau) in Lisbon (with Google Street layer as background). **b** Commuting shed to a suburban Freguesia (Almada) in Lisbo (with a terrain model and Google satellite image as background layers)

residences) can help researchers in different groups spot potential job-housing balance concerns and regional development bottlenecks and can incorporate these concerns into their scenario modelling/planning efforts. Also, it is noteworthy that two commuting shed maps are using different background Google map layers (street layer and satellite layer). This provides some degree of freedom on the choice of background map based on researchers' preference.

3.4 Flex-Based Visualization Application

The Flex-based application represents a different choice of visualization tool, which aims to provide rich visualization and interaction experience for the researchers. Thanks to the ArcGIS server, the ArcGIS REST API for Flex, and the open-source Flex Software Development Kit (SDK), the flex based application allows for much interactivity and flexibility. In a flex-based application, modelling results can be presented in a variety of forms including mapping, tabulating, and charting. Maps can be delivered to a flex-based application as a raster image or as vector geographic features. The former yields a static map generated by the ArcGIS server or geoprocessing services for each users' request, while the latter sends georeferenced data in a form that allows the Flex client to render a variety of maps based on user interaction without additional requests to the server, executing functions related to geographic features at the front-end Flex side. Meanwhile, Flex SDK can work with web services that define their interfaces in a Web Services Description Language 1.1 (WSDL 1.1) document, which is available as a URL.

3.4.1 Data Retrieving

When Flex is coupled with PHP and a database server, the HttpService component of Flex can call a PHP program with GET or POST function to perform a database query. The retrieved data are written as an XML file and fed to the Flex application. As a result, it is convenient for Flex to bind the modelling results related to map features from the PostgreSQL database server to corresponding vector features from the ArcGIS server at the client side. This strategy can improve the time efficiency of the application considerably when the number of map features is moderate, and help to realize semi-automatic communication between the server-side database, ArcGIS map services and the client-side Flex interface.

3.4.2 Application Examples

Figure 5 shows a screenshot of the Flex application. The upper map plots density of low-income households aggregated to the Freguesia level in 2011 under low income housing subsidy scenario. The values are simulated using UrbanSim and shown in the table on the right in the map. In Fig. 5b, the thematic map illustrates the differences in density of low-income households between the low income housing subsidy scenario in 2011 and the base year (year 2001). This provides a chance to comprehend the spatial differentiability of indicators under different

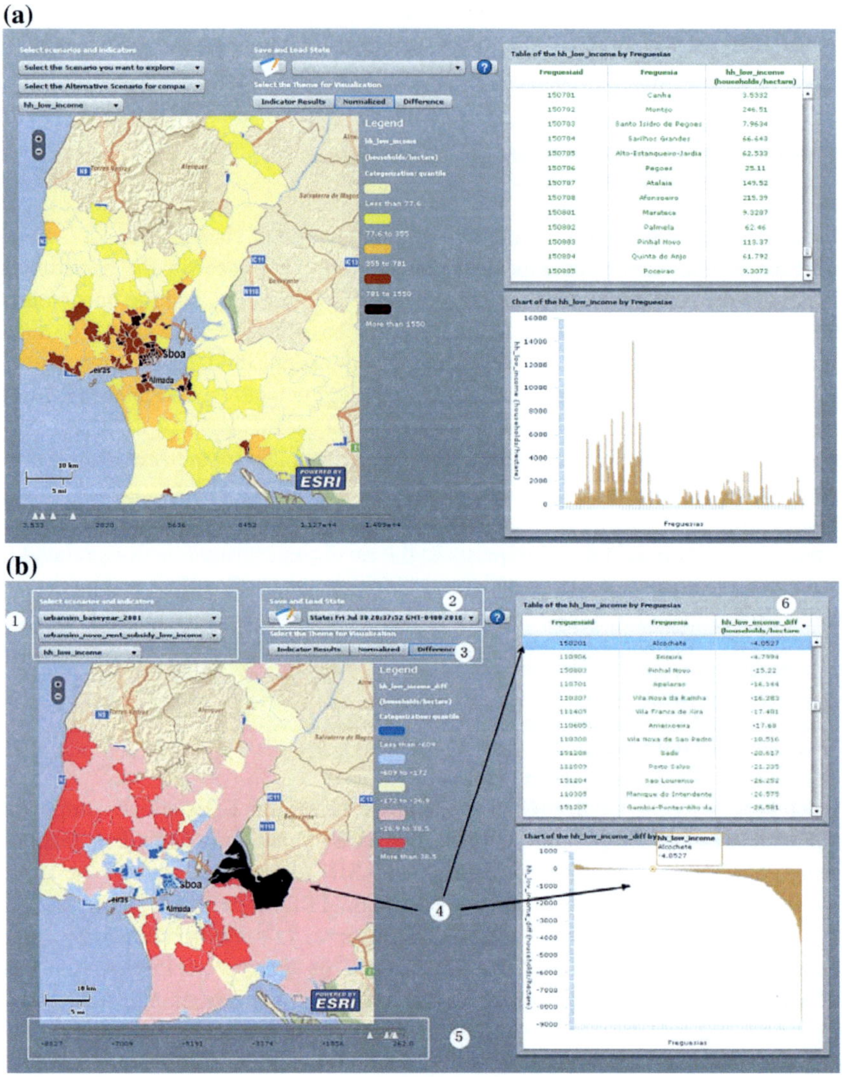

Fig. 5 The Flex visualization application for the urban modelling in Lisbon. **a** Lisbon low-income household density under the low-income housing subsidy (LIHS) scenario, 2011. **b** The difference in the estimated low-income household density between base year and the LIHS scenario

scenarios and hence helps to enhance the understanding of modelling results, scenarios and alternative urban futures.

3.4.3 Mouse-Over Interaction

Once the geographical features and the modelling results are loaded into the Flex application, the features on the map and the related attributes in the table and chart are all visually interlinked. When one moves the mouse across the chart, the corresponding Freguesia on the map will be highlighted in black and the relevant data in the table will also be highlighted.

Meanwhile, a tag will pop up on the chart showing the name and the picked indicator value of the selected Freguesia. Pairing geospatial maps with statistical charts can promote the sense-making of the scenarios and the interpretation of spatial patterns and relationships. Also, since the color scheme of the thematic map can disguise spatial differences and complicate scenario comparisons, users can change the break points of the shading scheme by dragging the five ticks on the symbology slide bar.

To improve the usability of the application, additional interactive effects are added. For example, when users click the head title of the columns in the table, the table and the chart will be sorted by the values in corresponding column in an ascending order. If the head title clicked again, the table and the chart will be sorted in a descending order.

3.4.4 State Saving and Sharing

Applications based on Flex can imbed code fragments in an extended URL in order to retain information about the view, such as map zoom-level bounding box, the indicator selected for exploration, etc. Such URLs can thereby describe an exact state of the visualization application. In this application, we use such extended URLs to allow users to save state by clicking the pad writing icon on the tool bar (see item #2 of Fig. 5b). The current state of the application will then be saved with a time stamp and automatically added to a drop down list of states any one of which can be restored by selecting from the drop-down list. The saved states last only for a current user session since the Flex application does not write any records back to the server-side database.

3.5 Summary

Overall, the two (Ajax- and Flex-based) visualization applications have their own advantages and shortcomings. The Ajax-based application provides on-the-fly visualization of the modelling results, which permits exhibiting modelling results at the first time, because the maps are delivered through WMS service as images. As a result, it enables mapping large dataset with satisfactory performances. For example, it only takes around 5 s to display the thematic map of a geospatial dataset with 35,000 features. By contrast, it is formidable to use Flex and ArcGIS server to map such a large dataset since it is very time consuming to download 35,000 features onto client-side application and usually triggers a request time-out error. However, Flex-based application tends to have richer interactive functionalities and hence more appropriate for users to manipulate data to have in-depth understanding of the results. The second difference between two applications is access control. The Ajax application is built on the intelligent middleware and has its own login/logout system, while the Flex application is stand-alone and can only be protected by HTTPs. However, these two techniques together represent our efforts in exploring different visualization techniques and delivering fast and rich visualization applications to the users.

4 System Implementation and Use

Version two of the GeoPortal started to be deployed for the SOTUR project of the MIT-Portugal Program (MPP) in 2011. In the process of test, our modelling developers found the primary values of the system to be in modeling diagnose and simulation results exploration which facilitates cross-group dialogue. Besides, they also found the wiki useful to catalog data and other information related to the SOTUR model, and the Druple-based content management system convenient to navigate between different components of the modelling. In this section, we will use accessibility results as an example to demonstrate the types of inter-group modelling collaboration and knowledge transfer supported by the GeoPortal.

4.1 Measurement of Accessibility

The outcomes of the SOTUR modelling are comprehensive, ranging from the spatial distribution of population, jobs and activities to the performance metrics of lands, transportation networks and urban environment. Among these results, one indicator worth particular attention is accessibility, which is usually abstract and rich in information but serves as the main linkage between land use and transportation components in many urban modelling frameworks. For a transportation system, accessibility is a composite indication of the easiness of access to opportunities, taking into account of travel impedance factors like travel time, travel costs and transfers of all available transportation mode options. For urban land use, accessibility is considered as an important attribute influencing the attractiveness of places. In addition, accessibility is an important criterion for evaluating the equity and economic vitality of the scenarios.

In the SOTUR model, the accessibility of an individual is operationalized as the expected maximum utility of engaging a given type of activity in all potential destinations taking into account the utility of traveling to those destinations from the origin place, i.e., housing location in this case (Dong et al. 2006). The resulting accessibility measure for each person given housing location is then:

$$Access_{i,h} = \frac{1}{\mu_d} \ln \left\{ \sum_{j \in D} \exp\left[\mu_d(\beta_l L_{ij} + \beta X)\right] \right\}$$

$$L_{ij} = \ln\left(\sum_{m \in M} \exp(\mu_m V_{ijm}) \right)$$

where $Access_{i,h}$ is the accessibility measure of place i and activity type h. L_{ij} is the approximated log-sum of travel utility (V_{ijm}) of all available transportation modes M from location i to j. X contains a vector of factors influencing the destination choice of a particular agent like population, employment and land use properties associated with origin and destinations as well as decision maker's own characteristics. β denotes the estimated coefficients of explanatory variables of a multinomial logit destination choice model.

4.2 Collaborative Modelling Support

4.2.1 Scenario Comparison

GeoPortal has the capabilities that not only allow exploring the spatial variation but also facilitate comparison of accessibility measures for different categories of decision makers, simulation years and various scenarios as shown in Fig. 6.

Each thematic map in the Fig. 6a and b presents five different levels of accessibility with the same breaking points for categories. The darker the shade, the greater the accessibility. In Fig. 6a, it appears that persons with car access have greater accessibility than the persons having no car access for the home-based work trips. Besides, for the same zone, persons within the high income category perceive less accessibility than the low income individuals because travel time tends to be a greater disutility for them than for the low income individuals. But in general, the zones with the highest accessibility for all categories of decision makers are located around the city center of the Lisbon. Figure 6b plots the time-lapse change of the accessibility measure of the high income group with car access under a hypothetical cordon charging scenario, which charges 8 euro for each car entering the downtown Lisbon. The temporal trend of accessibility in the central city of Lisbon suggests the cordon charging scenario will lead to an increase over time in accessibility for the high income group.

One of the primary purposes of developing SOTUR model is to simulate the ripple effects of a set of hypothetical policies on the future evolution of people's behaviors,

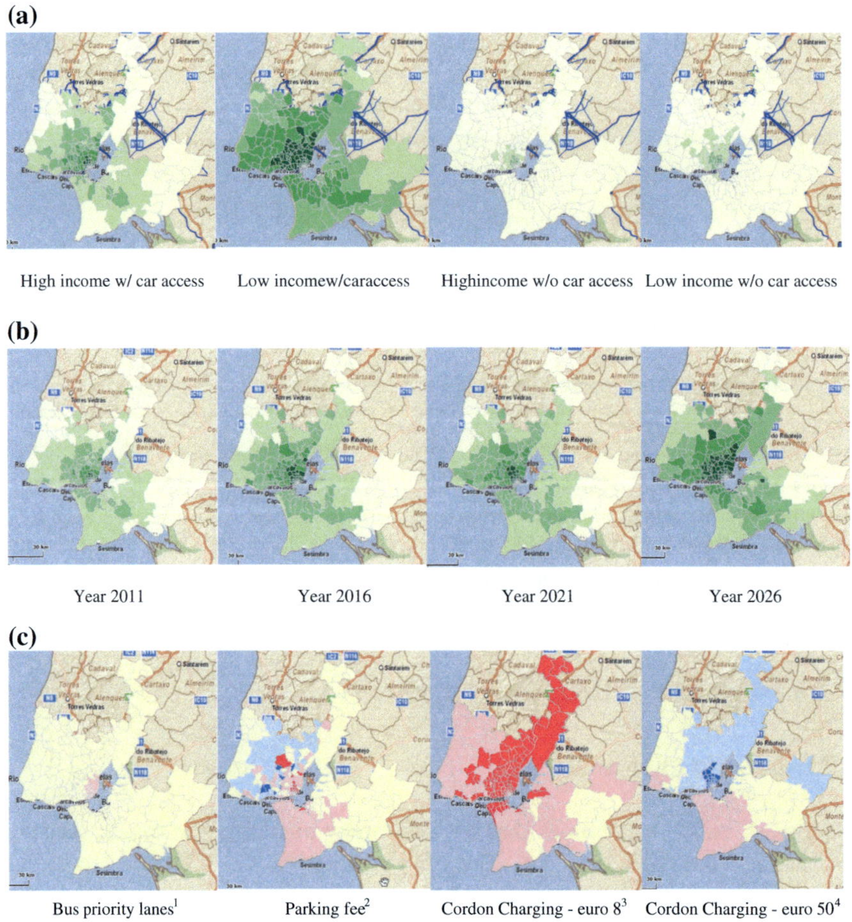

[1] Bus priority lanes: four major corridors are identified for the improved bus services and travel time for transit trips between freguesia pairs along the designated routes are decreased.

[2] Parkingfee: parking fee zones are specified by freguesia and encompass the most heavily developed areas of the metropolitan, The parking fee for trips ending in the parking fee zone will be increased from 2-5 euro/day to 15euro/day.

[3] A 8 euro fee is charged for auto trips ending in the zone defined as the city center.

[4] A 50 euro fee is charged for auto trips ending in the zone defined as the city center.

Fig. 6 Screenshots of the examples of resulting accessibility measures for various categories of decision makers, scenarios and simulation years. **a** Home-based work (HBW) accessibility measure by categories of decision makers. **b** HBW accessibility measure of the cordon charging scenario for high income group with car access by simulation years. **c** Difference in the home-based work accessibility measure between business as usually and selected scenarios for high-income, car-accessible individuals

urban land use and transportation system. In this sense, it is essential for the visualization applications of the GeoPortal to support comparing among scenarios. Figure 6c shows the accessibility difference between the selected policy scenario simulation and the "business as usual" scenario with the red shade representing the gain in accessibility and the blue shade meaning the loss. The maps suggest that the accessibility measure for the high-income and car accessible individuals are less sensitive to the bus service improvement but more responsive to the increased parking fees or cordon charging fees, which indicates the potential of achieving more sustainable urban transportation by increasing the cost of driving in Lisbon.

4.2.2 Modeling Diagnose

In the SOTUR model, the accessibility of places is the logsum derived from the nested transportation mode choice and destination choice model. However, the accessibility is also influenced by the spatial distribution of people and firms. Thus, the exploration and diagnose of accessibility requires model developers from both land use and transportation sides, and synthesis of relevant information from different sources, e.g., housing location choice, firm location choice model and destination choice model. Since the web-based interactive visualization tools of the GeoPortal are directly linked to the PostgreSQL database server, it can conveniently retrieve the inputs of outputs of all models and visualize them on the dynamic web pages, which provide views of not only accessibility measures, but also the factors influencing accessibility measurement and the factors being influenced by accessibility. In additional to the thematic map, tools like sorting tables and histogram charts provided by the Flex visualization are useful for pinpointing exceptions and outliers.

Further, because many researchers lack of local knowledge of the study area, the assumptions made in the modelling process may not be consistent with the local realities. For example, in the destination choice model, model developers initially assumed that travel options and frictions, build environment of the places and the characteristics of individuals are the main driving factors of the choices, but in this case other factors like the type of places and the south or north side of the Lisbon river also play important roles. Lack of this knowledge could result in biased accessibility measures, poor simulation results and wrong policy suggestions. In this context, having the inputs from the researchers and participants with local knowledge is also very necessary to help correct inappropriate assumptions and improve the soundness of the model. As researchers distributed in different places lack of effective means of contact when facing unfamiliar spatial distributions, web based interactive visualization provides a channel to support broader discussions among researchers with different emphases when diagnosing the modelling results collectively. Meanwhile, this also leads to improved cross-group learning.

4.2.3 View Sharing

In the context of asynchronous sharing, where communication takes place in different location and at different time, it is very important for participants to agree on the content and object they are talking about before they can have a productive discussion (Heer et al. 2008). Without a common ground, researchers may refer to

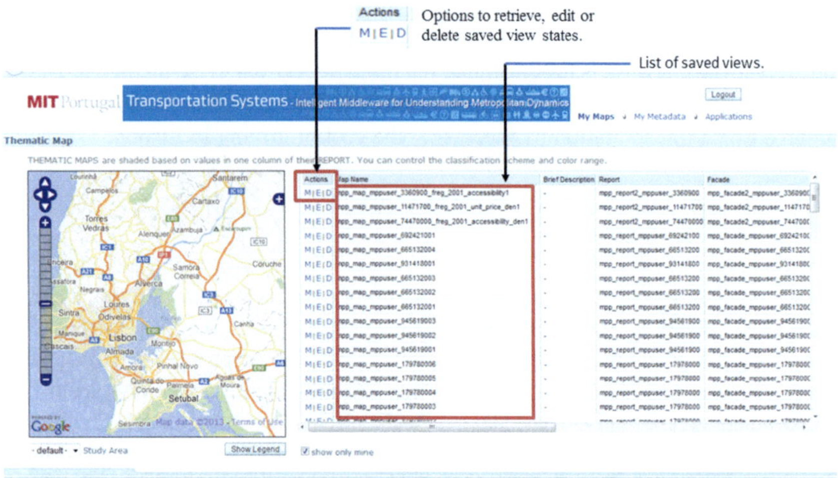

Fig. 7 View sharing of the accessibility measures in Ajax-based visualization tools

different things, which can lead to confusion in communication and discourage collaboration. While static maps or screenshots fails to support further drilling down for details and view sharing through video only works for synchronous collaboration, if the interactive visualization tool is able to save necessary states of the applications, it will be very convenient to share exploratory results with remote researchers through URLs or save unfinished investigation for future analysis.

As introduced in Sects. 3.3 and 3.4, both Ajax (Fig. 7) and Flex-based visualization tools support some forms of state saving and sharing. Web services and URL links can be saved on wiki or discussion pages or passed to others via emails or RSS feeds since any user with the requisite permission can copy and paste them into a browser to reproduce the saved visualization state. This allows the information and knowledge on modelling to be shared asynchronously and to be examined collectively. We also expect this capability to facilitate collaborative scenario analysis and engage more researchers into discussion.

In summary, sharing, synthesis and sensing-making of the data are critical for distributed and collaborative urban modeling. Version two of the GeoPortal is designated to support such activities so as to push the large-scale land use, transportation and environmental modeling effort carried out by distributed research teams to an integrated end.

5 Conclusion

This chapter presents a modelling infrastructure supporting the loosely-coupled large scale Land Use, Transportation and Environment modelling work with distributed research groups, and public agencies. The infrastructure centers on the

PostgreSQL database structure and two visualization tools, which can be used to coordinate the results from a variety of analytical models, to provide web-based visualization of data, and to save and share useful spatial patterns via WMS records or URL links. We focus specifically on the usability, automation, and interactivity of the visualization tools to make sure that urban modellers and planners would be comfortable using them.

At the top level, a web content management system Drupal was employed to orchestrate key components to attain structural integrity and easy navigation among components. The infrastructure could be easily adjusted to accommodate other GeoPortal components such as a wiki and FTP (Ferreira et al. 2010) to improve its usability. More importantly, the GeoPortal sustains multiple levels of connections among different modelling efforts, thereby enhancing research collaboration.

As the modelling infrastructure evolves, some emergent issues need to be addressed. First, the database structure and the visualization tools need to be more scalable and flexible to accommodate new types of data (e.g., time-series data, matrix), new units of analysis and new levels of aggregation. Second, the visualization applications are mainly used for thematic mapping and charting of indicators at this point. Their support for other visualization types like flow visualization, surface visualization and even 3D visualization remains inadequate. In order to portray the abstract relationships among variables from different perspectives, it is necessary to develop more powerful visualization capabilities in the future so that the model and the urban system can be understood in its entirety. Third, we need to follow the principle of minimum programming and flexible functionalities. Coordinating and integrating these tools for consistent but differentiated access control and convenient navigation is another task. At this stage, we are collecting feedbacks from researchers and proceeding to consolidate the existing system. We expect such infrastructure to enable government agencies, public and private sectors, planners and developers to be more tightly integrated into urban modelling efforts.

Acknowledgments We acknowledge the MIT Portugal Program for supporting this work on information infrastructure and geoprocessing. We also acknowledge the contributions of our collaborators in Portugal and at MIT including Nuno Pinto, Joao Zeferino, Dr. Jorge Tavares and Prof. Antonio Antunes at the University of Coimbra; Luis Martinez and Prof. Jose Viegas at the Institute Superior Tecnico, Lisboa; and Lisa Rayle, C. Angelo Guevara and Prof. Chris Zegras at MIT; and Prof. Marianne Hatzopoulou at the McGill University. In addition, we appreciate the IT assistance of MIT's Computer Resource Network (CRON) staff and Dr. Raj Singh of OGC, plus the support of the Urban Markets Initiative of the Brooking Institution on earlier elements of the middleware services.

References

Batty M (2007) Fifty years of urban modeling: macro-statics to micro-dynamics. In: Albeverio S, Andrey D, Giordano P, Vancheri A (eds) The dynamics of complex urban systems: an interdisciplinary approach. Physica-Verlag Heidelberg, pp 1–20

Bowman JL, Bradley MA (2008) Activity-based models: approaches used to achieve integration among trips and tours throughout the day. Presented at the 2008 European transport conference, Leeuwenhorst, The Netherlands, Oct 2008

Dong X, Ben-Akiva M, Bowman J, Walker J (2006) Moving from trip-based to activity-based measures of accessibility. Transp Res Part A 40(2):163–180

Ferreira J, Diao M, Zhu Y, Li W, Jiang S (2010) Information infrastructure for research collaboration in land use, transportation, and environmental planning. Transp Res Record J Transp Res Board 2183:85–93

Geertman S, Stillwell J (2009) Planning support systems: content, issues and trends. In Planning support systems best practice and new methods. Springer, Netherland, pp 1–26

Guevara CA (2009) Qualitative analysis of UrbanSim as a modelling test-bed for the SOTUR project. MIT Portugal Working Paper # TSI-SOTUR-09-02

Heer JM, Viegas FB, Wattenberg M, Agrawala M (2008) , ″Point, Talk, Publish: Visualization and the Web,″ in Trends in Interactive Visualisation: State of the Art Survey, Zudilova-Seinstra E, Adriaansen T, van Liere R (eds)., Advanced Information and Knowledge Processing, Berlin, Germany: Springer, ch. pp 269–283

Martínez LM, Viegas JM (2009) Effects of transportation accessibility on residential property values: a hedonic price model in the Lisbon, Portugal, metropolitan area. Transp Res Record J Transp Res Board 2115:127–137

Martínez LM, de Abreu e Silva J, Viegas JM (2010) Assessment of residential location satisfaction in the Lisbon metropolitan area. In: Proceedings of the 89th annual meeting of the transportation research board, Washington, D.C

The MIT/Portugal Program (2010) Research within the transportation systems area is described at: http://www.mitportugal.org/transportation-systems/research.html

Pinto N, Antunes AP (2007) A cellular automata model for the study of small urban areas. 15th European colloquium on theoretical and quantitative geography, Montreux, Switzerland

Waddell P, Ševcíková H, Socha D, Miller E, Nagel K (2005) Opus: an open platform for urban simulation. Presented at the computers in urban planning and urban management conference, London, U.K., June 2005

Waddell P, Wang L, Liu X (2008) UrbanSim: an evolving planning support system for evolving communities. In: Brail R (ed) Planning support systems for cities and regions. Lincoln Institute for Land Policy, Cambridge, pp 103–138

The Participatory Cube: A Framework for Analysis of Online Participation Platforms

Alenka Poplin, Gilberto Corso Pereira and Maria Célia Furtado Rocha

Abstract The main goal of this book chapter is to present a framework for analysis of online participation platforms. Recently, the whole range of various participation platforms emerged and there is a need for a model, which would enable to analyse their specific characteristics. The framework presented in this chapter, the participatory cube, is based on models proposed by Fung (2006) and Ferber et al. (2007). It consists of three axes: interactive communication, access to space of participation, and decision power. These categories play a major role in the analysis of the implemented study cases. The study cases were taken from Germany and Brazil. We concentrated on the selection of a variety of technologies that support civic engagement. The participatory cube served as the model for the comparison of the selected cases. We conclude the article with a discussion about the presented model and further research directions.

1 Introduction

The public sphere is changing in many ways due to the use of new technologies; especially Internet plays a major role in altering our daily lives. The concept of a public sphere was first defined by Jurgen Habermas as "a realm of our social life, in which something approaching public opinion can be formed" (Habermas 1974: 49).

A. Poplin (✉)
HafenCity University, Hamburg, Germany
e-mail: alenka.poplin@hcu-hamburg.de

G. C. Pereira
Federal University of Bahia, Salvador, Brazil
e-mail: corso@ufba.br

M. C. F. Rocha
PRODEB, Salvador, Brazil
e-mail: mariacelia.rocha@prodeb.ba.gov.br

S. Geertman et al. (eds.), *Planning Support Systems for Sustainable Urban Development*, Lecture Notes in Geoinformation and Cartography,
DOI: 10.1007/978-3-642-37533-0_23,

It presents a domain of social life in which public opinions can be expressed. Its ultimate goal is public discourse and debate related to a variety of issues relevant for the society and individuals. The dialogue can focus on the unsustainability of the urban development or on the needs to implement more sustainable planning alternatives as discussed in Wheeler and Beatley (2004) and Wheeler (2004). Ebenezer "Howard exemplifies how some of the most revolutionary ideas in city planning have come from concerned citizens rather than from professional planners and architects" (Wheeler and Beatley 2004: 11). Amado et al. (2010) report about the development of urban planning activities in protected landscape areas of the Sintra-Cascais Natural Park in Portugal and the implementation of participatory process that can support sustainable development of these protected landscape areas.

Internet can enable such dialogues and public participation leading to more sustainable planning. However, it also has some disadvantages. Papacharissi (2002) identified three issues related to the capacity of Internet to improve democratic exchange of ideas; access to information, the possibility of meeting people with different backgrounds and the future considering the commercialization of this environment. Regarding the access to information, she stressed that greater access to information is useful, but not enough to expand the public sphere. With respect to the possibility of meeting people, she claims that the Internet may become more fragmented with the virtual mass of people being subdivided into small groups of interest. The mere existence of a virtual space does not guarantee democratic and rational discourse. She concludes that Internet can provide a forum for political deliberation, facilitate discussions and promote the democratic exchange of ideas and opinions, but it does not necessarily guarantee a sustainable public sphere.

Gomes (2006) highlights the exciting hopes about the possibilities of democratizing Internet until the second half of the 90s of the last century. He also lists problems encountered by the use of Internet, which are: the qualification of political information, which is generally biased once produced by political agents; inequality of access, and the prevalent political culture in the sense of apathy. According to him, (Gomes 2006: 75) "a one-dimensional view of the internet which was seen only as an instrument for progress and democracy" can be a possible problem.

In our research we focus on a multi-dimensional view of the Internet and a constant growth of different online applications aim to facilitate online civic engagement. The main goal of this chapter is to design a model that can enable to analyze this growing number of online participatory platforms. Based on the work of Fung (2006) and Ferber et al. (2007) we designed a framework called participatory cube. The participatory cube in general can be used for the analysis of online participatory platforms. We tested its feasibility and usability on some selected examples of participatory platforms developed in Brazil and Europe. We are particularly interested in analyzing online participatory applications that allow citizens, either as an individual or as member of civil society organizations, to participate in public debates, to express their opinions, suggestions, and informations, or try new ways and solutions to problems and issues raised by contemporary urban life.

2 Previous Work

2.1 The Virtual Sphere 2.0

In her book chapter, Papacharissi (2009) discusses the democratizing powers of online media. She questions whether the idea of a virtual public sphere is indeed the best way to describe the democratizing potential of the internet. She argues that the current uses have little in common with the Habermasian public sphere. According to her, the interactions enabled by web 2.0 reinforce contemporary values of self-expression and personalization, contemplating, the organization of information based on a subjective order of importance, and the proliferation of personal/private space online such as blogs and social networks. Blogs and forums may encourage the proliferation of voices that could expand the public sphere reaching the goals that are beyond simply connecting with others. According to Papacharissi (2009), in a contemporary democracy, the citizen can act from a private sphere, whereas previously their engagement would have been activated through the public sphere. In this private sphere, even alone, this citizen is part of the public sphere, and not completely isolated.

2.2 Urban Sensing

Recent developments in technologies enable citizens to share their knowledge and information with other interested users of participation platforms. Campbell et al. (2006) refer to such processes as people centric urban sensing in which citizens act as sensors. Campbell et al. (2008) claim that "the users are the key architectural system component, enabling a host of new application areas such as personal, public, and social sensing". Participatory sensing is according to Goldman et al. (2009) defined as "a new collective capacity…in which people participate in sensing and analysing aspects of their lives that were previously invisible." Resch et al. (2009) refer to it as "Live Geography". In their chapter they present their approach as a standards-based distributed architecture combining current sensor data with complex event processing mechanisms. The sensors in this case are part of the geo-sensor network including measurement devices, Global Navigation Satellite System (GNSS) connectivity and basic processing capabilities. Their architecture specifically identifies citizens as sensors.

2.3 Volunteered Geographic Information

Goodchild in his earlier publications (Goodchild 2007a, b) focuses on a particular use of geographic information and citizens collecting and contributing geographic information via online platforms. Examples of such applications are Wikimapia and OpenStreetMap which enable the citizens to create global network of "mappers". Goodchild observes "the widespread engagement of large numbers of private citizens, often with little in the way of formal qualifications, in the creation of geographic information, a function that for centuries has been reserved to official agencies" (Goodchild 2007c). He termed this phenomena volunteered geographic information (VGI) as a special case of a more general Web phenomenon of the "user generated content" where the users voluntarily contribute their knowledge, information, data and ideas with the help of online participatory platforms. Coleman et al (2009) investigated why people volunteer and contribute geographic information on the platforms available online.

Lee (2007) in her work investigates the shift from a knowledge repository approach to a collaborative foundation of knowledge management. She refers to it as a collaborative intelligence in which the parties involved reach each other through collaboration. Collaborative intelligence is defined as the measure of the collaborative ability of an entity or a group. Knowledge derived from collaborative efforts is proportionally increasing to the magnitude of the World Wide Web.

2.4 Dimensions of Participation

Fung (2006) in his article "Varieties of Participation in Complex Governance" describes three dimensions of direct participation. The first dimension relates to those who participate in a participatory process. The second dimension specifies how the participants share information and make decisions. The third dimension describes the link between discussions and political/public actions. These three dimensions—scope of participation, mode of communication and decision-making, and extent of authority—constitute a three-dimensional space in which any specific mechanism of public decision can be located.

Fung (2006) additionally remarks that although the vast majority of public participation is open to all who wish to participate, those who choose to participate are often not the representatives of a large audience. He describes a scale of participatory mechanisms that runs from the "only presence of experts or elected representatives", towards the "mini-publics" (among them are representatives selected from the general population and/or an audience interested in the subject), up to the "macro-public" composed by diffuse public sphere of mass media, and local associations.

In the communicative and decision making dimension, the scale proposed by Fung (2006) runs from the less to the more intense level of investment, knowledge and commitment required by participants. According to him, the vast majority of people, who participate in events such as public hearings and community meetings, does not present their own views. They participate as observers who receive information about policies or projects from others. They represent interest groups or organizations in a public dialogue. Other participatory environments, however, are organized to allow participants to explore, develop, and perhaps transform their preferences and perspectives. In this case, participants usually discuss with others, often in small groups, rather than simply listen to the experts, politicians or lawyers.

In some cases, there is an attempt to develop a collective choice by combining various methods of decision-making. The most common one, according to Fung (2006), is "aggregation and negotiation". In such situations, participants know what they want, and the decision-making is carried on by the aggregation of preferences. "Deliberation and negotiation" is the second mode of decision making in which participants discover what they want by sharing experiences and reasons, developing their views and discovering their interests in a group context.

3 Designing a Framework for the Analysis: The Participatory Cube

3.1 The Main Goal of the Model

The concepts such as volunteered geographic information, participatory sensing, and user generated content are not just academic, theoretical models. In the last few years, they resulted in a variety of applications and online platforms designed for the citizens or organizations with the aim to share their knowledge, information or ideas. There is a need for a framework which would enable to analyze these applications in a systematic way. Based on the work of Fung (2006) and Ferber et al. (2007) we propose the participatory cube as a framework for the analysis of a variety of participatory applications available online. It consists of three axes which form a cube and represent three dimensions identified as the most relevant for the analysis. These dimensions are: decision power, interactivity of communication, and the access to space of communication.

3.2 Decision Power

Power is often defined as the ability to influence the behavior of others with or without resistance. Lukes (1974) in his book *Power: A Radical View* discusses three dimensions, three faces of power in which a government controls people. These dimensions include decision-making power, non decision-making power and ideological power. Arnstein (1969) discusses the rungs of power in her ladder of citizen participation. The lower levels represent less power and no participation, including what she calls "manipulation" and "therapy". Higher rungs of decision power include informing, consultation, placation, partnership, delegated power and citizen‘ control. The third rung, informing, represents the first step to legitimate citizen participation. Only on the sixth rung, delegated power, the effective participation can be initiated. On the seventh rung, citizens can deliberate and negotiate with government or other stakeholders. On the last rung, citizen control, the political power can be transferred to the citizens in a form of a direct democracy.

The Arnstein's work as the subsequent of Carole Pateman's work—Participation and Democratic Theory—are still widely cited publications. In 1984, another influential author, Benjamin Barber, strengthened the participations' perspective in his book Strong Democracy. Later, from 1989 to 1992, Habermas (1989) converted his notion of public sphere into the idea of public deliberation which becomes the starting point of the contemporary democratic theory called deliberative democracy (Gomes 2008). When analyzing applications of e-participation we shall pay tribute to these models of democracy.

The term digital democracy is often used to refer to the experience with the Internet, online allplications and devices aimed at increasing the potential for civil participation in conducting public affairs and political decisions. Gomes (2005a, b) lists five degrees of popular participation offered by the Internet. Access to information is at the most elemental degree. In the second degree, citizens are consulted about the public agenda and eventually promoting it on these two levels. In the third one, the state provides information and services. The fourth degree combines participative democracy and representative democracy models; the public can intervene in the political decisions. Finally, following the direct democracy model, the professional political sphere vanishes and the public controls political decisions. According to Gomes (2005a, b) there are no examples of efficient implementations of the third, fourth and fifth degree.

3.3 Interactivity of Communication

Ferber et al. (2007) in their article "Cyberdemocracy and Politics Online: A New Model of Interactivity" presented an interaction model that can be applied to electronic communication on the Internet. It enhances structures and practices that encourage deliberation and allow users to become participants. The authors suggested a six-part model of cyber-interactivity. Their model is built on the four-part model of cyber-interactivity first introduced by McMillan (2002a, b). McMillan and Hwang (2002), Hwang and McMillan (2002) distinguishes between one- and two way of communication and low- or high level of the receiver control. He proposes variations of one-way communication: *monologue* with lower receiver control and *feedback* with the user's high level of control over the communication process. According to McMillan (2002a, b) in a two-way communication, *responsive dialogue* represents the lower level of receiver control while *mutual discourse* indicates the high level of the user's control. According to Ferber et al. (2007) communication in both directions, two-way communication, is primarily interpersonal.

Ferber et al. (2007) add a third way of communication which they call three-way communication describing the situations of *controlled response* and *public discourse*. In the case of controlled response, the online applications give users the opportunity to participate. Examples are questionnaires, online forums or bulletin boards. In these cases, the users do not have a significant control over the content. In a questionnaire, the e-participation platform determines both the questions as well as the presentation of the results. Online forums usually require a moderator who controls and moderates the comments published by the participants. In the case of a public discourse, as enabled in some forums and chats, participants have the opportunity to determine and submit almost unrestricted content. The site control is limited to some actions such as deleting the comments because of defamation, obscenity or violation of rules of the site/application. In this case, participants have a high level of control.

3.4 Access to Space of Participation

"Access" in the context of digital participation can reflect technical, social or political aspects (Smith and Craglia 2003). In this chapter we discuss the type of public that has an access to the participatory platforms. Based on prior work done by the blogging community Global Voices, Fung et al. (2010) examined seven cases that successfully enhance accountability in many different political contexts in some developing countries, among them the Brazilian platform Cidade Democrática (www.cidadedemocratica.org.br). The authors noted, that the technological interventions should be defined according to the audience they hope to reach which may consist of mass users, government, media, political elites and Non-Governmental Organizations (NGOs). They found that it is useful to differentiate between mass users (citizens, consumers, residents) and organizational users (journalist, NGOs, governments and corporations) to the extent that the large majority of examined platforms have increased accountability when more centralized users acted upon available information (Fung et al. 2010). In our model we distinguish among access restricted to specific organizations, restricted to organizations devoted to specific themes, or enjoying the freedom of access.

3.5 The Participatory Cube

For the purpose of our analysis we created a model—the participatory cube. We designed a simplified three-dimensional space that enables to analyze the selected examples. The participatory cube is visualized in Fig. 1. We selected *access to the space of participation* as the first dimension due to its importance for placing and shaping public opinion through the democratic exchange of ideas. It includes requirements from belonging to specific organization through belonging to organizations devoted to a specific theme, up to the total freedom to participate, either as citizens or as civil society organizations of any kind.

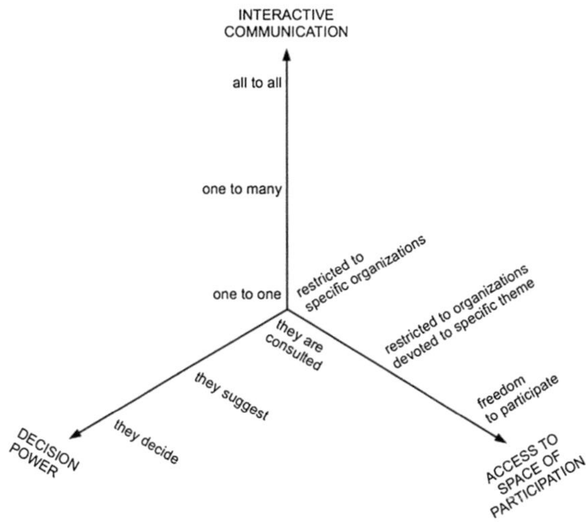

Fig. 1 Participatory cube adapted from Fung (2006) and Ferber et al. (2007)

The second dimension involves modes of *interactivity based* on Ferber et al. (2007). At the lowest level of interactivity, the user (citizens and/or civil society organization) interacts in a one-to-one way with anyone who sends or receives the message. At the intermediate level, it sends message to many, receives messages from many, and interacts with each of the participants. At the highest level of interaction, all users can comment and exchange views with all who wish to participate (all-to-all as in online forums with unrestricted content). Figure 2 represents directions and different ways of communication.

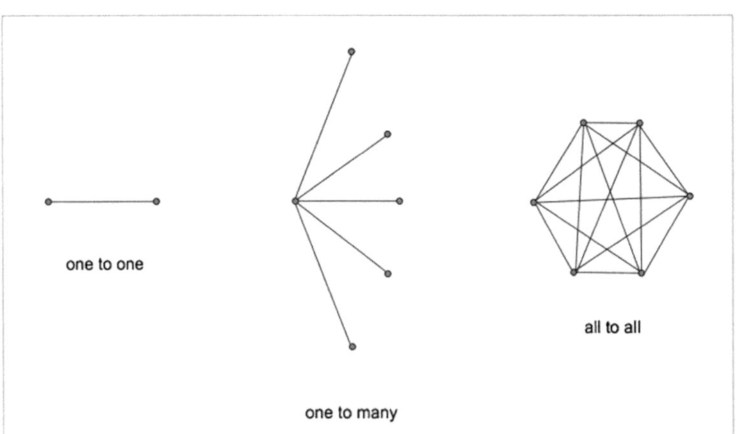

Fig. 2 Direction and ways of communication

The third dimension was inspired by the model of Fung (2006) and is related to the capacity to transform opinions in decisions and actions. At the lowest level participants can only submit their opinions, if consulted. At the medium level of power they may share suggestions of changes or alternatives. At the highest level they can participate in the decision-making. In the first level they can choose between previously selected options offered to them by the application or other participants. In the second level they can propose their own ideas and solutions, and contribute their knowledge, opinions and information. On the highest level the participants have the power to decide.

The participatory cube as presented in Fig. 1 was used as the analytical framework for our comparison of a variety of online participatory platforms. The selected study cases as examples of such platforms were chosen because of their variety in their goals and implementations. They come from two very different countries; Germany and Brazil.

4 Presentation of Case Studies from Germany and Brazil

The case studies selected were the applications in which geographical information may be included explicitly in the form of interactive maps, or implicitly in the form of geographical references (addresses, location, pictures). Aspects related to the content and the technology used for their presentation may be associated on the axis of "access" or the axis which presents the "interactive communication".

4.1 Leehrstandsmelder

The public participation platform Leehrstandsmelder (www.leerstandsmelder.de) is freely accessible online to everybody. Its main goal is to collect the information about empty apartments or offices in Germany. Figure 3 shows the entrance page with a dominant online interactive map available on the left side. On the right side the users can review the news relevant to the topic of empty real estates in different towns in Germany. The platform serves as a central hub for the information collected about empty real estate, and enables a quick search for an empty real estate in any of the cities in Germany. The platform is regularly used and there is a high number of entered empty sites in some of the Germany cities. Examples: 642 inserted empty real estates in Hamburg, followed by Bremen wit 391 and Berlin with 319 in November 2012.

Leehrstandsmelder was developed by an association Gängeviertel e.V. The rights and obligations of the users are clearly described on the website. The users of the platform have to officially register in order to be able to enter the empty real estate on the interactive map. The metadata describing the empty real estate include the street, a picture of the real estate and an additional description of the real estate. Figure 4 demonstrates the situation in Hamburg and a description of one of the empty real estates.

Fig. 3 Lerstandsmelder online

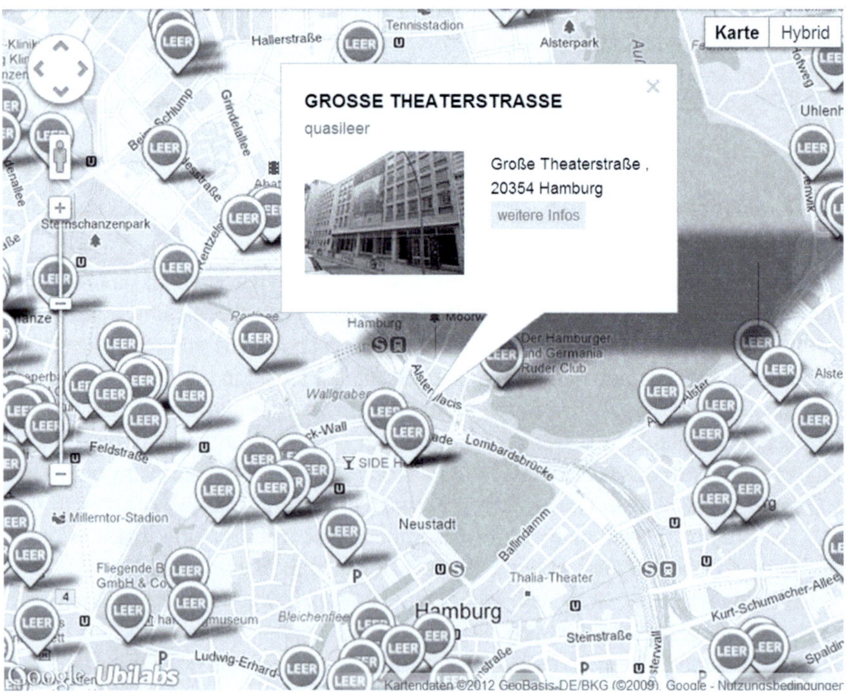

Fig. 4 Empty real estate in Hamburg

4.2 Maerker Brandenburg and Berlin

Maerker Brandenburg (http://maerker.brandenburg.de) is the official website of two states in Hamburg; Brandenburg and Berlin. It aims at getting information about the infrastructure in the commune that needs to be fixed. Figure 5 shows the comments for the commune Bad Belzig in which there are currently 70 contributions. The website also enables quick search according to the given categories, ID or text.

The comments posted on the website by the citizens are marked in different colors. Figure 5 shows a citizen‘s report about the damage on the street; the address of the street and the data are entered precisely. There is no photo attached, instead, the user can click on the viewer and see the situation on the map. Figure 6 demonstrates the location of the damage on the interactive map. Status is reported for every of the comments. In this case the website notifies that this comment has been forwarded to the responsible office which belongs to the city authority.

4.3 Climate Change in Elmshorn

Online public participation platform which enables discussions about climate change was developed for Elmshorn, a city district in Hamburg, Germany. Figure 7 presents the user interface with the interactive map. It is available to everybody under the following address: http://www.elmshorn-klimaanpassung.de. The platform enables the users to participate in a discussion forum, enter their suggestions for the protection against climate change on the interactive map, and vote for other’s suggestions. A short registration is needed for the active users that

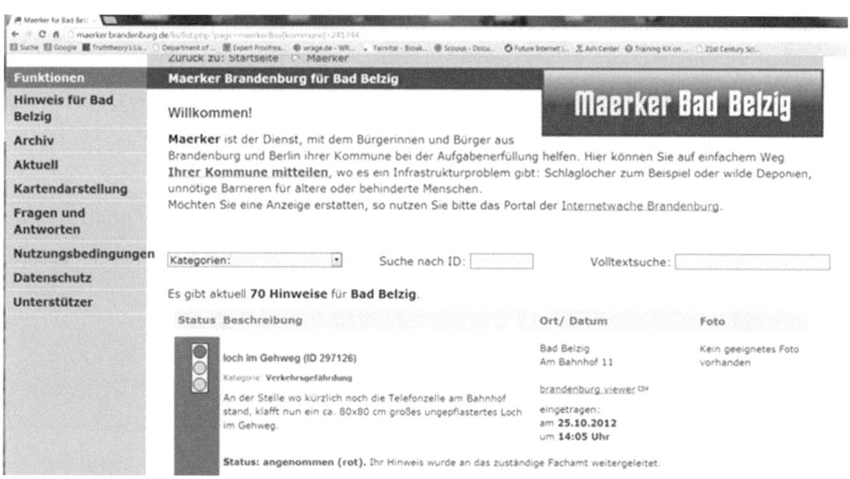

Fig. 5 Maerker Brandenburg

Fig. 6 Maerker Brandenburg viewer

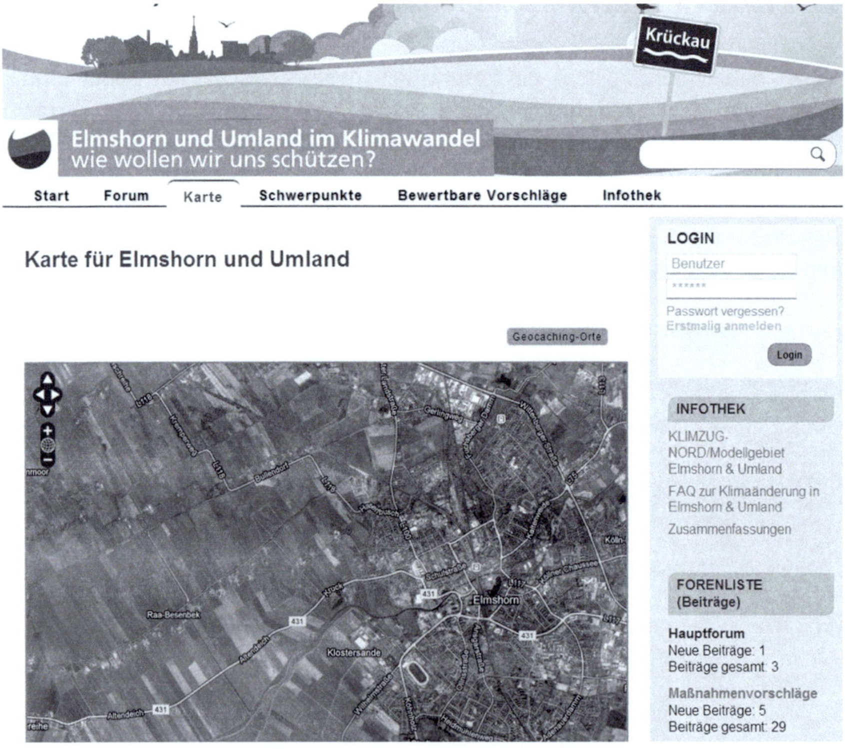

Fig. 7 Map-based discussion about climate change

want to actively contribute to the discussion. The discussion was open to the public in a limited time window between November 5th and 20th 2012. Discussion topics concentrated on the climate change in this particular district, selected as a case study in the project KLIMZUG-NORD (www.klimzugnord.de). A moderator, at the end of the discussion time, summarized the issues addressed by the citizens.

4.4 Urbanias

Urbanias (www.urbanias.com.br) aims to serve as a link between the citizens and the government. Its goal is to help finding solutions for the citizen's demands and needs. The citizen registered on the website can report about a problem in the city of São Paulo, classify it according to the predetermined themes—transit, transport, garbage, noise, etc. She can enter the date of the submission, the address of the problem, and can attach photos or videos. The complaint will then be forwarded to the responsible public agency by the site's team. The application was designed and is managed by a private company. The company received a financial subsidy for the development of the platform from the federal agency for innovation (Fig. 8).

Urbanias is also present on the Facebook through Urbaville application, a kind of a game that allows users to post problems, propose solutions, ask for support and include other messages/demands. It is also possible to post photos, evaluate places and access a list of posts. Using an avatar, it is possible to identify points of interest and participate in a quest where the user is challenged to post sites and certain problems through photos (Fig. 9).

Fig. 8 Urbanias website

Fig. 9 Urbanias on the Facebook

4.5 Portoalegre.cc

The platform Portoalegre.cc (www.portoalegre.cc) enables discussions about the history, present and future of the city of Porto Alegre, capital of State of Rio Grande do Sul. It encourages the creation of issues/problems in order, not only to solve problems and improve the city, but also to promote places positively evaluated. The user has to register on the site and connect to it using their Twitter, Facebook or Google account. The idea is to get civic engagement through spreading the information via social networks. The content is moderated in order to prevent abusive, untrue or offensive information. Launched in 2011, the project had some support from Porto Alegre city hall, but does not aim to be the channel of communication between the users and the city hall officials. Maps are used as the interface for locating issues created by the users (Fig. 10).

4.6 Cidade Democrática

Cidade Democrática (www.cidadedemocratica.org.br) is a platform that enables political participation. It allows the citizens and organizations to express their opinions, communicate and mobilize towards a sustainable city. The platform was designed and implemented by the institute Instituto Seva. Citizens, organizations

Fig. 10 Portoalegre.cc: a suggestion to mobility

or their representatives registered on the platform Cidade Democrática can communicate any problem or suggestion they observed in their city. They can provide locations of these problems with the help of a listing of the states, cities or places in Brazil. The users can classify the problem in a list of predefined categories which are defined by the platform and can be selected in a pull-down menu. The most relevant problems as well as the most active cities, posting the majority of the problems, are listed on the main page. Maps are used in specific cases: see plan for bicycle paths in Jundiaí. The suggestions and/or problems can be also shared through the online social networks connected to the platform. (https://maps.google.com/maps/ms?hl=pt-BR&ie=UTF8&msa=0&msid=108164969110847667788.00047892f01234a45aa92&z=12) (Fig. 11).

5 Results of the Analysis

Our analysis is based on the participatory cube presented in Sect. 3 and the selected case studies of the platforms described in Sect. 4. The participatory cube enables to understand the differences among the variety of participatory platforms and offers a framework for a comparison. The three axes represent the criteria for the comparison, and we comment all three axes on the selected examples.

In the case of the Lehrstandsmelder we observe about freedom of participation; a simple registration of the user enables everybody to participate. This is the highest level in the access to space of participation in the participatory cube. The communication is interactive and can be described as one-to-many. In this case we cannot really talk about power. The participants are not interested in expressing

Fig. 11 Cidade democrática: city of Jundiaí (state of São Paulo)

any power through this platform; it enables an exchange among the users, or simply to gather the information about empty real estate for everybody who needs this information.

Maerker Brandenburg and Berlin also enables freedom to participate, however, the freedom is limited by the knowledge of the citizens. Only the citizens that live either in Brandenburg or in Berlin will most likely have enough knowledge about their environment to be able to report about infrastructural damage. Sometimes tourists would notice such damages, but might not be willing to make an effort to report about them or might have difficulty in expressing themselves in German language. The access is also restricted to a specific theme; in this case, the focus is on the infrastructure in the selected states and communes. The communication is one to many as one user communicates about the damage, and the posts can be seen and read by many users. The power of deciding about the reported damage is hold by the responsible office, which is an example of "they suggest" in the participatory cube.

Climate change in Elmshorn is devoted to a specific theme, but it is not restricted to any organization. In principle everybody can participate even though the participants are limited by their knowledge of this particular area and the knowledge of the topics relevant for the analysis of climate change. The communication can be described as "all to all" because of the online forum included in the platform and the possibility to vote for the best suggestion. In this way everybody can communicate with everybody. The category of power can be described as "they are consulted". This is a research project which aims to enable the dialogue between the experts on climate change and the citizens of the city district Elmshorn. The comments and suggestions provided by the citizens can serve as information to the researchers and experts on climate change.

Urbanias provides free access to any citizen, who registers with an e-mail and a password. It fits the dimensions "one to many" and "they suggest" in the participatory cube. Posts are visible to all users through the selection of topics and are presented on a map. The platform does not enable the users to establish interactive communication with other users outside their social network. Urbanias intends to promote citizenship, sustainability and quality of life through civic engagement and communication between people. It promises to mediate between citizens and government as a way to help solving problems, which seems to encourage a more passive attitude rather than an active engagement and participation. The platform presents a set of issues/problems, and some of them have been resolved. It is not clear whether Urbanias contributed to solving the reported problems. The site has recently changed and the original link redirects users to the Urbanias' Facebook page. There is an ongoing attempt to improve the interaction between the users of the platform. In the future it will be possible to evaluate the success of the platform capabilities in incorporating network locality and stimulate civic engagement with new functionalities.

The platform Portoalegre.cc arises from an experience gained by Unisinos University of Rio dos Sinos Valley in the project related to Redenção Park, a large park located in downtown Porto Alegre. The project Redenção.cc aimed to support the preservation of the historic and cultural public places. Called wikiparque it enables discussions about the history, the reality and the future of specific territories (http://redencao.cc/#/about). The projects Redenção.cc and Portoalegre.cc are based on the creative commons license 3.0 (http://creativecommons.org/licenses/by-nc-sa/3.0/). Portoalegre.cc allows access to all citizens, but it requires connecting to their platform through an online social network (Facebook, Twitter or Google). For the purpose of framing its position in the participatory cube, the platform fits in the "restricted access" devoted to specific themes, allows interaction one-to-many and the participants might suggest, but they are not consulted, nor can decide about interventions in the city.

Cidade Democrática is a mature platform (Fung et al. 2010). It is implemented as an open space for people to point out problems, propose solutions and share opinions about their city. The platform, which was opened to the public in November 2009, hosted the Digital Cities program for building public agendas for collaboration between local governments and society. In December 2011, a technical cooperation agreement between the City Council Chamber of São Paulo and Cidade Democrática was signed in order to change the relationship between representatives and civil society ("São Paulo Democrática" project). Cidade Democrática aims to act as an organizer of civil society and a facilitator between citizens and public authorities enabling discussions of public interest. It has a clear focus on political participation and on finding and sharing solutions to improve the city. It seems to establish itself as a kind of forum for public deliberation for the exchange of ideas which may have consequences for the political decisions—the goal of a deliberative public sphere. In the participatory cube, the initiative is open access to all; interaction is in the one-to-many way since comments are always directed to given user but there are no discussions among users as in a forum.

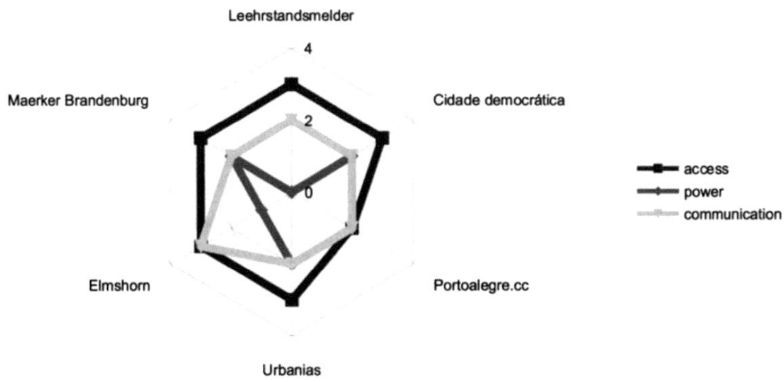

Fig. 12 Results of the analysis

Participants can propose but not decide, except in competitions such as "Cidadonos" where people could vote on proposals about Jundiaí city. The Fig. 12 summarises the results of our analysis and shows that the most common to all platforms is a wide access, centralized communication (one to many) and average power of influence.

6 Conclusions and Further Research Directions

The growing amount of research literature in the areas of volunteered geographic information, crowdsorcing, citizen science, urban sensing and live geography demonstrates the interest of researchers to define and evaluate the novel concepts, ideas and technological implementations that can facilitate participation. In this book chapter we suggested the concept of the participatory cube as a framework for analysis of a variety of implementations following the concepts of participation. The participatory cube concentrates on the aspects of interactive communication with the online participatory platform, the ways the users can access the information published on the platform and its participatory functions, and the decision power that the users can/or cannot gain through their online participatory activities. The use of the participatory cube enabled us to structure our analysis around three suggested axes and provided the framework for the analysis of the selected study cases, online participatory platforms that were designed, developed and implemented in Germany and Brazil.

All of the analyzed platforms, except Portoalegre.cc, enable high level of access. The producers and organizers of participatory processes recognized the importance of access to the information and possibilities to participate. We expect that this trend will continue in the future. We observe initiatives for free and open data available, and a growing need for people to express their opinion freely and through accessible media. The use of maps is still very limited; only some of the participatory platforms include an option of interacting with an online map. These include

Leehrstandsmelder, Climate change in Elmshorn and Urbanias. Interaction with online maps still seems to be rather complex and not all producers value the possibility to locate the problems on a map. We can only assume the reasons for that the lack of map inclusion, which could be the missing knowledge in geographic information systems (GIS) or possible difficulties in maps and digital data acquisition. The online maps can be used for variety purposes in different stages of participatory processes. Inclusions of maps in participatory applications, argumentation with the help of maps, and communication with interactive maps are still challenging and interesting research areas. In our future research we aim to concentrate on the specific characteristics of interaction with online maps, the design of map-based participation platforms and the user's experiences with online maps.

In our future work we will consider the impact of technology on the presented content, the tentative users and the design of the participatory applications. This is also one of the disadvantages of the model presented by Fung (2006); he does not include technology in his analysis. We will concentrate on the characteristics of technologies and their accessibility for different user's groups. Additional research is needed to understand which approaches can enable satisfactory participation processes and how can this success and satisfaction be measured. The main idea, we believe, is to enable platforms and forms that would enable interested citizens to contribute their knowledge and opinions and create sustainable living for everybody.

References

Amado MP, Santos CV, Moura EB, Silva VG (2010) Public participation in sustainable urban planning. Int J Human Social Sci 5(2):102–108

Arnstein SR (1969) A ladder of citizen participation. Am Inst Planners J 35:216–224

Campbell AT, Eisenman SB, Lane ND, Miluzzo E, Peterson RA (2006) People centric urban sensing. 2nd International WS on Wireless Internet (WICON'06)

Campbell AT, Eisenman SB, Lane ND, Miluzzo E, Peterson RA, Lu H, Zheng X, Musolesi M, Fodor K, Ahn GS (2008) The rise of people-centric sensing. Internet Comp 12(4):12–21

Coleman DJ, Georgiadou Y, Labonte J (2009) Volunteered geographic information: the nature and motivation of produsers. Int J Spatial Data Infrastruct Res 4:332–358

Ferber P, Foltz F, Pugliese R (2007) Cyberdemocracy and online politics: A New Model of Interactivity. Bull Sci Technol Soc 27(5):391–400. doi: 10.1177/0270467607304559

Fung A (2006) Varieties of participation in complex governance. Public administration review, pp 66–75, Dec 2006

Fung A, Gilman HR, Shkabatur J (2010) Impact case studies from middle income and developing countries. New technologies. Open Society Foundation. http://www.transparency-initiative.org/reports/impact-case-studies-from-middle-income-and-developing-countries. Accessed in 9 Nov 2012

Goldman J, Shilton K, Burke J, Estrin D, Hansen M, Ramanathan N, Reddy S, Samanta V, Srivastiva M (2009) Participatory sensing: a citizen-powered approach to illuminating the pattern that shape our world. White paper, Woodrow Wilson international center for scholars

Gomes W (2005a) Internet e participação política em sociedades democráticas. Revista FAMECOS, n. 27, ago 2005, Porto Alegre, pp 58–78. http://revistaseletronicas.pucrs.br/ojs/index.php/revistafamecos/article/view/3323. Accessed 31 May 2012

Gomes W (2005b) A democracia digital e o problema da participação civil na decisão política. Revista Fronteiras 7(3):214–222. http://200.144.189.42/ojs/index.php/fronteiras/issue/view/205. Accessed 21 Oct 2012
Gomes W (2006) Apontamentos sobre o conceito de esfera pública política. In: Maia R. e CASTRO, M. C. S (org.) Mídia, esfera pública e identidades coletivas. Belo Horizonte: Editora UFMG, p 49–61
Gomes W, Maia RCM (2008) Comunicação e democracia. São paulo, Paulus. Coleção Comunicação
Goodchild MF (2007a) Citizens as sensors: Web 2.0 and the volunteering of geographic information. Geofocus 7:8–10
Goodchild MF (2007b) Citizens as voluntary sensors: spatial data infrastructure in the world of Web 2.0. Int J Spatial Data Infrastruct Res 2:24–32
Goodchild MF (2007c). Citizens as sensors: the world of volunteered geography. GeoJournal 69(4): 211–221. Reprinted in M. Dodge, R. Kitchin, and C. Perkins, editors, The Map Reader: Theories of Mapping Practice and Cartographic Representation. Wiley, Hoboken, pp 370–378
Habermas J (1974) The public sphere: an encyclopedia article. New Ger Crit 3:49–55
Habermas J (1989) The structural transformation of the public sphere, Massachusetts Institute of technology, Cambridge, MA
Hwang J-S, McMillan SJ (2002) The Role of Interactivity and Involvement in Attitude toward the Web Site. In: Abernathy A (ed.) Proceedings of the 2002 conference of the american academy of advertising. Auburn University, Auburn, pp 10–17
Lee MR (2007) From Web 2.0 to conversational knowledge management: towards collaborative intelligence. J Entrep Res 2(2):47–62
Lukes S (1974) Power: a radical view. Macmillan Press, London
McMillan SJ, Hwang J-S(2002) Measures of perceived Interactivity: an exploration of the role of direction of communication, user control, and time in shaping perceptions of interactivity. Journal of Advertising, 31(3):41-54
McMillan SJ (2002a) Exploring models of interactivity from multiple research traditions: users, documents, and systems. In Lievrouw Livingstone S (Eds.), Handbook of New Media. Sage, London, pp 162–182
McMillan SJ (2002b) A four-part model of cyber-interactivity: some cyber-places are more interactive than others. New Media Soc 4(2):271–291
Papacharissi Z (2002) The virtual sphere. The internet as a public sphere. Sage Publications, London, Thousand Oaks, CA and New Delhi, 4(1):9–27
Papacharissi Z (2009) The virtual sphere 2.0: the internet, the public sphere, and beyond, in the book. Handbook of Internet Politics. In: Chadwick A, Howard P (eds) Routledge, Taylor & Francis Group, London and New York, pp 230–245
Resch B, Mittlboeck M, Girardin F, Britter R, Ratti C (2009) Live geography—Embedded sensing for standardised urban environmental monitoring. Int J on adv in syst and meas 2(2 and 3), ISSN 1942-261x, S.156–167
Smith RS, Craglia M (2003) Digital participation and access to geographic information: a case study of local government in the united Kingdom. URISA J 15, APA II
Wheeler SM (2004) Planning for sustainability: creating livable, equitable and ecological communities. Routledge, New York
Wheeler SM, Beatley T (eds) (2004) The sustainable urban development reader, The Routledge Reader Series, by Routledge, London

Part V
Planning and Policy Support

Application of Socio-Technical Research Methods in Understanding the Genesis and Potential Sustainability of Planning Support Systems

Wayne Williamson and Bruno Parolin

Abstract This research compares two methodologies to gain an insight into which may best fit the research of Planning Support Systems (PSS) used in planning practice. The approach taken by this research is twofold; firstly using data collected through case studies and the application of Actor-Network Theory (ANT), and secondly, an online questionnaire of staff in government and private practice. The questionnaire data was analysed using the Unified Theory of Acceptance and Use of Technology (UTAUT). Results of applying ANT can provide useful insights into the social and technical interactions that are relied upon to build and implement a PSS. Moreover, the UTAUT results found that in order for Information and Communication Technology (ICT) applications to be widely accepted by planners, the organizations in which they work need to address *performance expectancy* and *facilitating conditions* as priorities. Although the methods used in this research are vastly different, results have been found to be somewhat complimentary.

1 Introduction

Research suggests that a mismatch exists between the supply and demand of Planning Support Systems (PSS) that result in an underutilization of software by planners (Geertman 2008). Further, Vonk et al. (2005) argues that because demand is lagging behind the supply, PSS are not reaching maturity due to a slow product life cycle that does not benefit from an information feedback loop from real-world examples, which may be referred to as the PSS *implementation gap*. While research addresses the utility of PSS in research-based scenarios, there is limited

W. Williamson (✉) · B. Parolin
Faculty of the Built Environment, University of New South Wales, Sydney, Australia
e-mail: wayne.williamson@unswalumni.com

S. Geertman et al. (eds.), *Planning Support Systems for Sustainable Urban Development*, Lecture Notes in Geoinformation and Cartography,
DOI: 10.1007/978-3-642-37533-0_24, © Springer-Verlag Berlin Heidelberg 2013

research on the demand side factors involved in the adoption and use of PSS in practice (Geertman 2008).

PSS are typically designed to support the planning process and are usually based on several technologies, while using a common interface (Harris 1989; Harris and Batty 1993; Klosterman 2001; Geertman and Stillwell 2004; Geertman 2006). There are three general types of PSS distinguished by the literature; informing, communicating and analysing PSS (Geertman and Stillwell 2004; Klosterman and Pettit 2005). Informing PSS aim to make information accessible through a flow of information to the user. Communicating PSS aim to facilitate communications and discussions amongst participants in the planning process, while analysing PSS attempt to facilitate advanced processing of data for simulation and evaluation purposes.

Klosterman (2001) predicted that increased use of PSS would be aided by the rapid development of computer hardware and software; however, an inventory of PSS conducted by Geertman and Stillwell (2004), concluded that the majority of PSS had not progressed past the prototype stage, with little evidence of PSS reaching stages of maturity and use in planning practice. The situation is considered somewhat surprising as a small number of PSS have reached a level of maturity that allows them to be sold as off-the-shelf software tools, typically as ArcGIS plug-ins (Klosterman and Pettit 2005; Geertman 2008). Couclelis (2005) argues that the role of land use modelling remains problematic, especially in respect to future orientated scenario planning. Furthermore, PSS developers have some role to play in this situation as they have not provided the tools that planners really need. Geertman and Stillwell (2004) concluded that PSS should meet user and context requirements, including multiple levels of expertise, interdisciplinary perspectives, effective outputs for users, flexibility to focus on different problems and finally a need for more real-world experiences.

Vonk et al. (2005) found in their initial survey results that planning practitioners have little awareness of PSS type tools, which leads to low intentions to start using PSS among possible users. Vonk et al. (2007a) then built on these initial findings with follow-up interviews and a literature review, which concluded that PSS is in an early growth stage, with evidence of this found in some positive experiences by planners with PSS, and more critically, PSS tools are quite complex, while planners are calling for simple tools.

Vonk et al. (2007b) turned their attention to applying an information systems research theory known as the Technology Acceptance Model (TAM) to investigate the formal and informal paths of adoption in a select group of planning organizations in the Netherlands. Their conclusions highlighted that informal employee innovation pathways are used for geographic and information-based software tools in these organizations. Staff are reluctant to use formal pathways because they perceive the formal strategy as less rational than their own ideas; secondly, they do not experience enough social pressure to pursue formal pathways and finally, their own ideas do not have a great influence on formal pathways. The authors concluded by suggesting that organizations need to adopt a learning culture using knowledge management and employ staff with innovation management skills.

Pettit et al. (2008) also found that planners struggle to understand the outputs of PSS and that the introduction of PSS into planning education may help.

More recently, te Brömmelstroet and Schrijen (2010) suggests that PSS developers do not understand the relevance of PSS to daily planning activities; therefore, planners must be heavily involved in the design and analysis phases of the software development life cycle in order to provide planner relevant software products. te Brömmelstroet and Schrijen's (2010) contribution has addressed Geertman's (2008) research agenda by providing information on the development of four real-world software development examples using close collaboration with planning practitioners. These examples effectively transfer detailed knowledge of the planning practitioner's demands, professional skills and knowledge to the supply side developers of PSS.

Batty (2008) summarises that the future of PSS is somewhat inconsistent, with computer hardware and software allowing PSS to be proliferated at greater rates, the ability of software products to be linked together like never before and advances in visualization software that is drifting into an Internet based context will allow for a more distributed use of PSS. However, planning has become fragmented in practice and software tools built to support it have also become unique to specific contexts. This situation does not allow for a coherent framework within which PSS can mature. Ultimately the directions taken by land use planning will determine what direction PSS will take in the future.

This chapter seeks to address the following question: How does the application of socio-technical research methods enhance our understanding of the genesis and potential sustainability of Planning Support Systems. The approach taken by this research to data collection and analysis is twofold; firstly by using the data collected from interviews and the application of Actor-Network Theory (ANT), and secondly, an online questionnaire of planners and GIS staff working in local and state government and private practice. The questionnaire data was analysed using a structural equation modelling (SEM) technique under the Unified Theory of Acceptance and Use of Technology (UTAUT) framework. The results of both analysis techniques are then compared.

This chapter assumes that the majority of PSS attempt to implement sustainable urban development principals. The case studies used in this chapter all approach urban development through the consolidation of new housing in close proximity to existing town centres and transportation, which promotes sustainable transport usage, housing close to employment and more liveable mixed use town centres.

2 Theoretical Frameworks

This section describes the theoretical frameworks employed by this research. These theories have been chosen for their widespread use and their tested ability to be used in the research of socio-technical factors in the adoption of information systems.

2.1 Actor-Network Theory Concepts and Components

This section describes the fundamentals of ANT which was utilized to organize and present the data collected from interviews. ANT was developed in the 1980s with its origins in the sociology of science and technology. ANT is a constructivist theory that follows the strategies used by actors to mobilize other actors and resources to ultimately construct a heterogeneous network, and is usually chosen by researchers because it focuses on the relationships between non-humans and humans, as a central part of the production of scientific knowledge (Rodger et al. 2009). Moreover, ANT is concerned with the processes by which scientific disputes become closed, ideas accepted and tools and methods adopted. The theory places a strong focus on how decisions are made about what is known. These decisions are often temporary, but closing the *black box* on disputes allows people to take the work of others as a resource and move on, rather than continually reproducing and questioning it (Latour 1987), which is an issue observed by Geertman (2008) that is hampering the progression of PSS development.

ANT has two major approaches to data collection. One is to "follow the actor," via interviews and ethnographic research, the other is to examine inscriptions. Inscriptions include texts, but also images of many sorts, databases, and anything else that can be considered central to knowledge creation.

Actors become enrolled in a network through the translation process depicted in Fig. 1. Callon (1986) describes translation as a multi-step process of *problemization, obligatory passage point (OPP), intéressement, enrolment, mobilization* and *black boxing*, which was both introduced and followed by Callon's (1986) study of the scallop industry in North Western France. This important text in ANT literature followed the construction of a diverse actor-network by three researchers which eventually failed once other actors dissented from the network (Callon 1986).

Problemization is the first moment of translation, during which a focal actor frames the problem in its own terms, identifies other relevant actors, and highlights how the problem affects the other actors (Law 1986). Rodger et al. (2009) explains

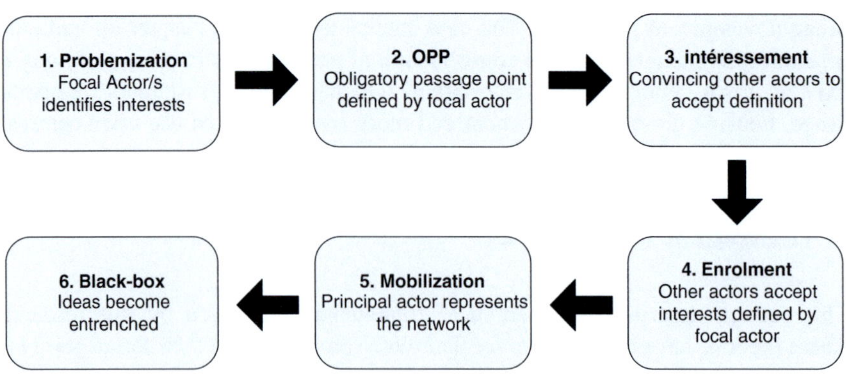

Fig. 1 Translation process (after Rodger et al. 2009)

that ANT is predominantly concerned with how ideas are conceived and the resistance experienced during their development. During *problemization* the focal actor will form an *Obligatory Point of Passage* (OPP) which is a network node acting as an intermediary between networks or network components. A strong OPP will control resources and claim responsibility for a network. The focal actor defines the OPP through which the other actors must pass through and by which the focal actor becomes indispensable (Law and Callon 1992).

The third moment of translation involves a process of convincing other actors to accept the definition of the focal actor (Callon 1986). Actors attempt to impose their priorities on other human and non-human actors. This stage of translation is reliant on the focal actor re-enforcing the identities and associations identified in the *problemization* stage. *Intéressement* is a crucial phase of the process (Callon 1986) and is a competitive part of the building of an actor-network since actors are often implicated in other networks.

In the fourth moment of translation, actors in the network accept or become aligned to interests defined for them by the focal actor. *Enrolment* seeks to reinforce the alliances found earlier in the *intéressement* phase, usually through negotiation (Callon 1986). *Enrolment* can be thought of as the moment that other actors accept the interests defined by the focal actor and become a functioning node in the network.

The fifth stage of translation is *mobilization*, in which actors are given the tools of communication and are able to create an interest in the network or to create sub-networks. *Mobilization* occurs as the proposed solution gains wider acceptance and an even larger network of absent entities is created through some actors acting as spokespersons for others (Tatnall and Burgess 2002). The sixth and final moment of translation is when the network reaches the *black box* stage and is identified by all actors being enrolled and the network is completely stabilized. An indication of a project reaching the *black box* phase is when managers can be seen using derived knowledge for management purposes (Rodger et al. 2009).

Rodger et al. (2009) explains that the strength of ANT is the attention paid to how networks emerge, who and what each network involves and how the network is maintained. Furthermore, ANT explores why some networks become established for a significant period of time, while others fail. Callon (1986) also identifies that well formed networks can become unstable, since the translation process is an ongoing process that can never be considered permanent. New actors, the departure of existing actors and changes in alliances can result in networks being reconsidered, reconfigured or dismantled completely (Callon 1986; Tatnall and Gilding 1999).

Conversely, there are numerous criticisms of ANT. These criticisms are focused on the theories attempt to distance itself from scientifically based methods. ANTs creators (Latour 1987; Callon 1986; Law and Callon 1992) argue that only a descriptive technique should be used, while critics argue that much is lost by describing only what the researcher sees. Criticism is also aimed at the theories biased towards non-human actors over humans, as well as the theory treating organizations as networks, when they could be open to alternative descriptions of the research context. Furthermore, critics argue that the real world cannot be reduced to "responsibility descriptions" in a network, as explanations and causes are equally important for researchers to uncover (Miles 2012).

2.2 Unified Theory of Acceptance and Use of Technology

The UTAUT model is aimed at explaining a user's intention to use information systems and the user's subsequent usage behavior. UTAUT was developed by Venkatesh et al. (2003) who systematically tested and verified nine information systems adoption models and then consolidated the most consistent elements into a unified theory. UTAUT uses four constructs to determine usage behavior; the first three constructs of *performance expectancy, effort expectancy* and *social influence*. A fourth unique construct for the UTAUT model is *facilitating conditions*, which is defined as the degree to which an individual believes that organizational and technical infrastructure exists to support the use of a system. Table 1 lists the four UTAUT constructs and their descriptions.

The theory also uses four moderators, being *gender*, *age*, *experience* and *voluntariness of use*, of an intended user's behavior and their subsequent use behavior. The introduction of demographic moderators such as *age* and *gender* have proven to be a positive aspect for UTAUT in explaining variations of intended use and behavior of information systems users, Venkatesh et al. (2003) notes that these two moderators have been given little attention in other theories. Furthermore, *age* and *experience* were identified as limitations of previously developed adoption models when analysing data related to a users level of experience with a given information system. The UTAUT model is shown in Fig. 2. The four boxes on the left hand side of Fig. 2 represent the constructs outlined in Table 1. The arrows leading across Fig. 2 to behavior intention and use behavior represent the intended influence on a user's behavior. The dashed arrows leading up from the bottom of the figure represent the moderating influence of *gender*, *age*, *experience* and *voluntariness of use* on the constructs. Essentially, the arrows between constructs and moderating influences represent hypotheses that can be tested to understand the likelihood of PSS being used and supported in an organization.

Table 1 Constructs in UTAUT

UTAUT construct	Definition
Performance expectancy	The degree to which an individual believes that using the system will help him or her to attain gains in job performance
Effort expectancy	The degree of ease associated with the use of the system
Social influence	The degree to which an individual perceives that important others believe he or she use an information system
Facilitating conditions	The degree to which an individual believes that an organizational and technical infrastructure exists to support use of the system

Source Venkatesh et al. (2003)

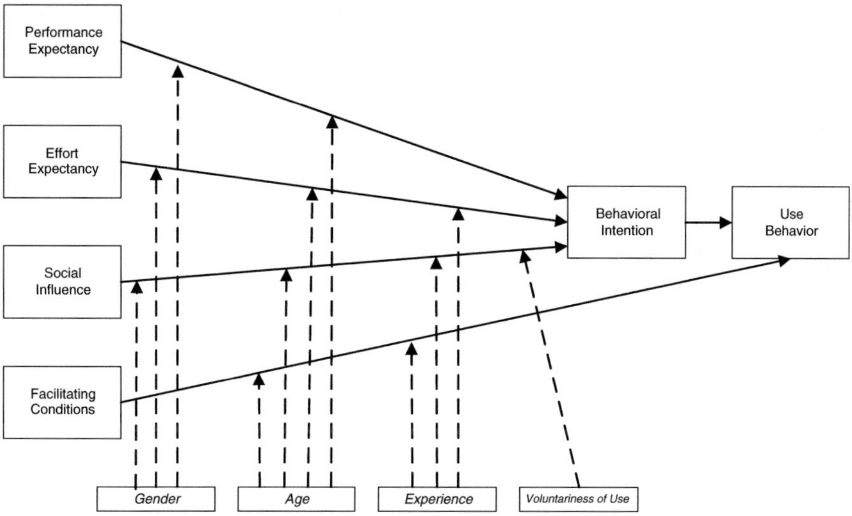

Fig. 2 UTAUT research model (*Source* Venkatesh et al. 2003)

2.3 Comparison of ANT and UTAUT

The main difference between the two is in the approach and definition of social-technical. In ANT, social systems and technology are considered inseparable. The only separation between them is when a technology has been successfully adopted, and therefore, is considered to have reached the *black box* stage and there is evidence that the actor-network is stable. In UTAUT, social systems and technology are regarded as separate. From this perspective, the diffusion of an innovation equates to the adoption of a technology by a social system. UTAUT considers a technology as a set of attributes that have to be demonstrated to potential adopters. Once a technology is distributed, it should be just a matter of time before all, except the most immovable, recognise its advantages (McMaster et al. 1997; McMaster and Wastell 2005; Mitev 2005).

3 Interview Methodology

Seven potential case studies were investigated. Two potential case studies were not used because they only used spreadsheets; a third case study was abandoned due to potential interviewees not being willing to participate; and one remaining case study was abandoned due to a lack of access to interviewees.

The case studies were identified by reviewing recent literature published about the New South Wales (NSW) planning system (Australia). In two cases, descriptions of the projects were found in a NSW Department of Planning publication

(Department of Planning 2007), while the third was discovered through word-of-mouth. The case studies that proceeded included; a *Community Viz* implementation; the METRIX sub-regional planning tool that was used to guide the distribution of in-fill development to meet metropolitan sub-regional dwelling targets; and thirdly, a custom built public participation PSS that generated electronic maps according to criteria entered by public users (Williamson and Parolin 2012).

Due to the difficulties encountered, there was no pre-determined case study strategy, other than to locate PSS and people willing to discuss their implementation. This situation concurs with the literature (Geertman and Stillwell 2004; Vonk et al. 2005) that finds a lack of communication through feedback loops between PSS supply and demand.

The interviewer did not use a standard set of questions or a script to conduct interviews, because each project was markedly different, and the people interviewed were from varying backgrounds. However, the interviews followed the translation process; hence the interviewer informally structured the interviews in the following manner.

After providing an initial overview of the research project and data collection technique, the interviews started with questions related to the *Problemization* phase. The discussions were aimed at teasing out what the interviewee believed to be the problem, how this problem was conceived, how a potential solution was identified and the main actors that were involved. Discussion was also directed on funding sources, planning rules and regulations, the use of consultants, internal management structures and external agencies involved. The interviewer started sketching an association diagram with the interviewee at this early stage as the interviewee nominated the relevant actors involved. The association diagram concept was adopted from Martin (2000) GIS-based research. This idea evolved through the data collection phase as a simple way of explaining how ANT was being used. It was found that interviewees were able to understand the visual concepts of ANT depicted in an association diagram with relative ease.

The initial stage of the interview also attempted to identify the Obligatory Point of Passage (OPP) being the actor who operates as a network node acting as an intermediary between all network components. Effectively, a strong OPP will control resources and claim responsibility for a network. In all three case studies, the OPP was interviewed.

The interview then moved into the *intéressement* phase where discussion refocused on how the OPP initially gains the interest of other actors and persuades them to accept the problem definition. Interview discussion was focused on any incentives or alike offered to staff to participate and the reasons given by potential actors for not accepting offers to participate in a project.

From the *intéressement* phase, the interviews then moved towards discussing how the OPP formally *enrolled* actors into the network as the project begins to function. This part of the interview needs to focus on the stage of the project where actors accept the problem defined and become functioning nodes in the network. Interview discussion should focus on the roles and responsibilities of all people involved in the project.

The fifth stage of translation is *mobilization*. The goal of the interviewer is now to gather information regarding the communication of a project and any evidence of sub-networks emerging. These are signs of wider acceptance by actors in the network that result in the distribution of information and knowledge to new actors in other networks. Discussion should focus on communication activities and conferences or industry presentations, as well as any activities of others enrolled in the network.

Information that has been gathered during the interview will lead the final stages of the interview regarding the *black box* stage of the project. The interviewer needs to focus on two specific aspects of the project; firstly, how stable the actors are in the network, for instance, has there been many personnel and/or management changes, and secondly; is there evidence of information being derived from the project as tangible output. These are considered important signs that a project has reached the *black box* stage and is functioning satisfactorily.

4 Questionnaire Methodology

The questionnaire was distributed via email to council customer service personnel, asking for the email to be distributed to the strategic and development assessment managers within the council. Targeted respondents were local government staff whose primary role is in strategic planning or development assessment and staff whose primary role involves the use or support of GIS. The email contained a participant information sheet and http link to the online questionnaire. It is acknowledged that bias is introduced by the snowball method that reduces the likelihood of a good representation of the population. Bias is most likely to occur as a result of the over-representation of planners in metropolitan areas. The snowball method was chosen for this research because;

- the researcher is targeting a particular population group;
- the sample population size is unknown, but relatively small;
- the sample population is geographically dispersed; and,
- there is no publicly available listing of contact details.

The questionnaire collected 167 useable responses. The response rate between males and females was 51.5 and 48.5 % respectively. The average age of respondents is 35. Nearly two thirds of the respondents are between 26 and 35 years old. The average number of years work experience is 10.4. Almost 48 % of respondents had between 6 and 10 years experience; 25.7 % of respondents had less than 5 years experience. The majority of respondents work in management and planning officer positions. Responses were also collected from Geographic Information Systems (GIS) and Information Technology (IT) staff. Respondents were drawn from all regions of the state of NSW. The highest response rate was from the Sydney metropolitan area at almost 40 %, Coastal NSW was the second highest response rate at 20.5 %.

5 Compare and Contrast Research Method Results

The results generated from the UTAUT analysis and shown in Fig. 3 could be interrupted as planners do not seek out ICT as a tool and are not influenced by direct or indirect exposure to the use of ICT in their work places. This apparent attitude also shows itself in the *effort expectancy* result, where *ease of use* is considered a less important influencing factor.

The path diagram in Fig. 3 presents the R^2 statistic for *behavioural intention* and *use behaviour*. The R^2 refers to the fraction of variance explained by the model. The R^2 value for *behavioural intention* is 0.154, which shows that *performance expectancy*, *effort expectancy* and *social influence* together account for 15 % of the variance of *behavioural intention*. This result suggests that there are other factors at play which could further explain what influences the *behavioural intention* of planners to use ICT. Secondly, the R^2 value for *use behaviour* is 0.512, which shows *facilitating conditions* and *behavioural intention* together account for 51 % of the variance of *use behaviour*. This suggests that *behavioural intention* and to a greater extent *facilitating conditions* are significant influencing factors on *use behaviour* of planners in regards to ICT usage.

Performance expectancy was revealed as a moderate influence on behavioral intention. The perceived usefulness of ICT as a tool for obtaining job performance improvements is a factor that emerged from the ANT case studies in the form of *voluntary* versus *involuntary use*. The situation for PSS in all case studies was *voluntary use* and, as such, a PSS would need to display clear signs of usefulness to persuade a potential user to switch from their current practice. This insight

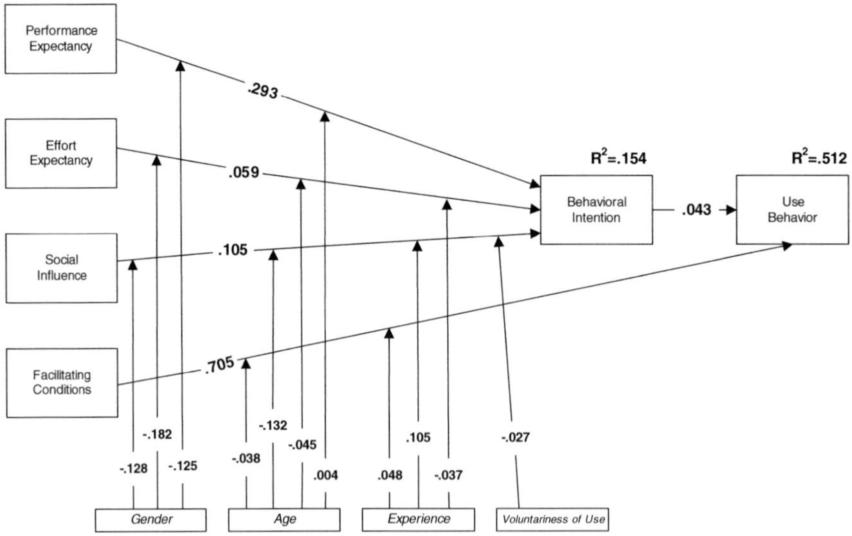

Fig. 3 Path diagram of UTAUT results

provides evidence of the *implementation gap* that exists between PSS and its potential users.

The UTAUT analysis revealed that *effort expectancy* is not a significant influence for potential users. Conversely, the ANT case studies demonstrated a situation where the usability of the PSS interface was a major deterrent of sustained use by planning staff. The ANT case studies also exhibited a clear failure to take user comments into account and attempt to improve the user interface. While UTAUT provided a clear result, ANT results took it a step further by providing potential reason for the situation.

Facilitating conditions is one of the standout results from the UTAUT analysis and there is no doubt that a supportive environment will aid the adoption of any software. However, this result also reveals that the sample contains experienced users who understand the value of a supportive environment. Conversely, the ANT case studies reveal that users have a low level of experience with PSS and little regard for offers of support. Research shows that *facilitating conditions* fluctuate with a user's level of experience.

Social influence was found to be a weak influence for use intention, which suggests *social influence* does not play a significant role in the adoption of ICT in the workplace of planners. The results were found to be conflicting on this subject, as the ANT case studies uncovered evidence of diffusion across organizations. However, the ANT case studies presented detailed evidence of unsuccessful staff attempts to diffuse PSS across organizations. Examples from all case studies illustrated the failed attempts to gain *enrolment* from planning staff and management. Moreover, the failed diffusion to planning staff and management was the ultimate weakness for all case studies. It is not fully understood why *social influence* is not stronger for respondents in this sample. Organizational structure and work place culture could be contributing factors and would be appropriate avenues for further research.

The *gender*, *age*, *experience* and *voluntariness* moderators have yielded a mixed bag of results for this sample. Factors that may have contributed to this include a sample that represents a fairly small range of *age* and *experience*. The similarity of planners' education attainment could also play a significant role in results showing little difference between *gender*.

Although the ANT and UTAUT methods used different approaches to analysis, the results are somewhat complimentary. In particular, the UTAUT areas of *facilitating conditions* and *social influence* demonstrate clear indicators of user behaviour towards using PSS while the more detailed case studies using ANT point to actual examples of diffusion and adoption failures. The case studies provided examples of various organisational entry points for PSS applications. The least successful entry point was via GIS staff, while the PSS with the longest continuous operation was introduced by management.

The case studies also demonstrate that even though ideas can be conceived by a planner and stakeholders were involved throughout the design process, as recommended by te Brömmelstroet and Schrijen (2010), a PSS implementation may only be a semi-permanent success. This suggests there is a need for sustainable funding for the ongoing maintenance and human resources to support of the PSS.

Furthermore, the case studies demonstrated a dependence on a single actor. Although two of the three case studies achieved some form of network stability, referred to as black boxing in ANT, it cannot be considered as a permanent situation. The main reasons for network instability stem from actors exiting, the reliance of an entire network on a single actor, or an inability to secure consistent funding to stabilise the network. It is simple to suggest that the best way to eliminate this issue would be to take the dependence away from the single actor. However, these case studies suggest that many occurrences of PSS are attributed to a single person's idea, who usually takes responsibility for driving the project to implementation.

6 Discussion

The insights gained from UTAUT and ANT analysis suggests that the introduction of software is not a value neutral innovation that is just adopted by an organization, but rather, software can be viewed as a non-human actor that translates and is translated throughout the adoption process. Secondly, there is a need for an ongoing translation process which will determine the success or failure of the adoption process. The process of introducing ICT tools into an organization is a process of network formation in which actors seek to persuade others to become enrolled and promote the acceptance of their view of the way the problem can best be solved (Tatnall 2000). In this process, the PSS itself is one of the prominent actors involved in shaping the network. The ongoing translation process is exemplified by Aslıgül Göçmen and Ventura (2010) who found that the awareness of GIS has increased, access to geospatial data and staff training has improved and costs have fallen, but planning agencies still face barriers to GIS use. The issues of organizational coordination and conflicts, management support and data standards and integration have given way to issues related to keeping up with the technology.

The ANT case studies suggest that the design and implementation of PSS consists of ideas that may or may not be accepted by potential users. These ideas are established by the focal actor and shape the functionality of the software. During the translation the content may change according to influences of different actors based on their own ideas. Once translated the use of the software may also be different to the initial idea. Just as human actors in a network bring their own ideas, politics, definitions of roles, responsibilities and agendas, so does software.

The success of the translation process may also be the result of failures. Latour (1986) argues that innovations travel through time in the hands of actors who may accept it, modify it, deflect it, betray it, add to it, appropriate it or let it drop. What is absolute is that faithful acceptance involving no change is rare (Tatnall 2000). An innovation requires an ongoing effort to keep the innovation alive; without the ongoing translation process, the notion will cease to exist (Monteiro 2000). Furthermore, Law and Callon (1992) argue that an innovation employed only once is a myth. The translation process is ongoing, rather than being achieved once and for all. A clear success or failure of a translation process is also a bit to simple, as software may not have a dramatic immediate effect, but a gradual effect over time. It is unrealistic that all actors will pick up and align their interests with a new method. The perceived failures of past implementation can be considered building blocks in the gradual process of adoption. Essentially, it is not a matter of adopting a tool, but the translation of ideas and techniques packaged into the PSS that will became an active member of an actor-network.

ANT suggests researchers should view the translation process as an ongoing process that changes over time as requirements evolve, actors interests change and different actors enter and exit the network. This implies that sustained use of PSS in planning practice may be provided through the deployment of the software as a service (SaaS) model in which a provider centrally locates the software and associated data services, while potential users access the software on a demand basis (Simon 2010; Brail 2009). The SaaS model could enable the realignment of planning support services to individual customers, while maintaining a stable core network of actors. This may provide PSS that achieve the most stable long-term use, rather than the packaged software and training approach that has been pursued, but lacks the feedback loop identified by Vonk et al. (2005, 2007a).

When software such as PSS is introduced into a planning agency, the effect is to impose a method or technique of land use planning on the planning professionals. In a somewhat voluntary environment for using ICT tools, this method has not proved to be successful (Vonk et al. 2005). By following the translation phases, the researcher can gain insights into where and when barriers appear, and how the network negotiates them or is damaged by them. People attach themselves to a practice for good reason. Moving to change these practices requires a process whereby current practices are critically evaluated, including the habitual aspects and alternatives imagined to transform the practice (Binder and Boldero 2012). This does not criticize the methods introduced by PSS, on the contrary; it questions the habitual practices of planners using current techniques, with little more reason than they have just always done things that way (Moore 2008).

What may work for one organization, may not work for another and vice versa. What is clear is the lack of literature focused on day-to-day planning practice, in Australia at least, how technologies may assist these practices and how they may be introduced. The introduction and use of GIS is well advanced compared to PSS, but even with funding and management level support, these technologies have taken decades of continual translation to become recognized as functioning actors in the networks that make up the NSW planning system. As actors come and go the planning system has realigned itself on several occasions, and the staff interested in GIS have realigned as well. The attempts to introduce PSS into planning practice can be viewed as part of the ongoing translation process. If the knowledge gained from ANT research is to be taken onboard, two things are for certain; the process of introducing PSS into planning practice will never be complete and secondly, PSS is unlikely to be used exactly as envisaged by its creators.

7 Conclusion

Although the ANT and UTAUT methods used in this research have vastly different approaches to analysis, the results have been found to be somewhat complimentary. Especially for the UTAUT constructs of *facilitating conditions* and *social influence* where the sample demonstrates clear indicators of user intentions towards using ICT, while the more detailed project level information collected using ANT points to actual examples of diffusion and adoption failures. It can be concluded that a pluralist approach to PSS research is a positive approach.

This chapter makes suggestions on how to improve the implementation of PSS in planning practice; however, further research is needed on the organizational structure, workplace culture and day-to-day tasks undertaken by planners with links to PSS functionality, where possible. An understanding of these factors may result in a more sustained use of PSS in planning practice. Comparative examples of PSS adoption and use using socio-technical methods from several counties are also needed.

The ANT method provides the researcher with a satisfactory approach to understanding, what is commonly, a complex socio-technical environment in the planners' work place. For this reasons, this chapter concludes that ANT is the more likely research method to gain insights into the demand side issues with PSS adoption and use.

References

Aslıgül Göçmen A, Ventura SJ (2010) Barriers to GIS use in planning. J Am Plann Assoc 76(2):172–183

Batty M (2008) Planning support systems: progress, predictions, and speculations on the shape of things to come. In: Brail R (ed) Planning support systems for cities and regions. Lincoln Institute of Land Policy, Cambridge

Binder G, Boldero JM (2012) Planning for change: the roles of habitual practice and habitus in planning practice. Urban Policy Res 30(2):175–188

Brail RK (2009) Planning support systems: bridging the gap between technology and user. Paper presented at 11th international conference on computers in urban planning and urban management, University of Hong Kong, Hong Kong, 16–18 June

Callon M (1986) Some elements of sociology of translation: domestication of the scallops and the fishermen of St. Brieuc Bay. In: Law J (ed) Power, action, and belief: a new sociology of knowledge? Sociological review monograph no. 32. Routledge & Kegan Paul, London

Couclelis H (2005) Where has the future gone? Rethinking the role of integrated land use models in spatial planning. Environ Plan A 37(8):1353–1371

Department of Planning (2007) Improving the NSW planning system: discussion paper, Sydney

Geertman S (2006) Potentials for planning support: a planning-conceptual approach. Environ Plan B: Plan Des 33:863–880

Geertman S (2008) Planning support systems: a planner's perspective. In: Brail RK (ed) Planning support systems for cities and regions. Lincoln Institute of Land Policy, Cambridge

Geertman S, Stillwell J (2004) Planning support systems: an inventory of current practice. Comput Environ Urban Syst 28(2):291–310

Harris B (1989) Beyond geographic information systems-computers and the planning professional. J Am Plan Assoc 55(1):85–90
Harris B, Batty M (1993) Locational models geographic information and planning support systems. J Plan Educ Res 12(2):184–198
Klosterman RE (2001) Planning support systems: a new perspective on computer aided planning. In: Klosterman RE, Brail RK (eds) Planning support systems: integrated geographic information systems, models and visualization tools. ESRI Press, California
Klosterman RE, Pettit C (2005) Guest editorial: an update on planning support systems. Environ Plan B Plan Des 32(4):477–484
Latour B (1986) The powers of association. In: Law J (ed) Power, action and belief: a new sociology of knowledge? sociological review monograph 32. Routledge & Kegan Paul, London
Latour B (1987) Science in action: how to follow scientists and engineers through society. Harvard University Press, Cambridge
Law J (1986) The heterogeneity of texts. In: Callon M, Law J, Rip A (eds) Mapping the dynamics of science and technology. Macmillan Press, Basingstoke
Law J, Callon M (1992) The life and death of an aircraft: a network analysis of technical change. In: Bijker WE, Law J (eds) Shaping technology/building society: studies in sociotechnical change. MIT Press, Cambridge
Martin EW (2000) Actor-networks and implementation: examples from conservation GIS in ecuador. Int J Geogr Inf Sci 14(8):715–738
McMaster T, Wastell D (2005) Diffusion—or delusion? challenging an IS research tradition. Inf Technol People 18(4):383–404
McMaster T, Vidgen RT, Wastell DG (1997) Towards an understanding of technology in transition: two conflicting theories. Paper presented at information systems research in Scandinavia, IRIS20 conference, University of Oslo, Hanko
Miles JA (2012) Management and organizational theory. Jossey-Bass, San Francisco
Mitev NN (2005) Are social constructivist approaches critical? The case if IS failure. In: Howcroft D, Truath EM (eds) Handbook of critical information systems research: theory and application. Edward Elgar Publishing, Northampton
Monteiro E (2000) Actor-network theory and information infrastructure. In: Ciborra CU (ed) From control to drift: the dynamics of corporate information infrastructures. Oxford University Press, Oxford
Moore T (2008) Planning support systems: what are practicing planners looking for? In: Brail RK (ed) Planning support systems for cities and regions. Lincoln Institute of Land Policy, Cambridge
Pettit C, Keysers J, Bishop I, Klosterman RE (2008) Applying the what if? planning support system for better understanding urban fringe growth. In: Pettit C, Cartwright W, Bishop I, Lowell K, Puller D, Duncan D (eds) Landscape analysis and visualization: spatial models for natural resource management and planning. Springer, Berlin
Rodger K, Moore SA, Newsome D (2009) Wildlife tourism science and actor network theory. Ann Tourism Res 36(4):645–666
Simon P (2010) The next wave of technologies: opportunities in chaos. Wiley, New Jersey
Tatnall A (2000) Working paper: information systems innovation—two different models. School of Information Systems, Victoria University, Victory, Australia
Tatnall A, Burgess S (2002) Using actor-network theory to research the implementation of B–B portal for regional SMEs in Melbourne, Australia. Paper presented at 15th electronic commerce conference and reality: constructing the e-economy, Bled, Slovenia
Tatnall A, Gilding A (1999) Actor-network theory and information systems research. Paper presented at 10th Australasian conference in information systems
te Brömmelstroet MCG, Schrijnen PM (2010) From planning support systems to mediated planning support: a structured dialogue to overcome the implementation gap. Environ Plan B: Plan Des 37(1):3–20

Venkatesh V, Morris M, Davis G, Davis F (2003) User acceptance of information technology: toward a unified view. MIS Q 27(3):425–478
Vonk G, Geertman S, Schot P (2005) Bottlenecks blocking widespread usage of planning support systems. Environ Plan A 37(5):909–924
Vonk G, Geertman S, Schot P (2007a) A SWOT analysis of planning support systems. Environ Plan A 39:1699–1714
Vonk G, Geertman S, Schot P (2007b) New technologies stuck in old hierarchies: the diffusion of geo-information technologies in Dutch public organizations. Public Adm Rev 67(4):745–756
Williamson W, Parolin B (2012) Investigating E-planning in practice: an actor-network case study approach. Int J E-Plan Res 1(3):68–87

Governance Approaches in the Regeneration of Immigrant Communities: Potential Roles of Planning Support Systems (PSS)

Yanliu Lin and Stan Geertman

Abstract In recent decades, governance approaches have been widely advocated in the context of immigrant communities. The interplay between state, market, and civil society results in several modes of governance in these approaches. This chapter shows that each mode of governance has its positive and negative consequences for the regeneration of immigrant communities. Planning support systems (PSS) can assist in dealing with the negative consequences. This chapter suggests that three types of PSS—namely "informing PSS," "communicating PSS," and "analyzing PSS"—can play a distinctive role in dealing with the specific deficiencies of each mode of governance in the context of immigrant communities.

1 Introduction

Globalization has exacerbated social polarization between rich and poor through, for example, the generalization of flexible production processes (Kesteloot 2003). Many cities in developing countries are characterized by high rates of immigration and, consequently, social inequality. It is almost impossible for governments in the developing world to provide or subsidize sufficient housing for the increasing number of migrants who arrive in their cities. The presence of low-income migrant households in rapidly growing cities thus often contributes to the widespread proliferation of immigrant communities.

Although immigrant communities in the developing world provide low-income housing for migrants, conflicts of interest and a lack of resources usually result in poor infrastructure and dilapidated housing. There are high concentrations of

Y. Lin (✉) · S. Geertman
Faculty of Geosciences, Utrecht University, Utrecht, The Netherlands
e-mail: Y.Lin@uu.nl

S. Geertman
e-mail: S.C.M.Geertman@uu.nl

S. Geertman et al. (eds.), *Planning Support Systems for Sustainable Urban Development*, Lecture Notes in Geoinformation and Cartography,
DOI: 10.1007/978-3-642-37533-0_25,

poverty and social and economic deprivation, and possibly also economic, physical, and social exclusion. Thus, immigrant communities are often characterized by informal settlements that are segregated from formal urban areas. In order to achieve a balance between economic development and social equity, it is important to promote the sustainable development of immigrant communities. However, traditional planning approaches usually fail to sustainably regenerate these communities due to complex social, economic, and spatial issues, conflicts between key stakeholders. New approaches and tools are urgently needed to resolve this situation.

In recent decades, governance approaches that emphasize partnerships between state, market, and civil society have been widely advocated in the context of deprived communities (Douglass and Friedmann 1998; UN-Habitat 2003; Healey 2006). In this, immigrant communities are one of the key targeted areas. The roles of and relationships among key stakeholders are the most important elements of these approaches. Nevertheless, the actual application of these approaches is characterized by problems ranging from a lack of sufficient information, to a lack of knowledge of the expected outcomes and the consequences of potential measures.

Planning support systems (PSS) are considered to fulfill a role in reducing this level of complexity. These instruments, which are based on geo-information technology, can be employed to, for instance, support stakeholder participation in the planning process. They can be used to deal with specific planning tasks and problems, and to inform the planning process through analysis, prediction, and prescription. PSS can help stakeholders to articulate their thoughts, express their visions, and communicate with each other throughout the planning process. They can also promote sustainable urban development by including a wide range of social groups, and by stimulating economic and spatial integration. However, the literature contains no examples of the application of these kinds of support instruments for the regeneration of immigrant communities.

This chapter bridges that research gap. After presenting some results of a literature study on modes of governance and types of PSS (Sect. 2), it discusses case studies in which governance approaches in the regeneration of immigrant communities play a role (Sect. 3). In Sect. 4, a link is made between several modes of governance and different types of PSS. Finally, some conclusions are drawn. This chapter suggests that three types of PSS—namely informing PSS, communicating PSS, and analyzing PSS—can contribute to dealing with some of the new challenges and issues in the regeneration of immigrant communities.

2 Modes of Governance and PSS

2.1 Three Key Stakeholders and Modes of Governance

The state and the market are the two conventional stakeholders in urban regeneration. Governments play a key role in regenerating low-income communities, by subsidizing the upgrading of infrastructure and facilities for marginal social groups.

The government's role has recently shifted from that of initiator and decision-maker, to that of facilitator and enabler. Market stakeholders can help governments to relieve some of the pressure on their stagnating budgets, while also tackling some of the urgently needed renewal of the built environment (Loeckx 2009).

In recent decades, civil society has become a new stakeholder in the regeneration of deprived communities (Douglass and Friedmann 1998; UN-Habitat 2003; Healy 2006). Civil society is an intermediate associational realm between the state and households (White et al. 1996). It encompasses various associational forms based on kinship, ethnicity, culture or region, as well as formal and informal social networks. It is important for raising the living standards of the poor in slum communities, and for furthering processes of democratization in partnership with the state, rather than being seen as marginal to development or as an alternative to the state strategy for development (UN-Habitat 2003, p. 148). However, a clear framework to differentiate between modes of governance in the context of immigrant communities is lacking and our understanding of why different modes of governance lead to different results is limited.

The notion of "governance" implies that in addition to governmental organizations, also civil organizations, market parties, and households want to have a voice in spatial policymaking, and will try to influence decision-making (Healey 2006; Geertman 2012). A conceptual framework of modes of governance that is based on the relationships and roles among state, market, and civil society, has been discussed in environmental studies (Driessen et al. 2012). Driessen et al. identify five modes of governance: centralized governance, decentralized governance, public–private governance, interactive governance, and self-governance (Fig. 1). By analyzing and comparing actor features and institutional features concerning policy content, one can distinguish between the various modes of governance.

First, there are governance arrangements in which the actors are in the public sector. This mode can be subdivided into centralized and decentralized modes of

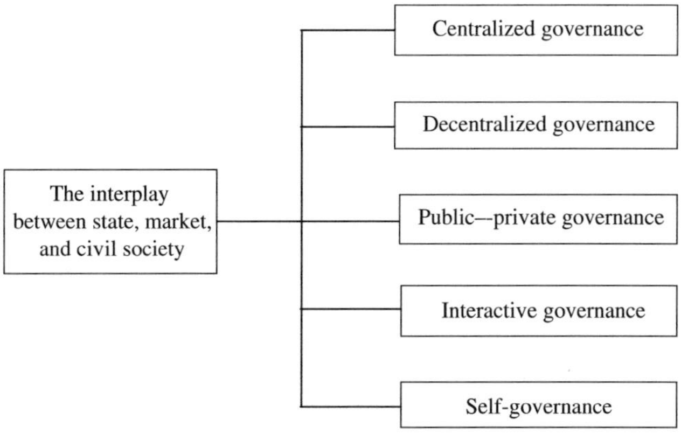

Fig. 1 Five modes of governance (based on Driessen et al. 2012)

governance. In both cases, either central or local governments take the lead, and the market and civil society are the recipients of the government's incentives.

Second, governance arrangements can be characterized by the joint effects of partners in the public and the private sector. These modes of governance include public–private governance—in which cooperation is mainly between government and market actors—and interactive governance, in which the actor base is broader and governments, market actors, and civil society collaborate on equal terms.

Third, self-governance is characterized by the fact that actors from the market and civil society enjoy far-reaching autonomy and are able to initiate new approaches themselves. These five modes of governance can coexist.

However, this conceptual framework on modes of governance is not sufficient to analyze the complex relationships between key stakeholders in the Chinese context. In Guangzhou, the "third realm" (Huang 1993), in which state and society merge, is playing a key role in the redevelopment of "villages in the city" (immigrant communities) (Lin and De Meulder 2012). The collective organizations are institutions in the third realm. Instead of "civil society," "society" could be a more suitable term to indicate the dimension beyond the state and the market, one that includes collective companies, the informal sector, households, etc. The interplay between these key stakeholders (state, market, and society) results in several modes of governance in the redevelopment of ViCs, for example, public–collective–private partnerships, collective–private cooperation, and private–private alliances. Collective–private cooperation and private–private alliances are characterized by self-governance. Nevertheless, public–collective–private governance is an addition to the five modes of governance identified by Driessen and colleagues. As public–private governance is not an important mode of governance in the present case studies, it has been replaced by public–collective–private governance.

2.2 Planning Support Systems

Planning support systems (PSS) are advocated to assist in dealing with the increasing complexity of present planning tasks. PSS are geographical information and communication technology instruments that support planning processes, particularly those in which intensive public participation is a key element (Geertman 2002). They bring together in the public realm the three components of traditional decision-support systems, namely information, models, and visualization. PSS deal with specific planning tasks and transform basic data into information that provides the driving force for modeling and design (Harris and Batty 1993). They can also help stakeholders to articulate their thoughts, express their visions, and discuss their ideas with others in the planning process.

There are three main application orientations for PSS in planning practice (Geertman 2012) (Fig. 2). First, "informing PSS" are predominantly designed to fulfill the task of information provision, for example, websites providing

Fig. 2 Three types of PSS (based on Geertman 2012)

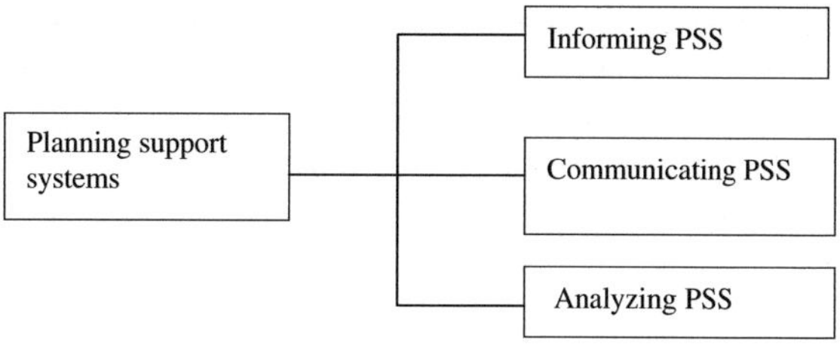

information about land uses and an area's associated land-use regulations. They furnish one-way communication from sender to recipient. Examples can be found on thousands of websites that provide information on spatial plans and developments.

Second, "communicating PSS" primarily support communication processes. An example of such a PSS is CommunityViz, a modular system built on the ArcGIS platform that can be used to integrate public values with geographic data and generate development alternatives (Lieske et al. 2009). Its "Scenario 360" module provides functionality for creating and assessing the potential impacts of proposed land-use actions, by facilitating scenario building and change monitoring in a series of associated indicators. In that sense, the module is also an "analyzing PSS" (see below). Another example is SoftGIS, which allows residents to contribute to their localized experiential knowledge (Kahila and Kyttä 2009). SoftGIS gathers data on how the everyday lives of the residents are organized, what kinds of place-based positive and negative experiences they have, and how they behave in their physical environment. This knowledge is collected through user-friendly internet-based applications. It is then used to create spatial layers in the GIS, which are subsequently used by experts such as urban planners and researchers.

Third, "analyzing PSS" are intended to accomplish analysis functions. Two examples are MIGMOD (ODPM 2002) and METRONAMICA (Van Delden and Hagen-Zanker 2009). MIGMOD allows the user to investigate the effects on migration of various policy scenarios, and to map various attributes of the migration system at the national level. The central features of this PSS are the separate modeling of out-migration from each area based on a set of determinant variables (Stage 1), and the distribution of migrants between destinations based on a set of determinants (Stage 2). The outputs of the Stage 1 and Stage 2 model calibrations can indicate which attributes of origins and destinations are associated with high rates of out-migration and in-migration, respectively.

The second example—METRONAMICA—can simulate and assess the integrated effects of planning measures on urban development. It models socioeconomic and physical planning aspects. It is able to integrate various scales and to capture a wide range of drivers (external factors, policy options, autonomous behavior, etc.). Van Delden and Hagen-Zanker (2009) developed a methodology linking qualitative storylines and quantitative land-use models. Stakeholders developed scenarios and storylines in three workshops, and between the workshops, modelers prepared model simulations based on the qualitative information derived from these storylines. The stakeholders and modelers then discussed how to improve the storylines and model applications in order to arrive at a creative, plausible, and consistent scenario.

In sum, these three types of PSS can play an important role in dealing with complexities that demand information transparency, involve various stakeholders, and require in-depth analysis. With their emphasis on stakeholder participation and complex planning tasks, these three types of PSS have some relationships with governance approaches in the context of immigrant communities. These relationships are explored in the following section through three case studies.

3 Case Studies

The case studies are the Favela–Bairro Program (Brazil), the Kampung Improvement Program (Indonesia), and the redevelopment of “villages in the city” (China). They were chosen because they are well known and representative in terms of the scale and effects of the approaches, and because the different political contexts and governance triangles would allow an examination of the impact of different modes of governance on the regeneration of immigrant communities. In addition, as all three cases concern high concentrations of migrants, they were likely to have some similar issues and opportunities.

3.1 The Favela-Bairro Program

Favelas are highly consolidated residential areas of self-construction on invaded public and private land on the edges of Brazil’s main cities. They provide housing for rural migrants and the urban poor, but no or few public services. In the 1990s, the municipality of Rio de Janeiro abandoned its policy of slum clearance, as it entailed very high social and economic costs, and launched the Favela–Bairro [“Slum-to-Neighborhood”] Program, which is aimed at integrating *favelas* into the urban fabric through infrastructure upgrading and service increases.

The program not only included sanitation systems and other basic infrastructure, but also emphasized the importance of opening up public space and improving housing. By promoting communication between several sectors of the municipal administration, the program also led to progress in establishing the procedures required for land ownership—one of the main demands of populations that live in subnormal settlements (UN-HABITAT 2003). Key to the success of the program was a committed and flexible municipal government, and the use of partnerships with NGOs, the private sector, professionals, and communities. The program was characterized by both decentralized governance and interactive governance.

The municipality cooperated with NGOs, the central government, and the private sector to finance the program. It first applied to the Inter-American

Development Bank (IDB) for a loan to finance public works in low-income settlements. The bank agreed to finance 60 % of the program; the remainder would be in matching funds from the municipality. The implementation of the social projects was funded by the relevant departments, although project resources paid for the construction of the buildings in which those services were to be delivered (Fiori et al. 2000). The construction of some public facilities also had financial backing from the central government. While the Favela–Bairro project funded the installation of the infrastructure, the utility providers used their own resources to provide those services. Although utility service fees and property taxes were collected from the residents of the improved *favelas* to repay the IDB loan and increase the municipality's revenue, Favela–Bairro remained a highly subsidized program (Fiori et al. 2000).

Community participation in the design and implementation processes contributed immensely to the success of the program. First, the leaders of communities played key roles in the program. The program was most successful where community organization was strong, and where active leaders were politically aware and therefore more independent of the government (Conde and Magalhaes 2004). Second, the households participated in meetings at the local community organization to learn about the program and plans, to express their concerns and demands, and to approve the final project plans. Third, grassroots-level infrastructure upgrading experts were employed as project managers, as they could work easily with both the government and the community members. Fourth, community participation was enriched, and in some cases promoted, by the support of university students who were involved in monitoring the work along with the residents. Thus, a dialogue was established between households, institutions, professions, and the public and private sectors in the search for effective solutions. Although the dialogue began when projects were designed, it was not limited to consultation processes, because the technicians built a base for future cooperation, as the residents would monitor the work while it was underway and be responsible for the physical maintenance and the implantation of social projects (Conde and Magalhaes 2004).

However, the program was criticized for its "perverse effects" regarding the dynamics of the land market and the consequences of regularization (Smolka and Iracheta 2000). Researchers found that the value of the properties had increased very little in relation to the realized public investment. They stated that the stigma of the *favela* had not been overcome, and that the program had encouraged further illegal occupation. The program was also criticized for its focus on spatial factors at the cost of dealing with complexities of poverty, and for failing to take action in other fields of social development and income generation (Vescina 2007).

In addition, although households were largely involved in informal economics based on social networks within *favelas* (Souto de Oliveira 2003), the program paid little attention to this. It was also not able to install a sustainable process, and the lack of continuity threatened the benefits already delivered. Integrated strategies (see 4.2) and strengthening the role of community in planning and implementation processes (see 4.4) are needed for the sustainable development of *favelas* and the city.

3.2 Kampung Improvement Program

Kampungs ("villages") are Indonesian rural settlements that became urbanized through a process of densification and transformation (Silas 2010). The main problem in most *kampungs* is flooding. Most of them lack water, sewer, and electrical infrastructure. Roads and footpaths are often in poor condition.

In response to this, in the 1930s the central government launched the Kampung Improvement Program (KIP). A committee was established to investigate the problems and possibilities and to propose a nationwide KIP; however, its proposals were never carried out because of a dispute over funding between central and local governments (Silas 1984). Since the 1960s, the KIP in Jakarta has some characteristics of decentralized governance. Local governance played an important role in initiating, subsidizing, and implementing the program (Silas 1984). This program was criticized for its comprehensive approach and lack of community involvement. Nevertheless, the approaches applied in Surabaya (see below) successfully promoted community participation in planning and implementation processes. The approaches were characterized by self-governance and interactive governance.

Cooperation between local governments and *kampung* communities resulted in three distinct types of *kampung* improvement, that is, the People Self-help Projects, the W.R. Soepratman Projects, and the Urban Kampung Improvement Program in Surabaya (Silas 1992). The first two types of improvement had some features of self-governance. The improvement was basically within the communities.

In the People Self-Help Projects, the community identified specific neighborhood needs or problems, and addressed them by bringing together local funds and labor, whilst the government provided the appropriate guidelines and standards for the proposed facilities. The W.R. Soepratman Projects were initiated and organized by the local communities themselves, and they financed more than half of the costs. Works were initiated by the local community, and technical assistance could be requested from the public works department to check that the improvements conform to municipal regulations, plans, and standards (Silas 1984). These projects were characterized by high levels of resourcing by the local community in terms of both money and direct labor, and the forging of an extremely effective partnership between low-income communities and the public sector.

In 1976, the KIP, which was more comprehensive than the other two types of projects, was increased in both scale and scope. It was funded by local, provincial, and central governments; funds from the World Bank were channeled through the provincial government. The community provided the land, and was responsible for moving dwellings and operating and maintaining facilities. This implied that the KIP shifted from self-governance to interactive governance, within which various stakeholders were involved and intensively interact with each other.

The three approaches were strongly based on intense community participation in planning and implementation processes. Households were given the opportunity to identify issues and problems in their neighborhoods, to comment on or make changes to proposed plans, and were encouraged to carry out the planning and improvement of their own plots and to take responsibility for the ongoing maintenance of *kampungs* (Silas 1984).

However, a survey by the Word Bank (1995) shows that the KIP needs to assess the dynamics of local real estate markets and to address the issue of compensating displaced low-income families, as many improved *kampungs* might be redeveloped by the private sector. Furthermore, the focus of the KIP has moved from inner city areas to the urban fringe, where villages have become centers for accommodating the growth in Surabaya's population. Implementing a modified form of the KIP in fringe areas at an early stage of development can pre-empt problems related to over-crowding in inner-city *kampungs* and create the basis for development into healthy urban residential areas with a formal status (Silas 1992). Although 70 % of low-income housing encompasses some kind of home industry (informal economies) (Santosa 2000), the KIP pays little attention to the relationships between these economic activities and spatial structures. Integrated strategies are consequently required to address the relationships between urban areas and *kampungs*, between inner city areas and the urban fringe area, and between the formal and the informal sector (see 4.4 and 4.5).

3.3 The Redevelopment of "Villages in the City"

In order to deal with the housing issues of middle and low-income households in Chinese cities, the central government subsidizes, and establishes regulations for, the construction of public rental housing for urban residents and migrants. Local government is mainly in charge of policy implementation and subsidizes a part of this project. This top-down approach is characterized by both central and decentralized governance.

This approach is currently facing many difficulties. There are conflicts of interest between the central government and local governments. Due to a lack of finance, most public rental housing is located in remote areas of Chinese cities, and it is not accessible to low-income migrants. In Beijing, the redevelopment of "villages in the city" (ViCs) is combined with the construction of public rental housing (Cai and Hu 2012). However, it has not proven successful due to conflicts of interest and limited budgets. While public rental housing has largely failed to meet the housing needs of low-income migrants, in many Chinese cities ViCs provide affordable rental homes for millions of migrants.

ViCs were initially rural settlements. During the rapid urbanization process, city governments usually requisition farmland for urban developments, while not touching the residential areas of villages where higher levels of compensation are required. As a result, many villages become ViCs with dual urban–rural structures. They provide informal housing for mass migrants, who are institutionally and economically excluded by formal urban systems. They also provide new incomes for indigenous villagers who lost their land as a result of the urbanization process.

There are complex relationships between urban areas and ViCs. The proximity of urban roads, industrial areas, commercial activities, and amenities contributes greatly to the development of ViCs, while the inferior quality developments on collective land in turn provide services for the city (Lin and De Meulder 2012). However, due to over construction and a lack of resources, ViCs usually have poor living environments. The infrastructure and facilities are inadequate, and the housing quality is poor.

The redevelopment of ViCs in most Chinese cities is mainly based on the "demolish and redevelop" model (Zhang 2005; Hao et al. 2011; Lin et al. 2011). Local governments, developers, and collective organizations are the key stakeholders in the demolition and redevelopment of ViCs. This approach is characterized by public–collective–private governance. Municipal governments are usually incapable of investing in the reconstruction of ViCs, and propose that developers and collective organizations should self-finance the redevelopment projects. They have made policies for the redevelopment of ViCs and have participated in some redevelopment projects as partners. The financial power has enabled collective organizations (shareholding cooperative companies) to negotiate with government authorities during redevelopment process, as they could use their own resources to facilitate redevelopment (Chung 2009).

Generally speaking, collective organizations represent the interests of villagers and act as mediators between villagers, developers, and governments. These public–collective–private partnerships, together with the high land value of the reclaimed land, made possible the demolition and redevelopment of several centrally located ViCs in Guangzhou and Shenzhen (Lin et al. 2011; Hao et al. 2011). However, these approaches demolish not only the low-income communities themselves, but also commercial and industrial projects on collective land that, although of inferior quality, are important for the urban economy. The lack of transparent information and efficient communication between stakeholders also creates a complex and difficult environment in which to reach an agreement on redevelopment (Hao et al. 2011). In many cases, these approaches have failed as a result of the complex issues and conflicts between key stakeholders.

There are also bottom-up approaches that restructure the space in a majority of ViCs in Guangzhou (Lin and De Meulder 2012). Collective organizations cooperate informally with market parties in the development of commercial and industrial projects on collective land. They also upgrade collective infrastructure and facilities (sports centers, primary schools, kinder gardens, gardens, medical centers, activity centers for the elderly, etc.). On the other hand, informal property management companies cooperate with villager households in upgrading small parts of ViCs. Many migrant households are involved in informal economic activities that are based on social networks. They cooperate with villager households in the gradual upgrading of villagers' houses and the village's main roads. However, due to profit maximization and the lack of regulation and cooperation, these self-governance approaches usually result in spatial fragments in ViCs (see 4.5).

4 Linking Governance, PSS, and the Regeneration of Immigrant Communities

The case studies show that there are possible relationships between modes of governance, PSS, and the regeneration of immigrant communities. Several modes of governance are based on the relationships among and the roles of three key stakeholders (state, market, society/civil society) in the regeneration of immigrant communities. These modes include centralized governance, decentralized governance, public–collective–private governance, interactive governance, and self-governance. Each mode has its positive and negative effects on the regeneration of immigrant communities and urban developments. Three types of PSS (informing PSS, communicating PSS, analyzing PSS) could play a role in dealing with specific problems and issues in the approach within each mode of governance (Table 1).

4.1 Centralized Governance

Within the mode of centralized governance, the central government takes the lead, and the market and civil society are the recipients of the government's incentives. The regeneration of immigrant communities, or the provision of public housing for the poor, is initiated and highly subsidized by the central government (see the KIP in Indonesia at the beginning of the twentieth century, and public rental housing in China). This approach can provide resources for the poor through a redistributive effect, for example, by subsidizing housing and facilities.

However, this top-down approach is unfeasible due to the limited resources and a conflict of interest between central and local governments. Communicating PSS (e.g., websites like IntelCities) can play a role in overcoming these problems, by acting as a platform for the communication and cooperation between multilevel governments and other stakeholders.

In addition, analyzing PSS—for instance, MIGMOD (e.g., see ODPM 2002)—can be used to map various attributes of the migration system at a national level. They can indicate which attributes of origins and destinations are associated with high rates of out-migration and in-migration, respectively, and thus contribute to policymaking for the allocation of redistributive resources for low-income migrants.

4.2 Decentralized Governance

The local government plays a key role in initiating, subsidizing, and implementing the regeneration of immigration programs (e.g., the KIP in Jakarta). It provides subsidized housing and facilities for the poor, and it has the capacity to allocate and redistribute resources at the local level, and to deal with specific issues and

Table 1 Linking governance, PSS, and the regeneration of immigrant communities

Modes of governance	Centralized governance	Decentralized governance	Public–collective–private governance	Interactive governance	Self-governance
Case studies	KIP in Indonesia in the 1930s, public rental housing in China	Favela–Bairro program, KIP in Jakarta, public rental housing in China	Demolition and redevelopment of ViCs in China	Favela–Bairro program, KIP in Surabaya	KIP in Surabaya, bottom-up approaches in ViCs of Guangzhou
Actions	Subsidized housing and facilities	Subsidized housing and facilities	Commercial redevelopment projects of immigrant communities	Upgrading of immigrant communities	Upgrading of immigrant communities
Key stake-holders	Central government	Local government	Local government, real estate companies, collective organizations (shareholding cooperative companies)	Multi-governments, formal and informal sectors, civil society (NGOs, community organizations and households)	Community organizations, leaders, households (or with support of governments and experts), formal and informal sectors
Positive aspects	Redistributive resources for the poor	Redistributive resources for the poor, dealing with specific issues and problems at local level	Benefits for key stakeholders, reclaimed land for urban development	Redistributive resources, empowerment of the poor, public participation, physical integration	Bottom-up, demand-led, empowerment of the poor, public participation, social networks (or informal economic activities)
Negative aspects	Top-down, unfeasible, limited resources, conflict of interests between central and local government	Top-down, little public participation, ineffective	Conflicts of interest, little public participation, destroying proximity and spatial structures that contains survival strategies of the poor	Limited boundary, little attention to the symbiotic relations between urban areas and immigrant communities, lack of economic and social integration	Limited scope and finance, little attention to spatial, economic and social relations between urban areas and immigrant communities; spatial fragment
Potential uses of PSS	Communicating PSS (e.g., websites like IntelCities), analyzing PSS	Communicating PSS (e.g., SoftGIS)	Informing PSS (e.g., websites), analyzing PSS (e.g., communityViz)	Informing PSS (e.g., websites), analyzing PSS (e.g., communityViz)	Informing PSS (e.g., websites), communicating PSS (e.g., What-if)

problems. However, due to the lack of stakeholder participation (particularly community participation) in the planning process, the approaches within this mode of governance are usually unable to establish a sustainable process.

Therefore, communicating PSS—for instance, SoftGIS (e.g. see Kahila and Kyttä 2009)—can provide a user-friendly internet platform for residents to evaluate their everyday living environment and communicate with professionals. SoftGIS methods allow residents to contribute to localized experiential knowledge. This knowledge is collected through user-friendly internet-based applications and used by professionals, such as urban planners and researchers. It can be applied to support the communications and collaboration between residents, professionals, and governments in the planning process.

4.3 Public-Collective-Private Governance

Public–collective–private governance is characterized by the cooperation between local governments, market parties, and collective organizations (see the redevelopment of "villages in the city" (ViCs) in China). This form of governance implies that there is almost no room for input from migrant households during the planning process. Villager households theoretically have their representatives (collective organizations), but they do not always benefit from the redevelopment of ViCs within the mode of public–collective–private governance, and they sometime conflict with local governments during the demolition process. This approach is criticized for destroying the symbiotic relationships between urban areas and ViCs, and for demolishing spatial structures that contain the survival strategies of migrants (Lin and De Meulder 2012).

Two types of PSS can contribute to overcoming the spatial, economic, and social problems caused by this approach. First, informing PSS (e.g. websites) can provide a wide range of stakeholders and users with information on spatial plans and development. This information transparency can contribute to democracy and empowerment. Second, analyzing PSS—for instance, CommunityViz (see Lieske et al. 2009)—can be used to analyze and evaluate the impact of urban policies and land-use actions on immigrant communities, by facilitating scenario building and change monitoring in a series of associated indicators.

4.4 Interactive Governance

Within the mode of interactive governance, there is intensive interaction between multi-governments, community organizations, households, designers, and the private sector. Since the 1970s, the partnership between state, market, and civil society has made a major contribution to the success of the Favela–Bairro Program and the KIP in Surabaya. However, this does not mean that it is an ideal approach to the regeneration of immigrant communities. The case studies show that attention is usually paid to physical integration, rather than to social and economic integration. Interventions usually have a limited scope and there is a lack of understanding of the symbiotic relationships between urban areas and immigrant communities.

Communicating PSS—for instance, SMURF (see Soutter and Repetti 2009)—can improve information exchange on urban projects in the surrounding urban areas, and can support the monitoring of urban developments and the exchange of information among the various stakeholders. In addition, analyzing PSS—for instance, METRONAMICA (see Van Delden and Hagen-Zanker 2009)—can be used to simulate and access the integrated effects of the planning measures on the regeneration of immigrant communities. They make it possible to integrate various scales and to capture a wide range of drivers (external factors, policy options, autonomous behavior), and can integrate socioeconomic and physical planning aspects.

4.5 Self-Governance

The mode of self-governance is characterized by stakeholders from the market and society/civil society. Projects within this mode of governance (e.g. the People Self-Help Projects, the W.R. Soepratman Projects, and the redevelopment of "villages in the city") are initiated by local communities and focus on the satisfaction of household needs. They successfully solve specific problems on the neighborhood scale by relying on the people's own resources. In Surabaya, community organizations and households are empowered in the implementation and planning process, and take responsibility for the ongoing maintenance of *kampungs*. In Guangzhou, bottom-up approaches in ViCs are characterized by partnerships between collective organizations and market parties, and between informal organizations and households. These two cases show that approaches within the mode of self-governance are flexible and adaptable. The case of Guangzhou also shows that social networks and informal economic activities play an important role in spatial and economic restructuring.

However, due to the limited scope (neighborhood scale) of the approach, a limited budget, and the lack of involvement of stakeholders from urban areas, it is not surprising that this approach pays little attention to other activities in the surrounding urban areas and can easily lead to spatial fragmentation. Communication and cooperation between stakeholders in immigrant communities and those in the surrounding urban areas may be required. Informing PSS (e.g. websites) can deliver information about planning rules, policies, and regulations. Analyzing PSS –for instance, What-If (Pettit and Wyatt 2009)—can be used to integrate public values with geographic data and generate development alternatives. They can also help the communication among various stakeholders from urban areas and immigrant communities, and the mutual coproduction of a common vision at the city level.

5 Conclusions and Discussions

Immigrant communities in developing contexts represent specific human settlements with similar social, economic, and spatial issues and challenges. In order to deal with the complex relationships among key stakeholders and intricate issues, governance approaches have been widely advocated in the regeneration of immigrant communities in developing countries. In these approaches, the interplay between state, market, and society/civil society results in several modes of governance, including centralized governance, decentralized governance, public–collective–private governance, interactive governance, and self-governance. Each mode has its positive and negative consequences, and its own contributions and deficiencies.

This chapter investigated governance approaches in the regeneration of immigrant communities. Three representative case studies allowed the detection and explication of the issues and problems related to each mode of governance. Three types of planning support systems—namely informing PSS, communicating PSS, and analyzing PSS—can contribute to dealing with the identified problems and issues within each mode of governance in the context of immigrant communities. These three types of PSS can play a role in countering the negative effects of each mode of governance, and can contribute to the development of integrated approaches for the sustainable regeneration of immigrant communities.

It should be noted that these potential contributions of distinctive types of PSS to the proper handling of governance approaches in the regeneration of immigrant communities, are purely hypothetical at the moment. In further research, the actual role of these distinctive PSS within diverse governance approaches will have to be investigated in greater detail.

References

Cai XX, Hu HD (2012) Design research on social housing for non-registered migrants in the urban fringe area with fast urbanization background (taking the southern suburban urban villages of Beijing as instance). Master thesis, Architecture, Urban Design and Regional Planning Department, K. U. Leuven

Chung H (2009) The planning of "villages in the city" in Shenzhen, china: the significance of the new state-led approach. Int Plann Stud 14(3):253–273

Conde LP, Magalhaes S (2004) Favela-Bairro: rewriting the history of Rio. ViverCidades, Rio de Janeiro

Douglass M, Friedmann J (1998) Cities for citizens. Wiley, New York

Driessen PJP, Dieperink C, Van Laerhoven F, Runhaar HAC, Vermeulen WJV (2012) Towards a conceptual framework for the study of shifts in modes of environmental governance—experiences from the Netherlands. Environ Policy Governance 22:143–160

Fiori J, Riley L, Ramirez R (2000) Urban poverty alleviation through environmental upgrading in Rio de Janeiro: Favela Bairro. Research report for the UK department for international development (DFID)

Geertman S (2002) Participatory planning and GIS: a PSS to bridge the gap. Environ Plan B: Plan Des 29:21–35

Geertman S (2012) Planning support systems as research instruments. J Urban Des Plan (accepted)
Hao P, Sliuzas R, Geertman S (2011) The development and redevelopment of urban villages in Shenzhen. Habitat Int 35(2):214–224
Harris B, Batty M (1993) Locational models geographic information and planning support systems. J Plan Educ Res 12:184–198
Healey P (2006) Collaborative planning: shaping places in fragmented societies. Palgrave Macmillan
Huang PCC (1993) Public sphere/civil society in China? The third realm between state and society. Mod China 19(2):216–240
Kahila M, Kyttä M (2009) SoftGIS as a bridge-builder in collaborative urban planning. In: Geertman G, Stillwell J (eds) Planning support systems best practice and new methods. Springer, New York, pp 389–412
Kesteloot C (2003) Urban social-spatial configurations and the future of European cities. Paper prepared for the Eurex online seminar 2003
Lieske SN, Mullen S, Hamerlinck JD (2009) Enhancing comprehensive planning with public engagement and planning support integration. In: Geertman G, Stillwell J (eds) Planning support systems best practice and new methods. Springer, New York, pp 295–316
Lin YL, De Meulder B (2012) A conceptual framework for the strategic urban project approach for the sustainable redevelopment of 'villages in the city' in Guangzhou. Habitat Int 36:380–387
Lin YL, De Meulder B, Wang SF (2011) Understanding the "village in the city" in Guangzhou: economic integration and development issue and their implications for the urban migrant. Urban Stud 48(16):3575–3590
Loeckx A (2009) Project and design, amending the project mode. In: Loeckx A (ed) Framing urban renewal in Flanders. SUN, Amsterdam
ODPM (2002) Development of a migration model, report prepared by the university of Newcastle upon Tyne, the university of Leeds, and the greater London authority/London research centre. ODPM, London
Pettit C, Wyatt R (2009) A planning support system toolkit approach for formulating and evaluating land use change scenarios. In: Geertman G, Stillwell J (eds) Planning support systems best practice and new methods. Springer, New York, pp 69–92
Santosa H (2000) Environmental management Surabaya with reference to national agenda 21 and the social safety programme. Environ Urbanization 12(2):175–184
Silas J (1984) The kampung improvement programme of Indonesia: a comparative case study of Jakarta and Surabaya. In: Payne GK (ed) Low-income housing in the developing world. Wiley Ltd
Silas J (1992) The Kampung improvement program, Surabaya. www.Mukinits.com
Silas J (2010) KIP (kampung improvement program) in Surabaya. In: d'Auria V, De Meulder B, Shannon K (eds) Human settlements: formulations and (re)calibrations. SUN Architecture Publishers, San Francisco
Smolka M, Iracheta A (2000) Mobilizing land value increments to provide service land for the poor. Colegio Mexiquense, Mexico city
Souto de Oliveira J (2003) Rethinking informal economy in Rio de Janeiro. http://mospi.nic.in/informal_paper_2003_06.htm
Soutter M, Repetti A (2009) Land management with the SMURF planning support system. In: Geertman G, Stillwell J (eds) Planning support systems best practice and new methods. Springer, New York, pp 369–388
UN-Habitat (2003) The challenge of slums: global report on human settlements 2003. Earthscan Publications Ltd, London
Van Delden H, Hagen-Zanker A (2009) New ways of supporting decision making: linking qualitative storylines with quantitative modelling. In: Geertman G, Stillwell J (eds) Planning support systems best practice and new methods. Springer, New York, pp 295–316

Vescina L (2007) Experiences from the South: a critical appropriation of the urban project: some cases in Brazil and Argentina, ISOCARP Review 03
White G, Howell J, Shang X (1996) In search of civil society in China. Clarendon press, Oxford
Zhang L (2005) Migrant enclaves and impact of redevelopment policy in Chinese cities. In: Ma LJC, Wu FL (eds) Restructuring the Chinese city: changing society, economy and space. Routledge, London

Supporting Planning Processes by the Use of Dynamic Visualisation

Stefano Pensa, Elena Masala and Isabella M. Lami

Abstract Large scale and long term decision making processes encounter a huge amount of difficulties. On the one hand, the uncertainty about the future represents a fundamental problem which is innate in long-term planning. On the other hand, the different interests and opinions regarding goals and strategies among the various actors are complex issues to be tackled. In order to build common knowledge and facilitate reasoning among the different actors, the following research proposes a method for sharing information through the use of dynamic maps. By means of the visual localisation of costs and benefits, the participants in the spatial decision processes are led to evaluate methods and objectives for a number of alternative development options. The system has been used in different case studies and has shown its effectiveness in creating awareness of spatial problems and enhancing discussions.

1 Introduction

Planning and decision making on spatial questions is recognised as a complex process which must deal with a large number of variables, interests and actors (Andrienko et al. 2007, 2011). Large scale and long term issues have a deep impact on individual and collective life, so discussing them generates a huge amount of

S. Pensa (✉) · E. Masala · I. M. Lami
SiTI—Higher Institute on Territorial Systems for Innovation,
Via Pier Carlo Boggio, 61 10138 Turin, Italy
e-mail: stefano.pensa@polito.it

E. Masala
e-mail: elena.masala@polito.it

I. M. Lami
e-mail: isabella.lami@polito.it

S. Geertman et al. (eds.), *Planning Support Systems for Sustainable Urban Development*, Lecture Notes in Geoinformation and Cartography,
DOI: 10.1007/978-3-642-37533-0_26, © Springer-Verlag Berlin Heidelberg 2013

elements to be taken in consideration. For these reasons the setting of goals and strategies requires a high level of awareness on the part of decision-makers. To allow knowledge building and achieve an informed decision, literature provides different tools and methods. In particular, many benefits can be gained from the use of spatial data visualisation (Bertin 1981; MacEachren and Taylor 1994; Thomas and Cook 2005), also known as geovisualisation, when applied to support spatial decision making and planning.

Tools for supporting planning and decisional processes, more generally known as Planning Support Systems (PSS), Decision Support Systems (DSS) and Spatial Decision Support Systems (SDSS), try to deal with a multitude of actors, opinions, interests, evaluation criteria and data. These diverse elements lead to a corresponding number of attempts to solve the issues involved, which have produced new research fields and technologies. As a result, PSS, DSS and SDSS represent a large part of scientific studies in spatial issues.

Numerous definitions as well as many concepts and classifications, have been produced, providing different, confusing point of views, especially regarding the classification of PSS, DSS and SDSS. As reported by Geertman and Stillwell (2003) citing Clarke (1990), PSS are conceived for "long-range problems and strategic issues", while SDSS "are generally designed to support short-term policy-making by isolated individuals and by business organisations". Also Lima et al. (2003) highlights that PSS are meant to include strategic issues while DSS and SDSS "are in general designed to support more specific or short-term policy-making processes". The first distinction outlines the two main branches of a family tree of definitions that have been assembled by many authors. In particular, two key classifications can be found for PSS. On the one hand, Gertman and Stillwell (2003), according to Harris (1989) and Batty (1995), refers to PSS as "a subset of geotechnology-related instruments that incorporate a suite of components (theories, data, information, knowledge, methods and tools) that collectively support all of or some parts of a unique planning task". On the other hand, Klosterman (1997) interprets PSS as an "information framework", in which technology is not the foremost priority, but rather a means to achieve planning purposes. Our experience confirms Klosterman's view, showing how this persistent research in building the most complex and realistic model is leading to the construction of scarcely comprehensible scientific axioms which policy makers and the public do not trust (Klosterman 2008, 2012; te Brömmelstroet 2010). This approach leads to models which are no longer an abstraction of reality but a one-to-one scaled map (Borges 1960). At the same time, the difficulty in accurately defining the problems inherent in spatial issues highlights how planning is not fully suitable to being converted into an automatic process (Andrienko et al. 2007), but the individual experience and expertise of the actors involved is essential in the building and assessment of plans.

In this continuous battle between a technology-centered approach and human skills, the application of PSS in real case studies encounters many difficulties (Couclelis 2005; Vonk et al. 2005; Geertman and Stillwell 2009; te Brömmelstroet 2010) especially due to a lack of communication between developers and end-users such as policy-makers, stakeholders and the public. Furthermore, each case study has specific requirements which can rarely be met by offering a single tool, so much time is spent on investigating existing tools and finding the most appropriate ones to fit the goals of the case study. The choice in technology often requires further customisation, which again takes up more time. For this reason a PSS should be very flexible when studying a problem, allowing for a change of scale, easy customisation, the analysis and evaluation of different goals as well as having a simple interface for the exchanging of information.

In time of financial resource constraints and in presence of actors who produce and manage the urban space, often with conflicting goals, it is evident the importance of having tools that can allow to better understand how spatial planning can act within the complex urban dynamics. In this sense, it could be very useful to find visualisation and evaluation tools which are able to manage this complexity at the different scales such as a district inside a city, the whole urban system or a European infrastructural corridor. For these reasons, the following research focuses on implementing planning process by using visualisation in order to stimulate communication and discussion among the actors involved.

Therefore, the goal of this research is to build a method for using an intuitive visual language which should enable communication among decision and policy makers, as well as other kinds of public or private actors for allowing the planning process to be effective in including participation and collaboration as approaches to a sustainable urban development. In accordance with Klosterman's concept of PSS as *information framework* (1997), the purpose of this study is to generate a methodology for organising, managing, relating and communicating data in a visual environment, in order to inform the actors and increase awareness within a planning process. The output of this research is the Interactive Visualisation Tool (InViTo), a visual method for managing spatial data in real-time, based on parametric three-dimensional modelling.

The plan of the chapter is the following. Section 2 contains an overview of the topic. Section 3 describes the methodology developed by the authors. Section 4 describes an application of the methodology in the field of wide area transformations. Section 5 contains the main findings and the evaluation of tool. The conclusion proposes some possible implementation of the research in the field under investigation.

2 Overview

Planning Support Systems require a high flexibility in order to integrate various disciplines and actors, and to be applicable in a large amount of cases, where different characteristics and needs must be met.

In the last two decades, many studies have been carried out in an effort to integrate support systems with other technologies, such as Geographic Information Systems (GIS), Multiple-criteria Decision Analysis (MCDA), Land-Use and

Transport Models (LUTM) and geovisualisation in order to increase their effectiveness in real case studies. More recently, the widespread use of virtual globes, such as Google Earth or Microsoft Virtual Earth, has increased the public curiosity towards the spatial location of data. This new interest has generated a huge amount of web applications but also a diffuse and simplistic approach to geography, even among professionals. Geo-referred data is widely used to locate information, but scarcely exploited to define goals and strategies for cities. Some attempts in this direction arrive from Boston MIT (SENSEable City Laboratory 2012), which proposes the visual analysis of urban areas by recording the spatial movements of personal devices such as mobile phones or bicycles. Even if the visualisation can be very eye-catching, the final purpose of such applications does not enter the real planning processes, remaining an interesting study which does not interact with the applied urban policy-making or strategy definition. With regard to the exploration of data, a good example is the "map tube project" (Centre for Advanced Spatial Analysis 2011). Here a free online resource allows the overlapping and meshing of different maps of the city of London. This initiative of UCL's Centre for Advanced Spatial Analysis is a Google Maps-based challenge to explore urban clusters within a free public interface. Besides these two famous examples, there is vibrant activity across the globe concerning the combined use of visualisation technologies and geo-based data, especially in interactive environments and customisable interfaces. However, the union between the latest technologies and planning practice does not appear to be particularly vibrant. Among professionals, a deep mistrust of digital support still remains. As reported by Te Brömmelstroet (2010), the main difficulty for the implementation of PSS is a lack of transparency due to which these instruments may be perceived as "sophisticated black boxes". Further important bottlenecks concern communication issues between models and users, such as "low communication value, lack of user friendliness" and lack of "interaction" (Te Brömmelstroet 2010).

Planning is a process in which people have to interact, communicate, exchange ideas, share information, but also defend their interests and carry out their reasoning. Effective communication is the basis for a well-performed activity in planning practice. Tools derived from complex mathematical formulas which do not employ eloquent images, can not aid this process, but rather help increase the mistrust in technological supports. Many PSS intend to reproduce the behaviour of a complete spatial system, with its inter-relations and connections at the different scales, expressed in various macro-scenarios, as a way to provide different forecasts for a number of planning options. As such, these tools place automatic processing before the whole set of expertise, ideas and opinions which animate the planning debate. The actors may perceive that they are misled by the tools instead of being assisted in building their own reasoning. In order to preserve a priority to human minds, the research field of geovisual analytics aims at supporting spatial problems through the integration of "the power of computational methods with human's background knowledge, flexible thinking, imagination, and capacity for insight" (Andrienko 2007, 2011). To achieve the enhancement of human capabilities towards building knowledge, interaction with data and interactive interfaces are recognised as the most effective means, because they allow data exploration and discovery (MacEachren and Taylor 1994; MacEachren et al. 2004).

3 Methodology

The following research proposes a visual method for managing spatial data in real time. This method, called InViTo, is based on Grasshopper, a free plug-in of McNeel's "Rhinoceros", a 3D modeling software most often used in architecture design. Grasshopper generates parametric shapes through the use of different kinds of input such as GIS data, databases, raster and vector files, allowing high compatibility with the largest number of software used by technicians involved in planning processes. InViTo makes use of scripts in Visual Basic language to customise Grasshopper components. The structure of Grasshopper allows the creation of a complex structure of links among different spatial shapes, which InViTo uses to build specific models for each case study. InViTo organises information in order to provide a visual outcome of relationships among spatial objects in real-time. It provides both 2D and 3D outputs such as dynamic maps or volumetric diagrams, which can show the behaviour of single elements or clusters of elements at both micro and macro scale. Data can be presented as an abstract visualisation offering a conceptual view of spatial dynamics, thus allowing users to analyse spatial information in a symbolic way. Nevertheless, photo-realistic representations are also possible, allowing final images to be defined by choosing from a wide range of visual techniques, in order to match the level of expertise of the actors involved in the debate. Moreover, visualisations can be displayed both in Rhinoceros environment, as well as in a virtual globe such as Google Earth. Thanks to the visual interactive framework, which enables users to work with information in real-time, InViTo is a tool conceived as Planning Support System for aiding the actors involved in sharing information and raising awareness of spatial issues. In particular, InViTo aims to provide a number of innovations in approaching planning practice.

First of all, InViTo uses visualisation as the basis for structuring the case study and exploits the properties of visual languages to create a common grammar among the participants to the planning processes.

Secondly, InViTo focuses on the possibility of interacting with data, thereby generating different advantages (Batty 2007). In fact, on the one hand, the opportunity to change input data in a model increases knowledge and awareness on spatial issues and project tasks, allowing actors to have a fruitful "dialogue" with the data displayed. On the other hand, it enhances discussions and debates, supporting users by means of shared images.

Thirdly, InViTo looks for models which are simple and transparent. It avoids the black-box approach and shuns complicated relationships among factors. In fact, these characteristics, which are common among many land use and transport simulation models, have a negative impact on planners' and policy-makers' trust in information tools. For this reason, InViTo looks for linear connections among spatial elements, as shown in the flowchart structure of Grasshopper. It allows users to access the model and evaluate, validate or customise it.

One further feature of InViTo is its flexibility. Common land-use and transport model are based on rigid frameworks which are difficult to adapt to the unique needs of each case study. These can often be used only for specific applications, such as land-use optimisation or accessibility analysis, and only at particular scales. InViTo, however, allows the production of specific models for each case study by a full customisation, which meets the purposes of application, its scale and planning requirements. Its flexibility also concerns the compatibility with different file formats so that it can employ databases, GIS data, raster and vector files.

Finally InViTo can be applied to the morphological features of spatial elements. Many models are based on the regular subdivision of areas by gridcells, which are used to calculate the values of land transformation. InViTo can work with discrete values of space, but it can also use the real shape of urban spaces, increasing comprehension of the localisation for displayed data.

4 Case Study

InViTo has been used as PSS, intended as an "information framework" (Klosterman 1997), in different applications. In the COST Action TU1002 on accessibility tools, it has been used to study the accessibility to public transport in urban areas. To achieve this task, its structure has been arranged to reproduce an integrated land use and transport model (LUTM) (Bertolini et al. 2012). In other applications, InViTo has been used to simulate the dynamics of rule-based land use models (Pensa et al. 2011; Marina et al. 2012) in order to asses a number of projects for residential purposes. In these cases, InViTo has been organised to relate different elements, as is usual in spatial models, though by means of a visual and interactive interface as *leitmotif* to approach simulation.

In this chapter, InViTo is presented with another application, which focuses on the visual support to the Analytic Network Process (ANP), a multi-criteria analysis technique for the assessment of alternative project or planning options (Saaty 2001, 2005; Saaty and Vargas 2006).

The ANP is a recent development of the well-known Analytic Hierarchy Process (AHP) (Saaty 1980). It represents a theory of relative measurement on absolute scales of both tangible and intangible criteria, which are based on the judgment of experts, and on existing measurements and statistics needed to make a decision. Recent applications of ANP to urban and territorial problems (Abastante et al. 2012; Abastante and Lami 2012; Abastante et al. 2011; Bottero and Lami 2010; Bottero et al. 2008) show that the theory considers the views of different actors, even with heterogeneous languages and may contribute to the construction and review of alternatives. In this sense it is important to underline that ANP allows the concept of participation to be developed, by means of focus groups where different actors and decision-makers involved can deal directly with each other.

In this version, InViTo has been used in different workshops and focus groups for studying a number of bottlenecks along the trans-European railway axis (TEN-T) 24 Genoa-Rotterdam, which is part of an Interreg IVB NWE Project called "CoDe24"(CoDe24 2010). The case study here presented is about a German section of corridor 24 between Frankfurt and Mannheim. The research aims at identifying strategies and goals for the area, which has been recognised as one of the most critical for the development of strategies for the Genoa-Rotterdam corridor (Masala 2012).

Through the collaboration between a research team of SiTI, which is a non-profit association between the Politecnico di Torino and Compagnia di San Paolo, and part of the Institut für Raum- und Landschaftsentwicklung (IRL) of ETH Zurich, the problem was structured in three different scenarios, which were discussed in a set of workshops and focus groups at the Value Lab of ETH-Zurich. These events were organised in order to involve both parties of the CoDe24 partnership and experts coming from the two cities of Frankfurt and Mannheim. The Analytic Network Process was used to analyse different areas at different scales, in particular from the urban to the trans-national, in order to assess transport, economic and environmental issues (Lami et al. 2011).

The ANP divides the spatial problems into simple elements and, through a pair comparison survey, asks actors to assign a weight to each element. Like other methods, ANP provides the ranking of alternatives as a final result. In this procedure, InViTo was used to implement discussion during the ANP questionnaire. By means of a visual and interactive representation of each element, ANP provides a visual support to compare the localisation of costs and benefits for each scenario.

4.1 ANP Questionnaire

The assessment procedure within ANP methodology consists of pairwise comparisons. This is essential to establish the relative importance of the different elements, with respect to a certain component of the network. In pairwise comparisons, a ratio scale of 1–9, namely Saaty's fundamental scale, is used to compare any two elements (Saaty 1980).

The structure of the pair comparison survey is divided into two distinct levels: the cluster, which is principally strategic, and the node (or element), which has a specific and detailed nature. The numerical judgments used in the pairwise comparison matrices are normally derived from a specific focus group made up of decision-makers and stakeholders. This group works together to evaluate the different aspects that characterise the problem with respect to the overall objective. Its main purpose is to reach a consensual decision on the weights and priorities. At the cluster level, the result of this phase is represented by the so-called cluster

With reference to the choice of the best alternative development scenario for the Rhein/Main-Rhein/Neckar area, which one of these two aspects do you think is more beneficial? And to what extent?

Economic and Transport Aspects	9	8	7	6	5	4	3	2	1	2	3	4	5	6	7	8	9	Environmental and Social Aspects

Fig. 1 Example of pair comparison between two clusters of elements

matrix, while at the node level, the ensemble of related vectors forms the unweighted supermatrix.

As an example, considering the aforementioned application, the questions that the focus group were asked to solve were similar to the one reported in Fig. 1.

Finally, according to the ANP theory, the cluster matrix is applied to the initial supermatrix as a cluster weight. The result is the weighted supermatrix, which is raised to a limiting power in order to obtain the limit supermatrix, where all columns are identical and each column gives the global priority vector. In the case of this specific application, the spatial problem has been structured into questions concerning the whole area, divided into benefits and costs, where nine elements shared among four clusters are identified. To evaluate the problem, 17 sets of questions have been given to participants who were asked to respond by providing a weight to each question. The pair comparison provided two limit supermatrices, one for each subnetwork of Costs and Benefits. Each column of the limit supermatrices, obtained from either subnetwork, provides the final priority vector of the elements under consideration. Finally, the raw priorities of the alternatives obtained from the limit supermatrices were normalised by cluster and synthesised. These priorities became the input values for the final aggregation and synthesis of the model results.

4.2 Visualisation of ANP Elements

In order to help the actors involved to understand spatial issues, each question of ANP was supported by the visualisation of the symbolic positioning of the expected effects. A map of expected consequences was hereby built for each ANP element (Tables 1 and 2) on the basis of the expertise of researchers in the fields of transport, economics, environment and spatial planning.

The wide scale of the case study corresponds to an approximation in the building of the maps. However, no precise detail is needed because the workshops were intended to aid the reasoning of experts and stakeholders in defining the key

Table 1 Maps used to represent the expected effects of each element under the sub-network "benefits"

Clusters	Elements	Maps		
		Scenario 1	Scenario 2	Scenario 3
Economic and transport aspects	Increase in level of attractiveness due to the improvement in speed/frequency/ capacity of passenger transport connections	Frankfurt and Mannheim 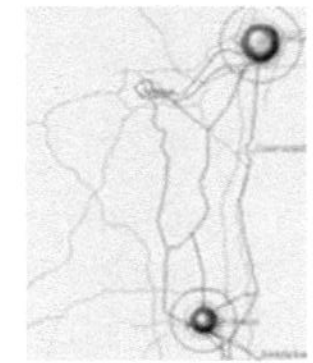	Railway stations along passenger tracks 	Railway stations along passenger tracks 
	Increase in level of attractiveness due to the improvement in speed/frequency/ capacity of freight transport connections	Freight transport railway tracks 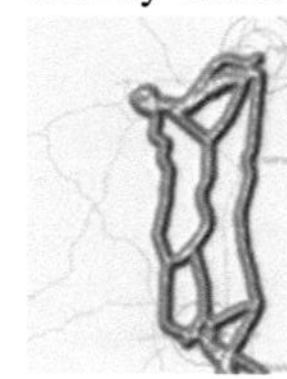	Freight transport railway tracks 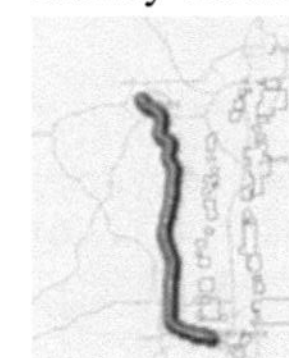	Freight transport railway and high capacity tracks 

(continued)

Table 1 (continued)

Clusters	Elements	Maps		
		Scenario 1	Scenario 2	Scenario 3
Environmental and urban planning aspects	Reduction in pollution due to the displacement on railway lines of a portion of road traffic	Motorway network 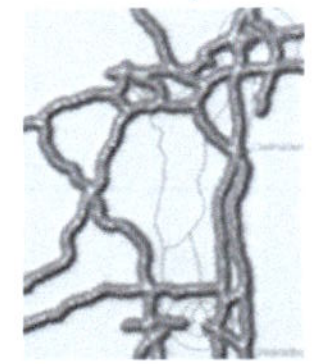	Motorway network 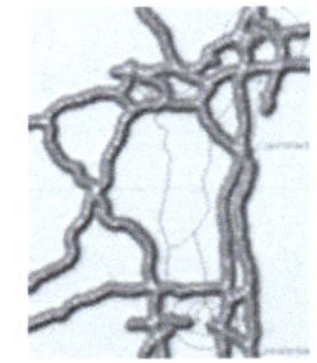	Motorway network 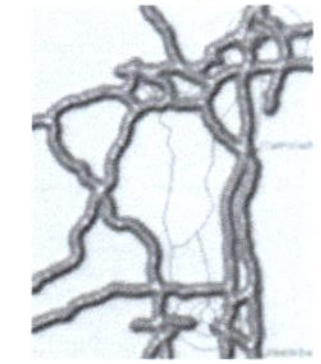
	Optimisation in soil consumption (widespread urbanisation is limited)	Frankfurt and Mannheim 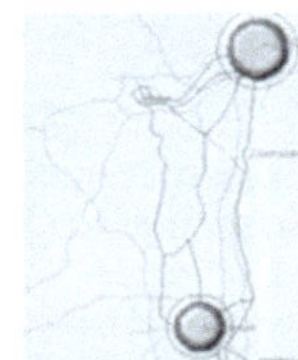	Brownfield on Mannheim area 	Brownfield on Mannheim area 
	Increase in level of services for the local population	Frankfurt and Mannheim 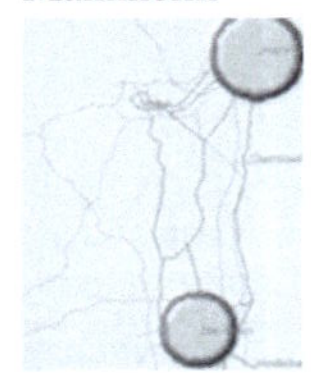	Settlements along passenger tracks 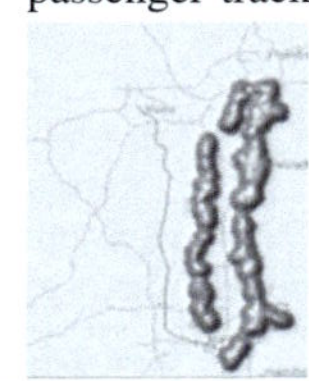	Settlements along passenger tracks 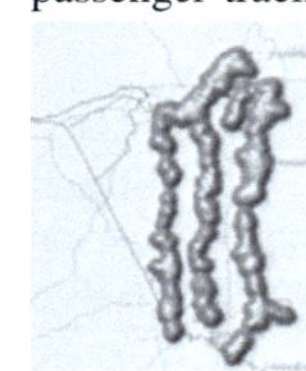

Table 2 Maps used to represent the expected effects of each element under the sub-network "costs"

Clusters	Elements	Maps		
		Scenario 1	Scenario 2	Scenario 3
Economic and transport aspects	Missing financial resources and construction costs (initial investments, reclamation costs)	Constant over the area 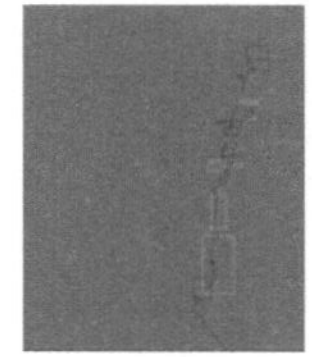	Constant over the area 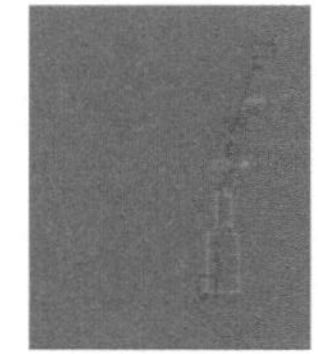	Constant over the area 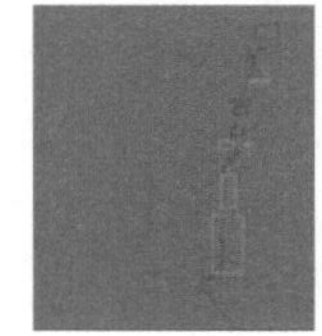
	Operational cost	High speed railway 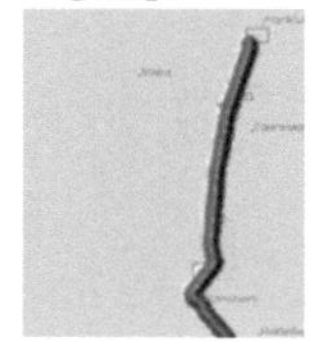	Existing railways 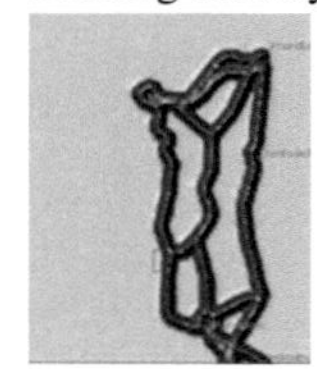	High capacity railway 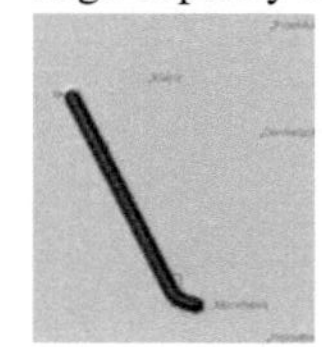

(continued)

Table 2 (continued)

Clusters	Elements	Maps		
		Scenario 1	Scenario 2	Scenario 3
Environmental and urban planning aspects	Negative impact (noise, vibrations and visual impact) due to the passage of trains	Settlements along high speed track 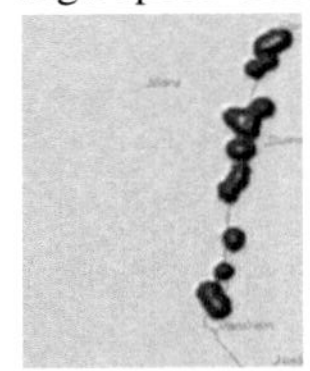	Settlements along freight track 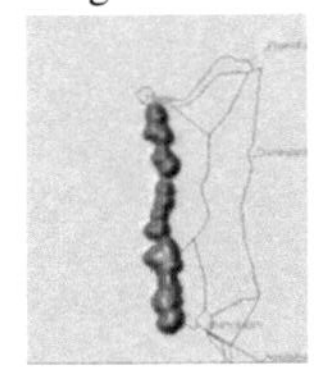	Settlements along high capacity track 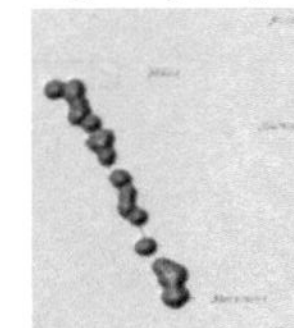
	Destruction of protected areas between Frankfurt and Mannheim and in the Mannheim region	Green areas along high speed track 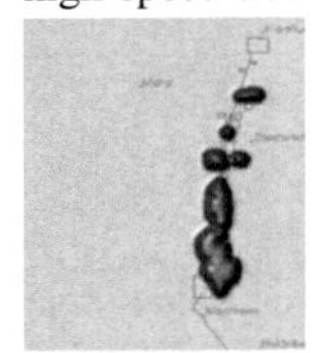	Green areas along freight track 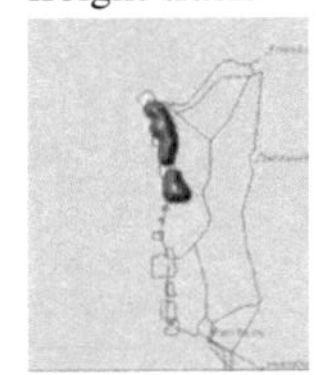	Green areas along high capacity track 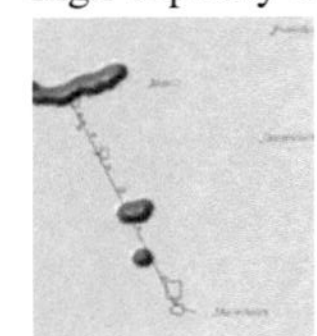

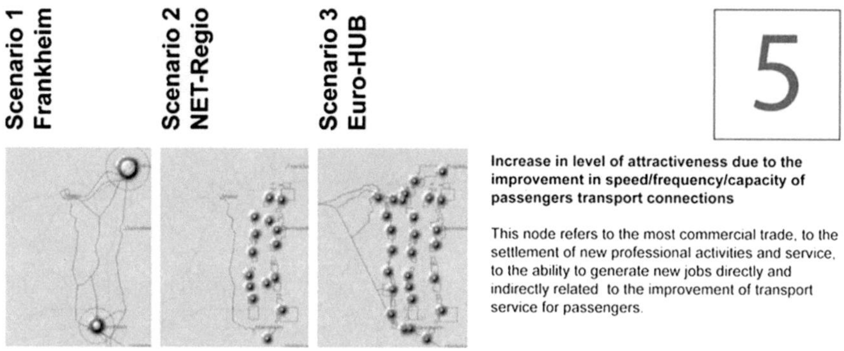

Fig. 2 Example of maps presented to participants for the evaluation of each single question of ANP. Each map represents the expected position of the effect of a specific ANP element depending on the scenario

elements for long term strategies on a very large scale. The maps should therefore give an indication of the effect rather than the actual position of an event.

During the discussion, a set of dynamic maps for each question (Fig. 2) were displayed to participants showing, in real-time, the behaviour of maps if the given weight was 1 or 9 for one element or another. Effectively, participants were guided in understanding where their choice might fall, so as to evaluate step by step the importance of their response.

InViTo was used to build all the visualisations used in this ANP assessment procedure. These visualisations are GIS-based and work with parametric features which have been set to reproduce the ANP scale of evaluation. They have been structured on a three-dimensional mesh where peaks change in height in correspondence to the expected effect, as defined by the weights assigned (Fig. 3).

This means that the mesh changes its shape on the basis of the values given by actors during the pair comparison of ANP elements and clusters. In particular, the interactive interface enables the use of maps to display specific values according to the requests of participants, as well as the identification of the areas with more benefits (coloured in light grey) or costs (painted in black). To allow a better understanding of the visualisations, the three-dimensional meshes are intersected by a slicing plane, which works as a cursor running up and down to visually select the areas included on a specific range of values. This plane (the transparent and white horizontal plane) has the same height in all scenarios so to allow a better comparison between them.

Once the whole set of questions has been answered, the weights resulting for each element are summed up as defined by ANP technique and summarised in two final maps, one for the Benefits and one for the Costs. These two maps overlap in order to compare the localisation and the amount of effects due to the actors' choices (Fig. 4).

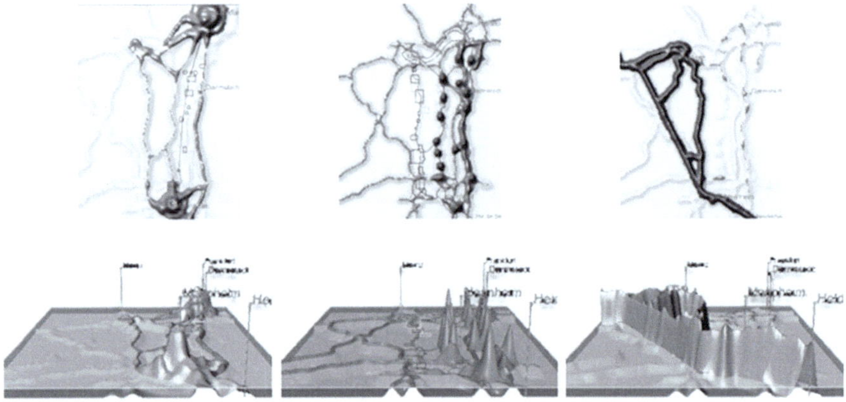

Fig. 3 Maps showing the total amount of the aggregate benefits for the three scenarios

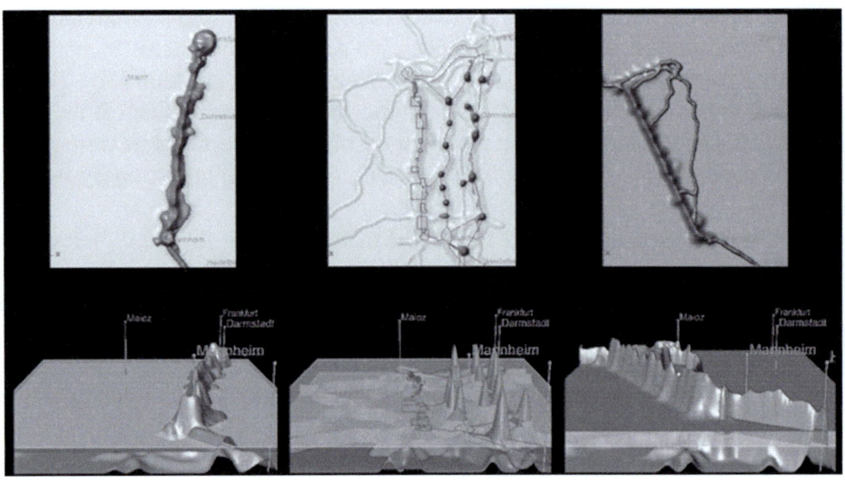

Fig. 4 Overlapping of the total amount of costs (*coloured in black*) and benefits (*coloured in light grey*)

The ANP questionnaire provided only one a ranking of scenarios, which showed the second as the best and the third as the worst. The reading of maps confirmed the ANP results and provided a lot of important points for the following discussion. The maps highlighted where and how much benefits overcome costs and vice versa, providing important elements for building reasoning and increasing awareness in participants. The positioning of the effects and their intensity added important contents to the debate so that some people changed their mind.

5 Methodology Evaluation

A survey about visualisation and its use in assessment sessions, distributed among the participants to one of the workshops at Value Lab of ETH Zurich, provided important feedbacks for the improvement of the support system. Although by and large comments were positive and participants satisfied, some participants requested further explanation about the language used by the visualisation. Dynamic maps seemed to be a good way to communicate spatial information, but the high level of abstraction has introduced new rules in the reading of maps. This proved to be a difficult obstacle to overcome especially in the case of participants with technical expertise in GIS management. In fact, the deterministic approach to the use of maps which distinguishes GIS technicians is far removed from the conceptual representation used in this case study. Therefore, a more detailed explanation of visualisation techniques have proven necessary.

In general, InViTo has been recognised by the majority of participants as being very useful in real planning processes while all the participants evaluated it as a useful tool for sharing information, knowledge building and as a support to the discussion.

6 Conclusion and Discussion

The case study described in this chapter presents a new approach to visualisation during the planning process. The goal of facilitating reasoning and awareness has been achieved to the point that many participants in the workshops confirmed InViTo as a being a useful PSS. The integration with ANP was effective in leading participants to build their own reasoning and sharing it with the others.

The visualisation used in this research introduced new languages for reading maps, overcoming the deterministic use and enhancing an abstract approach for clustering and selecting spatial data. InViTo has shown to be highly flexible and capable of being adapted to different case studies at diverse scales along the corridor 24, as well as functioning as a fully transparent interface between ANP technique and users.

The results obtained are encouraging, even if numerous issues remain to be solved. The foremost being that the interface with users is mediated by a technician. The values of weight given by users are input into the model by an operator as well as the movement of the slicing plane. On the one hand this is an advantage to leading the workshop because helps maintain a framework for the discussion. On the other it is a limit because the lack of an individual interface confines the interaction between users and data, therefore limiting the possibility of a comprehensive knowledge of the model. This entails that a graphic interface for direct interaction with the model is needed. First attempts have already been made by introducing the aid of laptops and tablet PCs, although they have not been tested in real decisional sessions.

Acknowledgments InViTo is a result of PhD research in Architecture at Politecnico di Torino, and it has been developed at SiTI (Higher Institute on Territorial Systems for Innovation), Turin, (IT). In particular, the authors would like to thank F. Corsico, D. Inaudi, G. Mondini, M. Arnone and M. Tabasso for their precious work and suggestions. Further thanks go to ETH Zurich (CH), University "Sts. Cyril and Methodius" in Skopje (MK), Institute of Architecture and new Media, Graz University of Technology (AT) for their collaboration on the project. The case study described in this chapter has been developed under CoDE24 project, financied by Intereg IVB NWE Project.

References

Abastante F, Lami IM (2012) Quality function deployment (QFD) and analytic network process (ANP): an application to analyze a cohousing intervention. J Appl Oper Res 4(1):14–27

Abastante F, Bottero M, Greco S, Lami IM (2011) The role of analytic network process and dominance-based rough set approach in strategic decisions for territorial transformations. Valori e Valutazioni 7(4):101–122

Abastante F, Bottero M, Lami IM (2012) Using the analytic network process for addressing a transport decision problem. Int J Anal Hierarchy Process 4:41–60

Andrienko G et al (2007) Geovisual analytics for spatial decision support: setting the research agenda. Int J Geogr Inf Sci 21(8):839–857

Andrienko G, Andrienko N, Keim D, MacEachren AM, Wrobel S (2011) Challenging problems of geospatial visual analytics. J Vis Lang Comput 22(4):251–256

Batty M (1995) Planning Support systems and the new logic of computation. Reg Dev Dialogue 16(1):1–17

Batty M (2007) Planning support system: progress, predictions, and spectaculations on the shape of things to come, CASA, working paper series. UCL, London, no. 122, 09/2007

Bertin J (1981) La graphique et le traitement graphique de l'information. de Gruyter (French), Berlin

Bertolini L, Halden D, Iltanen S, Pensa S, Santos B (2012) Cross-analysis of the accessibility instruments presented in chapter 3. Accessibility instruments for planning practice. In: Hull A, Silva C, Bertolini L (eds) Cost office, European science foundation, pp 195–203

Borges JL (1960) Del rigor de la ciencia. El Hacedor, Emecé ed., Buenos Aires

Bottero M, Lami IM (2010) Analytic network process and sustainable mobility: an application for the assessment of different scenarios. J Urbanism 3(3):275–293

Bottero M, Lami IM, Lombardi P (2008) Analytic network process, La valutazione di scenari di trasformazione urbana e territoriale. Alinea, Firenz

Centre for Advanced Spatial Analysis (2011) http://www.maptube.org. Accessed 11 2012

Clarke M (1990) Geographical information systems and model-based analysis: towards effective decision support systems. Geographical information systems for urban and regional planning. In: Scholten H, Stillwell J (eds) Kluwer, Dordrecht, pp 165–175

CoDe24 (2010) http://www.code-24.eu/. Accessed 11 2012

Couclelis H (2005) Where has the future gone? Rethinking the role of integrated land-use models in spatial planning. Environ Plann A 37(8):1353–1371

Geertman SCM, Stillwell J (2003) Planning support systems in practice. Springer, Heidelberg

Geertman SCM, Stillwell J (2009) Planning support systems: new methods and best practice (advances in spatial science). Springer, New York

Harris B (1989) Beyond geographic information systems: computers and the planning professional. J Am Plann Assoc 55(4):85–90

Klosterman RE (1997) Planning support systems: a new perspective on computer-aided planning. J Plann Educ Res 17:45–54

Klosterman RE (2008) A new tool for a new planning: the what if? Planning support system. Planning support systems for cities and regions. In: Brail RK (ed) Lincoln institute of land policy, Cambridge, pp 85–99

Klosterman RE (2012) Simple and complex models. Environ Plann B Plann Des 39(1):1–6

Lami IM, Masala E, Pensa S (2011) Analytic network process (ANP) and visualization of spatial data: the use of dynamic maps in territorial transformation processes. Int J Anal Hierarchy Process (IJAHP) 3(2):92–106

Lima RS, Silva ANR, Mendes JFG (2003) A SDSS for integrated management of health and education facilities at the local level: challenges and opportunities in a developing country. In: International conference on computers in urban planning and urban management, 8, Sendai, Japan, proceedings, (CD-ROM). Center for Northeast Asian Studies, Tohoku University, Sendai

MacEachren AM, Taylor DRF (1994) Visualization in modern cartography. Pergamon, Oxford

MacEachren AM, Gahegan M, Pike W, Brewer I (2004) Geovisualization for knowledge construction and decision-support. Comput Graph Appl 24(1):13–17

Marina O, Masala E, Pensa S, Stavric M (2012) Interactive model of urban development in residential areas in Skopje. Usage, usability and utility of 3D city models. In: Leduc T, Moreau G, Billen R (eds) EDP sciences

Masala E (2012) ETH March 20, 2012—Computational assessment workshop. https://www.youtube.com/watch?v=6ghuI0h1YRU&feature=BFa&list=PLA68D9CE96846CD66. Accessed 10 2012

Pensa S, Masala E, Marietta C (2011) The effects of decision-making on urban form: a tool for supporting planning processes. In: Pinto NN, Tenedorio JA, Santos M, Deus R (eds) Proceedings of the 7th international conference on virtual cities and territories, Lisbon, October 11th to 13th, 2011, Coimbra: department of civil engineering of the university of coimbra and e-GEO, research centre in geography and regional planning of the faculty of social sciences and humanities of the nova university of Lisbon

Saaty TL (1980) The analytic hierarchy process. McGraw Hill, New York

Saaty TL (2001) The analytic network process: decision making with dependence and feedback. RWS Publications, Pittsburgh

Saaty TL (2005) Theory and applications of the analytic network process. RWS Publications, Pittsburgh

Saaty TL, Vargas LG (2006) Decision making with the analytic network process. Springer, New York

SENSEable City Laboratory (2012). http://senseable.mit.edu/. Accessed 11 2012

te Brömmelstroet M (2010) Equip the warrior instead of manning the equipment: land use and transport planning support in the Netherlands. J Transport Land Use 3:25–41

Thomas JJ, Cook KA (2005) Illuminating the path: the R and D agenda for visual analytics. In: Thomas JJ, Cook KA (eds) IEEE Computer Society Press

Vonk G, Geertman S, Schot P (2005) Bottlenecks blocking widespread usage of planning support systems. Environ Plann A 37(5):909–924

Beauty and Brains: Integrating Easy Spatial Design and Advanced Urban Sustainability Models

Eduardo Dias, Marianne Linde, Azarakhsh Rafiee, Eric Koomen and Henk Scholten

Abstract This chapter contributes to the literature on the dyad creative design versus analytical assessment. It describes the challenges of creating a planning support system acceptable by designers, and reports the success of combining two planning support tools: *Phoenix*, an easy-to-use natural user-interface (using multitouch and physical objects input) for spatial design and *Urban Strategy,* a framework for integrating advanced environmental and socio-economic models for urban sustainability analysis. The systems were developed in different circumstances, contexts, by and for different groups, but their combination unlocked the key to user acceptance and adoption. Their extremely low learning curve attracts even the most technophobic stakeholders and makes use of recent advances in cloud computing for storage and processing to deliver immediate feedback. It allows anyone (especially designers) to sketch their solutions with natural movements and immediately receive feedback on key indicators of the sustainability performance, enabling iterative improvements of designs.

1 Introduction

1.1 Context

Our world is continuously evolving and we have been witnessing for the last centuries a continuous shift from a rural society to urban modes of living. This shift has been especially dramatic in the last five decades where a massive urbanization took place and cities grew to mega-cities. Nowadays, more than half

E. Dias (✉) · M. Linde · A. Rafiee · E. Koomen · H. Scholten
Department of Spatial Economics/SPINlab, Vrije Universiteit Amsterdam,
De Boelelaan 1105 1081 HV Amsterdam, The Netherlands
e-mail: edias@feweb.vu.nl

M. Linde
TNO, Princetonlaan 6 3584 CB Utrecht, The Netherlands

S. Geertman et al. (eds.), *Planning Support Systems for Sustainable Urban Development*, Lecture Notes in Geoinformation and Cartography,
DOI: 10.1007/978-3-642-37533-0_27,

of the world population live in cities and soon there will be more than two-thirds. Nevertheless, knowledge about city growth has not accompanied city development itself and not every city is growing in a desired fashion. Urban development is a complex dynamic process that is characterized by substantial spatiotemporal variation. Growth and decline coexist within neighbouring regions at short distances from each other. For example, population and employment are expected to grow in a few crucial regions in the Netherlands (the so-called main-, brain- and green-ports), whereas a decline is expected between these regions and, especially, in more peripheral parts of the country (Snellen et al. 2011). This makes steering urban expansion and intensification important policy issues in many regions, while preparing for decline and urban restructuring have become hot topics in others. These processes are driven by various interacting and sometimes even conflicting societal and economic forces that may impact regions differently.

Therefore, urban spatial planning is a complex process that goes much beyond government steering. Many stakeholders, each with their own interests, are involved as urban areas encapsulate different perspectives that range from economy, employment, mobility, noise, air quality, safety, aesthetics, health and climate change. It is common for spatial planning processes to recur to analytical frameworks to assess impacts in the different perspectives that are important for the stakeholders in order to minimize negative impacts and maximize benefits of the intended interventions.

1.2 Governance, Governing, Planning and Design

Also on the governing side, we witness opposing forces: societal concerns such as safety, accessibility and economic development call for active and preferably centralised government control, but at the same time central government is increasingly delegating its responsibilities to lower tiers of government. In fact, attention is shifting from government to governance and societal organisations (including businesses and non-governmental organisations) and individual citizens become more important in decision-making processes. This process of change is especially apparent in the renowned Dutch spatial planning system that is currently being stripped from its most prominent features: top–down restrictive zoning polices and urban concentration polices are being abolished. Also the underlying principles of distributive justice and solidarity between regions are being removed from spatial planning and the related allocation of funds. This is directly relevant for the management of regional population decline as it implies that classical, costly interventionist urban restructuring policies will not be feasible and calls for the development of other, more innovative strategies, and using participatory approach involving local industry, citizen groups and governing body around the table.

Although there are many actors involved, the planning (or the planning process) of cities, parts of cities or urban zones usually starts with designers and creative professions. And creativity delivers new solutions… there can be many new and different solutions, but in order to maximize positive impacts and minimize negative ones, also the initial designs should be evaluated in integrated models reflecting the stakeholders criteria, the impact directly visible and potentially (most likely) re-designed. So design is not a unilateral flow process, but becomes an interactive process that allows in-design feedback, external reviews and contributing to the thought process of defining a polygon or a policy decision.

1.3 Science and Creativity Integrated in Design

When we design (design as a verb) we attempt to imagine a future before creating it. The imagined future can be illustrated in a design (design as a noun). The verb implies purpose and intention, before the intervention. We should thus enable futures that are purposefully designed with full understanding of consequences. To understand the consequences of our designs, we need to integrate science (process and impact models) and creativity directly in the process of design. Dangermond (2010) refers to this integration as a systematic methodology for geographic planning and decision making now being called geodesign.

Instead of creating several alternatives to be tested and evaluated in a subsequent step, designers can get direct feedback of their options, understand the impacts and redesign to improve and achieve better results. These successive improvements can be compared to genetic algorithms from computer science (Goldberg and Holland 1988). These algorithms try to mimic the process of natural evolution by selecting the most fit combination of solutions to create a new and better set of solutions in subsequent interactions. Similarly the geodesign process allows the designers to visualize the fitness-index (less impacts) of solutions and consider maybe cross-solutions (or learn from the indicators) in the creation of succeeding and improved designs, trying to optimize the solutions to minimize impacts or fit a certain purpose.

The Geodesign name has just recently been coined and defined (Dangermond 2010; Ervin 2011; Flaxman 2010; Steinitz 2012; Zwick 2010). For example, according to Flaxman (2010), "Geodesign is a design and planning method that tightly couples the creation of design proposals with impact simulations informed by geographic context". Although the name is new, it is not a new concept. The concept has been placed in the 1960s, when a strong movement defined the integration of science where it would be possible to bring together scientific and regulatory information into the design phase as early as possible and this information could improve the design method by filtering unfeasible options in an early stage and allowing designers to focus on the options that the models suggested best outcomes. This was described in McHarg's (1971) seminal book *Design with Nature* where he elaborated on the original idea for an empirical layered world abstractions (which later would give birth to GIS with the layers concept) and this layers could indicate impacts to the designers. But for PSS scholars, this concept was described even in an earlier period, studied in the end of the 1950s, when Harris

(1960, 1965) proposed a revolutionary approach to planning by integrating "sketch planning" (as the drawing of alternatives) with state of the art analytical modeling to directly visualize the implications of the sketches/alternatives. Nevertheless, technology limitations (in data storage, processing capacity and uneasy human–computer interaction) limited these proposals to become reality until recently. Nowadays, the concept of direct interaction feedback with design can be implemented in real time, so we now have a new name to celebrate the new opportunities.

2 Research Questions and Method

Nevertheless, it is uncommon for design studios with landscape architects, designers, architects, to use analytical tools to inform and feedback on their design decisions. So we asked ourselves: "Why?" and found a twofold research. First we had to study how do these professionals work (which methodologies) and understand if it limits the analytical potential to inform and improve designs. Secondly, "How can we develop a tool that would really support and benefit the design process" (and be accepted by designers)? In order to understand what designers need, we cooperated in around 50 workshops (from 2009 to 2012) where we presented the sketching tool in an early stage, discussed and gathered feedback from landscape architects, regional planners, policy makers, governmental positions (international, national, regional, local), even with local farmers who used the tool to share local knowledge and support decision making. The extensive feedback collected allowed for the adaptation of the tool to the demands. A summary presenting the lessons learned will be discussed in the subsequent section.

3 Observations and Discussion

3.1 Bridging the Right and the Left

The left side of our brain is associated with analytical and objective functions, while the right is famous for creativity, intuition, and aesthetics. Let us consider this simplistic distinction within the spatial planning professionals when envisioning and creating new landscapes and new urban developments. Planners and geospatial researchers/professionals use analytical and objective tools to empirically describe the world and its processes (also known as spatial data and spatial models). Designers (defined broadly as all developing creative spatial solutions, such as architects, landscape architects, urban designers) perform by imagining and developing creative solutions in zoning problems. The two professions are generally seen separated in *time* and *space*. During the planning process (*time*), designers start with creative and intuitive solutions which are only later evaluated for feasibility and sustainability using concrete measurements. And physically (*space*), designers seat in the studio or zoning section, while the analytical or GIS professionals usually sit in the information technology division (e.g. in municipality offices). Different from analytical professionals, designers operate by generating many ideas and then deciding on presenting just few alternatives (chosen based on different criteria).

Even worse, when considering the construction of facility management or infrastructure planning, if design is separated/isolated from evaluation (due to professional and disciplinary divisions) that leads to a lack of information flow back to designers which ultimately can produce less advantageous design (or plainly bad design) or at least that important conditions or considerations might be ignored in the specific design alternatives that are selected for evaluation. This means slow and expensive workflows by deferring evaluation to a post phase.

It has been previously reported that GIS, PSS and SDSS and other analytical tools commonly fall short in efficiently supporting design efforts and decision making in implemented and established processes (Beukers et al. 2012; Geertman and Stillwell 2003; Te Brömmelstroet 2010; Te Brömmelstroet and Schrijnen 2010; Vonk and Geertman 2008). Corroborating with such previous research, we also observed in the past that such analytical tools have not been able to support the creative and dynamic process of design, a structural reason is that the data gathering, processing base data, modeling tools takes too much time and resources and cannot respond in the speed with which new ideas are generated from the creative professionals. The position for the analytical tools would then be relegated to evaluate a limited set of different options in a later stage independently.

But this lack of flexibility and unsupportive role of analytics in design is now changing. With the advent of (1) new tools to host and manage gigantic geo-spatial databases (cloud computing, big tables, hardware advances), (2) the recent achievements in computer processing performance and (3) ubiquitous "natural" human–computer interfaces, we are now observing new opportunities for geo-analytical tools to support the design professionals in real time. Base data is available for regional to national levels in high detail, models run on the fly and impacts calculated in real time as the lines in the sketch are designed.

From these recent computing developments, we felt that the widespread of the "Natural" user interface was a decisive development in planning support systems acceptability. Recently, several studies applied natural user interfaces for direct exploration of geo-information and spatial planning with reported success in the users appreciation (Arciniegas et al. 2011, 2012; Jacobs and Koomen 2010; Ligtenberg et al. 2004). Designers have always been used to tangible interfaces (such as large paper sheets, pens, markers or models/'maquettes'), but the indirect computing interfaces from the last decades (low resolution vertical screens, mouse and keyboards) implied very intrusive technology and a large usability threshold. So an important new paradigm in computer science that made computing more accessible are the natural user interfaces where the users can interact directly with the computer, for example in exploring geographic information, panning and zooming became intuitive and natural by just touching the screen.

3.2 More Lessons Learned

In the previous section we discussed two major limitations for PSS to be fully used in the design phases: (1) the design phase usually happens in an early stage of the planning, while analytical assessment is done after the designers present their designs proposals, which usually are just a few alternatives from the hundreds that could have been generated; (2) designers design many different alternatives in outbursts and it has been impossible for information systems to prepare data, process and display results in real time. But in our earlier developments and testing of the system with designers we learned more lessons; (3) drawing in GIS is different from designing: typically PSS and GIS develop from a technology perspective and the way drawings can be input in the system reflects a strict unnatural data structure (point, lines, polygons) and very specific drawing rules (polygons to be drawn point by point, clockwise and the last point has to be the same as the first), while most GIS professionals understand and tolerate such rules, designers need the flexibility and to be able to draw as they always did in unstructured ways (to draw a rectangle, sometimes they don't draw sequentially clockwise the sides of the rectangle), so we devised algorithms to be able to accommodate flexible drawings that mimic the way designers draw (they draw freely, and if they can close the polygon in the end, so that the algorithm understands the shaper regardless of the order. This was just a small change, but a great example illustrating how listening to and observing the users we can dramatically improve acceptance and usability, and (4) the black-box approach, this is a classic complaint from PSSs that we also confirmed. If the analytical models are predefined, hard to change, sometimes complex to understand, then stakeholders don't trust what they don't manage or understand. The calculated indicators had to be simple to understand and change and most important to be defined together with the stakeholders in the workshop session. This was often an interactive process itself also. Last but not least, (5) Independent work versus collaborative work: designers sometimes work together (on a table and they collaboratively draw in the same plan/paper) and we developed a multitouch environment to allow designers to collaborate on the same project without the need to a mouse input, just touching the screen which has the advantage of identical access to the system (instead of who controls the mouse and keyboard, controls the discussion, creating an hierarchical information flow) and the system was well evaluated initially! but... sometimes (many times) individual team members also want to seclude for inspiration, or just work on an idea thoroughly before presenting it to the colleagues. So, we needed to support in situ collaboration system where they can draw together (in turns or at the same time, although they never did it at the same time) on the large multitouch table/screen and we also feature individual design (design as a verb) so that when the designer is happy with the portrayal of the idea he can share it with the others in the collaborative large screen. For that we implemented a websocket based multi-user map so that individual team members can see the map data and individually draw designs and alternatives in personal devices (PC, Tablet, even smart phones). When needed/desired they would share their designs in the table version. Using websockets (which seem to be the next big thing for full-duplex communication on the web, but the future of technology is harder to predict than land use), it is most commonly implemented in "messaging/chat" clients, so we made sure that any sketched line and even the extent and individual user is looking at can be directly shared or the user can wait for a few lines to be ready and then share the alternative with all. This means that designers can cooperate co-situated, in different devices, or even in different continents. This multi-user PSS where you can actually see the map extent your colleague is looking at (what he is seeing) and together edit the map; think "google-docs" for planning/map editing.

4 Tools for Planners, Planning for Tools

4.1 Beauty

Phoenix, the beauty, where citizens, stakeholders or designers can sit around the table (see Fig. 1) and easily draw their ideas via a very intuitive touch interface, even without any geo-spatial expertise. They can also add information and interact with the system using physical objects that represent functions (see Fig. 2a) and easily access a repository of spatial data, a Spatial Data Infrastructure (SDI) in order to overlay and explore the environment. The base data from the SDI revealed

Fig. 1 Planning session around the "table" (early prototype in 2009), when the physical ink pot is placed on the table, the users can "draw" their ideas and plans

Fig. 2 Details of the user interaction: **a** sketching with a finger, **b** using a physical object to display a dynamic scale, **c** placing and rotating a small wooden block to define the center and size of a buffer/distance

to have a crucial role in the discussions because the users were able to easily turn layers on and off that were relevant for the discussion (e.g. suitability aspects). Phoenix is being developed by Geodan, a private Dutch company specialized in geo-information.

This tool had evolved as a sketching and layer viewer tool to support the design of digital spatial plans. Its human computer interface and easy to use functionality evolved during several years via discussions and iterations with landscape designers from a large planning department of a particular ministry. But it did not contain any advanced impact models, so it only became a geodesign tool when we decided to integrate it with *Urban Strategy*.

4.2 The Brains

Urban Strategy, is an interactive, decision support system developed by TNO (the Netherlands Organisation for Applied Scientific Research) that contains modeling expertise in local urban environments, involving aspects such as air quality, noise, accessibility, safety, urban energy production and consumption, greenhouse gases emission, and urban heat. Changes can be implemented in an interactive way (e.g. extra residential area, road closure, clean trucks only in environmental zone) and the effects of these on the quality of the surroundings are shown directly using the state-of-the-art urban calculation models. The current models include: Traffic (intensity and accessibility); Air quality; Noise (level, nuisance, sleep disturbance); External safety (group and place related risk); Groundwater flows and levels; Accessibility of parks and gardens; Shade and shadow; Sustainability (e.g. climate effects); and Cost/benefits of demolition/new development (Schelling et al. 2010). The input to the models was only accessible via a PC, with an interface accessible only to experts. In workshops, it is an operator/moderator who would manage the impact tool (see Fig. 3). The stakeholders could discuss and propose changes.

From a technical perspective, also conceptual, this tool introduced an innovation in the modeling infrastructure. The different models (air, noise, traffic, etc.)

Fig. 3 Impact outputs of *Urban Strategy*. *Left* indicators in 1D, *Centre* 2D map which also has editing functions, and 3D visualization that can show model outputs (such as noise or air pollution contours)

exist as separate entities (can even be deployed in separate devices), but they communicate with each other via a so called subscription communication framework (see Fig. 4). In this way, interconnected or inter-dependent models can subscribe to each other and if something changes on one, triggers computation and new results on the other.

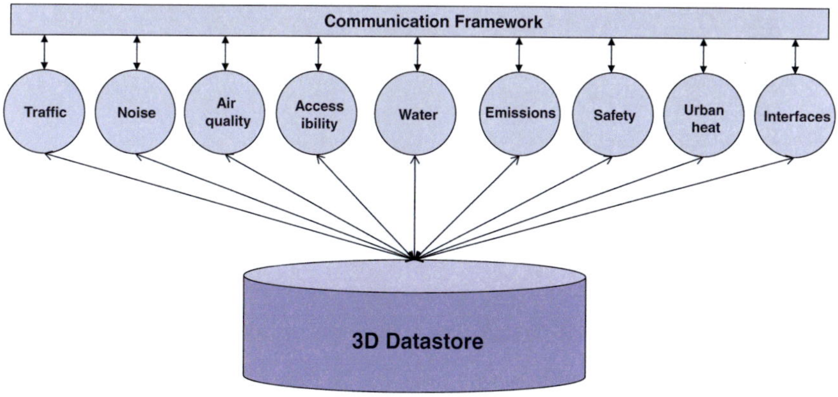

Fig. 4 Models and components of the analytical framework, focusing specially on urban sustainability (e.g. emissions of greenhouse gases, urban heat, air quality, traffic)

As an example, when a new building is designed in a certain street and 100 inhabitants ascribed to it, the traffic model would then be listening for change in the map and automatically calculate the changes in traffic (congestion) in that street due to the increased number of residents. The traffic model generates the traffic demand and assignment of the traffic based on an all-or-nothing assignment. The output consists of intensities per road segments. Based on this output the environmental models can run, which answer to the requirements of the National and European legislation for calculating these local effects. So in a similar fashion, air quality model is subscribed to the traffic model and when it notices, it changing automatically computes new air quality maps and indicators in accordance with the new traffic. Also the noise model is subscribed to both the traffic and 3D map module and will use the increased number of cars on the road and the position and size of the new building to compute the noise propagation map and the indicators. And so forth with all the subscribed models. This allows for direct feedback to the designer from the available indicators.

Both tools were extended so that they could interact with each other. So now the design sketched in workshops in *Phoenix* are sent to the analytical framework and the results (impact maps and indicator graphs) are provided back to the designer (Fig. 5).

Fig. 5 Latest interface that now calculates impacts of the plans right after finishing drawing (in this case noise disturbance from roads)

5 Past Experiences, Feelings, Reactions and Findings

Phoenix has evolved during several years 2007–2013 and it has been applied in more than 50 workshops with landscape architects, regional planners, policy makers, and different governmental authorities (local, regional, national and international). It has been used in diverse cases ranging from register ideas on spatial plans for road infrastructure with the "Dutch minister of infrastructure and environment", to gather local knowledge of the farmers in a climate adaptation workshop in a remote area of the Netherlands. *Urban Strategy* has been under development since 2006 and has been used in impact assessment studies on several projects, but due to its complexity, it is handled by an expert operator and results as shared in consultation sessions. This section tries to summarize the lessons learned for stakeholder acceptance and interaction with the interface tool (*Phoenix*). It proposes settings for an optimal utilization of the tools and describes recognized limitations in its applicability.

5.1 Less is More

While developing the tool, it was clear from the start that the drawing functionality had to be intuitive. In creative phases one needs to have all the expression options that one has when using paper and pen (e.g. colors, line thickness). The tool needs to provide these in an intuitive way, virtually without instruction. If the expression options are not provided or not intuitive, it will surely frustrate creativity. In addition, and also crucial, was the finding that designers want to grasp the "whole" application (as they do with their other tools), this meant we had to reject buttons with extra functionalities. It was a clear message of: less is more.

5.2 Flashy Gadget versus Black Box: Proceed with Care

We identified that a tool that looks new and flashy boosts enthusiasm and curiosity, which immediately gives a positive boost to a planning process. But beyond that it actually steered the process, in the beginning without modelling, since models are sometimes considered too much of a black box, also noticed in previous research (Geertman and Stillwell 2009; Vonk et al. 2005), so the models have to be introduced with care and sensitivity. For example, the tool was used in 2011 in a planning case in Bangladesh to discuss future interventions for regional climate adaptation. We used a land use model to predict future scenarios, not in real time, just as context background maps, but it lead to very intense discussions and difficulties to get those results accepted (even without real time modelling). But such reaction was not unique in the Bangladesh case. In many cases where modelling results were introduced we witnessed similar reactions.

It is important to note that planning is a very dynamic arena due to the many stakeholders involved. Each individual has his/hers own interests, positions, professional background and beliefs. But the predominant powers (authorities) in this arena often steer the choices in a planning process. Therefore, the participatory characteristics of such tools are crucial to allow more voices in the project. A lot of the "black-boxness" many stakeholders feel in complex planning processes disappears if they can freely navigate for themselves. This is one of the most important strengths of *Phoenix*.

5.3 Data Balance

When we present limited (little amount of) background data layers, the discussions become too vague and the planning workshop was inundated with questions. After hours we could not see advances on the process. But correcting that with too much context data, we witnessed increased and distracting discussions about the data itself (e.g. on data suitability, lineage, quality, portrayal choices). These are indeed relevant discussions, but they cost precious time of the planning workshop, which is a very limited resource as the stakeholders' workshops consisted of two hour sessions. In the end of the session, only data had been discussed.

5.4 The Right People at the Right Place (Phase)

Our feeling is that in order to successfully implement this tools, it should not be about the tool (or its data), but about the people and ideas. Our experience was that we needed to prepare the people (to make them open for the tool). On a workshop level this requires careful introduction of the tool and a willingness of all participants to interact with a focus on optimizing the overall outcome. On a broader planning process-level the fuzzy, ad-hoc, time and money constrained nature of most planning processes makes it very difficult to organize a planning process with the proper tools in the proper phases.

It is crucial to apply the tool in the right phase. We discovered the tool to be especially useful in the diversion phase, and not so much in the conversion phase. The "diverging phase" in planning cycles (Faludi and van der Valk 1994) is when stakeholders attempt to identify appropriate and different solutions. And our attempts to use the tools in the "conversion phase", when a choice needs to be made (choosing the best from three alternatives), it was not so useful. For that, in depth analysis and reporting was a preferred method.

6 The Next Case Studies: Public Participation and Smartness Modeling

Two new studies are planned to explore the use of this system from different aspects.

At this moment the most important channel for citizens to influence decision-making is the traditional public consultation and participation meeting. These meetings have important disadvantages as participation is restricted in time and space, meetings are often dominated by vocal minorities and the average citizen does not understand the jargon of planners. We will explore if the ease of use map interface allows and motivates citizens to contribute ideas to the planning process, then we can overlay and compare the hundreds of thousands of ideas and try to merge them into a democratic proposal (using common features and medians of the different plans). The second case study is to model city smartness, as in knowledge-work-force presence and creative industry attractiveness which are precursors of innovation and economic growth.

Globalisation is a dominant socioeconomic force that is associated with changes in the production and employment structure of countries and regions. It brings business opportunities and foreign investment to, for example, the Brainport region of Eindhoven in the south of the Netherlands, a high-tech cluster of industry and academia focused on innovation and technology. But at the same time, globalisation is partially responsible for rapid reductions in employment and outflows of high-skilled people in other regions. We are working together with this growing region and it was clear from the initial requirements assessment that a major priority for this urban zone is the economic development via the capture of knowledge workers and high skilled work force (including creative industry). The advantages have been previously described: Spatial concentration of activities, involving spatial and social proximity, increases the opportunities for interaction and knowledge transfer, while the resulting spill-over effects reduce the cost of obtaining and processing knowledge. In addition, knowledge workers preferably interact with each other in agglomerated environments to reduce interaction costs, and they are more productive in such environments (Florida 2002). But knowledge workers feel attracted to environments with particular characteristics (cultural offer, environmental quality, heritage) so the next challenge is to incorporate as an indicator the impact in the knowledge workforce attractiveness in the planning tool.

7 Concluding Remarks

The world evolves and changes. Frequently these changes result from uncoordinated human interventions. Our landscapes are becoming more fragmented and depleted and many cities are growing in unsustainable pace and fashion. It seems we are not consciously designing our future! We are not evaluating and accounting for long-term impacts of our actions on our communities, our cities, our society. In order to minimize negative impacts, the Geodesign concept is proposed as it provides feedback to designers directly on the planned interventions. We implemented the concept of Geodesign by integrating an ease-of-use digital sketching tool with a complex and comprehensive modelling framework.

The design is facilitated by a digital natural user interface that transparently records the creative ideas in sketches and provides feedback, immediately. This new interactive process allows for immediate improvements and more targeted designs which are evaluated and measured against intended future outcomes. The interactive tool took many years with iterative improvements with the design professionals. The lessons learned for the development and implementations of such tools can be summarized into 5 aspects: (1) design phase is typically separate and earlier than the assessment phase, designers need to be motivated to explore the analytical tools, but one should be aware of the different creative method from designers who, (2) develop tens or hundreds of alternatives to later remorselessly scratch out most of them to the benefit of a very few (sometimes chosen only from its innovative or aesthetics), so the tool should support the sketching and fast outburst by being flexible in inputs and fast in results, since (3) to be flexible in input is only achieved by observing and listening to the needs of the designers who are used to tools as ease as pen and paper and will not accept technical limitation in the creative input, (4) but they are not only interested in creative input, to use the quantitative impact models, it is clear that the models need to be changeable during the process and understandable, otherwise the we fall in the classic black-box pit. Last lesson learned was that collaborative planning is a key word, but does not mean designers can or desire to always work on the same group interface, (5) sometimes they need solitude moments where they can explore an idea before presenting it to the colleagues or process, so the system has to be flexible to link to additional interfaces (such as tablet devices) that allow users to individually retreat and create.

But naturally there are also limitations to the use and implementation of these tools. It proves to be especially useful in "diverging planning" (start or preparation) phases, usually at the start of a planning process, where the intention is to create many possible and divergent alternatives. It was not so welcome in the conversion phase when a final decision (where the "best" alternative) should be selected. A recognized risk in using the exploration and sketching tool, *Phoenix,* lies in trying to include all the relevant background data for a specific planning objective. Our experience shows that too many maps (which require too much introduction) often saturate the planning process. Humans only have a limited capacity to handle information. A crucial step that we have not observed in geodesign tools or in planning support systems is to procedurally limit and simplify the available data to the core, but without neglecting any crucial parts. In addition, for the individual (or the combination of tools) to work in practice, one needs to consciously and carefully introduce them in the planning process. On a workshop level this requires careful introduction of the tool and a willingness of all participants to interact with a focus on optimizing the overall outcome. On a broader planning process-level the fuzzy, ad-hoc, time and money constrained nature of most planning processes makes it very difficult to organise a planning process with the proper tools in the proper phases.

Acknowledgments This research is funded by the research programme Urban Regions in the Delta (URD), part of the VerDuS-programme ('Verbinding Duurzame Steden') of the Netherlands organisation for scientific research (NWO). The authors would specially like to thank Tim Ebben (from Geodan), main developer of the Phoenix tool for his efforts and patience. And to Walter Lohman (from TNO), main developer of Urban Strategy, for sharing his knowledge and his efforts in connecting the tools. Last but not least, to all the policy makers, researchers, experts and citizens who participated in planning workshops using these tools (whose feedback allowed us to improve): thank you for believing in designing a better future!

References

Arciniegas G, Janssen R, Omtzigt N (2011) Map-based multicriteria analysis to support interactive land use allocation. Int J Geogr Inf Sci 25(12):1931–1947

Arciniegas G, Janssen R, Rietveld P (2012) Effectiveness of collaborative map-based decision support tools: results of an experiment. Environ Model Softw

Beukers E, Bertolini L, Te Brömmelstroet M (2012) Why Cost Benefit Analysis is perceived as a problematic tool for assessment of transport plans: a process perspective. Trans Res Part A Policy Pract 46(1):68–78

Dangermond J (2010) GeoDesign and GIS: designing our future. In: Buhmann E, Pietsch M, Kretzler E (eds) Proceedings of digita landscape architecture, Wichmann, pp 502–514

Ervin S (2011) A system for GeoDesign. In: Buhmann E et al (eds) Proceedings of digital landscape architecture, LE:NOTRE, pp 145–154

Faludi A, van der Valk AJ (1994) Rule and order Dutch planning doctrine in the twentieth century, Kluwer Academic Publishers, Dordrecht and Boston

Flaxman M (2010) Fundamentals of Geodesign. In: Buhmann E, Pietsch M, Kretzler E (eds) Proceedings of digital landscape architecture, Wichmann, pp 28–41

Geertman S, Stillwell J (2003) Planning support systems in practice. Springer, New York

Geertman S, Stillwell J (2009) Planning support systems: content, issues and trends. In: Geertman S, Stillwell J (eds) Planning support systems best practice and new methods, Springer, pp 1–26

Goldberg DE, Holland JH (1988) Genetic algorithms and machine learning. Mach Learn 3(2–3):95–99

Harris B (1960) Plan or projection: an examination of the use of models in planning. J Am Inst Planners 26(4):265–272

Harris B (1965) New tools for planning. J Am Inst Planners 31(2):90–95

Jacobs C, Koomen E (2010) A GIS-assisted method for defining an integrated vision of a climate proof Groningen. In: Painho M, Santos MY, Pundt H (eds) Proceedings of the 13th AGILE international conference on geographic information science, Guimarães, Portugal: AGILE

Ligtenberg A, Wachowicz M, Bregt AK, Beulens A, Kettenis DL (2004) A design and application of a multi-agent system for simulation of multi-actor spatial planning. J Environ Manage 72(1–2):43–55

McHarg IL (1971) Design with Nature Doubleday/American Museum of Natural History

Schelling A, van Meijeren J, Borst J, Huybregts R, van Campenhout I, Geerlings H (2010) Urban strategy koppelt modellen. Verkeerskunde 4:34–39

Snellen D, van Eck JR, Hilbers H (2011) Nederland in 2040: een land van regios. Ruimtelijke verkenning 2011. In Planbureau voor de Leefomgeving (ed) Colloquium Vervoersplanologisch Speurwerk

Steinitz C (2012) A framework for Geodesign: changing geography by design. ESRI Press, Redlands

Te Brömmelstroet M (2010) Equip the warrior instead of manning the equipment: land use and transport planning support in the Netherlands. J Transp Land Use 3(1):25–41

Te Brömmelstroet M, Schrijnen PM (2010) From planning support systems to mediated planning support: a structured dialogue to overcome the implementation gap. Environ Plan B: Plan Des 37(1):3–20

Vonk G, Geertman S (2008) Improving the adoption and use of planning support systems in practice. Appl Spat Anal Policy 1(3):153–173

Vonk G, Geertman S, Schot PP (2005) Bottlenecks blocking widespread usage of planning support systems. Environ Plan A 37(5):909–924

Zwick P (2010) The world beyond GIS. Planning 76(6):20–23

MIX
Papier aus verantwortungsvollen Quellen
Paper from responsible sources
FSC® C105338
FSC
www.fsc.org

If you have any concerns about our products,
you can contact us on
ProductSafety@springernature.com

In case Publisher is established outside the EU,
the EU authorized representative is:
Springer Nature Customer Service Center GmbH
Europaplatz 3, 69115 Heidelberg, Germany

Printed by Libri Plureos GmbH
in Hamburg, Germany